PMP®

Project Management Professional E

Study Guide
2021 Exam Update

Tenth Edition

Kim Heldman, PMP

SYBEX®
A Wiley Brand

To BB, my forever love.

Acknowledgments

Thank you for buying *PMP®: Project Management Professional Exam Study Guide, Tenth Edition*, to help you study and prepare for the PMP® exam. Thousands of readers worldwide have used previous editions of this book to help them study for and pass the exam. Because of their success and recommendations to friends and co-workers, we've been able to keep this study guide up to date to reflect the changes made in *A Guide to the Project Management Body of Knowledge (PMBOK® Guide), Sixth Edition* (PMI®, 2017).

I would also like to thank the countless instructors who use my book in their PMP® prep classes. Thank you for your continued interest in using it in your classes. A big thanks goes to all the PMI® chapters who use this book in their classes as well.

A huge thank-you goes to Neil Edde, former vice president and publisher at Sybex, for taking a chance way back when on the first edition of this book. I can't thank him enough for having the foresight at that time to believe in this little-known exam. Almost 20 years later, the exam is well known and instructors and readers all over the globe have used this Study Guide to help them pass the PMP® exam.

Publishing this book clearly fits the definition of a project, and the team at Sybex is one of the best project teams you'll ever find. I appreciate all the hard work and dedication everyone on the team put into producing this book. A special thanks goes to Kenyon Brown, the acquisitions editor, for giving me the opportunity to update the book and for his ideas on making it stronger.

Next, I'd like to thank John Sleeva, the project editor, for his diligent work in helping me make this edition the best it can be. A big thanks also goes to Christine O'Connor, the production editor, for all her help on the book. It was great to work with her again. And thanks also to Liz Welch, the copyeditor, for all her help.

There were many folks involved behind the scenes who also deserve my thanks, including Louise Watson, the proofreader, and Johnna VanHoose Dinse, the indexer.

Next, I'd like to thank Vanina Mangano, technical editor, for her wealth of suggestions and ideas for new topics I should add to the text. It is always a pleasure to work with Vanina. She is a true professional. Vanina is an instructor and consultant in project management. I highly recommend Vanina's classes and videos if you're looking for additional study materials for the exam. She has been a joy to work with and I so appreciate everything she brings to this book. Last, but always the first on my list, is my best friend for a few decades and counting, BB. I love you, and I would never have accomplished what I have to date without your love and support. You're the best! And I'd be remiss if I didn't also thank Jason and Leah, Noelle, Amanda, and Joe, and of course the two best granddaughters on the planet, Kate and Juliette, for their support and understanding.

About the Author

Kim Heldman, MBA, PMP®, is the Senior Manager/CIO for the Regional Transportation District in Denver, Colorado. Kim directs IT resource planning, budgeting, project prioritization, and strategic and tactical planning. She directs and oversees IT design and development, enterprise resource planning systems, IT infrastructure, application development, cybersecurity, the IT program management office, intelligent transportation systems, and datacenter operations.

Kim oversees the IT portfolio of projects ranging from those that are small in scope and budget to multimillion-dollar, multiyear projects. She has over 25 years of experience in information technology project management. Kim has served in a senior leadership role for over 20 years and is regarded as a strategic visionary with an innate ability to collaborate with diverse groups and organizations, instill vision, improve morale, and lead her teams in achieving goals they never thought possible.

Kim wrote the first edition of *PMP®: Project Management Professional Study Guide*, published by Sybex, in 2002. Since then, thousands of people worldwide have used the study guide in preparation for the PMP® exam. Kim is also the author of *CompTIA Project+ Study Guide: Exam PK0-004, Second Edition*, *Project Management JumpStart, Third Edition*, and *Project Manager's Spotlight on Risk Management*. Kim has also published several articles and is currently working on a leadership book.

Most of the Real-World Scenarios in this study guide are based on Kim's real-life experiences. The names and circumstances have been changed to protect the innocent.

Kim continues to write on project management best practices and leadership topics, and she speaks frequently at conferences and events. You can contact Kim at Kim.Heldman@ gmail.com. She loves hearing from her readers and personally answers all her email.

Contents at a Glance

Contents

**Chapter   12   Controlling Work Results and Closing
                   Out the Project   683**

Appendices   **757**

Appendix A   **Answers to Review Questions**   **759**

Introduction

This book was designed for anyone thinking of taking the Project Management Professional (PMP®) exam sponsored by the Project Management Institute (PMI®). This certification is in high demand in all areas of business. PMI® has experienced explosive growth in membership over the last few years, and more and more organizations are recognizing the importance of project management certification.

Although this book is written primarily for those of you taking the PMP® exam, you can also use this book to study for the Certified Associate in Project Management (CAPM®) exam. The exams are similar in style, and the information covered in this book will help you with either exam.

This book has been updated to reflect the latest edition of *A Guide to the Project Management Body of Knowledge (PMBOK® Guide), Sixth Edition* (PMI®, 2017), and the new exam domains introduced in 2020. It assumes you have knowledge of general project management practices, although not necessarily specific to the *PMBOK® Guide*. It's written so that you can skim through areas you are already familiar with, picking up the specific *PMBOK® Guide* terminology where needed to pass the exam. You'll find that the project management processes and techniques discussed in this book are defined in such a way that you'll recognize tasks you've always done and be able to identify them with the *PMBOK® Guide* process names or methodologies.

PMI® offers the most recognized certification in the field of project management, and this book deals exclusively with its procedures and methods. Project management consists of many methods, each with its own terminology, tools, and procedures. If you're familiar with another organized project management methodology, don't assume you already know the *PMBOK® Guide* processes. I strongly recommend that you learn all of the processes—their key inputs, tools and techniques, and outputs. Take the time to memorize the key terms found in the Glossary as well. Sometimes just understanding the definition of a term will help you answer a question. It might be that you've always done that particular task or used the methodology described but called it by another name. Know the name of each process and its primary purpose.

The process names, inputs, tools and techniques, outputs, and descriptions of the project management process groups and related materials and figures in this book are based on content from *A Guide to the Project Management Body of Knowledge (PMBOK® Guide), Sixth Edition* (PMI®, 2017). The references to adaptive and hybrid methodologies, related materials, and figures in this chapter are based on content from the *Agile Practice Guide* (PMI®, 2017).

What Is the PMP® Certification?

PMI® is the leader and the most widely recognized organization in terms of promoting project management best practices. PMI® strives to maintain and endorse standards and ethics in this field and offers publications, training, seminars, chapters, special interest groups, and colleges to further the project management discipline.

PMI® was founded in 1969 and first started offering the PMP® certification exam in 1984. PMI® is accredited as an American National Standards Institute (ANSI) standards developer and also has the distinction of being the first organization to have its certification program attain International Organization for Standardization (ISO) 9001 recognition.

PMI® boasts a worldwide membership of more than a half a million members with more than 200 countries and territories around the globe. Local PMI® chapters meet regularly and allow project managers to exchange information and learn about new tools and techniques of project management or new ways to use established techniques. I encourage you to join a local chapter and get to know other professionals in your field.

Why Become PMP® Certified?

The following benefits are associated with becoming PMP® certified:

- It demonstrates proof of professional achievement.
- It increases your marketability.
- It provides greater opportunity for advancement in your field.
- It raises customer confidence in you and in your company's services.

Demonstrates Proof of Professional Achievement

PMP® certification is a rigorous process that documents your achievements in the field of project management. The exam tests your knowledge of the disciplined approaches, methodologies, and project management practices as described in the *PMBOK® Guide*.

You are required to have several years of experience in project management before sitting for the exam, as well as 35 hours of formal project management education. Your certification assures employers and customers that you are well grounded in project management practices and disciplines. It shows that you have the hands-on experience and a mastery of the processes and disciplines to manage projects effectively and motivate teams to produce successful results.

Increases Your Marketability

Many industries are realizing the importance of project management and its role in the organization. They are also seeing that simply proclaiming a head technician to be a "project manager" does not make it so. Project management, just like engineering, information technology, and a host of other trades, has its own specific qualifications and skills. Certification tells potential employers that you have the skills, experience, and knowledge to drive successful projects and ultimately improve the company's bottom line.

A certification will always make you stand out above the competition. If you're a PMP® credential holder and you're competing against a project manager without certification, chances are you'll come out as the top pick. As a hiring manager, all other things being equal, I will usually opt for the candidate who has certification over the candidate who doesn't have it. Certification tells potential employers you have gone the extra mile. You've spent time studying techniques and methods as well as employing them in practice. It shows dedication to your own professional growth and enhancement and to adhering to and advancing professional standards.

Provides Opportunity for Advancement

PMP® certification displays your willingness to pursue growth in your professional career and shows that you're not afraid of a little hard work to get what you want. Potential employers will interpret your pursuit of this certification as a high-energy, success-driven, can-do attitude on your part. They'll see that you're likely to display these same characteristics on the job, which will help make the company successful. Your certification displays a success-oriented, motivated attitude that will open up opportunities for future career advancements in your current field as well as in new areas you might want to explore.

Raises Customer Confidence

Just as the PMP® certification assures employers that you've got the background and experience to handle project management, it assures customers that they have a competent, experienced project manager at the helm. Certification will help your organization sell customers on your ability to manage their projects. Customers, like potential employers, want the reassurance that those working for them have the knowledge and skills necessary to carry out the duties of the position and that professionalism and personal integrity are of utmost importance. Individuals who hold these ideals will translate their ethics and professionalism to their work. This enhances the trust customers will have in you, which in turn will give you the ability to influence them on important project issues.

How to Become PMP® Certified

You need to fulfill several requirements in order to sit for the PMP® exam. PMI® has detailed the certification process quite extensively at its website. Go to www.pmi.org and click the Certifications tab to get the latest information on certification procedures and requirements.

As of this writing, you are required to fill out an application to sit for the PMP® exam. You can submit this application online at the PMI® website. You also need to document 35 hours of formal project management education. This might include college classes, seminars, workshops, and training sessions. Be prepared to list the class titles, location, date, and content.

In addition to filling out the application and documenting your formal project management training, there is one set of criteria you'll need to meet to sit for the exam. The criteria in this set fall into two categories. You need to meet the requirements for only one of these categories:

- Category 1 is for those who have a baccalaureate degree. You'll need to provide proof, via transcripts, of your degree with your application. In addition, you'll need to complete verification forms—found at the PMI® website—that show 4,500 hours of project management experience that spans a minimum of three years. You'll also need 35 hours of project management education/training.

- Category 2 is for those who do not have a baccalaureate degree but do hold a high school diploma or associate's degree. You'll need to complete verification forms documenting 7,500 hours of project management experience that spans a minimum of five years. You'll also need 35 hours of project management education/training.

As of this writing, the exam fee is $405 for PMI® members in good standing and $555 for non-PMI® members. Testing is conducted at Prometric testing centers. You can find a center near you on the Prometric center website, but you will not be able to schedule your exam until your application is approved by PMI®. You have one year from the time PMI® receives and approves your completed application to take the exam. You'll need to bring two forms of identification, such as a driver's license and a credit card in your name, with you to the Prometric testing center on the test day. You will not be allowed to take anything with you into the testing room and will be provided with a locker to store your personal belongings. You will be given a calculator, pencils, and scrap paper. You will turn in all scrap paper, including the notes and squiggles you've jotted during the test, to the center upon completion of the exam.

The exam is scored immediately, so you will know whether you've passed at the conclusion of the test. You're given four hours to complete the exam, which consists of 200 randomly generated questions. Only 175 of the 200 questions are scored. Twenty-five of the 200 questions are "pretest" questions that will appear randomly throughout the exam. These 25 questions are used by PMI® to determine statistical information and to determine whether they can or should be used on future exams. You will receive a score of Proficient, Moderately Proficient, or Below Proficient for each exam domain, as well as a Pass or Fail score. Because PMI® uses psychometric analysis to determine whether you have passed the exam, a passing score is not published. The questions on the exam cover three domains with roughly half the questions involving agile or hybrid methodologies. There will also be questions regarding professional responsibility. You'll answer questions on the following domains:

- People
- Process
- Business Environment

Questions pertaining to professional responsibility on the exam will be inter-mixed with questions for all the process groups. You won't see a section or set of questions devoted solely to professional responsibility, but you will need to understand all the concepts in this area. I've devoted a good portion of the last chapter of this book to discussing professional responsibility topics.

All unanswered questions are scored as wrong answers, so it benefits you to guess at an answer if you're stumped on a question.

After you've received your certification, you'll be required to earn 60 professional development units (PDUs) every three years to maintain certification. Approximately one hour of structured learning translates to one PDU. The PMI® website details what activities constitute a PDU, how many PDUs each activity earns, and how to register your PDUs with PMI® to maintain your certification. As an example, attendance at a local chapter meeting earns one PDU.

Who Should Buy This Book?

If you are serious about passing the PMP® exam (or the CAPM® exam for that matter), you should buy this book and use it to study for the exam. This book is unique in that it walks you through the project processes from beginning to end, just as projects are performed in practice. When you read this book, you will benefit from the explanations of specific *PMBOK® Guide* processes and techniques coupled with real-life scenarios that describe how project managers in different situations handle problems and the various issues all project managers are bound to encounter during their careers. This study guide describes in detail the exam objective topics in each chapter and has attempted to cover all of the important project management concepts.

Interactive Online Learning Environment and Test Bank

The interactive online learning environment that accompanies *PMP® Project Management Professional Exam Study Guide, Tenth Edition* provides a test bank with study tools to help you prepare for the certification exam—and to increase your chances of passing it the first time! The test bank includes the following tools:

Sample Tests   All of the questions in this book are provided, including the **Assessment Test**, which you'll find at the end of this Introduction, and the **Chapter Tests** that include the review questions at the end of each chapter. In addition, there are two **Bonus Exams** and two practice **CAPM Exams**. Use these questions to test your knowledge of the study guide material. The online test bank runs on multiple devices.

If you are interested in more practice exams, check out the *PMP Project Management Professional Practice Tests, 2nd Edition,* co-authored by me and Vanina Mangano.

Flashcards Questions are provided in digital flashcard format (a question followed by a single correct answer). You can use the flashcards to reinforce your learning and for last-minute test prep before the exam.

Other Study Tools Several bonus study tools are included:

> **Glossary** The key terms from this book and their definitions are available as a fully searchable PDF.
>
> **Bonus Questions** Supplement the topics in the book with bonus questions that'll help you test your knowledge and understanding of concepts.
>
> **Audio Instruction and Review** With over two hours of audio instruction, you can listen to the author review essential concepts, helping you to increase your understanding and fine-tune critical skills.

Go to www.wiley.com/go/sybextestprep to register and gain access to this interactive online learning environment and test bank with study tools.

How to Use This Book and the Test Bank

I've included several study tools, both in the book and in the test bank at www.wiley.com/go/sybextestprep. Following this Introduction is an assessment test that you can use to check your readiness for the actual exam. Take this test before you start reading the book. It will help you identify the areas you may need to brush up on. The answers to the assessment test appear after the last question of the test. Each answer includes an explanation and a note telling you in which chapter this material appears.

Exam Essentials appear at the end of every chapter to highlight the topics you'll most likely find on the exam and help you focus on the most important material covered in the chapter so that you'll have a solid understanding of those concepts. However, it isn't possible to predict what questions will be covered on your particular exam, so be sure to study everything in the chapter.

Like the exam itself, this study guide is organized in terms of process groups and the natural sequence of events a project goes through in its life cycle. This is in contrast to other study guides, where material is organized by Knowledge Area (Human Resource Management, Communications Management, and so on); such organization can make mapping the processes in each Knowledge Area to process groups confusing when you're studying for the exam.

Review questions are also provided at the end of every chapter. You can use them to gauge your understanding of the subject matter before reading the chapter and to point out the areas in which you need to concentrate your study time. As you finish each chapter, answer the review questions and then check to see whether your answers are right—the correct answers appear in Appendix A. You can go back to reread the section that deals with each question you got wrong to ensure that you answer the question correctly the next time you are tested on the material. If you can answer at least 80 percent of the review questions correctly, you can probably feel comfortable moving on to the next chapter. If you can't answer that many correctly, reread the chapter, or the section that seems to be giving you trouble, and try the questions again. You'll also find more than 200 flashcard questions on the website for on-the-go review.

 Don't rely on studying the review questions exclusively as your study method. The questions you'll see on the exam will be different from the questions presented in the book. There are 200 randomly generated questions on the PMP® exam and 150 on the CAPM®, so it isn't possible to cover every potential exam question in the review questions section of each chapter. Make sure you understand the concepts behind the material presented in each chapter and memorize all the formulas as well.

In addition to the assessment test and the review questions, you'll find bonus exams online. Take these practice exams just as if you were actually taking the exam (that is, without any reference material). When you have finished the first exam, move on to the next exam to solidify your test-taking skills. If you get more than 85 percent of the answers correct, you're ready to take the real exam.

Finally, you will notice various Real-World Scenario sidebars throughout each chapter. They are designed to give you insight into how the various processes and topic areas apply to real-world situations.

The Exam Objectives

Behind every certification exam, you can be sure to find exam objectives—the broad topics in which the exam developers want to ensure your competency. PMP® exam objectives are listed at the beginning of every chapter in this book.

Exam objectives are subject to change at any time without prior notice and at the sole discretion of PMI®. Please visit the Certifications page of PMI®'s website, www.pmi.org, for the most current listing of exam objectives.

How to Contact the Author

I welcome your feedback about this book or about books you'd like to see from me in the future. You can reach me at Kim.Heldman@gmail.com. For more information about my work, please visit my website at KimHeldman.com.

PMP®: Project Management Professional Exam Study Guide, 10th Edition

Updated for the 2020 Exam

Exam Objectives

PMP Exam Domains

Domain	Chapter
1. People	**All Chapters**
1.1 Manage conflict	10
1.2 Lead a team	1, 2, 3, 4, 5, 6, 9, 10
1.3 Support team performance	9
1.4 Empower team members and stakeholders	2, 3, 4, 8, 9, 11, 12
1.5 Ensure team members/stakeholders are adequately trained	9
1.6 Build a team	3, 4, 5, 6, 9,
1.7 Address and remove impediments, obstacles, and blockers from the team	9, 10

Domain	Chapter
1.8 Negotiate project agreements	8, 10
1.9 Collaborate with stakeholders	2, 3, 4, 6, 10
1.10 Build shared understanding	3, 4, 6, 8,
1.11 Engage and support virtual teams	9
1.12 Define team ground rules	10
1.13 Mentor relevant stakeholders	3, 6
1.14 Promote team performance through the application of emotional intelligence	9
2. Process	All Chapters
2.1 Execute project with urgency required to deliver business value	3, 5, 6, 8, 9
2.2 Manage communications	1, 3, 6, 10, 11, 12
2.3 Assess and manage risks	7, 10, 11
2.4 Engage stakeholders	3, 6, 11
2.5 Plan and manage budget and resources	6, 8, 12
2.6 Plan and manage schedule	5, 12
2.7 Plan and manage quality of products/deliverables	8, 10, 11
2.8 Plan and manage scope	4, 12
2.9 Integrate project planning activities	2, 8, 11, 12
2.10 Manage project changes	11
2.11 Plan and manage procurement	2, 8, 10, 11
2.12 Manage project artifacts	5, 6, 7, 10, 11, 12
2.13 Determine appropriate project methodology/ methods and practices	1, 2, 3, 4, 5, 6, 7, 8, 9, 10
2.14 Establish project governance structure	2, 8
2.15 Manage project issues	7, 11
2.16 Ensure knowledge transfer for project continuity	12
2.17 Plan and manage project/phase closure or transitions	12

Domain	Chapter
Business Environment	All Chapters
3.1 Plan and manage project compliance	2, 3, 4, 7, 11
3.2 Evaluate and deliver project benefits and value	2, 3,4, 6, 8, 11
3.3 Evaluate and address external business environment changes for impact on scope	3, 11
3.4 Support organizational change	3, 4, 5, 8, 11

Assessment Test

1. You work for Writer's Block, a service that reviews and critiques manuscripts for aspiring writers. You were assigned to be the project manager for a new computer system that logs, tracks, and electronically scans and files all submitted manuscripts along with the editor's notes. You hired a vendor to perform this project, and they used an agile methodology to manage the project. You are documenting how well the tailoring processes and project integration worked for this project. Which of the following information will you document regarding project integration?

    A. You documented the project life cycle and development life cycle you used to manage the project.

    B. You documented the management approaches used on the project.

    C. You documented the expected benefits to ensure the intended benefits were brought about on the project.

    D. You documented how project knowledge was managed.

    E. B, C, D

    F. A, B, C, D

2. The project sponsor has approached you with a dilemma. At the annual stockholders' meeting, the CEO announced that the project you're managing will be completed by the end of this year. The problem is that this is six months prior to the scheduled completion date. It's too late to go back and correct her mistake, and now stockholders expect implementation by the announced date. You must speed up the delivery date of this project. Your primary constraint before this occurred was the budget. Choose the best action from the options listed to speed up the project.

    A. Hire more resources to get the work completed faster.

    B. Ask for more money so that you can contract out one of the phases you had planned to do with in-house resources.

    C. Utilize negotiation and influencing skills to convince the project sponsor to speak with the CEO and make a correction to her announcement.

    D. Examine the project management plan to see whether there are any phases that can be fast tracked, and then revise the project management plan to reflect the compression of the schedule.

3. These types of dependencies can create arbitrary total float values and limit your scheduling options.

    A. Discretionary

    B. External

    C. Mandatory

    D. Hard logic

4. Project managers spend what percentage of their time communicating?

   A. 90

   B. 85

   C. 75

   D. 50

5. Match the following agile measurements with their definitions.

Agile measurements and definitions

Agile measurements	Definitions
A. Definition of done	1. The time a task waits before work starts
B. Empirical measure	2. A type of in-the-moment measure
C. Lead time	3. Describes the specifics of the tasks planned for the iteration before the team begins work
D. Definition of ready	4. The time it takes for a task to go from request to completion
E. Response time	5. The time it takes to complete work on a task from the time work starts
F. Cycle time	6. A checklist of elements needed to ensure the deliverable is ready for the customer to use
G. Capacity measures	7. Typically expressed as deliverables, functionality, or features

6. Your project has a high degree of certainty, firm requirements, a stable team, and low risk. Which of the following life cycle methodologies does this describe?

   A. Flow-based agile

   B. XP

   C. Six Sigma DMAIC

   D. Predictive

7. Which of the following statements regarding configuration management is not true?

   A. Configuration management involves managing changes to the project baselines.

   B. Change control systems are a subset of the configuration management system.

   C. Configuration management focuses on the specifications of the deliverables of the project.

   D. Configuration management validates and improves the project by evaluating the impact of each change.

8. Name the difference between the agile iterative approach and the agile incremental approach as discussed in the *Agile Practice Guide* (PMI®, 2017).

    A. Incremental uses prototypes and iterative helps in performing the work faster and in speeding up the project.

    B. Incremental focuses on learning optimization and iterative focuses on speed of delivery.

    C. Iterative plans the work at the beginning of the project before starting work, and incremental plans the work at the beginning of each iteration.

    D. Iterative uses prototypes and incremental helps in performing the work faster and speeding up the project.

9. Your project has a high degree of uncertainty, high risk, evolving requirements, and cross-functional teams. Which of the following life cycle methodologies does this describe?

    A. Agile

    B. Hybrid

    C. Predictive

    D. Waterfall

10. During your project meeting, a problem was discussed, and a resolution to the problem was reached. During the meeting, the participants started wondering why they thought the problem was such a big issue. Sometime after the meeting, you received an email from one of the meeting participants saying they've changed their mind about the solution reached in the meeting and need to resurface the problem. The solution reached during the initial project meeting is a result of which of the following conflict resolution techniques?

    A. Collaborate

    B. Forcing

    C. Smoothing

    D. Storming

11. According to the *PMBOK® Guide*, which of the following names all the components of an interactive communication model?

    A. Encode, transmit, decode

    B. Encode, transmit, decode, acknowledge, feedback/response

    C. Encode, transmit, decode, feedback/response

    D. Encode, transmit, acknowledge, decode

12. What are benefit measurement methods?

    A. Project selection criteria

    B. Project selection methods

    C. Project selection committees

    D. Project resource and budget selection methods

13. During your project meeting, a problem was discussed, and the project sponsor described the resolution they wanted the project team to implement. The project manager tried to interject with another idea that might solve the issues but the sponsor didn't want to discuss the idea. Which of the following conflict resolution techniques does this question describe?

A. Collaborate

B. Forcing

C. Smoothing

D. Storming

14. Which of the following factors are changes that occur to the external business environment that may impact the organization and your project?

A. Mergers and acquisitions

B. Geopolitical and marketplace

C. Regulatory and technology

D. Social and economic

E. A, C, D

F. A, B, C, D

15. You've been assigned as a project manager on a research and development project for a new dental procedure. You're working in the Project Scope Management Knowledge Area. What is the purpose of the scope management plan?

A. The scope management plan describes and documents a scope baseline to help make future project decisions.

B. The scope management plan decomposes project deliverables into smaller units of work.

C. The scope management plan describes how project scope will be developed and how scope changes will be managed.

D. The scope management plan describes how cost and time estimates will be developed for project scope changes.

16. Which of the following statements regarding Ishikawa diagrams are true?

A. Ishikawa diagrams are also called cause-and-effect diagrams.

B. Ishikawa diagrams are also called fishbone diagrams.

C. Ishikawa diagrams help identify the root cause of the problem.

D. Ishikawa diagrams are also known as why-why diagrams.

E. A, B, C, D

F. A, B, D

17. What is one of the most important skills a project manager can have?

A. Negotiation skills

B. Influencing skills

C. Communication skills

D. Business skills

18. Which of the following terms are other names for inspections?

   A. Reviews

   B. Assessment

   C. Walk-through

   D. Audits

   E. A, C, D

   F. A, B, C, D

19. You are the project manager for Xylophone Phonics. It produces children's software programs that teach basic reading and math skills. You're performing cost estimates for your project and don't have a lot of details yet. Which of the following techniques should you use?

   A. Analogous estimating techniques, because this is a form of expert judgment that uses historical information from similar projects

   B. Bottom-up estimating techniques, because this is a form of expert judgment that uses historical information from similar projects

   C. Monte Carlo analysis, because this is a modeling technique that uses simulation to determine estimates

   D. Parametric modeling, because this is a form of simulation used to determine estimates

20. Project managers have the highest level of authority and the most power in which type of organizational structure?

   A. Project-oriented

   B. Simple

   C. Functional

   D. Hybrid

21. This process is concerned with determining the engagement levels of the stakeholders.

   A. Plan Communications Management

   B. Control Communications

   C. Plan Stakeholder Engagement

   D. Manage Stakeholder Engagement

22. All of the following statements are true regarding risk events except which one?

   A. Project risks are uncertain events.

   B. If risks occur, they can have a positive or negative effect on project objectives.

   C. Unknown risks can be threats to the project objectives, and nothing can be done to plan for them.

   D. Risks that have more perceived rewards to the organization than consequences should be accepted.

23. Which of the following describes the key focus or purpose of the Manage Project Knowledge process?

    A. Gathering, creating, storing, distributing, retrieving, and disposing of project information

    B. Managing communications, resolving issues, engaging others on the project, managing expectations, improving project performance by implementing requested changes, and managing concerns in anticipation of potential problems

    C. Sharing organizational and project knowledge and creating new knowledge that can be shared in the future

    D. Performing systematic activities to determine which processes should be used to achieve the project requirements, and to ensure that activities and processes are performed efficiently and effectively

24. You are the project manager for Xylophone Phonics. This company produces children's software programs that teach basic reading and math skills. You are ready to assign project roles, responsibilities, and reporting relationships. On which project Planning process are you working?

    A. Estimate Activity Resources

    B. Plan Resource Management

    C. Acquire Project Team

    D. Plan Organizational Resources

25. Match the following types of tests that are used on agile-based software projects to their description.

Testing at all levels

Name of test	Description
A. Integration test	1. Testing the software from the start to the end to ensure the application is working correctly.
B. End-to-end test	2. A high-level test designed to identify simple failures that could jeopardize the software program.
C. Regression test	3. This test is performed after changes are made to the code or when maintenance activities are performed on the hardware the code resides on to ensure the software works the same way it did before the change.
D. Unit test	4. This test combines software modules and tests them as a group.
E. Smoke test	5. This test is performed on individual modules or individual components of source code.

26. These types of meetings are associated with the agile project management methodologies. They occur at the beginning of an iteration. Team members choose items from the backlog list to work on during the iteration. What is this meeting called?

 A. Review planning meeting

 B. Planning meeting

 C. Retrospective planning meeting

 D. Daily stand-up planning meeting

27. These diagrams rank-order factors for corrective action by frequency of occurrence. They are also a type of histogram.

 A. Control charts

 B. Process flowcharts

 C. Scatter diagrams

 D. Pareto diagrams

28. You are a project manager who has recently held a project team kickoff meeting where all the team members were formally introduced to each other. Some of the team members know each other from other projects and have been working with you for the past three weeks. Which of the following statements is not true?

 A. Team building improves the knowledge and skills of team members.

 B. Team building builds feelings of trust and agreement among team members, which can improve morale.

 C. Team building can create a dynamic environment and cohesive culture to improve productivity of both the team and the project.

 D. Team building occurs throughout the life of the project and can establish clear expectations and behaviors for project team members, leading to increased productivity.

29. You are a project manager for the Swirling Seas Cruises food division. You're considering two different projects regarding food services on the cruise lines. The initial cost of Project Fish'n for Chips will be $800,000, with expected cash inflows of $300,000 per quarter. Project Picnic's payback period is six months. Which project should you recommend?

 A. Project Fish'n for Chips, because its payback period is two months shorter than Project Picnic's

 B. Project Fish'n for Chips, because the costs on Project Picnic are unknown

 C. Project Picnic, because Project Fish'n for Chips's payback period is four months longer than Project Picnic's

 D. Project Picnic, because Project Fish'n for Chips's payback period is two months longer than Project Picnic's

30. Which of the following compression techniques increases risk?

    A. Crashing

    B. Resource leveling

    C. Fast-tracking

    D. Lead and lag

31. You are the project manager for a construction company that is building a new city and county office building in your city. You recently looked over the construction site to determine whether the work to date conformed to the requirements and quality standards. Which tool and technique of the Control Quality process were you using?

    A. Defect repair review

    B. Inspection

    C. Sampling

    D. Quality audit

32. You have been assigned to a project that will allow job seekers to fill out applications and submit them via the company website. You report to the VP of human resources. You are also responsible for screening applications for the information technology division and setting up interviews. The project coordinator has asked for the latest version of your changes to the online application page for his review. Which organizational structure do you work in?

    A. Functional organization

    B. Weak matrix organization

    C. Virtual organization

    D. Balanced matrix organization

33. The primary function of the Closing processes is to perform all of the following except which one?

    A. Formalize lessons learned and distribute this information to project participants.

    B. Complete all activities associated with closing out the project.

    C. Validate that the deliverables are complete and accurate.

    D. Ensure all project work is complete and accurate.

34. You are the project manager for Lucky Stars Candies. You've identified the requirements for the project and documented them where?

    A. In the requirements documentation, which will be used as an input to the Create WBS process

    B. In the project scope statement, which is used as an input to the Create WBS process

    C. In the product requirements document, which is an output of the Define Scope process

    D. In the project specifications document, which is an output of the Define Scope process

35. What is the purpose of the project charter?

    A. To recognize and acknowledge the project sponsor

    B. To recognize and acknowledge the existence of the project and commit organizational resources to the project

    C. To acknowledge the existence of the project team, project manager, and project sponsor

    D. To describe the selection methods used to choose this project over its competitors

36. Which of the following are tools and techniques of the Identify Stakeholders process you can use to categorize stakeholders? (Choose two.)

    A. Salience model

    B. Power/interest grid

    C. Stakeholder register

    D. Stakeholder engagement assessment

37. You are a project manager working on a software development project. You've developed the risk management plan, identified risks, and determined risk responses for the risks. A risk event occurs, and you implement the response. Then, another risk event occurs as a result of the response you implemented. What type of risk is this?

    A. Trigger risk

    B. Residual risk

    C. Secondary risk

    D. Mitigated risk

38. All of the following are a type of project ending except for which one?

    A. Extinction

    B. Starvation

    C. Desertion

    D. Addition

39. You are working on a project that will upgrade the phone system in your customer service center. You have considered using analogous estimating, parametric estimating, bottom-up estimating, and three-point estimating to determine activity costs. Which process does this describe?

    A. Estimating Activity Resources

    B. Estimate Costs

    C. Determine Budget

    D. Estimating Activity Costs

40. Failure costs are also known as which of the following?

A. Internal costs

B. Cost of poor quality

C. Cost of keeping defects out of the hands of customers

D. Prevention costs

41. Feeding buffers and the project buffer are part of which of the following Develop Schedule tool and technique?

A. Critical path method

B. Schedule network analysis

C. Applying leads and lags

D. Critical chain method

42. You are the project manager for a construction company that is building a new city and county office building in your city. Your CCB recently approved a scope change. You know that scope change might come about as a result of all of the following except which one?

A. Schedule revisions

B. Product scope change

C. Changes to the agreed-on WBS

D. Changes to the project requirements

43. You are working on a project that will upgrade the phone system in your customer service center. You have used bottom-up estimating techniques to assign costs to the project activities and have determined the cost baseline. Which of the following statements is true?

A. You have completed the Estimate Cost process and now need to complete the Determine Budget process to develop the project's cost baseline.

B. You have completed the Estimate Cost process and established a cost baseline to measure future projects against.

C. You have completed the Determine Budget process and now need to complete the Schedule Development process to establish a project baseline to measure future project performance against.

D. You have completed the Determine Budget process, and the cost baseline will be used to measure future project performance.

44. Each of the following options describes an element of the Develop Project Management Plan process except for which one?

A. Project charter

B. Outputs from other planning processes

C. Configuration management system

D. Organizational process assets

45. This type of leader leads the team in learning and maturing agile practices. They promote emotional intelligence and self-awareness, they are good listeners, put the needs of others first, help team members improve their skills, they coach and mentor, encourage safety, encourage respectful behaviors, build trust among the team, and promote the skills and intelligence of others. Which of the following leadership styles does this question describe and what are the three steps, in order, they use to help the team learn and mature agile processes? (Choose two.)

A. This question describes a servant leader.

B. This type of leader takes these steps in this order: people, purpose, and process.

C. This question describes a democratic leader.

D. This type of leader takes these three steps in this order: purpose, people, process.

46. Monte Carlo analysis can help predict the impact of risks on project deliverables. It is an element of one of the tools and techniques of one of the following processes. The other tools and techniques of this process include sensitivity analysis, decision tree analysis, and influence diagrams.

A. Plan Risk Responses

B. Perform Quantitative Risk Analysis

C. Identify Risks

D. Perform Qualitative Risk Analysis

47. You know that PV = 470, AC = 430, EV = 460, EAC = 500, and BAC = 525. What is VAC?

A. 65

B. 20

C. 25

D. 30

48. Which of the following contracts should you use for agile projects that will be priced based on user stories?

A. Multitiered structure

B. Dynamic scope

C. Graduated time and materials

D. Fixed-price increments

49. Every status meeting should have time allotted for reviewing risks. Which of the following options are true?

A. Risk identification and monitoring should occur throughout the life of the project.

B. Risk audits are performed during the Monitoring and Controlling phase of the project.

C. Risks should be monitored for their status and to determine whether the impacts to the objectives have changed.

D. Technical performance measurement variances may indicate that a risk is looming and should be reviewed at status meetings.

E. A, C, D

F. A, B, C, D

50. Name the two types of agile approaches discussed in the *Agile Practice Guide* (PMI®, 2017).

A. Iteration-based and flow-based

B. Hybrid and incremental-based

C. Incremental-based and agile

D. Predictive-based and release planning–based

51. Name the ethical code you'll be required to adhere to as a PMP® credential holder.

A. Project Management Policy and Ethics Code

B. PMI® Standards and Ethics Code of Conduct

C. Project Management Code of Professional Ethics

D. PMI® Code of Ethics and Professional Conduct

52. According to the *PMBOK® Guide*, the project manager is identified and assigned during which process?

A. During the Develop Project Charter process

B. At the conclusion of the Develop Project Charter process

C. Prior to beginning the Planning processes

D. Prior to beginning the Define Scope process

53. The project manager is responsible for all of the following regarding business value except which one?

A. Delivering the project so business value can be realized

B. Work with the team to subdivide tasks into the minimum viable product whenever possible

C. Measuring business value

D. Delivering business value at the end of the project

54. Which of the following statements are true regarding risks?

A. Risks might be threats to the objectives of the project.

B. Risks are certain events that may be threats or opportunities to the objectives of the project.

C. Risks might be opportunities to the objectives of the project.

D. Risks have causes and consequences.

E. A, C, D

55. Shu Ha Ri is a technique used to develop an agile team. Which of the following are true about this model?

   A. Shu stage means to obey or protect.

   B. Ha stage means to separate or leave.

   C. Ri stage means to break free or digress.

   D. Shu Ha Ri comes from Aikido, a Japanese martial art form.

   E. Shu Ha Ri is practiced in a progressive fashion.

   F. A, D, E

   G. A, B, D, E

56. Match the following scaling agile frameworks with their description.

Scaling agile frameworks

Framework	Description
A. LeSS	1. Two or more Scrum teams work on the project together. Each Scrum team focuses on a portion of the work.
B. SAFe	2. This consists of up to eight Scrum teams with up to eight members each who all work on the project together.
C. Enterprise Scrum	3. This extends Scrum practices to all aspects of the organization.
D. Scrum of Scrums	4. This combines several agile best practices and includes information from functional areas of the business.
E. DA	5. This is an interactive knowledge base consisting of technical guidance, knowledge, and information on agile.

57. Which performance measurement tells you what the projected total cost of the project will be at completion?

   A. ETC

   B. EV

   C. AC

   D. EAC

58. Which of the following statements is true regarding the Project Management Knowledge Areas?

   A. They include Initiation, Planning, Executing, Monitoring and Controlling, and Closing.

   B. They consist of 10 areas that bring together processes that have things in common.

   C. They consist of five processes that bring together phases of projects that have things in common.

   D. They include Planning, Executing, and Monitoring and Controlling processes because these three processes are commonly interlinked.

59. What are the Define Scope process tools and techniques?

    A. Cost–benefit analysis, scope baseline, expert judgment, and facilitated workshops

    B. Product analysis, alternatives generation, and expert judgment

    C. Product analysis, alternatives analysis, expert judgment, multicriteria decision analysis, and facilitation

    D. Alternatives generation, stakeholder analysis, and expert judgment

60. You are the project manager for Heartthrobs by the Numbers Dating Services. You're working on an updated website that will display pictures as well as short bios of prospective heartbreakers. You have your activity list and resource requirements in hand. You are using an adaptive methodology to manage the project. Which of the following is true?

    A. A Kanban Board is capacity based.

    B. A Scrum Board is time or velocity based.

    C. A burndown chart will show the remaining work of the sprint.

    D. All of the above.

61. Which performance measurement tells you the cost of the work that has been authorized and budgeted for a WBS component?

    A. PV

    B. EV

    C. AC

    D. BCWP

62. Your team is developing the risk management plan. Which tool and technique of this process is used to develop risk cost elements and schedule activities that will be included in the project budget and schedule?

    A. Meetings

    B. Data analysis

    C. Information-gathering techniques

    D. Risk data quality assessment

63. You are the project manager for Xylophone Phonics. It produces children's software programs that teach basic reading and math skills. You are performing the Plan Quality Management process and are identifying operational definitions. Which of the following does this describe?

    A. The quality metrics

    B. The quality management plan

    C. The project documents update

    D. The cost of quality

64. You need to convey some very complex, detailed information to the project stakeholders. What is the best method for communicating this kind of information?

   A. Verbal

   B. Vertical

   C. Horizontal

   D. Written

65. You have just prepared an RFP for release. Your project involves a substantial amount of contract work detailed in the RFP. Your favorite vendor drops by and offers to give you and your spouse the use of their company condo for your upcoming vacation. It's located in a beautiful resort community that happens to be one of your favorite places to go for a getaway. What is the most appropriate response?

   A. Thank the vendor, but decline the offer because you know this could be considered a conflict of interest.

   B. Thank the vendor, and accept. This vendor is always offering you incentives like this, so this offer does not likely have anything to do with the recent RFP release.

   C. Thank the vendor, accept the offer, and immediately tell your project sponsor so they're aware of what you're doing.

   D. Thank the vendor, but decline the offer because you've already made another arrangement for this vacation. Ask them whether you can take a rain check and arrange another time to use the condo.

66. Directing project work on an agile project consists of several steps. Match the following steps with their descriptions.

Directing agile teamwork

Steps	Description
A. First step	1. Review meetings are held to examine the work of the iteration and provide and receive feedback.
B. Second step	2. Daily stand-ups are conducted to examine what was worked on yesterday, what will be worked on today, and what obstacles are standing in the way.
C. Third step	3. The product backlog is defined.
D. Fourth step	4. Retrospectives are held at the end of the iteration to determine what went well, what improvements can be made to the process, and what didn't go well.
E. Fifth step	5. Planning meetings are held at the beginning of the iteration to pull user stories into the iteration backlog.

67. You are a project manager for Waterways Houseboats, Inc. You've been asked to perform a cost–benefit analysis for two proposed projects. Project A costs $2.4 million, with potential benefits of $12 million and future operating costs of $3 million. Project B costs $2.8 million, with potential benefits of $14 million and future operating costs of $2 million. Which project should you recommend?

- **A.** Project A, because the cost to implement it is cheaper than with Project B
- **B.** Project A, because the potential benefits plus the future operating costs are less in value than the same calculation for Project B
- **C.** Project B, because the potential benefits minus the implementation and future operating costs are greater in value than the same calculation for Project A
- **D.** Project B, because the potential benefits minus the costs to implement are greater in value than the same calculation for Project A

68. Louis R. Pondy, a professor of business administration and author on organizational management and other topics, developed the stages of conflict. Which of the following are stages of conflict according to Pondy? (Choose three.)

- **A.** Collaborate
- **B.** Latent
- **C.** Avoid
- **D.** Felt
- **E.** Perceived

69. You are performing alternatives analysis as part of the Define Scope process. Which of the following options is not true?

- **A.** Alternatives analysis is a component of the data analysis tool and technique.
- **B.** Alternatives analysis is used in the Plan Scope Management process and the Define Scope process.
- **C.** Alternatives analysis involves unanimity, plurality, majority, and autocratic voting methods.
- **D.** Brainstorming and lateral thinking are types of alternative analysis.

70. The project manager has the greatest influence over quality during which process?

- **A.** Plan Quality Management
- **B.** Manage Quality
- **C.** Control Quality
- **D.** Monitor Quality

71. What type of organization experiences the least amount of stress during project closeout?

- **A.** Project-oriented
- **B.** Functional
- **C.** Weak matrix
- **D.** Strong matrix

72. You are working on the product description for your company's new line of ski boots. Your customers have been asking for changes in style. New advances in the manufacturing process allows you to make these changes quickly and get the new line on the shelves before the next ski season. Your organization hopes to increase revenues and market share by offering this new line of boots. Which of the following are true? (Choose two.)

    A. One KPI used to measure business value might be improving the organization's business relationship with the manufacturer.

    B. The business value for this project is increasing revenues and market share.

    C. This project came about due to an organizational need to add a new style of boots and get them to market quickly.

    D. This project came about as a result of a customer request and technological advance.

73. The business need or demand that brought about the project, high-level scope description, analysis of the problem or opportunity the project presents, recommendation, and an evaluation statement together describe elements of which of the following?

    A. Organizational process assets

    B. The feasibility study

    C. The business case

    D. The project charter

74. Which of the following statements is true regarding constraints and assumptions?

    A. Constraints restrict the actions of the project team, and assumptions are considered true for planning purposes.

    B. Constraints are considered true for planning purposes, and assumptions limit the options of the project team.

    C. Constraints consider vendor availability and resource availability to be true for planning purposes. Assumptions limit the project team to work within predefined budgets or timelines.

    D. Constraints and assumptions are inputs to the Initiation process. They should be documented because they will be used throughout the project Planning process.

75. People are motivated by the need for achievement, power, or affiliation according to which theory?

    A. Expectancy Theory

    B. Achievement Theory

    C. Contingency Theory

    D. Theory X

76. You are a project manager working in a foreign country. You observe that some of your project team members are having a difficult time adjusting to the new culture. You provided them with training on cultural differences and the customs of this country before they arrived, but they still seem uncomfortable and disoriented. Which of the following statements is true?

A. This is the result of working with teams of people from two different countries.

B. This condition is known as culture shock.

C. This is the result of jet lag and travel fatigue.

D. This condition is known as global culturalism.

77. As a result of a face-to-face meeting you recently had to discuss the items in your issue log, you have resolved issues, managed expectations, and come away with an action plan that will improve project performance and will also require an update to the communications management plan. Which process does this describe?

   A. Manage Stakeholder Engagement

   B. Control Communications

   C. Manage Project Communications

   D. Manage Team

78. Which of the following are a type of agile project management methodology primarily used for information technology projects? (Choose two.)

   A. Scrum

   B. Six Sigma

   C. XP

   D. Kaizen

79. What is the definition of free float?

   A. The amount of time you can delay the earliest start of a task without delaying the ending of the project

   B. The amount of time you can delay the start of a task without delaying the earliest start of a successor task

   C. The amount of time you can delay the latest start of a task without delaying the ending of the project

   D. The amount of time you can delay the start of a task without delaying the earliest finish of a successor task

80. Generational diversity is an important component of diversity and inclusion when building your team. Which of the following are true statements about the five generations in the workplace today? (Choose two.)

   A. Baby Boomers experienced rationing of food, gas, and other everyday items while growing up. They are often frugal and strong savers.

   B. Millennials are also known as the "latchkey" generation.

   C. Gen X are also known as the "MTV" generation.

   D. Gen Z grew up with technology as a way of life and are heavily influenced by social media.

81. Your project involves the research and development of a new food additive. You're ready to release the product to your customer when you discover that a minor reaction might occur in people with certain conditions. The reactions to date have been very minor, and no known long-lasting side effects have been noted. As project manager, what should you do?

   A. Do nothing because the reactions are so minor that very few people will be affected.

   B. Inform the customer that you've discovered this condition and tell them you'll research it further to determine its impacts.

   C. Inform your customer that there is no problem with the additive except for an extremely small percentage of the population and release the product to them.

   D. Tell the customer you'll correct the reaction problems in the next batch, but you'll release the first batch of product to them now to begin using.

82. You are a project manager working on gathering requirements and establishing estimates for the project. Which process group are you in?

   A. Planning

   B. Executing

   C. Initiating

   D. Monitoring and Controlling

83. According to the *PMBOK® Guide*, which of the following names all the components of an interactive communication model?

   A. Encode, transmit, decode

   B. Encode, transmit, decode, acknowledge, feedback/response

   C. Encode, transmit, decode, feedback/response

   D. Encode, transmit, acknowledge, decode

84. Who is responsible for performing and managing project integration when using an agile project management approach?

   A. Team members

   B. Product owner

   C. Scrum master

   D. Project manager

85. The Plan Procurement process applies evaluation criteria to bids and proposals and selects a vendor. It also uses independent estimates to compare vendor prices. This is also known as which of the following?

   A. Independent comparisons

   B. Analytical techniques

   C. Should cost estimates

   D. Expert judgment

86. All of the following statements are true of the project Closing process group except for which one?

  A. Probability for success is greatest in the project Closing process group.

  B. The project manager's influence is greatest in the project Closing process group.

  C. The stakeholders' influence is least in the project Closing process group.

  D. Risk occurrence is greatest in the project Closing process group.

87. Which of the following can you use in addition to the probability and impact matrix to prioritize risks?

  A. Urgency

  B. Manageability

  C. Propinquity

  D. Detectability

  E. PESTLE

  F. A, B, C, D

  G. A, B, C, D, E

88. As a PMP® credential holder, one of your responsibilities is to ensure integrity on the project. When your personal interests are put above the interests of the project or when you use your influence to cause others to make decisions in your favor without regard for the project outcome, this is considered which of the following?

  A. Conflict of interest

  B. Using professional knowledge inappropriately

  C. Culturally unacceptable

  D. Personal conflict issue

Answers to Assessment Test

1. F. All of the options are considered when examining whether the tailoring processes and project integration management produced the results you were expecting.

2. D. Fast tracking is the best answer in this scenario. Budget was the original constraint on this project, so it's unlikely the project manager would get more resources to assist with the project. The next best thing is to compress phases to shorten the project duration. For more information, please see Chapter 1.

3. A. Discretionary dependencies can create arbitrary total float values, and they can also limit scheduling options. For more information, please see Chapter 5.

4. A. Project managers spend about 90 percent of their time communicating through status meetings, team meetings, email, verbal communications, and so on. For more information, please see Chapter 10.

5. A-6, B-7, C-4, D-3, E-1, F-5, G-2. Agile measurements should focus on customer value. For more information, please see Chapter 12.

6. D. This question describes a predictive life cycle methodology. Options B, C, and D are all adaptive methodologies, which have high degrees of uncertainty, high risk, evolving requirements, and cross-functional teams. For more information, see Chapter 7.

7. A. Change control systems are a subset of the configuration management system. Change control systems manage changes to the deliverables and/or project baselines. For more information, please see Chapter 11.

8. D The iterative approach uses prototypes and mockups produced in time-bound periods such as sprints. The incremental approach produces usable deliverables at the end of the workflow, which helps in performing the work faster and speeds up the project. The incremental approach focuses on speed of delivery and the iterative approach focuses on learning optimization. For more information, please see Chapter 6.

9. A. This question describes an agile life cycle methodology. Hybrid life cycles have some degree of uncertainty and risk, but not at the level the question describes. Option C and D describe predictive methodologies which have high degrees of certainty, firm requirements, and stable teams.

10. C. The smoothing technique (also known as accommodate) does not usually result in a permanent solution. The problem is downplayed to make it seem less important than it is, which makes the problem tend to resurface later. For more information, please see Chapter 10.

11. B. The components of the interactive communication model are encode, transmit, decode, acknowledge, and feedback/response. The basic communication model consists of the sender, message, and receiver elements. For more information, please see Chapter 6.

12. B. Benefit measurement methods are project selection methods that use benefit cost ratio and other financial analysis to select projects. For more information, please see Chapter 2.

13. B. This question describes the forcing technique because the project sponsor insisted on implementing their solution. The forcing technique occurs when one party forces a solution on others. For more information, please see Chapter 10.

14. F. These factors, and others, may impact the organization and/or your project and you should continually monitor and review both the internal and external environment for changes that can impact the project. For more information, please see Chapter 11.

15. C. The scope management plan outlines how project scope will be managed and how scope changes will be incorporated into the project. For more information, please see Chapter 4.

16. E. Cause-and-effect diagrams—also called *Ishikawa, fishbone diagrams,* and *why-why diagrams*—show the relationship between the effects of problems and their causes. Kaoru Ishikawa developed cause-and-effect diagrams. For more information, please see Chapter 10.

17. C. Negotiation, influencing, and business skills are all important for a project manager to possess. However, good communication skills are the most important skills a project manager can have. For more information, please see Chapter 1.

18. E. Inspections are also called reviews, peer reviews, walkthrough, and audits. For more information, please see Chapter 11.

19. A. Analogous estimating—also called *top-down estimating*—is a form of expert judgment. Analogous estimating can be used to estimate cost or time and considers historical information from previous, similar projects. For more information, please see Chapter 5.

20. A. Project managers have the highest level of power and authority in a project-oriented organization. They also have high levels of power and authority in a strong matrix organization. For more information, please see Chapter 3.

21. C. Plan Stakeholder Engagement is concerned with determining the engagement levels of the stakeholders, understanding their needs and interests, and understanding how they might impact the project or how the project may impact them. For more information, please see Chapter 6.

22. C. Unknown risks might be threats or opportunities to the project, and the project manager should set aside contingency reserves to deal with them. For more information, please see Chapter 7.

23. C. Sharing knowledge and creating knowledge are the focus of this process. Option A describes the Manage Communications process, option B describes the Manage Stakeholder Engagement process, and option D describes the Manage Quality process. For more information, please see Chapter 10.

24. B. The Plan Resource Management process identifies project resources, documents roles and responsibilities of project team members, and documents reporting relationships. For more information, please see Chapter 8.

25. A-4, B-1, C-3, D-5, E-2. Testing at all levels is a concept used in Extreme Programming, and other agile methodologies, to expose issues and problems early in the coding process. For more information, see Chapter 9.

26. B. The planning meeting occurs at the beginning of an iteration or sprint. Team members choose the items from the backlog that they will work on in the upcoming sprint. For more information, please see Chapter 5.

27. D. Pareto diagrams rank-order important factors for corrective action by frequency of occurrence. For more information, please see Chapter 11.

28. D. Team building does occur throughout the life of the project, but ground rules are what establish clear expectations and behaviors for project team members. For more information, please see Chapter 9.

29. D. The payback period for Project Fish'n for Chips is eight months. This project will receive $300,000 every three months, or $100,000 per month. Project Fish'n for Chips has the shortest payback period and should be chosen over Project Picnic. For more information, please see Chapter 2.

30. C. Fast-tracking is a compression technique that increases risk and potentially causes rework. Fast-tracking is performing two activities previously scheduled to start one after the other in parallel. For more information, please see Chapter 5.

31. B. Inspection involves physically looking at, measuring, or testing results to determine whether they conform to your quality standards. For more information, please see Chapter 11.

32. B. Functional managers who have a lot of authority and power working with project coordinators who have minimal authority and power characterizes a weak matrix organization. Project managers in weak matrix organizations are sometimes called *project coordinators*, *project leaders*, or *project expeditors*. For more information, please see Chapter 3.

33. C. The deliverables are validated and accepted during the Validate Scope process. For more information, please see Chapter 12.

34. A. The requirements documentation contains a list of requirements for the project along with other important information regarding the requirements. For more information, please see Chapter 4.

35. B. The purpose of a project charter is to recognize and acknowledge the existence of a project and commit resources to the project. The charter names the project manager and project sponsor, but that's not its primary purpose. For more information, please see Chapter 2.

36. A, B. Identify Stakeholders tools and techniques are expert judgment, data gathering, data analysis, data representation, and meetings. The Salience model and power/interest grid are two of the data representation techniques you can use to categorize and show stakeholder information. The stakeholder register is where the information is recorded. For more information, please see Chapter 3.

37. C. Secondary risk events occur as a result of the implementation of a response to another risk. For more information, please see Chapter 7.

38. C. The four types of project endings are addition, integration, starvation, and extinction. For more information, please see Chapter 12.

39. B. Estimate Costs is where activity costs are estimated using some of the tools and techniques listed in the question. The remaining tools and techniques of this process are expert judgment, data analysis, project management information system, and decision-making. For more information, please see Chapter 6.

40. B. Failure costs are associated with the cost of quality and are also known as cost of poor quality. For more information, please see Chapter 8.

41. D. The critical chain is a resource-constrained critical path that adds duration buffers to help protect schedule slippage. For more information, please see Chapter 5.

42. A. Scope changes will cause schedule revisions, but schedule revisions do not change the project scope. Project requirements are part of the project scope statement; therefore, scope change might come about as a result of changes to the project requirements, as stated in option D. For more information, please see Chapter 12.

43. D. The Determine Budget process establishes the cost baseline, which is used to measure and track the project throughout the remaining process groups. For more information, please see Chapter 6.

44. C. The inputs to Develop Project Management Plan include project charter, outputs from other processes, enterprise environmental factors (EEF), and organizational process assets (OPA). The tools and techniques of this process are expert judgment, data gathering, interpersonal and team skills, and meetings. For more information, please see Chapter 4.

45. A, D. This question describes a servant leader. They take three steps to ensure the team learns and matures the agile process: purpose, people, process. For more information, please see Chapter 9.

46. B. Monte Carlo analysis is a simulation technique that is part of a simulation tool and technique performed in the Perform Quantitative Risk Analysis process. For more information, please see Chapter 7.

47. C. VAC is calculated this way: VAC = BAC – EAC. Therefore, 525 – 500 = 25. For more information, please see Chapter 12.

48. D. Fixed-price increments are contracts used on agile projects that are based on breaking down the work into user stories, rather than pricing the contract as a whole. For more information, please see Chapter 8.

49. E. Risk audits should be performed throughout the life of the project, and you are specifically interested in looking at the implementation and effectiveness of risk strategies. For more information, please see Chapter 11.

50. A. The two types of agile approaches discussed in the *Agile Practice Guide* (PMI®, 2017) are iteration-based (like Scrum) and flow-based (like Kanban). For more information, please see Chapter 6.

51. D. The *PMI® Code of Ethics and Professional Conduct* is published by PMI®, and all PMP® credential holders are expected to adhere to its standards. For more information, please see Chapter 12.

52. A. According to the *PMBOK® Guide*, the project manager should be assigned during the development of the project charter, which occurs in the Develop Project Charter process. For more information, please see Chapter 2.

53. D. The project manager is responsible for delivering business value incrementally throughout the project, not just at the end of the project. For more information, please see Chapter 3.

54. E. Risks are uncertain events that may be threats or opportunities to the objectives of the project. For more information, please see Chapter 7.

55. F. Shu Ha Ri is a technique that comes from Aikido. Shu means to obey or protect, Ha means to break free or digress, and Ri means to separate or leave. For more information, please see Chapter 9.

56. A-2, B-5, C-3, D-1, E-4. Scaling agile frameworks is a technique used to scale agile practices to the organization and incorporate multiple teams using agile methodologies. For more information, please see Chapter 10.

57. D. Estimate at completion (EAC) estimates the total cost of the project at completion based on the performance of the project to date. For more information, please see Chapter 12.

58. B. The project management Knowledge Areas bring together processes that have commonalities. For example, the Project Quality Management Knowledge Area includes the Plan Quality Management, Manage Quality, and Control Quality processes. For more information, please see Chapter 2.

59. C. The tools and techniques of the Define Scope process include product analysis, alternatives analysis, expert judgment, multicriteria decision analysis, and facilitation. For more information, please see Chapter 4.

60. D. All of the options are true in relation to an adaptive methodology. For more information, please see Chapter 5.

61. A. Planned value is the cost of work that has been authorized and budgeted for a schedule activity or WBS component. For more information, please see Chapter 12.

62. A. The Plan Risk Management process contains three tools and techniques: data analysis (stakeholder analysis), expert judgment, and meetings. Meetings are used to determine the plans for performing risk management activities. One of the key components of these meetings is to determine risk cost elements, along with schedule activities, and definitions of terms, and the development or definition of the probability and impact matrix. For more information, please see Chapter 7.

63. A. Operational definitions are quality metrics. They describe what is being measured and how it will be measured during the Control Quality process. For more information, please see Chapter 8.

64. D. Information that is complex and detailed is best conveyed in writing. A verbal follow-up would be good to answer questions and clarify information. Vertical and horizontal are ways of communicating within the organization. For more information, please see Chapter 10.

65. A. The best response is to decline the offer. This is a conflict of interest, and accepting the offer puts your own integrity and the contract award process in jeopardy. For more information, please see Chapter 12.

66. A-3, B-5, C-2, D-1, E-4. Agile projects typically follow a workflow that consists of defining the product backlog, holding planning meetings, conducting daily stand-ups, holding review meetings, and conducting retrospectives. For more information, see Chapter 9.

67. C. Project B's cost–benefit analysis is a $9.2 million benefit to the company, compared to $6.6 million for Project A. Cost–benefit analysis takes into consideration the initial costs to implement and future operating costs. For more information, please see Chapter 2.

68. B, D, E. Pondy identified five stages of conflict including latent, perceived, felt, manifest, and aftermath. Options A and C describe conflict resolution techniques. For more information, please see Chapter 10.

69. C. Option C describes voting methods that are used in the decision-making tool and technique. They are not part of alternatives analysis. For more information, please see Chapter 4.

70. B. Manage Quality is the process where project managers have the greatest amount of influence over quality. For more information, please see Chapter 10.

71. C. Weak matrix organizational structures tend to experience the least amount of stress during the project closeout processes. For more information, please see Chapter 12.

72. B, D. Business value brings short- or long-term benefits to the organization. Business value for this project is stated in the question as increasing revenues and market share. KPIs are usually numeric metrics used to examine whether business value was achieved. This project was requested by customers and the new manufacturing processes describe a technological advancement. Organization need usually entails projects focused on internal organizational needs such as upgrading a software system or finding a new building to lease. For more information, please see Chapter 3.

73. C. These elements are part of the business case used as an input (through the business documents input) to the Develop Project Charter process. For more information, please see Chapter 2.

74. A. Constraints limit the options of the project team by restricting action or dictating action. Scope, time, and cost are the three most common constraints, and each of these has an effect on quality. Assumptions are presumed to be true for planning purposes. Always validate your assumptions. For more information, please see Chapter 4.

75. B. Achievement Theory conjectures that people are motivated by the need for achievement, power, or affiliation. For more information, please see Chapter 9.

76. B. When people work in unfamiliar environments, culture shock can occur. Training and researching information about the country you'll be working in can help counteract this. For more information, please see Chapter 12.

77. A. The clues in this question are the face-to-face meetings resolving issues, managing expectations, and improving project performance, which are the primary purposes of the Manage Stakeholder Engagement process. Project management plan updates include both the communications management plan and stakeholder engagement plan and are an output of this process. For more information, please see Chapter 10.

78. A, C. Scrum and Extreme Programming (XP) are agile project management methodologies that are used in the information technology field. For more information, please see Chapter 1.

79. B. Option A describes total float. Options C and D are incorrect. For more information, please see Chapter 5.

80. C, D. Option A describes the Silent Generation and option B describes the Gen X generation. For more information, please see Chapter 9.

81. B. Honesty and truthful reporting are required of PMP® credential holders. In this situation, you would inform the customer of everything you know regarding the problem and work to find alternative solutions. For more information, please see Chapter 12.

82. A. The Planning process group is where requirements are fleshed out and estimates on project costs and time are made. For more information, please see Chapter 1.

83. B. The components of the interactive communication model are encode, transmit, decode, acknowledge, and feedback/response. The basic communication model consists of the sender, message, and receiver elements. For more information, please see Chapter 6.

84. D. The project manager is responsible for performing and managing project integration while the project team members are responsible for planning, control, and delivery of the product. For more information, please see Chapter 4.

85. C. Independent estimates are also known as should cost estimates. For more information, please see Chapter 8.

86. D. Risk occurrence is lowest during the Closing process group because you've completed the work of the project at this point. However, risk impacts are the greatest in the Closing process because you have much more at stake. For more information, please see Chapter 12.

87. F. PESTLE is used to assist in identifying risks. The other factors you can consider that are not listed in the options are proximity, dormancy, controllability, connectivity, and strategic impact. For more information, please see Chapter 7.

88. A. A conflict of interest is any situation that compromises the outcome of the project or ignores the impact to the project to benefit yourself or others. For more information, please see Chapter 12.

Chapter

1

Building the Foundation

THE PMP® EXAM CONTENT FROM THE PEOPLE DOMAIN COVERED IN THIS CHAPTER INCLUDES THE FOLLOWING:

✓ **Task 1.2: Lead a team**

  ▪ Task 1.2.1 Set a clear vision and mission

✓ **Task 1.9 Collaborate with stakeholders**

  ▪ 1.9.2 Optimize alignment between stakeholder needs, expectations, and project objectives

THE PMP® EXAM CONTENT FROM THE PROCESS DOMAIN COVERED IN THIS CHAPTER INCLUDES THE FOLLOWING:

✓ **Task 2.2: Manage communications**

  ▪ 2.2.3 Communicate project information and updates effectively

✓ **Task 2.13: Determine appropriate project methodology/ methods and practices**

  ▪ 2.13.4 Recommend a project methodology/approach

Congratulations on your decision to study for and take the Project Management Institute (PMI®) Project Management Professional (PMP)® certification exam (PMP® exam). This book was written with you in mind. The focus and content of this book revolve around the principles of sound project management as outlined in the three domains contained in the *Project Management Professional (PMP)® Examination Content Outline* published in June 2019. The three domains include People, Process, and Business Environment. I will cover each of these domains in depth in this book.

PMI® has also published a body of work called *A Guide to the Project Management Body of Knowledge (PMBOK® Guide), Sixth Edition.* This guide covers the process aspects of project management and I will refer to this information during the course of this book. Keep in mind that the PMP® exam is primarily principle focused, rather than process focused, but all topics are potential exam questions, so don't skip anything in your study time. When possible, I'll pass on hints and study tips that I collected while studying for the exam. I will also be referencing another publication from PMI® called the *Agile Practice Guide* (PMI®, 2017). This guide outlines practices related to agile and hybrid development methodologies. A large portion of the exam will cover agile and hybrid techniques, and I'll cover those in detail throughout the book.

PMI® is the de facto standard for project management principles, techniques, and processes. To become familiar with the material and principles they've developed, I recommend first familiarizing yourself with the terminology used in the *PMBOK® Guide.* Volunteers from differing industries from around the globe worked together to come up with the standards and terms used in the guide. These folks worked hard to develop and define project management terms, and the terms are used interchangeably among industries. For example, *resource planning* means the same thing to someone working in construction, information technology, or healthcare. You'll find many of the *PMBOK® Guide* terms explained throughout this book. Even if you are an experienced project manager, you might find that you use specific terms for processes or actions you regularly perform but that the *PMBOK® Guide* calls them by another name. So, the first step is to get familiar with the terminology.

The next step is to become familiar with the processes as defined in the *PMBOK® Guide.* The process names are unique to PMI®, but the general principles and guidelines underlying the processes are used for most projects across industry areas.

This chapter starts with the Process domain and lays the foundation for building and managing your project. We'll focus on this domain through Chapter 5. I'll start this chapter with an overview of projects versus operations and a discussion on how projects come about. We'll take a look at the PMI® process groups and their purposes and end the chapter with an overview of project management life-cycle methodologies. We'll examine the various project methodologies such as predictive (also known as waterfall), agile methodologies (of which there are several), and the hybrid approach. These project management methodologies will be discussed and used in examples in the remainder of this book. I'll continue to build on these topics, and others, throughout the book. Good luck!

> The process names, inputs, tools and techniques, outputs, and descriptions of the project management process groups and related materials and figures in this chapter are based on content from *A Guide to the Project Management Body of Knowledge (PMBOK® Guide), Sixth Edition* (PMI®, 2017). The references to adaptive and hybrid methodologies, related materials, and figures in this chapter are based on content from the *Agile Practice Guide* (PMI®, 2017).

Establishing the Foundation

Consider the following scenario: The VP of marketing approaches you with a fabulous idea—"fabulous" because he's the big boss and because he thought it up. He wants to live-stream the organization's board meetings and live-stream town hall meetings with the CEO and allow employees to ask questions in real time. He tells you that the board of directors has already cleared the project, and he'll dedicate as many resources to this as he can. He wants the live-streaming solution in place by the end of this year. The best news is he has assigned you to head up this project.

Your first question should be "Is it a project?" This might seem elementary, but projects are often confused with ongoing operations. Projects are temporary in nature; have definite start and end dates; produce a unique product, service, or result; and are completed when their goals and objectives have been met and signed off by the stakeholders or when the project is terminated.

Exam Spotlight

Projects may be terminated for any number of reasons. Examples include: the objectives have been met, the objectives cannot be met, funding is no longer available or was spent, the project is no longer needed, or it was terminated for legal cause or convenience.

A common characteristic of projects is that they often initiate change in an organization by moving the business from one state to another. For example, when an organization undergoes a new back office software implementation or upgrade, changes will occur within the organization to accommodate the new business process and software. The business moves from current state—where they are now—to the future state—where they will be once the software is implemented.

Projects also bring about business value creation. Business value can be tangible (revenue, goods, market share) or intangible (goodwill, recognition, public benefit).

When considering whether you have a project on your hands, you need to keep some issues in mind. First, is it a project or an ongoing operation? If so, what characteristics distinguish this endeavor as a project? We'll look at each of these next.

Projects vs. Operations

Projects are temporary in nature and have definitive start dates and definitive end dates. The project is completed when its goals and objectives are accomplished (by producing deliverables) to the satisfaction of the stakeholders. Sometimes projects end when it's determined that the goals and objectives cannot be accomplished or when the product, service, or result of the project is no longer needed and the project is canceled. Projects exist to bring about a product, service, or result that didn't exist before. This might include tangible products, components of other products, services such as consulting or project management, and business functions that support the organization. Projects might also produce a result or an outcome, such as a document that details the findings of a research study. In this sense, a project is unique. However, don't be confused by the term *unique*. For example, Ford Motor Company is in the business of designing and assembling cars. Each model that Ford designs and produces can be considered a project when it is first introduced to the marketplace. The models differ from one another in their features and are marketed to people with various needs. An SUV serves a different purpose and clientele than a luxury sedan or a hybrid. The initial design and marketing of these three models are unique projects. However, the actual assembly of the cars is considered an operation—a repetitive process that is followed for most makes and models.

Determining the characteristics and features of the different car models is carried out through what the *PMBOK® Guide* terms *progressive elaboration*. This means the characteristics of the product, service, or result of the project are determined incrementally and are continually refined and worked out in detail as the project progresses. More information and better estimates become available the further you progress in the project. Progressive elaboration improves the project team's ability to manage greater levels of detail and allows for the inevitable change that occurs throughout a project's life cycle. This concept goes along with the temporary and unique aspects of a project because when you first start the project, you don't know all the minute details of the end product. Product characteristics typically start out broad-based at the beginning of the project and are progressively elaborated into more and more detail over time until they are complete and finalized.

> **Exam Spotlight**
>
> Progressive elaboration is most often used when creating the project or product scope, developing requirements, determining human resources, scheduling, and defining risks and their mitigation plans.

Operations are ongoing and repetitive. They involve work that is continuous without an ending date, and you often repeat the same processes and produce the same results. One way to think of operations is the transforming of resources (steel and fiberglass, for example) into outputs (cars). The purpose of operations is to keep the organization functioning, whereas the purpose of a project is to meet its goals and to conclude. At the completion of a project, or at various points throughout the project, the deliverables may get turned over to the organization's operational areas for ongoing care and maintenance. For example, let's say your company implements a new human resources software package that tracks employees' time, expense reports, benefits, and so on. Defining the requirements and implementing the software is a project. The ongoing maintenance of the site, updating content, and so on are ongoing operations.

It's a good idea to include some members of the operational area on the project team when certain deliverables or the end product of the project will be incorporated into their future work processes. They can assist the project team in defining requirements, developing scope, creating estimates, and so on, helping to ensure that the project will meet their needs. The process of knowledge transfer to the team is much simpler when they are involved throughout the project; they gain knowledge as they go and the formal handoff is more efficient. This isn't a bad strategy in helping to gain buy-ins from the end users of the product or service either. Often, business units are resistant to new systems or services, but getting them involved early in the project rather than simply throwing the end product over the fence when it's completed may help gain their acceptance.

Managing projects and managing operations require different skill sets. Operations management involves managing the business operations that support the goods and services the organization is producing. Operations managers may include line supervisors in manufacturing, retail sales managers, and customer service call center managers. The skills needed to manage a project include general management skills, interpersonal skills, planning and organization skills, and more. The remainder of this book will discuss project management skills in detail.

According to the *PMBOK® Guide*, several examples exist where projects can extend into operations up until and including the end of the product life cycle:

- Concluding the end of each phase of the project
- Developing new products or services

- Upgrading and/or expanding products or services
- Improving the product development processes
- Improving operations

The preceding list isn't all inclusive. Whenever you find yourself working on a project that ultimately impacts your organization's business processes, I recommend you get people from the business units to participate on the project.

Project Characteristics

You've just learned that a project has several characteristics:

- Projects are unique.
- Projects are temporary in nature and have a definite beginning and ending date.
- Projects are completed when the project goals are achieved or it's determined the project is no longer viable.
- A successful project is one that meets the expectations of your stakeholders.
- Projects initiate change in the organization.
- Projects bring about business value creation.

Using these criteria, let's examine the assignment from the VP of marketing to determine whether it is a project:

Is it unique?   Yes, because the board conference room does not currently have live-streaming equipment or software.

Does the project have a limited time frame?   Yes, the start date of this project is today, and the end date is the end of this year. It is a temporary endeavor.

Is there a way to determine when the project is completed?   Yes, the live-streaming equipment and software will be installed. Once the system is intact and operating, the project will come to a close.

Is there a way to determine stakeholder satisfaction?   Yes, the expectations of the stakeholders will be documented in the form of deliverables and requirements during the Planning processes. These deliverables and requirements will be compared to the finished product to determine whether it meets the expectations of the stakeholders.

Is it driving change?   Yes, this is a new way of communicating with their shareholders and employees.

Will business value creation be realized?   Yes, this project will improve communication among shareholders and employees.

Houston, we have a project.

What Is Project Management?

You've determined that you indeed have a project. What now? The notes you scratched on the back of a napkin during your coffee break might get you started, but that's not exactly good project management practice.

We have all witnessed this scenario: an assignment is made, and the project team members jump directly into the project, busying themselves with building the product, service, or result requested. Often, careful thought is not given to the project-planning process. I'm sure you've heard co-workers toss around statements like, "That would be a waste of valuable time" or "Why plan when you can just start building?" Project progress in this circumstance is rarely measured against the customer requirements. In the end, the delivered product, service, or result doesn't meet the expectations of the customer. This is a frustrating experience for all those involved. Unfortunately, many projects follow this poorly constructed path.

Project management brings together a set of tools and techniques—performed by people—to describe, organize, and monitor the work of project activities. *Project managers* are the people responsible for managing the project processes and applying the tools and techniques used to carry out the project activities. All projects are composed of processes, even if they employ a haphazard approach. There are many advantages to organizing projects and teams around the project management processes endorsed by PMI®. However, process is only one side of the equation. A project manager is a business-savvy leader who is able to link project benefits with business objectives and ensure that the project produces business value. They will do this in part with project management processes and by utilizing their business acumen and leadership skills. We'll be examining both of these concepts in depth throughout the remainder of this book.

According to the *PMBOK® Guide*, project management involves applying knowledge, skills, tools, and techniques during the course of the project to meet requirements. It is the responsibility of the project manager to ensure that project management techniques are applied and followed.

Exam Spotlight

Remember that according to the *PMBOK® Guide*, the definition of project management is applying tools, techniques, skills, and knowledge to project activities to bring about successful results and meet the project requirements.

Project management is a collection of processes that includes initiating a new project, planning, putting the project management plan into action, and measuring progress and performance. It involves identifying the project requirements, establishing project objectives, balancing constraints, and taking the needs and expectations of the key stakeholders into consideration. Planning is one of the most important functions you'll perform during the course of a project. It sets the standard for the remainder of the project's life and is used to track future project performance. Let's look at some of the ways the work of project management is organized.

Programs

According to the *PMBOK® Guide*, programs are groups of related projects, subsidiary programs, and other activities that are managed using similar techniques in order to capitalize on benefits that wouldn't be feasible if you managed the projects individually. When projects are managed collectively as programs, it's possible to capitalize on benefits that wouldn't be achievable if the projects were managed separately. This would be the case where a very large program exists with many subsidiary projects under it—for example, building an urban live-work-shopping development. Many subsidiary projects exist underneath this program, such as design and placement of living and shopping areas, architectural drawings, theme and design, construction, marketing, facilities management, and so on. Each subsidiary project is a project unto itself. Each subsidiary project has its own project manager, who reports to a project manager with responsibility over several of the areas, who in turn reports to the head project manager (often called a program manager) who is responsible for the entire program. All the projects are related and are managed together so that collective benefits are realized and controls are implemented and managed in a coordinated fashion. Sometimes programs involve aspects of ongoing operations as well. After the shopping areas in our example are built, the management of the buildings and common areas becomes an ongoing operation. The management of this collection of projects—determining their interdependencies, managing among their constraints, and resolving issues among them—is called *program management*. Program management also involves centrally managing and coordinating groups of related projects to meet the objectives of the program.

A new type of project referenced in the *PMBOK® Guide* is called a *megaproject*. These projects, as implied by the term, are very large and take multiple years to complete. The total investment of these multiyear projects must be $1 billion or more (in US dollars) and the project must affect 1 million people or more. A few examples might include the construction of a new interstate highway or the implementation of a national healthcare system.

Portfolios

Portfolios are collections of programs, subsidiary portfolios, operations, and projects that support strategic business goals or objectives. Let's say our company is in the construction business. Our organization has several business units: retail, single-family residential, and multifamily residential. Collectively, the projects within all of these business units make up the portfolio. The program I talked about in the preceding section (the collection of projects associated with building the new live-work-shopping urban area) is a program within our portfolio. Other programs and projects could be contained within this portfolio as well. Programs and projects within a portfolio are not necessarily related to one another in a direct way. However, the overall objective of any program or project in this portfolio is to meet the strategic objectives of the portfolio, which in turn should meet the strategic objectives of the department and ultimately the business unit or corporation.

Portfolio management encompasses centrally managing the collections of programs, projects, other work, and sometimes other portfolios. It involves guiding the investment decisions of the organization and ensures that the investments are appropriately applied to the projects and programs so that they advance the organization's strategic goals. The project management office (discussed in the next section) is typically responsible for managing portfolios. This gives a central, enterprise view to the projects for the organization as a whole and leads to more effective management of the programs and projects within the portfolio. Maximizing the portfolio is critical to increasing success. This includes weighing the value of each project, or potential project, against the business's strategic objectives, selecting the right programs and projects, eliminating projects that don't add value, and prioritizing resources among programs and projects. It also concerns monitoring active projects for adherence to objectives, balancing the portfolio among the other investments of the organization, and ensuring the efficient use of resources. Portfolio managers also monitor the organizational planning activities of the organization to help prioritize projects according to fund availability, risk, the strategic mission, and more. As you can see, portfolio management (and program management) has a different objective, life cycle, and benefits than project management. Portfolio management is generally performed by a senior manager who has significant experience in both project and program management.

Exam Spotlight

Projects or programs within a portfolio are not necessarily related to or dependent on each other.

Table 1.1 compares the differences between projects, programs, and portfolios according to the *PMBOK® Guide*.

TABLE 1.1   Projects, programs, and portfolios

	Purpose	Area of focus	Manager	Objectives	Success
Project	Applies and uses project management processes, knowledge, and skills	Delivery of products, services, or results	Project manager	Detailed objectives for the project	Objectives are met and stakeholders are satisfied.
Program	Collections of related projects, subsidiary programs, or work managed in a coordinated fashion	Project interdependencies	Program manager	Coordinated objectives and interdependencies across the program to realize benefits	Collective objectives and benefits are realized.
Portfolio	Aligns projects/ programs/ portfolios/ subsidiary portfolios/ operations to the organization's strategic business objectives	Optimizing efficiencies, objectives, costs, resources, risks, and schedules	Portfolio managers. Project and program managers may report to the portfolio manager.	Align with the organization's strategic business objectives. Prioritizes the right programs and projects, prioritizes work, and ensures resources are available.	Performance and benefit realization of the portfolio

Exam Spotlight

Project and program management both focus on performing the projects and programs in the right way. Portfolio management is concerned with working on the right projects and the right programs for the organization.

Organizational Project Management

Organizational project management (OPM) ensures that projects, programs, and portfolios are aligned and managed according to the organization's strategic business objectives. It optimizes the organization's capabilities by correlating projects, programs, and portfolios to perform efficiently and align with the strategic goals. The focus of OPM is to ensure that the organization performs the right projects, that critical resources are available and assigned appropriately, and that the strategic objectives and business value are first and foremost.

Project Management Offices

The *project management office (PMO)* is usually a centralized organizational unit that oversees the management of projects and programs throughout the organization. The most common reason a company starts a project management office is to establish and maintain procedures and standards for project management methodologies and to manage resources assigned to the projects in the PMO. PMOs are often tasked with establishing an organizational project management (OPM) framework. OPM helps ensure that projects, programs, and portfolios are managed consistently and that they support the overall goals of the organization. OPM is used in conjunction with other organizational practices, such as human resources, technology, and culture, to improve performance and maintain a competitive edge.

According to the *PMBOK® Guide*, the key purpose of a PMO is to provide support for project managers. This may include the following types of support:

- Providing an established project management methodology, including templates, forms, and standards

- Mentoring, coaching, and training project managers

- Facilitating communication within and across projects

- Managing resources

Not all PMOs are the same. Some PMOs may have a great deal of authority and control, whereas others may only serve a supporting role. According to the *PMBOK® Guide*, there are three types of PMOs: supportive, controlling, and directive. Table 1.2 describes each type of PMO, its roles, and its levels of control.

A PMO might have full authority to manage projects, including the authority to cancel projects, or it might serve only in an advisory role. PMOs might also be called *project offices*, *program management offices*, or *Centers of Excellence*.

TABLE 1.2 PMO organizational types

PMO type	Role	Level of control
Supportive	Consulting: Templates, project repository, training	Low
Controlling	Compliance: Project management framework/Conformance to methodologies/Conformance to governance frameworks/Use of specific templates and tools	Moderate
Directive	Controlling: PMO manages projects	High

The PMO usually has responsibility for maintaining and archiving project artifacts for future reference. This office compares project goals with project progress and gives feedback to the project teams and management. It ensures that projects are aligned with the strategic objectives of the organization, and it measures the performance of active projects and suggests corrective actions. The PMO evaluates completed projects for their adherence to the project management plan and asks questions like "Did the project meet the time frames established?" and "Did it stay within budget?" and "Was the quality acceptable?" and "Did we bring about business value?"

Project managers are typically responsible for meeting the objectives of the project they are managing, delivering business value, controlling the resources within the project, and managing the individual project constraints. The PMO is responsible for managing the objectives of a collective set of projects, managing resources across the projects, and managing the interdependencies of all the projects within the PMO's authority.

Project management offices are common in organizations today, if for no other reason than to serve as a collection point for project artifacts. Some PMOs are fairly sophisticated and prescribe the standards and methodologies to be used in all project phases across the enterprise. Still others provide all these functions and also offer project management consulting services. However, the establishment of a PMO is not required in order for you to apply good project management practices to your next project.

There Ought to Be a Law

The importance of practicing sound project management techniques has grown significantly over the past several years. In 2015, President Obama signed the *Program Management Improvement and Accountability Act of 2015 (PMIAA)*. This act intended to enhance accountability in project management and ensure that best practices are used for projects performed by the federal government. It developed an official career path for project managers and recognized the importance of the role of the project sponsor and executive management on the project. This act enforced the development of a standards-based program management policy across the federal government and set up an interagency council on project management so that knowledge sharing could occur among agencies.

Understanding How Projects Come About

Your company's quarterly meeting is scheduled for today. You take your seat, and each of the department heads gets up and gives their usual "We can do it" rah-rah speech, one after the other. You sit up a little straighter when the CEO takes the stage. She starts this part of the program pretty much the same way the other department heads did, and before long, you find yourself drifting off. You are mentally reviewing the status of your current project when suddenly your daydreaming trance is shattered. You perk up as you hear the CEO say, "And the new phone system will be installed by Thanksgiving."

Wait a minute. You work in the telecom department and haven't heard a word about this project until today. You also have a funny feeling that you've been elected to manage this project. It's amazing how good communication skills are so important for project managers but not for. . .well, we won't go there.

Project *initiation* is the formal recognition that a project, or the next phase in an existing project, should begin and resources should be committed to the project. Unfortunately, many projects are initiated the way the CEO did in this example. Each of us, at one time or another, has experienced being handed a project with little to no information and told to "make it happen." The new phone system scenario is an excellent example of how *not* to initiate a project.

Taking one step back leads you to ask, "How do projects come about in the first place? Do CEOs just make them up like in this example?" Even though your CEO announced this new project at the company meeting with no forewarning, no doubt it came about as a result of a legitimate need. Believe it or not, CEOs don't just dream up projects just to give you something to do. They're concerned about the future of the company and the needs of the business and its customers.

According to the *PMBOK® Guide,* projects are initiated by business leaders due to four categories or factors that influence the organization:

- Regulatory compliance, legal requirements, or social requirements
- Stakeholder needs and requests
- Changing technology needs of the organization
- Creation or improvement of processes, services, or products

There are several needs or demands that bring about the creation of a project within these categories. I'll cover this topic next.

Needs and Demands and Other Factors That Lead to Project Creation

Organizations exist to generate profits, serve the public, and create business value. To stay competitive, organizations are always examining new ways of creating business, gaining efficiencies, or serving their customers. Sometimes laws are passed to force organizations to make their products safer or to make them less harmful to the environment, for example. Projects might result from any of these needs as well as from business opportunities or problems. According to the *PMBOK® Guide*, most projects will fit one of the seven needs and demands described next. Let's take a closer look at each of these areas:

Market Demand   The demands of the marketplace can drive the need for a project. For example, a bank initiates a project to offer customers the ability to apply for mortgage loans using a simple phone app.

Organizational Need   The new phone system talked about earlier that was announced at the quarterly meeting came about as a result of a business need. The CEO, on advice from her staff, was advised that call volumes were maxed on the existing system. Without a new system, customer service response times would suffer, and that would eventually affect the bottom line.

Customer Request   Customer requests run the gamut. Generally speaking, most companies have customers, and their requests can drive new projects. Keep in mind that customers can be internal or external to the organization. For example, government agencies don't have external customers per se (we're captive customers at any rate), but there are internal customers within departments and across agencies.

Technological Advance   Many of us own a smartphone that keeps names and numbers handy along with a calendar, a to-do list, and a plethora of other apps to help organize our day or add a little fun in between meetings. I couldn't live without mine. However, a newer, better version is always coming to market. Satellite communications, bigger screens, thinner bodies, touch screens, video streaming, and more are all examples of technological advances (I can't wait for a working foldable screen, but I digress). Electronics manufacturers are continually revamping and reinventing their products to take advantage of new technology (thank you!).

Legal Requirement   Private industry and government agencies both generate new projects as a result of laws passed during every legislative season. For example, new sales tax or healthcare laws might require changes to the computer programs that support these systems. The requirement that food labels on packaging describe the ingredients in the product, the calories, and the recommended daily allowances is another example of legal requirements that drive a project.

Ecological Impacts   Many organizations today are undergoing a "greening" effort to reduce energy consumption, save fuel, reduce their carbon footprint, and so on. Another example might include manufacturing or processing plants that voluntarily remove their waste products from water prior to putting the water back into a local river or stream to prevent contamination. These are examples of environmental considerations that result in projects.

Social Need   The last need is a result of social demands. For example, perhaps a developing country is experiencing a fast-spreading disease that's infecting large portions of the population. Medical supplies and facilities are needed to vaccinate and treat those infected with the disease.

Exam Spotlight

Understand the needs and demands that bring about a project.

In addition to the needs and demands that may bring about a project, other factors are outlined in the *PMBOK® Guide* that may also lead to project creation. Let's take a look at those next.

Strategic Opportunity or Business Need   An example here might include the acquisition or merger of two business entities in order to expand market penetration. The business may need to purchase or build a new warehouse, they may have a strategic opportunity to expand into a new market, and so on. There are an unlimited number of business needs that could bring about a project.

Competitive Forces   We've all likely seen the better burger commercials and two-for-one promotions, or maybe stood in line all night in the cold waiting to be one of the first to buy the next new technology gadget. When one business develops new pricing structures or new products, or offers more for the same, competing businesses must do something similar in order to stay competitive. This can bring about the creation of projects.

Political Changes   I have witnessed this example countless times in my career. A new government official is elected, or perhaps a change in power occurs in a legislative body, and once they take office, they kill the projects that the previous office held dear and initiate their own projects.

New Technology   This category is like the technological advance category described earlier. It can also include technology that will improve efficiencies within the organization, provide a better customer experience, save money, increase revenues, and so on.

Stakeholder Demands   Stakeholder demands can create an infinite number of projects. There is no limit to what they might need in order to support the organization or their customers.

Business Process Improvements   This category of project creation can also bring about numerous requests. For example, perhaps the organization currently uses a manual process for travel authorization requests. To improve the business process, they request an automated workflow and electronic signature so that travel requests can be routed electronically rather than using paper.

Material Issues   This need refers to issues that may occur with existing assets that require a project to replace or fix. For example, a warehouse holding products ready to ship is hit by a tornado. The building will need to be demolished and replaced.

All of these needs and demands represent opportunities, business requirements, or problems that need to be solved. Each may also introduce risk to the project. Management must decide how to respond to these needs and demands, which will more often than not initiate new projects.

 Real World Scenario

Project Initiation

Corey is an information technology manager who works for the National Park Service. One warm spring Sunday morning, he is perusing the local online news and comes across an article about new services being offered at some of the national parks. He perks up when he sees his boss's name describing the changing nature of technology and how it impacts the types of services the public would like to see at the parks. Corey knows that the public has been asking when wireless Internet services will be available in some of the larger parks and has also been involved in the discussions about the resources needed to make this happen, but the project never has enough steam to get off the ground. It seems that a higher-priority project always takes precedence. However, all that changes when Corey sees the next sentence in the article: his boss promising wireless access in two of the largest parks in their region by July 4. It looks like the customer requests have finally won out, and Corey has just learned he has a new project on his hands.

Skills Every Good Project Manager Needs

Many times, organizations will knight their technical experts as project managers. The skill and expertise that made them stars in their technical fields are mistakenly thought to translate into project management skills. This is not necessarily so.

Project managers are generalists with many skills in their repertoires. They are also problem solvers who wear many hats. They are also leaders who deliver business value to the organization. Project managers might indeed possess technical skills, but technical skills are not a prerequisite for sound project management skills. Your project team should include a few technical experts, and these are the people on whom the project manager should rely for technical details. Understanding and applying good project management techniques, along with a solid understanding of strategic and business management skills, leadership skills, as well as interpersonal skills, are career builders for all aspiring project managers.

Project managers have been likened to small-business owners. They need to know a bit about every aspect of management. General management skills, also known as business acumen, include every area of management, from accounting to strategic planning, supervision, personnel administration, and more. Interpersonal skills are often called soft skills and include, among others, communications, leadership, and decision making. General management and interpersonal skills are called into play on every project. But some projects require specific skills in certain application areas. Application areas consist of categories of projects that have common elements. These elements, or application areas, can be defined several ways: by industry group (automotive, pharmaceutical), by department (accounting, marketing), and by technical (software development, engineering) or management (procurement, research and development) specialties. These application areas are usually concerned with disciplines, regulations, and the specific needs of the project, the customer, or the industry. For example, most governments have specific procurement rules that apply to their projects but that wouldn't be applicable in the construction industry. The pharmaceutical industry is acutely interested in regulations set forth by the Food and Drug Administration. The automotive industry has little or no concern for either of these types of regulations. Having experience in the application area you're working in will give you a leg up when it comes to project management. Although you can call in the experts who have application area knowledge, it doesn't hurt for you to understand the specific aspects of the application areas of your project.

Project managers are not the sole performers on a project and are not expected to know everything or to perform every task. But they should have sound project management skills, adequate technical skills, and sufficient experience to manage the size, complexity, and risk of the project they'll undertake. They also serve as champions for the value of project management, help to socialize and gain acceptance of project management concepts, and advance the effectiveness and advantages of the PMO.

The *PMBOK® Guide* outlines the skills every project manager needs in what they call the *PMI® Talent Triangle™*, which is made up of technical project management skills, leadership skills, and strategic and business management skills. I will discuss each of those next, along with other skills that I consider the foundation of good project management practices. Your mastery of them (or lack thereof) will likely affect project outcomes. We'll look at an overview of these skills now, and I'll discuss each in more detail in subsequent chapters.

 Once you obtain the PMP® credential, you will need to complete a certain number of professional development units (PDUs) every three years. PMI® requires you to record the PDUs according to the Talent Triangle™. That is, PDUs are earned and recorded in the leadership, strategic, and technical categories.

Technical Project Management Skills

Technical skills, as the term relates to project management, refer to the technical aspects of performing the role. This incorporates skills like applying project management knowledge in order to deliver the objectives of the projects, defining the critical success factors of the project, developing a project schedule, and knowing when you don't know and need to ask for help. As I noted earlier, the project manager isn't expected to be the technical expert (from the perspective of the product of the project) and should have sufficient subject matter experts on the project team to address technical concerns. The project manager is concerned with using the right tools and techniques for the project, planning appropriately, and managing the schedule, budget, resources, and risks. Technical project management skills are included in the PMI® Talent Triangle™ skills.

Business Management and Strategic Skills

Project managers should be able to describe the business needs of the project and how they align to the organization's goals, including elements such as operations, market conditions, competition, and strategy. This also means that the project manager should have a basic understanding of how the goals of the project relate to various business functions in the organization, such as finance, marketing, customer service, and operations. Having an understanding of the business needs and the strategic vision of the organization helps you to recognize what aspects of the project may need to be closely monitored or which deliverables have higher priority than others.

Business skills also involve understanding the risks and issues involved in bringing about the results of the project, the financial impacts and their effects, how business value is maximized through a successful project, and how to manage the scope and schedule.

Another skill you might not equate to a business skill is understanding the politics of the organization and, most importantly, understanding who has the power, responsibility, and authority to make things happen for your project. Business management and strategic skills are included in the PMI® Talent Triangle™ skills.

Communication Skills

One of the single most important characteristics of a first-rate project manager is excellent communication skills. Written and oral communications are the backbone of all successful projects. Many forms of communication will exist during the life of your project. It's your job, as the creator or manager of most of the project communication (project documents, meeting updates, status reports, and so on), to ensure that the information is explicit, clear, and complete so that your audience will have no trouble understanding what has been communicated. Once the information has been distributed, it is the responsibility of the people receiving the information to make sure they understand it.

Many forms of communication and communication styles exist. I'll discuss them more in depth in Chapter 10, "Sharing Information."

Organizational and Planning Skills

Organizational and planning skills are closely related and probably the most important skills, after communication skills, a project manager can possess. Organization takes on many forms. As project manager, you'll have project documentation, requirements information, memos, project reports, personnel records, vendor quotes, contracts, and much more to track and be able to locate at a moment's notice. You will also have to organize meetings, put together teams, and perhaps manage and organize media-release schedules, depending on your project.

Time management skills are closely related to organizational skills. It's difficult to stay organized without an understanding of how you're managing your time. I recommend that you attend a time management class if you've never been to one. They have great tips and techniques to help you prioritize problems and interruptions, prioritize your day, and manage your time.

I discuss planning extensively throughout the course of this book. There isn't any aspect of project management that doesn't first involve planning. Planning skills go hand in hand with organizational skills. Combining these two with excellent communication skills is almost a sure guarantee of your success in the project management field.

Conflict Management Skills

Show me a project, and I'll show you problems. All projects have some problems, as does, in fact, much of everyday life. Isn't that what they say builds character? But I digress.

Conflict management involves solving problems. Problem solving is really a twofold process. First, you must define the problem by separating the causes from the symptoms. Often when defining problems, you end up just describing the symptoms instead of getting to the heart of what's causing the problem. To avoid that, ask yourself questions like, "Is it an internal or external problem? Is it a technical problem? Are there interpersonal problems between team members? Is it managerial? What are the potential impacts or consequences?" These kinds of questions will help you get to the cause of the problem.

Next, after you have defined the problem, you have some decisions to make. It will take a little time to examine and analyze the problem, the situation causing it, and the alternatives available. After this analysis, the project manager will determine the best course of action to take and implement the decision. The timing of the decision is often as important as the decision itself. If you make a good decision but implement it too late, it might turn into a bad decision.

Negotiation and Influencing Skills

Effective problem solving requires negotiation and influencing skills. We all utilize negotiation skills in one form or another every day. For example, on a nightly basis I am asked, "Honey, what do you want for dinner?" Then the negotiations begin, and the fried chicken versus swordfish discussion commences. Simply put, negotiating is working with others to come to an agreement.

Negotiation on projects is necessary in almost every area of the project, from scope definition to budgets, contracts, resource assignments, and more. This might involve negotiation one on one or with teams of people, and it can occur many times throughout the project.

Influencing is convincing the other party that swordfish is a better choice than fried chicken, even if fried chicken is what they want. It's also the ability to get things done through others. Influencing requires an understanding of the formal and informal structure of all the organizations involved in the project.

Power and politics are techniques used to influence people to perform. *Power* is the ability to get people to do things they wouldn't do otherwise. It's also the ability to change minds and the course of events and to influence outcomes.

Politics involve getting groups of people with different interests to cooperate creatively even in the midst of conflict and disorder.

These skills will be utilized in all areas of project management. Start practicing now because, guaranteed, you'll need these skills on your next project.

Leadership Skills

Leaders and managers are not the same. *Leaders* impart vision, gain consensus for strategic goals, establish direction, and inspire and motivate others. They guide and direct the team in accomplishing the project's objectives. *Managers* focus on results and are concerned with getting the job done according to the requirements. Even though leaders and managers are not the same, project managers must exhibit the characteristics of both during different times on the project. Understanding when to switch from leadership to management and then back again is a finely tuned and necessary talent. Leadership is one of the Talent Triangle™ skills.

Team-Building and Motivating Skills

Project managers will rely heavily on team-building and motivational skills. Teams are often formed with people from different parts of the organization. These people might or might not have worked together before, so some component of team-building groundwork might involve the project manager. The project manager will set the tone for the project team and will help the members work through the various stages of team development to become fully functional. Motivating the team, especially during long projects or when experiencing a lot of bumps along the way, is another important role the project manager fulfills during the course of the project.

An interesting caveat to the team-building role is that project managers many times are responsible for motivating team members who are not their direct reports. This scenario has its own set of challenges and dilemmas. One way to help this situation is to ask the functional manager to allow you to participate in your project team members' performance reviews. Use the negotiation and influencing skills I talked about earlier to make sure you're part of this process.

Multiple Dimensions

Project managers are an interesting bunch. They know a little bit about a lot of topics and are excellent communicators. They have the ability to motivate people, even those who have no reason to be loyal to the project, and they can make the hard-line calls when necessary. Project managers can get caught in sticky situations that occasionally require making decisions that are good for the company (or the customer) but that aren't good for certain stakeholders. The offended stakeholders will then drag their feet, and the project manager has to play the heavy in order to motivate them and gain their cooperation again. Some organizations hire contract project managers to run their large, company-altering projects just because they don't want to burn out a key employee in this role. Fortunately, that doesn't happen often.

Role of a Project Manager

Project managers are responsible for ensuring that the objectives of the project are met. Projects create value, which in turn increases the business value of the organization. Business value is the total value of all the assets of the organization, including both tangible and intangible elements. The project manager must be familiar with the organization's strategic plan in order to marry those strategic objectives with the projects in the portfolio. This requires all the skills we covered in the preceding sections.

Now that you've been properly introduced to some of the skills you need in your tool kit, you'll know to be prepared to communicate, solve problems, lead, and negotiate your way through your next project.

Understanding Project Management Process Groups

Project management processes organize and describe the work of the project. The *PMBOK® Guide* describes five process groups used to accomplish this end. These processes are performed by people and are interrelated and dependent on one another.

These are the five project management process groups that the *PMBOK® Guide* documents:

- Initiating
- Planning
- Executing
- Monitoring and Controlling
- Closing

All these process groups have individual processes that collectively make up the group. For example, the Initiating process group has two processes called Develop Project Charter and Identify Stakeholders. Collectively, these process groups—including all their individual processes—make up the project management process. Projects, or each phase of a project, start with the Initiating process and progress through all the processes in the Planning process group, the Executing process group, and so on, until the project is successfully completed, or it's canceled. All projects must complete the Closing processes, even if a project is killed.

Let's start with a high-level overview of each process group. If you want to peek ahead and see the complete list, Appendix B, "Process Inputs and Outputs," lists each of the process groups, the individual processes that make up each process group, and the Knowledge Areas in which they belong. (I'll introduce Knowledge Areas in the next chapter.)

Initiating   The *Initiating* process group, as its name implies, occurs at the beginning of the project and at the beginning of each project phase for large projects. Initiating acknowledges that a project, or the next project phase, should begin. This process group grants the approval to commit the organization's resources to working on the project or phase and authorizes the project manager to begin working on the project. The outputs of the Initiating process group, including the project charter and identification of the stakeholders, become inputs into the Planning process group.

Planning   The *Planning* process group includes the processes for formulating and revising project goals and objectives and creating the project management plan that will be used to achieve the goals the project was undertaken to address. The Planning process group also involves determining alternative courses of action and selecting from among the best of those to produce the project's goals. This process group is where the project requirements are fleshed out. Planning has more processes than any of the other project management process groups. To carry out their functions, the Executing, Monitoring and Controlling, and Closing process groups all rely on the Planning processes and the documentation produced during the Planning processes. Project managers will perform frequent iterations of the Planning processes prior to project completion. Projects are unique and, as such, have never been done before. Therefore, planning must encompass all areas of project management and consider budgets, activity definition, scope planning, schedule development, risk identification, staff acquisition, procurement planning, and more. The greatest conflicts a project manager will encounter in this process group are project prioritization issues.

Executing   The *Executing* process group involves putting the project management plan into action. It's here that the project manager will coordinate and direct project resources to meet the objectives of the project management plan. The Executing processes keep the project on track and ensure that future execution of project plans stays in line with project objectives. This process group is typically where approved changes are implemented. The Executing process group will consume the most project time and resources, and as a result, costs are usually highest during the Executing processes. Project managers will experience the greatest conflicts over schedules in this cycle.

Monitoring and Controlling   The *Monitoring and Controlling* process group is where project performance measurements are taken and analyzed to determine whether the project is staying true to the project management plan. The idea is to identify problems as soon as possible and apply corrective action to control the work of the project and ensure successful outcomes. For example, if you discover that variances exist, you'll apply corrective action to get the project activities realigned with the project management plan. This might require additional passes through the Planning processes to adjust project activities, resources, schedules, budgets, and so on.

Monitoring and Controlling is used to track the progress of work being performed and to identify problems and variances within a process group as well as the project as a whole.

Closing   The *Closing* process group is probably the most often skipped process group in project management. Closing brings a formal, orderly end to the activities of a project phase or to the project itself. Once the project objectives have been met, most of us are ready to move on to the next project. However, Closing is important because all the project information is gathered and stored for future reference. The documentation collected during the Closing process group can be reviewed and used to avert potential problems on future projects. Formal acceptance and approval are obtained from project stakeholders.

Exam Spotlight

The project manager and project team are responsible for determining which processes within each process group are appropriate for the project on which you're working. This is called *tailoring*. You should consider the size and complexity of the project and the various inputs and outputs of each of the processes when determining which processes to implement and perform. Small projects might not require all of the processes within a process group or the same level of rigor as a large project. Every process should be addressed, and it should be determined whether the process is appropriate for the project at hand and, if so, what level of implementation is required. Use your judgment when deciding which processes to follow, particularly for small projects.

Characteristics of the Process Groups

The process groups have several characteristics. The first is that costs are lowest during the Initiating processes, and few team members are involved. Costs and staffing increase in the Executing process group and then decrease as you approach the Closing process group. The chances for success are lowest during Initiating and highest during Closing. The chances for risks occurring are higher during Initiating, Planning, and Executing, but the impacts of risks are greater during the later processes. Stakeholders have the greatest influence during the Initiating and Planning processes and less and less influence as you progress through Executing, Monitoring and Controlling, and Closing. For a better idea of when certain characteristics influence a project, refer to Table 1.3.

TABLE 1.3   Characteristics of the project process groups

	Initiating	Planning	Executing	Monitoring and Controlling	Closing
Costs	Low	Low	Highest	Lower	Lowest
Staffing levels	Lowest	Medium	High	High	Low
Chance for successful completion	Lowest	Low	Medium	High	Highest
Stakeholder influence	Highest	High	Medium	Low	Lowest
Risk probability of occurrence	Highest	High	Medium	Low	Lowest

The Process Flow

You should not think of the five process groups as onetime processes that are performed as discrete elements. Rather, these processes interact and overlap with one another. They are *iterative* and might be revisited and revised several times as the project is refined throughout its life. The *PMBOK® Guide* calls this process of going back through the process groups an iterative process. The conclusion of each process group allows the project manager and stakeholders to reexamine the business needs of the project and determine whether the project is satisfying those needs—and it is another opportunity to make a go or no-go decision.

Figure 1.1 shows the five process groups in a typical project. Keep in mind that during phases of a project, the Closing process group outputs can provide inputs to the Initiating process group. For example, if you performed a feasibility study as the first phase of a project, once it's accepted or closed, it becomes an input to the Initiating process group.

It's important to understand the flow of these processes for the exam. If you remember the processes and their inputs and outputs, it will help you when you're trying to decipher an exam question. The outputs of one process group may in some cases become the inputs into the next process group (or the outputs might be a deliverable of the project). Sometimes just understanding which process the question is asking about will help you determine the answer.

As I stated earlier, each process group contains several individual processes. For example, the Closing process group consists of two processes: Close Project or Phase and Close Procurements. Each process takes inputs and uses them in conjunction with various tools and techniques to produce outputs.

FIGURE 1.1   Project management process groups

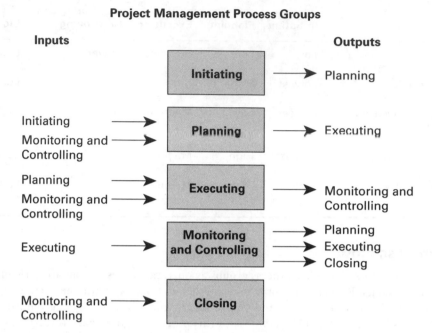

Project Management Process Groups

Exam Spotlight

Understand each project management process group and all the processes that make up these groups. Appendix B contains a table of all the processes, their inputs, their tools and techniques, their outputs, and the Knowledge Area in which they each belong. I'll cover Knowledge Areas in Chapter 2, "Assessing Project Needs."

You may see test questions regarding inputs, tools and techniques, and outputs of many of the processes within each process group. One way to keep them all straight is to remember that tools and techniques usually require action of some sort, be it measuring, applying some skill or technique, planning, or using expert judgment. Outputs are usually in the form of a deliverable. Remember that a deliverable is characterized by results or outcomes that can be verified. Last but not least, outputs from one process often serve as inputs to another process.

It's outside the scope of this book to explain all the inputs, tools and techniques, and outputs for each process in each process group (although they are in Appendix B). You'll find all the inputs, tools and techniques, and outputs detailed in the *PMBOK® Guide*, and I highly recommend you become familiar with them.

Process Interactions

We've covered a lot of material, but I'll explain one more concept before moving on to project management methodologies. As stated earlier, project managers must determine the processes that are appropriate for effectively managing a project based on the complexity and scope of the project, available resources, budget, and so on. As the project progresses, the project management processes might be revisited and revised to update the project management plan as more information becomes known. Underlying the concept that process groups are iterative is a cycle the *PMBOK® Guide* describes as the Plan-Do-Check-Act cycle, which was originally defined by Walter Shewhart and later modified by Edward Deming. The idea behind this concept is that each element in the cycle is results oriented. The results from the Plan cycle become inputs into the Do cycle, and so on, much like the way the project management process groups interact. The cycle interactions can be mapped to work with the five project management process groups. For example, the Plan cycle maps to the Planning process group. Before we go any further, here's a brief refresher:

- Project phases describe how the work required to produce the product of the project will be completed.

- Project management process groups organize and describe how the project activities will be completed in order to meet the goals of the project.

- The Plan-Do-Check-Act cycle is an underlying concept that shows the integrative nature of the process groups.

Figure 1.2 shows the relationships and interactions of the concepts you've learned so far. Please bear in mind that a simple figure can't convey all the interactions and iterative nature of these interactions; however, I think you'll see that the figure ties the basic elements of these concepts together.

FIGURE 1.2   Project management process groups interactions

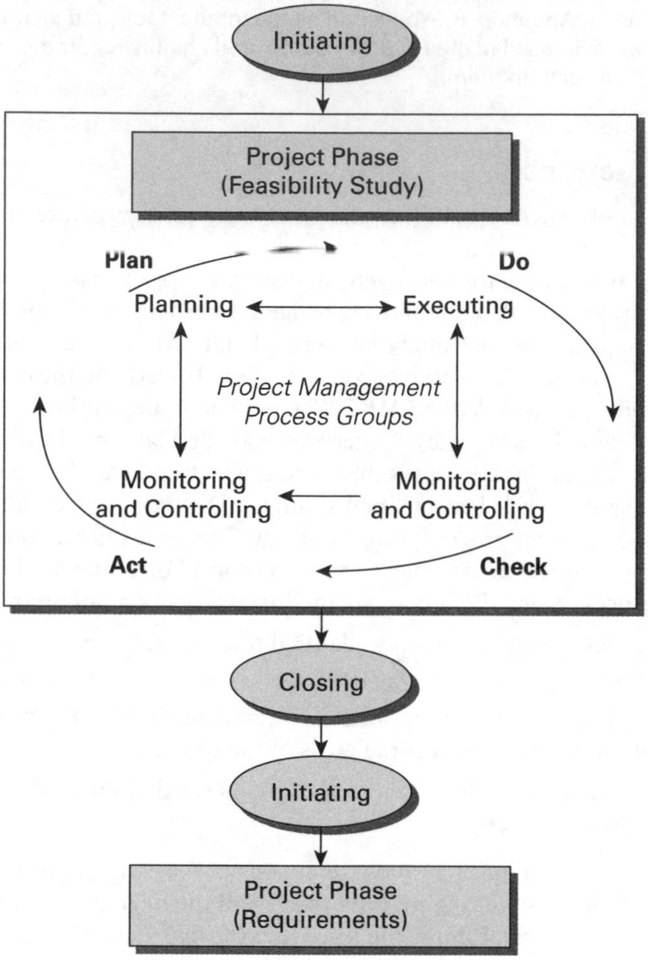

Determining a Project Methodology or Approach

Project life cycles are similar to the life cycle that parents experience raising their children to adulthood. Children start out as infants and generate lots of excitement wherever they go. However, not much is known about them at first. So, you study them as they grow, and you assess their needs. Over time, they mature and grow until one day the parents' job is done.

Projects start out just like this and progress along a similar path. Someone comes up with a great idea for a project and actively solicits support for it. The project, after being approved, progresses through the intermediate phases to the ending phase, where it is completed and closed out.

All the collective phases the project progresses through from the start of the project until the end are called the *project life cycle*. Project life cycles are similar for all projects regardless of their size or complexity.

The phases that occur within the project life cycle are sequential or may sometimes overlap each other. Most projects consist of the following life-cycle structure:

- Beginning the project
- Planning and organizing the work of the project
- Performing the work of the project
- Closing out the project

A *development life cycle* consists of the phases of the project associated with producing the product, service, or result of the project. A development life cycle is performed within the project life cycle. We'll look at categories of development life cycles next.

 Don't confuse a *product life cycle* with a project life cycle. A product life cycle consists of the phases that represent the development of a *product* idea from its inception, through market delivery, and eventually to retirement of the product. A project life cycle is the series of phases from the start of a *project* until the end. Product and project life cycles are independent of each other.

Life Cycle Categories

According to the *PMBOK® Guide*, there are four categories of development life cycles and two categories of project life cycles. Each contains the phases we just discussed (beginning, planning, performing, closing). Let's look at the development life cycles first.

Predictive Development Life Cycle (Also Known as Plan-Driven Life Cycle or Waterfall)   The predictive development life cycle approach defines the scope or deliverable at the beginning of the life cycle. Changes are monitored closely and typically permitted only if they are essential for completing the product of the project. The entire life cycle is completed before moving on to the remaining project phases.

Iterative Development Life Cycle   Project deliverables are defined early in the development life cycle and progressively elaborated as the project, or life cycle, progresses. Schedule and cost estimates are continually modified as the final product, service, or result of the project becomes clear. The product of the project continues to evolve by adding functionality during repeating, iterative cycles. Iterative life cycles may take more time than an incremental approach because this approach is optimized for learning rather than speed. This is a great development life cycle to use when the team is new to agile approaches because of the optimization for learning.

Incremental Development Life Cycle   The incremental development life cycle is similar to iterative, but the incremental approach uses predetermined periods of time called *iterations* (not to be confused with the iterative life cycle) to complete the deliverable. During each iteration, new functionality is introduced. The deliverable is considered complete after the final iteration is finished. Incremental development life cycles are best when speed of delivery is necessary.

Iterative and incremental development life cycles are perfect choices for large projects, for complex projects (these life-cycle processes will reduce the complexity), for projects where changing objectives and scope are known ahead of time, or for projects where deliverables need to be delivered incrementally.

Hybrid Development Life Cycle   This development life cycle is a combination of a predictive and adaptive life cycle (*adaptive* in the development life cycle refers to an iterative or incremental life cycle). Typically, the elements of the deliverable that are well known at the beginning of the life cycle will follow a predictive approach whereas the elements that are not clear at that stage will follow either an incremental or an iterative approach.

Project life cycles are either predictive or adaptive. Remember that a development life cycle can be performed within the project life cycle, so you may use more than one life cycle approach during the course of the project. For example, the development life cycle might be incremental whereas the project life cycle might be predictive. We discussed predictive in the context of the development life cycle, so next we'll look at it from a project life cycle approach.

We will look at the differences in development life cycle approaches in more depth in Chapter 6, "Developing the Project Budget and Engaging Stakeholders." If you want to peek ahead, take a look at Table 6.2.

Predictive Life Cycle Methodology

A predictive or *waterfall* approach is a step-by-step methodology whereby each stage of the project is completed in order. Typically, you don't move from one stage to another until the previous stage is completed, although there are exceptions that we'll get to later in this section. For example, once the project is kicked off, requirements are defined in detail and approved before any work begins. In this approach, it is very difficult to change requirements or incorporate new ideas as the project progresses. Sometimes months, if not years, of planning go into getting every detail of the planning before actual work begins. The customer may not see any deliverables, functionality, or understand if business value is being created until the project is close to the end of its life cycle.

If changes are made, you must revisit and modify project plans and formally accept the changes to the scope and subsequent project management plan. Schedules and budgets are defined early in the life cycle as well. Predictive life cycles are an excellent choice for projects where the requirements are well understood, for low-risk projects, and for projects where the team is well established and stable.

Waterfall projects are sometimes divided into phases. A project *phase* generally consists of segments of work that allow for easier management, planning, and control of the work and generally produce at least one deliverable by the end of the phase. The work and the deliverables produced during a phase are typically unique to that phase. The work of each phase is usually distinct and not repeated in other phases. Each phase has an emphasis on a different portion of the project activities, and different project management process groups are performed during each phase. You can use a phased approach with adaptive methodologies, but they are more commonly used in a waterfall approach.

Project phases are determined in any number of ways. The type of project itself and the industry you're in may drive the phases of the project, as could decision points such as milestone completion or go/no-go decisions. The number of phases in the project life cycle depends on the project complexity and the industry you're in. According to the *PMBOK® Guide*, phases are described by such attributes as name, number, duration, resource requirements, and entrance and exit criteria. For example, information technology projects might progress through phases named this way: requirements, design, development, test, and implement.

Many industries use a *feasibility study* as one of the first phases of the project (this is also another potential phase name). A feasibility study is used to determine whether the project is worth undertaking and whether the project will be profitable to the organization. It's a preliminary assessment of the viability of the project; the viability or perhaps marketability of the product, service, or result of the project; and the project's value to the organization. It might also determine whether the product, service, or result of the project is safe and meets industry or governmental standards and regulations. The completion and approval of the feasibility study triggers the beginning of the requirements phase, where requirements are documented and then handed off to the design phase, where blueprints are produced, and so on through the phases. The feasibility study might also show that the project is not worth pursuing and the project is then terminated; therefore, the next phase never begins.

The group of people conducting the feasibility study should not be the same people who will work on the project. Project team members might have built-in biases toward the project and will tend to influence the feasibility outcome toward those biases.

Phase Reviews

Project phases evolve through the life cycle in a series of phase sequences called *handoffs*, or technical transfers. For projects that consist of sequential phases, the end of one phase typically marks the beginning of the next.

A *phase review* should be held at the end of each phase. This review allows the project manager, stakeholders, and project sponsor the opportunity to determine whether the project should continue to the next phase and whether business value has been achieved. They will examine the progress performance to date against the project charter, the business case, the project management plan, and the benefits management plan. As each phase is completed, it's handed off to the next phase.

For example, as we've discussed, a deliverable produced in the beginning phase of a project might be the feasibility study. Producing, verifying, and accepting the feasibility study will signify the ending of this phase of the project. The successful conclusion of one phase does not guarantee authorization to begin the next phase. Let's say our feasibility study was performed for a construction project. The study showed that environmental impacts of a serious nature would result if the construction project were undertaken at the proposed location. Based on this information, a go or no-go decision can be made at the end of the phase. Phase reviews give the project manager the ability to discover, address, and take corrective action against errors discovered during the phase.

The *PMBOK® Guide* states that phase reviews are also known by a few other names: *phase gate, phase entrance, phase exit, stage gate,* and *kill points.*

Phase Completion

You will recognize phase completion because each phase has a specific deliverable, or multiple deliverables, that marks the end of the phase. A *deliverable* is an output that must be produced, verified, and approved to bring the phase, life-cycle process, or project to completion. Deliverables are unique and verifiable and may be tangible or intangible, such as the ability to carry out a service. Deliverables might also include things such as design documents, project budgets, blueprints, project schedules, prototypes, and so on.

Multiphased Projects

Projects may consist of one or more phases. The phases of a project are often performed sequentially, but there are situations where performing phases concurrently, or overlapping the start date of a sequential phase, can benefit the project. Multiple phases allow the team to analyze project performance and take action in concurrent and later phases to correct or prevent the problems that occurred previously. According to the *PMBOK® Guide*, there are three ways project life-cycle phases could be performed:

Sequential   One phase must finish before the next phase can begin.

Iterative   More than one phase is being performed at the same time.

Overlapping   One phase starts before the prior phase completes.

Sometimes phases are overlapped to shorten or compress the project schedule. This is called *fast tracking*. Fast tracking means that a later phase is started prior to completing and approving the phase, or phases, that comes before it. This technique is used to shorten the overall duration of the project.

Project Phases vs. Project Management Process Groups

Don't confuse project phases and project life cycles with the project management process groups. Project phases and life cycles describe how the work associated with the product of the project will be completed. For example, a construction project might have phases such as feasibility study, design, build, inspection, and turnover. The five project management process groups (Initiating, Planning, Executing, Monitoring and Controlling, and Closing) organize and describe how the project activities will be conducted in order to meet the project requirements. These processes are generally performed for each phase of a large project. The five process groups are the heart of the *PMBOK® Guide*. Be certain you understand each of these processes as they're described in the *PMBOK® Guide*.

Agile Methodologies

Agile project management is a method of managing projects in small, incremental portions of work that can be easily assigned, easily managed, and completed within a short period of time.

An agile project management approach allows the team to quickly adapt to new requirements and allows for the continual assessment of the goals, deliverables, and functionality of the product.

You'll want to choose an agile, or adaptive, methodology when active participation of your stakeholders is required throughout the project, when you are not certain of all the requirements at the beginning of the project, or when you work in a changing environment.

 Agile is a highly iterative approach where requirements can be continually defined and refined based on continuous feedback from the product owner. This allows the development team to quickly adapt to changes and accommodate new or modified functionality requests.

Agile has been around for many years in one form or another. In 2001, several software developers converged to formalize the agile approach. They published the *Manifesto for Agile Software Development* (agilemanifesto.org) and identified 12 principles that are the focus of any agile approach. These principles include factors such as daily interactions between the business and agile teams, frequent deliverables, and self-motivated teams, with a focus on continuous improvement.

One of the key principles in the *Agile Manifesto* is the focus on the value to the customer. Instead of measuring how efficiently a certain process runs or the quality of a deliverable, attention is given to the value the customer perceives. For example, this value may include a tangible deliverable that contains functionality that's critical to the success of the project; business value is thereby created during each period of work. Rather than measuring project success based on the "on time and on budget" approach, success is measured by the progress made in incremental steps and the value that the functionality or deliverables bring to the stakeholders as the project is progressively elaborated.

There are several agile methodologies to choose from. Some are more suited for software development projects and some more suited for manufacturing processes. All of them have elements that can be applied to your next project. We'll look at each next.

Scrum

Scrum is a form of agile project management. Scrum project teams consist of cross-functional team members and are self-organized and self-directed. Scrum emphasizes daily communication and the flexible reassessment of plans, which are carried out in short, iterative phases of work called *sprints*. Sprints are always time-bound and generally consist of two-week time periods, but they can consist of any short period of time defined and agreed on by the team. The goal of the sprint is to produce a deliverable, or a tangible portion of a deliverable, by the end of the sprint. Sprint is a term that is specific to the Scrum methodology. Other agile methodologies use this same time-bound approach and call these short work periods *iterations*.

Scrum is most commonly used in software development projects for similar reasons to those just stated. Using an agile approach, a development team can assess results and adjust processes in order to meet new or modified requirements during or after each sprint. Prototypes can be delivered early and tangible progress is made in each sprint.

Scrum is customer focused, meaning that the project team gathers the wish list from the customers (sometimes called end users), listing all the elements that will make the product great. These are called *user stories* or, simply, *stories*. User stories contain features that are written from the perspective of the end user.

The user stories are kept in the *product backlog*. Keep in mind that the product backlog contains all the user stories that define the features and elements that make up the requirements of the project. Once they are all documented, they are prioritized by the customer in order of importance. Then, a few user stories are chosen from the backlog and worked on during each sprint. The Scrum team estimates the amount of work involved to complete each user story at the beginning of the sprint. The completed results are released at the end of the sprint for the customer to validate. User stories that were not completed in the previous sprint, or those that require changes, along with new user stories, are worked on in the next sprint, and the cycle repeats until the project is completed.

Kanban

Kanban is a lean scheduling agile methodology that was developed by the Toyota Motor Corporation. Kanban is an agile project management methodology that is typically seen in manufacturing projects, but it also has a presence in the information technology field. With Kanban, the work is balanced against available resources or available capacity for work. It's a pull-based concept where work progresses to the next step only when resources are available. It's also considered an on-demand scheduling methodology because the work is pulled through the system according to demand.

Kanban means "billboard" or "sign." Using Kanban, you construct a board that represents your project. The Kanban board can be physical, like a whiteboard, or you can use software to manage the board. The Kanban board is simple to construct and looks somewhat like Figure 1.3 in the beginning of the project.

FIGURE 1.3   Kanban board at the start of the project

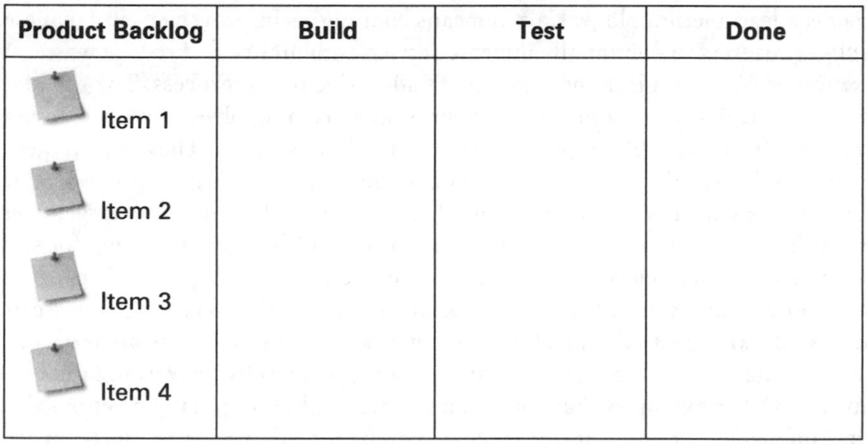

Product Backlog	Build	Test	Done
Item 1			
Item 2			
Item 3			
Item 4			

Any number of columns can exist in between the Product Backlog and Done columns. It's up to the team to define the stages of work represented by the columns on the board.

Kanban consists of one or more product owners, who are responsible for creating the work list. Each of the tasks are called user stories, tasks, or cards. In Figure 1.3, each of the sticky notes represents a task. Initially the product backlog contains all the notes, but as the project begins and team members start to work on tasks, the notes are moved from the product backlog to the next column to the right. As a sticky note task is completed in a given column, it is moved to the next column to the right, and so on. Once a note has vacated a column, a new note can move into its place. You can use the terms *story*, *task*, or *card* interchangeably. They all mean the same thing: time-bound modularized tasks with discrete deliverables.

In this regard, Kanban is like Scrum. Both are called *pull systems*, meaning that as a task moves from one column to the next, a new task is pulled from the previous column. However, unlike Scrum, which uses sprints that typically consist of two to four weeks' worth of work, Kanban is a continuous system. The work does not start and stop but continues through to completion. There are no sprints in Kanban. The Kanban methodology may involve more than one team working on different functional aspects of the work, although they all work from the same Kanban board in a continuous manner.

Lean

Lean is another agile methodology that is concerned with making work processes as efficient as possible while also assuring that the quality of the output is excellent. Lean thinking aims to reduce or eliminate waste. Much of lean thinking is derived from Toyota executives such as Taiichi Ohno. Although thinking lean is primarily focused on manufacturing, it does not mean that the principles cannot be used in other project management efforts.

Kaizen is a lean methodology. Kaizen means continuous improvement in Japanese. The idea behind Kaizen is to continually improve service and quality and reduce waste. Waste is anything the project team is doing that doesn't add value to the process. The foundational belief of Kaizen is that everything can be improved. Kaizen involves every person in the organization—from the CEO to project managers to line workers. They are charged with looking at their jobs and activities in a new light and finding ways to improve productivity and decrease waste in small steps over time. This isn't about looking for a large, onerous problem (although if you find one, you should deal with it) but about looking for small things that produce inefficiencies or waste in the process. For example, perhaps your PMO requires approvals and sign-offs on every document and nearly every task performed on the project. This is likely overkill, and eliminating many of the unnecessary approvals can free up the project managers to work on actual project tasks, thereby delivering project value more quickly. Other examples that come to mind are holding too many meetings that don't add value, implementing tools or processes that aren't effective and that the team avoids using, and requiring too much movement because team members are in different physical locations across the city or the globe. This philosophy could be applied to any number of activities.

Using the Kaizen approach, workers look for places where the seven wastes may appear and then take steps to reduce or eliminate them. The seven wastes are listed here. Reducing or eliminating them by changing your work to make it more efficient is the essence of Kaizen.

Motion   This concerns the movement, or amount of motion, employees go through while performing their work. Examine whether they move too much or too little.

Waiting   Examine whether workers have times where they are simply waiting for the next task and find ways to decrease that wasted time.

Transportation   Moving items or elements of work takes time; put the items needed to complete a task next to the employee.

Storage   Storing materials for tasks or storing completed items for shipment can create waste. Keep materials and supplies organized and easily accessible to employees to reduce waste.

Defects   Defects may be introduced by manufacturing defective parts or by making mistakes in the work. Examine the causes of defects and determine ways to eliminate them.

Processing   This could involve creating too much or too little effort when processing goods or services. Reducing efforts in processing will reduce waste.

Overproduction   This involves making too much of something. This could impact waiting times or storage as well.

Extreme Programming

Extreme Programming (XP) is another agile methodology used in software development. Technology changes rapidly and development teams today do not usually have the luxury of taking years to develop new products. XP involves delivering the software that's needed when the customer needs it. Consider using XP under these conditions:

- When there are dynamically changing user requirements. This typically happens when the customer, or end user, doesn't have a clear idea of what they need.
- When a high amount of risk is associated with the project.
- When you have small development teams of between two and twelve programmers.

XP requires that all project team members work together collaboratively to create the product of the project. They need to work side by side in order to bring about cohesiveness and be able to collaborate instantly. XP also requires automated unit and functional tests. Unit tests are tests on small, whole units of code to determine whether the code is functional. Integration testing involves testing several pieces of the code together to see if they perform as expected. Functional testing is an end-to-end test to ensure the code works throughout the entire process.

You will learn more about these agile techniques throughout the remainder of the book. I will also explain other agile methods when appropriate as we come to them.

Six Sigma

Six Sigma (also represented as the number six, and the Greek letter sigma—6σ) consists of a set of management techniques designed to do away with defects and increase productivity to its maximum. Six Sigma practitioners are highly regarded, well paid, and leaders in the lean manufacturing project management industry.

Six Sigma analysts typically look at one process at a time to evaluate the process for defects or errors. They correlate that information using statistical techniques and look for ways to improve the process. Six Sigma is discipline oriented and focuses on developing and practicing disciplines that improve and maintain quality.

Hybrid

A *hybrid* methodology is just like the name implies: it's a combination of one or more methodologies to create what works best for your team. This combination could include aspects of predictive and adaptive methodologies, or a combination of adaptive methodologies. Hybrid does not mean that there is an equal distribution of methodologies. You may work on a project where a waterfall approach is used to document requirements and it accounts for a third of the project time. The remainder of the project will utilize an agile approach, but the combination of approaches means this is technically a hybrid development life cycle. A good friend of mine calls the combination of waterfall and agile "Wagile."

Project Life Cycles

Life cycles, as you'll recall from an earlier section in this chapter, are all the phases a project progresses through from the beginning of the project to the end.

Table 1.4 highlights the differences in the life cycles between an agile approach (highly adaptive) and a waterfall approach (highly predictive). The hybrid of the two methodologies is listed in the middle of the chart. This hybrid column could read "It depends" for every entry, because the more the organization leans toward predictive (versus adaptive), the more the hybrid approach will behave as predictive does and will produce similar results. If the approach is more predictive, the planning will be more methodical in the beginning of the cycle with few changes. If it's more adaptive, the approach will be iterative and incremental. The bottom portion of the table highlights the differences between these life cycles and their interaction with each project process group.

TABLE 1.4   Life cycle differences

	Highly predictive (waterfall)	**Hybrid**	**Highly adaptive (agile)**
Require-ments	Detailed specifications	High-level planning and iterative refinements	Progressively elaborated during each iteration
Risks	A good deal of time is spent at the beginning of the project identifying risks. They are continually identified throughout the project.	Risks are iteratively identified throughout the project.	Risks are identified at the beginning of each iteration.

	Highly predictive (waterfall)	**Hybrid**	**Highly adaptive (agile)**
Costs	A good deal of time is spent at the beginning of the project documenting costs. Once the budget is established, there is little room for change.	Costs may be identified at the beginning of the project with some consideration for changes as the project progresses.	Costs are identified with each of the iterations. A high-level budget may be established at the beginning of the project but cost and time estimates are performed at the beginning of the iteration.
Stakeholders	Heavily involved in gathering and documenting requirements. Their involvement tapers off as the project progresses.	More involvement than highly predictive approach and less involvement than highly adaptive approach	Continuous involvement and frequent feedback because the stakeholders work beside the project team
Schedule	Created once for the project or phase	May be created once with high-level milestones and further defined as the project progresses.	Each iteration is its own schedule. The work of the iteration is defined at the beginning of each sprint.
Planning	Once the plan is approved, changes that impact scope, time, or budget are controlled and minimized.	A high-level plan is developed at the beginning of the project and further elaborated as the project progresses.	There is progressive elaboration of scope based on continuous feedback.

Understanding How This Applies to Your Next Project

As you can tell from this first chapter, managing projects is not for the faint of heart. You must master multiple skills and techniques in order to complete projects successfully.

I talked about the definition of a project in this chapter. You'd be surprised how many people think ongoing operations are projects. Projects create business value and move the organization from one state to another. This can be subtle or significant, depending on the project.

Projects come about for many reasons. Most of the time, understanding the reason it came about will give you some insight into its purpose. For example, if a new law is passed that requires anyone applying for a driver's license to show two forms of identification but the existing system has the space to record verification of only one document, you immediately have a firm grasp on the purpose of the project—you'll have to update the system to include additional space for recording the second document.

I've made the mistake of thinking the project management process groups are overkill for a small project. My team once embarked on a small project and thought that within a matter of weeks we'd have it wrapped up and delivered. We neglected to get signatures from the project requestor on the agreed-on scope, and, you guessed it, the scope grew and grew and changed several times before we were able to get the project back under control. If you're reading between the lines here, you can also tell we didn't have adequate change control in place. As you progress through the book, I'll highlight the important processes you'll want to include on all projects, large and small, so you don't get caught in this trap.

It's important to understand the differences in project management methodologies. Some projects need a waterfall approach whereas other projects would benefit from an agile approach where continuous feedback is provided. I work in the information technology industry and we often employ agile methodologies to our projects. It allows us to get continuous feedback from the customer and keeps us from developing something that isn't useful for our customers at the end of the project.

Summary

Phew! We covered a lot of ground in this chapter. You learned that projects exist to bring about a unique product, service, or result. Projects are temporary in nature and have definite beginning and ending dates.

I detailed how projects are initiated. Projects come about as a result of one of seven needs or demands: market demands, organizational needs, customer requests, technological advances, legal requirements, ecological impacts, or social needs. Other factors that bring about projects are strategic opportunity or business need, competitive forces, political changes, new technology, stakeholder demands, business process improvements, and material issues.

Project management is a discipline that brings together a set of tools and techniques to describe, organize, and monitor the work of project activities. Project managers are the ones responsible for carrying out these activities. Projects might be organized into programs or portfolios and might be managed centrally by a PMO.

All projects have one or more phases, and the progression of phases from the beginning to the end of the project is known as a project life cycle. There are also development life cycles that can occur within a project life cycle.

The project management process groups are performed throughout the project's life cycle. The process groups described in the *PMBOK® Guide* are Initiating, Planning, Executing, Monitoring and Controlling, and Closing.

The adaptive, or agile, methodology is a method of managing projects in small, incremental portions of work that can be easily assigned, easily managed, and completed within a short period of time. Predictive methodologies, also known as waterfall, typically require one process or phase of the project to be completed before moving to the next. A hybrid methodology is a combination of both waterfall and agile practices.

Scrum, Kanban, and XP are agile methodologies that are used in the information technology industry. Lean, Kaizen, and Six Sigma are primarily used in the manufacturing industry, and Kanban has some applications in this industry as well. Keep in mind that there are aspects of any of these methodologies that may apply to your project.

Exam Essentials

Be able to describe the difference between projects and operations.   A project is temporary in nature with a definite beginning and ending date. Projects produce unique products, services, or results. Operations are ongoing and use repetitive processes that typically produce the same result over and over.

Be able to distinguish between the seven needs or demands that bring about project creation.   The seven needs or demands that bring about project creation are market demand, organizational need, customer requests, technological advances, legal requirements, ecological impacts, and social needs.

Other factors that bring about projects are strategic opportunity or business need, competitive forces, political changes, new technology, stakeholder demands, business process improvements, and material issues.

Be able to describe a feasibility study.   A feasibility study is used to determine the viability of the project, the probability of success, and the viability of the product, service, or result of the project.

Be able to name the five project management process groups.   The five project management process groups are Initiating, Planning, Executing, Monitoring and Controlling, and Closing.

Be able to denote the skills listed in the PMI® Talent Triangle™.   The three skills are technical project management, leadership, and strategic and business management.

Be able to denote some of the skills every good project manager should possess. Communication, organizational, problem solving, negotiation and influencing, leading, team building, technical, and business knowledge are skills a project manager should possess.

Be able to name the three types of PMO organizations.   The three types of PMO organizations are supportive, controlling, and directive.

Be able to name the three types of development life cycles.   They are predictive (known as waterfall), adaptive (known as agile), and hybrid.

Be able to describe the agile project management methodology. A method of managing projects in small, incremental portions of work that can be easily assigned, easily managed, and completed within a short period of time. Agile involves continuous stakeholder involvement and feedback.

Be able to describe the difference between Scrum and Kanban. Scrum and Kanban are both agile methodologies. Scrum teams complete work in short, time-bound periods called sprints. Kanban is a continuous system. The work does not start and stop but continues through to completion. Kanban is also known as an on-demand scheduling system. Both Scrum and Kanban are known as pull systems.

Review Questions

You can find the answers to the review questions in Appendix A. Be sure to download the Bonus Exams and Bonus Questions so that you'll have a broader exposure and more experience answering questions related to the topics in this chapter.

1. Which organization has set the de facto standards for project management principles, processes, and techniques?

   A. PMBOK®

   B. PMO

   C. PMI®

   D. PBO

2. You work for a textile manufacturing firm. Your organization is introducing a new color line for their drapery materials. Changing colors for these materials during the manufacturing process is straightforward. This is considered which of the following?

   A. Project initiation

   B. Ongoing operations

   C. A project

   D. Project execution

3. Your company manufactures small kitchen appliances. It is introducing a new product line of appliances in designer colors with distinctive features for kitchens in small spaces. These new products will be offered indefinitely starting with the spring catalog release. Which of the following is true?

   A. This is a project because this new product line has never been manufactured and sold by this company before.

   B. This is an ongoing operation because the company is in the business of manufacturing kitchen appliances. Introducing designer colors and features is simply a new twist on an existing process.

   C. This is an ongoing operation because the new product line will be sold indefinitely. It's not temporary.

   D. This is not a project or an ongoing operation. This is a new product introduction not affecting ongoing operations.

4. Your company manufactures small kitchen appliances. It is introducing a new product line of appliances in designer colors with distinctive features for kitchens in small spaces. This project was approved by the stakeholders and you have been appointed the project manager. These new products will be offered starting with the spring online release. To determine the characteristics and features of the new product line, you will have to perform which of the following? (Choose two.)

   A. Defining business value for the new product line

   B. Consulting with the stakeholders about the characteristics and features of the new product line

    C. Planning the project life cycle for the project

    D. Progressively elaborating the characteristics and features of the new product line

5. You've been hired as a manager for the adjustments department of a nationwide bank based in your city. The adjustments department is responsible for making corrections to customer accounts. This is a large department, with several smaller sections that deal with specific accounts, such as personal checking or commercial checking. You've received your first set of management reports and can't make heads or tails of the information. Each section appears to use a different methodology to audit their work and record the data for the management report. You request that a project manager from the PMO come down and get started right away on a project to streamline this process and make the data and reports consistent. This project came about as a result of which of the following?

    A. Technological advance

    B. Organizational need

    C. Customer request

    D. Legal requirement

6. Which of the following applies a set of tools and techniques used to describe, organize, and monitor the work of project activities to meet the project requirements?

    A. Project managers

    B. The *PMBOK® Guide*

    C. Project management

    D. Stakeholders

7. Which of the following are true regarding multiphased relationships? (Choose three.)

    A. Planning for an iterative phase begins while the work of other phases is progressing.

    B. Overlapping phases occur when more than one phase is being performed at the same time.

    C. During sequentially phased projects, the previous phase must finish before the next phase can begin.

    D. Phase reviews should occur at the end of every phase.

8. Agile project life cycle methodologies are characterized by which of the following? (Choose three.)

    A. Dividing tasks into small deliverables that can be completed in a short time frame.

    B. Using a step-by-step process where one task is completed followed by another.

    C. This methodology is used primarily in the software development industry but can be applied across other industry areas.

    D. This methodology allows the project team to quickly adapt to new requirements and receive continuous feedback.

9. Your company sells Internet of Things appliances. They've just learned that another company is going to offer the same appliances you offer with new, updated features and functionality. Your project came about to incorporate similar features in your product line. What is the business reason or need that brought this project about?

   A. Technological advance

   B. Competitive forces

   C. Market demand

   D. Business process improvements

10. All of the following statements are true except for which one?

    A. Programs are groups of related projects.

    B. Project life cycles are collections of sequential, iterative, and overlapping project phases.

    C. A project may or may not be part of a program.

    D. Portfolios are collections of interdependent projects or programs.

11. What are the five project management process groups, in order?

    A. Initiating, Executing, Planning, Monitoring and Controlling, and Closing

    B. Initiating, Monitoring and Controlling, Planning, Executing, and Closing

    C. Initiating, Planning, Monitoring and Controlling, Executing, and Closing

    D. Initiating, Planning, Executing, Monitoring and Controlling, and Closing

12. During which project management process group are risk and stakeholders' ability to influence project outcomes the highest?

    A. Planning

    B. Executing

    C. Initiating

    D. Monitoring and Controlling

13. Which of the following are true about a PMO? (Choose three.)

    A. There are three types of PMOs: supportive, controlling, and collaborative.

    B. The PMO is often responsible for implementing the OPM.

    C. The key purpose of the PMO is to provide support to project managers.

    D. The PMO facilitates communication within and across projects.

14. Which of the following agile methodologies describes the seven wastes?

    A. Scrum

    B. Six Sigma

    C. Kaizen

    D. Kanban

15. Which of the following agile methodologies are a type of pull system? (Choose two.)

    A. Kaizen

    B. Six Sigma

    C. Lean

    D. XP

    E. Scrum

    F. Kanban

16. Which of the following agile methodologies relies heavily on statistical data?

    A. Kaizen

    B. Six Sigma

    C. Lean

    D. XP

    E. Scrum

    F. Kanban

17. Which of the following are characteristics of a predictive methodology? (Choose three.)

    A. The results of the work of the project are often not delivered until the end of the project.

    B. This methodology requires continuous feedback from your stakeholders throughout the project.

    C. Changes to the project require a review of project plans and documenting the changes in the project plan.

    D. Predictive methodologies might use a phased approach where deliverables are produced at the end of each phase.

    E. The project team reviews the work of the project with the stakeholders in an iterative fashion so that they can incorporate modifications to functionality in the next phase.

18. Which of the following are true regarding the *Agile Manifesto*? (Choose two.)

    A. It is concerned with the quality of the deliverable.

    B. Success is measured in incremental steps.

    C. The focus is on the value to the customer.

    D. It measures how efficiently the process was performed.

    E. It is concerned with business process improvements.

19. Match the following terms with the statements that describe them.

Term	Description
A. Project and program management	1. Concerned with working on the right projects at the right time
B. Portfolio management	2. Centralized unit that oversees the management of projects throughout the organization
C. Organization project management	3. Focus on performing the projects in the right way
D. Project management office	4. Ensures projects are aligned with the organization's strategic business objectives

20. Your project team has a solid idea of the requirements for the project up front. Some specific elements of the deliverables are known at this point. Not all deliverables have been completely defined yet. The team would like to start with the known requirements and specific deliverables and then change their approach later in the development phase to incrementally deliver results. What development life cycle does this describe?

A. Incremental

B. Predictive

C. Hybrid

D. Agile

Chapter

2

Assessing Project Needs

THE PMP® EXAM CONTENT FROM THE PEOPLE DOMAIN COVERED IN THIS CHAPTER INCLUDES THE FOLLOWING:

✓ **Task 1.2 Lead a team**

- 1.2.1 Set a clear vision and mission

✓ **Task 1.4: Empower team members and stakeholders**

- 1.4.3 Evaluate demonstration of task accountability
- 1.4.4 Determine and bestow levels of decision-making authority

✓ **Task 1.9: Collaborate with stakeholders**

- 1.9.1 Evaluate engagement needs for stakeholders
- 1.9.2 Optimize alignment between stakeholder needs, expectations, and project objectives
- 1.9.3 Build trust and influence stakeholders to accomplish project objectives

THE PMP® EXAM CONTENT FROM THE PROCESS DOMAIN COVERED IN THIS CHAPTER INCLUDES THE FOLLOWING:

✓ **Task 2.9 Integrate project planning activities**

- 2.9.2 Analyze the data collected

✓ **Task 2.12 Manage project artifacts**

- 2.12.1 Determine the requirements (what, when, where, who, etc.) for managing the Project artifacts
- 2.12.2 Validate that the project information is kept up to date (i.e., version control) and accessible to all stakeholders

✓ **Task 2.13 Determine appropriate project management methodology**

- 2.13.4 Recommend a project methodology/approach

THE PMP® EXAM CONTENT FROM THE BUSINESS ENVIRONMENT DOMAIN COVERED IN THIS CHAPTER INCLUDES THE FOLLOWING:

✓ **Task 3.1 Plan and manage project compliance**

- 3.1.1 Confirm project compliance requirements

✓ **Task 3.2 Evaluate and deliver project benefits and value**

- 3.2.1 Investigate that benefits are identified

- 3.2.2 Document agreement on ownership for ongoing benefit realization

Now that you're armed with a detailed overview of project management, you can easily determine whether your next assignment is a project or an ongoing operation. And one of the first skills you will put to use will be your communication skills. Are you surprised? Of course, you're not. It all starts with communication. You can't begin defining the project until you've first talked to the project sponsor, key stakeholders, and management personnel. All good project managers have honed their communication skills to a nice sharp edge.

You'll remember from Chapter 1, "Building the Foundation," that *Initiating* is the first process group in the five project management process groups. You can think of it as the official project kickoff. Initiating acknowledges that the project, or the next phase in an active project, should begin. This process group culminates in the publication of a project charter and a stakeholder register. We'll cover the project charter in this chapter and the stakeholder register in Chapter 3, "Delivering Business Value." But before we dive into these processes, we have one more preliminary topic to cover regarding the 10 Knowledge Areas. Then, I'll close out the chapter with a discussion of business value, what it entails, and how you know whether you're achieving it.

At the end of this chapter, I'll introduce a case study that will illustrate the main points of the chapter. I'll expand on this case study from chapter to chapter, and you'll begin building a project using each of the skills you learn.

The process names, inputs, tools and techniques, outputs, and descriptions of the project management process groups and related materials and figures in this chapter are based on content from *A Guide to the Project Management Body of Knowledge (PMBOK® Guide), Sixth Edition* (PMI®, 2017). The references to adaptive and hybrid methodologies, related materials, and figures in this chapter are based on content from the *Agile Practice Guide* (PMI®, 2017).

Exploring the Project Management Knowledge Areas

We talked about the five process groups in Chapter 1. They are Initiating, Planning, Executing, Monitoring and Controlling, and Closing. Each process group is made up of a collection of processes used throughout the project life cycle. *A Guide to the Project*

Management Body of Knowledge (PMBOK® Guide), Sixth Edition groups these processes into 10 categories that it calls the *Project Management Knowledge Areas.* These groupings, or Knowledge Areas, bring together processes that have characteristics in common. For example, the *Project Cost Management Knowledge Area* involves all aspects of the budgeting process, as you would suspect. Therefore, processes such as Estimate Costs, Determine Budget, and Control Costs belong to this Knowledge Area. Here's the tricky part: these processes don't belong to the same project management process groups (Estimate Costs and Determine Budget are part of the Planning process group, and Control Costs is part of the Monitoring and Controlling process group). Think of it this way: Knowledge Areas bring together processes by commonalities, whereas project management process groups are more or less the order in which you perform the project management processes (although remember that you can come back through these processes more than once). The *PMBOK® Guide* names the 10 Knowledge Areas as follows:

- Project Integration Management
- Project Scope Management
- Project Schedule Management
- Project Cost Management
- Project Quality Management
- Project Resource Management
- Project Communications Management
- Project Risk Management
- Project Procurement Management
- Project Stakeholder Management

Let's take a closer look at each Knowledge Area so that you understand how they relate to the process groups. Included in each of the following sections are tables that illustrate the processes that make up the Knowledge Area and the project management process group to which each process belongs. This will help you see the big picture in terms of process groups compared to Knowledge Areas. I'll discuss each of the processes in the various Knowledge Areas throughout the book, but for now, you'll take a high-level look at each of them with the exception of the Project Integration Management Knowledge Area. Since this chapter kicks off the processes in this Knowledge Area, we'll go a little deeper explaining the purpose of integration in the next section.

Exam Spotlight

The PMP® exam may have a question or two regarding the processes that make up a Knowledge Area. Remember that Knowledge Areas bring together processes by commonalities, so thinking about the Knowledge Area itself should tip you off to the processes that belong to it. Projects are executed in process group order, but the

Knowledge Areas allow a project manager to think about groups of processes that require specific skills. This makes the job of assigning resources easier because team members with specific skills might be able to work on and complete several processes at once. To broaden your understanding of the Knowledge Areas, cross-reference the purposes and the processes that make up each Knowledge Area with the *PMBOK® Guide*.

 The descriptions, process names, and project management process groups in the 10 Knowledge Areas described in this section are based on content from the *PMBOK® Guide*.

Project Integration Management

The *Project Integration Management Knowledge Area* consists of seven processes, as shown in Table 2.1.

TABLE 2.1 Project Integration Management

Process name	Project management process group
Develop Project Charter	Initiating
Develop Project Management Plan	Planning
Direct and Manage Project Work	Executing
Manage Project Knowledge	Executing
Monitor and Control Project Work	Monitoring and Controlling
Perform Integrated Change Control	Monitoring and Controlling
Close Project or Phase	Closing

The Project Integration Management Knowledge Area is concerned with coordinating all aspects of the project management plan and is highly interactive. This Knowledge Area involves identifying and defining the work of the project and combining, coordinating, unifying, and integrating the appropriate processes. Integration of these processes might occur once, or multiple times as needed, or continuously throughout the project. For example, obtaining resources for the project may occur at specific times when those skills or abilities are needed so this process is repeated periodically. Or processes that involve identifying risks may occur throughout the life of the project until it's closed out or risks no longer pose a threat, so these processes are performed continuously throughout the project. It also involves aligning deliverable due dates with the project life cycle and benefits management plan. This Knowledge Area is also concerned with choosing among alternative approaches, managing resource assignments, tailoring the processes, and performing trade-offs among the competing demands of several projects. Phase transitions are carried out in this Knowledge Area.

An example of integration across the Knowledge Areas is that project planning, executing, monitoring, and change control occur throughout the project and are repeated continuously while you're working on the project. Project planning and executing involve weighing the objectives of the project against the alternatives to bring the project to a successful completion. The project manager also makes choices about how to effectively use resources and how to coordinate the work of the project on a continuous basis. Monitoring the work of the project involves anticipating potential problems and issues and dealing with them before they reach the critical point. Change control can impact the project schedule, which in turn impacts the work of the project, which in turn can impact the project management plan, so you can see that these processes are tightly integrated.

According to the *PMBOK® Guide,* performing integration is a critical skill that all project managers should possess.

The project manager is responsible for managing the processes within the Project Integration Management Knowledge Area. These processes should not be delegated. The elements covered in this Knowledge Area include activities such as developing the project management plan, performing change control, monitoring and reporting on project work, directing and managing project work, managing project knowledge, and formally closing out the project. It is also concerned with satisfactorily meeting the requirements of the customer and stakeholder. These activities are all directly related to the processes found within this Knowledge Area.

Integration processes can be enhanced for both the project manager and stakeholders by using automated project management tools, visual aids, and sharing project management knowledge with others so that the knowledge or project and techniques are retained.

Integration involves processes, knowledge, and people. According to the *PMBOK® Guide,* integration occurs on three levels: the process level, cognitive level, and context level.

Process Level Integration   The process level is what we just discussed earlier in this section. The integration of project processes may occur once, multiple times, or continuously throughout the project, and processes may overlap one another.

Cognitive Level Integration   The cognitive level factors in elements such as the complexity of the project, the culture of the organization, the project size, and the complexity of the organization itself. Project managers' adeptness at understanding these elements, along with their technical project management skills, experience, and the other skills. will guide them in integrating the appropriate processes from the various Knowledge Areas in a way that leads to the fulfillment of the project objectives.

Context Level Integration   The context level refers to the business environment and its culture and the environment or culture the business serves, including elements such as technology, social networks, team makeup, and more. It's important that the project manager understands these factors so that integration is performed in a way that relates to the context level.

 Although all three levels of integration may involve process, knowledge, and people, as you can see from these descriptions, the process level leans more toward process integration, the cognitive level is more focused on knowledge, and the context level emphasizes people.

There isn't one right way to integrate processes. It is up to project managers, using their knowledge and consideration of the business environment, to determine and tailor which project processes will interact with one another and integrate them appropriately. Complex, high-risk projects require more rigor than smaller, low-risk projects will need. When tailoring, consider the life-cycle process, development and management approaches, change control and how changes will be managed, governance processes, lessons learned, and the benefits gained using the tailored approach.

Integration is managed and performed by the project manager when using an agile project management methodology. However, the planning, control, and delivery of the product is managed by the agile team. Changes can be made during the iteration as well. The project manager's role in performing integration within an agile approach entails the following:

- Providing an atmosphere of collaboration among the team members

- Assisting them in decision-making

- Providing an environment that allows them to respond to change

Project Integration Management uses visual tools to help present information about the project and automated tools to assist in integrating processes and document project information. The *project management information system (PMIS)* is an automated system used to document the project management plan and subsidiary plans, to facilitate the feedback process, and to revise the documents. It also incorporates the configuration management system and the change control system. Later in the project, the PMIS can be used to control changes to any of the plans. When you're thinking about the PMIS as an input to a process (generally it's listed as part of the enterprise environmental factors input), think of it as a collection and distribution point for information as well as an easy way to revise and update documents. When you're thinking about the PMIS as a tool and technique, think of it just that way—as a tool to facilitate the automation, collection, and distribution of data and to help monitor processes such as scheduling, resource leveling, budgeting, and web interfaces.

Project Scope Management

The *Project Scope Management Knowledge Area* has six processes, as shown in Table 2.2.

TABLE 2.2 Project Scope Management

Process name	Project management process group
Plan Scope Management	Planning
Collect Requirements	Planning
Define Scope	Planning
Create WBS	Planning
Validate Scope	Monitoring and Controlling
Control Scope	Monitoring and Controlling

Project Scope Management is concerned with defining all the work of the project and only the work needed to successfully achieve the project goals. These processes are highly interactive. They define and control what is and what is not part of the project. Each process occurs at least once—and often many times—throughout the project's life.

Project Scope Management encompasses both product scope and project scope. *Product scope* concerns the characteristics of the product, service, or result of the project. It's measured against the product requirements to determine successful completion or fulfillment. The application area usually dictates the process tools and techniques you'll use to define and manage product scope. *Project scope* involves managing the work of the project and only the work of the project. Project scope is measured against the project management plan. The scope baseline is made up of the project scope statement, the work breakdown structure (WBS), and the WBS dictionary.

To ensure a successful project, both product and project scope must be well integrated. This implies that Project Scope Management is well integrated with the other Knowledge Area processes.

Collect Requirements, Define Scope, Create WBS, Validate Scope, and Control Scope involve the following:

- Defining and detailing the deliverables and requirements of the product of the project
- Creating a WBS
- Validating deliverables using measurement techniques
- Controlling changes to the scope of the project

The steps outlined previously are used primarily in a predictive life-cycle methodology. In an agile methodology, these processes are performed before, during, and after the iteration. For example, when using Scrum, requirements (they are called *user stories* in Scrum) are defined at the beginning of the project. When a sprint is ready to begin, user stories are pulled from the list and worked on during short periods of time called sprints. The tasks related to the user stories are broken down by the team members; the work is completed, validated, and controlled; and changes are made as needed during the sprint. I will discuss the Scrum process in much more detail throughout the remainder of this book.

Project Schedule Management

The *Project Schedule Management Knowledge Area* has six processes, as shown in Table 2.3.

TABLE 2.3   Project Schedule Management

Process name	Project management process group
Plan Schedule Management	Planning
Define Activities	Planning
Sequence Activities	Planning
Estimate Activity Durations	Planning
Develop Schedule	Planning
Control Schedule	Monitoring and Controlling

This Knowledge Area is concerned with estimating the duration of the project activities, devising a project schedule, and monitoring and controlling deviations from the schedule. Collectively, this Knowledge Area deals with completing the project in a timely manner. Time management is an important aspect of project management because it concerns keeping the project activities on track and monitoring those activities against the project management plan to ensure that the project is completed on time.

Although most processes in this Knowledge Area occur at least once in every project (and sometimes more), in many cases—particularly on small projects—Sequence Activities, Estimate Activity Durations, and Develop Schedule are completed as one activity. Only one person is needed to complete these processes for small projects, and they're all worked on at the same time.

Schedule management occurs in an iterative fashion when using an agile project management methodology. For example, Scrum uses two- to four-week periods of time called *sprints* to perform work. There isn't a schedule per se, but some estimates could be made at the beginning of the project to determine an approximate length of time to complete the overall project.

Project Cost Management

As its name implies, the *Project Cost Management Knowledge Area* centers around costs and budgets. Table 2.4 shows the processes that make up this Knowledge Area.

TABLE 2.4   Project Cost Management

Process name	Project management process group
Plan Cost Management	Planning
Estimate Costs	Planning
Determine Budget	Planning
Control Costs	Monitoring and Controlling

The activities in the Project Cost Management Knowledge Area establish cost estimates for resources, establish budgets, and keep watch over those costs to ensure that the project stays within the approved budget. The earlier you can develop and agree on the scope of the project, the earlier you can estimate costs. The benefit of this practice is that costs are more easily influenced early in the project.

This Knowledge Area is primarily concerned with the costs of resources, but you should think about other costs as well. For example, be certain to examine ongoing maintenance and support costs for software or equipment that may be handed off to the operations group at the end of the project.

Depending on the complexity of the project, these processes might need the involvement of more than one person. For example, the finance person might not have expertise about the resources documented in the staffing management plan, so the project manager will need to bring in a staff member with those skills to assist with the activities in this process.

Two techniques are used in this Knowledge Area to decide among alternatives and improve the project process: life-cycle costing and value engineering. The life-cycle costing technique considers the total of all the project costs that might be incurred during the life of the project. This includes the initial investment; the costs to produce the product, service, or result of the project; the costs to operate the project once it's turned over to production; and the disposal costs at the project's end of life. The value engineering technique helps improve project schedules, profits, quality, and resource usage and optimizes life-cycle costs, among others. Value engineering, in a nutshell, involves optimizing project performance and cost and is primarily concerned with eliminating unnecessary costs. By examining the project processes, performance, or even specific activities, the project manager can identify those elements that are not adding business value and could be eliminated, modified, or reduced to realize cost savings. These techniques can improve decision-making, reduce costs, reduce activity durations, and improve the quality of the deliverables. Some application areas require additional financial analysis to help predict project performance. Techniques such as payback analysis, return on investment, and discounted cash flows are a few of the tools used to accomplish this.

Project Quality Management

The *Project Quality Management Knowledge Area* is composed of three processes, as shown in Table 2.5.

TABLE 2.5   Project Quality Management

Process name	Project management process group
Plan Quality Management	Planning
Manage Quality	Executing
Control Quality	Monitoring and Controlling

The Project Quality Management Knowledge Area ensures that the project meets the requirements that it was undertaken to produce. This Knowledge Area focuses on product quality as well as on the quality of the project management processes used during the project. These processes measure overall performance and monitor project results and compare them to the quality standards set out in the project-planning process to ensure that the customers will receive the product, service, or result they commissioned. It's easy to think that this Knowledge Area applies only to manufacturing or construction type projects that produce a tangible product. According to the *PMBOK® Guide,* Project Quality Management applies to all projects.

Quality is assessed iteratively during each iteration when using an agile approach like Scrum. All of the Project Quality Management processes are performed during an iteration. You'll plan, manage, and control quality during the sprint and adjust as needed during the iteration, or in the upcoming iteration, to meet quality standards.

Project Resource Management

The *Project Resource Management Knowledge Area* consists of six processes, as shown in Table 2.6.

Project Resource Management involves all aspects of people management and personal interaction, including leading, coaching, dealing with conflict, conducting performance appraisals, and more. It also involves the management of physical resources. These processes ensure that the physical and human resources assigned to the project are used

TABLE 2.6   Project Resource Management

Process name	Project management process group
Plan Resource Management	Planning
Estimate Activity Resources	Planning
Acquire Resources	Executing
Develop Team	Executing
Manage Team	Executing
Control Resources	Monitoring and Controlling

in the most effective way possible. Some of the project participants on whom you'll practice these skills are stakeholders, team members, and customers. Each requires the use of different communication styles, leadership skills, and team-building skills. A good project manager knows when to apply certain skills and communication styles based on the situation.

Projects are unique and temporary, and project teams usually are too. Teams are built based on the skills and resources needed to complete the activities of the project, and many times, project team members might not know one another. The earlier you can involve team members on the project, the better, because you'll want the benefit of their knowledge and expertise during the planning processes. Their participation early on will also help ensure their buy-in to the project. Whether using a predictive, adaptive, or hybrid methodology, the makeup of the project team may change as you progress through the project. The stakeholders involved in the various stages of the project might change as well, so you'll use different techniques at different times throughout the project to manage the processes in this Knowledge Area.

Agile teams are made up of team members from cross-functional areas of the business. They are self-directed and self-managed. Agile teams should also be built based on the skills needed and the types of resources needed to complete the project work.

Project Communications Management

The processes in the *Project Communications Management Knowledge Area* are related to general communication skills, but they encompass much more than an exchange of information. Communication skills are considered interpersonal skills that the project manager uses on a daily basis. The processes in the Project Communications Management Knowledge Area seek to ensure that all project information—including plans, risk assessments, meeting notes, and more—is collected, organized, stored, and distributed to stakeholders, management, and project members at the proper time. When the project is closed, the information is archived and used as a reference for future projects. This is referred to as *historical information* in several project processes. These activities involve developing and maintaining a solid communication strategy.

Everyone on the project has some involvement with this Knowledge Area because all project members will send and/or receive project communication throughout the life of the project. It is important that all team members and stakeholders understand how communication affects the project. The communications management plan details how, when, what, and with whom you'll communicate throughout the project. Communication is an activity that occurs on all projects, whether you're using a predictive, adaptive, or hybrid approach in performing the work of the project.

Three processes make up the Project Communications Management Knowledge Area, as shown in Table 2.7.

TABLE 2.7   Project Communications Management

Process name	Project management process group
Plan Communications Management	Planning
Manage Communications	Executing
Monitor Communications	Monitoring and Controlling

Time to Communicate

Project Communications Management is probably the most important Knowledge Area on any project, and most project managers understand the importance of good communication skills and making sure stakeholders are informed of project status. I know a project manager who had difficulties getting time with the project sponsor. The project sponsor agreed to meet with the project manager, even set up the meetings himself, and then canceled them or simply didn't show up. The poor project manager was at her wit's end about how to communicate with the sponsor and get some answers to the questions she had. Her desk was not far outside the project sponsor's office. One day as she peeked around the corner of her cube, she decided if the sponsor wouldn't come to her, she would go to him. From then on, every time the sponsor left his office, she would jump up from her chair and ride with him on the elevator. He was a captive audience. She was able to get some easy questions answered and finally convince him, after the fourth or fifth elevator ride, that they needed regular face-to-face meetings. She understood the importance of communication and went to great lengths to make certain the sponsor did too.

Project Risk Management

The *Project Risk Management Knowledge Area,* as shown in Table 2.8, contains seven processes.

TABLE 2.8   Project Risk Management

Process name	Project management process group
Plan Risk Management	Planning
Identify Risks	Planning
Perform Qualitative Risk Analysis	Planning
Perform Quantitative Risk Analysis	Planning
Plan Risk Responses	Planning
Implement Risk Responses	Executing
Monitor Risks	Monitoring and Controlling

Risks include both threats to and opportunities for the project. The processes in this Knowledge Area are concerned with identifying, analyzing, and planning for potential risks, both positive and negative, that might impact the project. This means minimizing the probability and impact of negative risks while maximizing the probability and impact of positive risks. These processes are also used to identify the positive consequences of risks and exploit them to improve project objectives or discover efficiencies that might improve project performance. All projects have risk, and the processes in this Knowledge Area should not be skipped in order to speed up project progress.

Project managers will often combine several of these processes into one step. For example, Identify Risks and Perform Qualitative Risk Analysis might be performed at the same time. The important factor of the Project Risk Management Knowledge Area is that you should strive to identify all the risks and develop responses for those with the greatest consequences to the project objectives. In a predictive approach to project management, you'll perform these processes in a more detailed fashion than you would in an adaptive or hybrid approach. If you are using an agile methodology to manage your project, these processes will be performed before, during, and after every iteration.

Project Procurement Management

There are three processes in the *Project Procurement Management Knowledge Area*, as shown in Table 2.9.

TABLE 2.9 Project Procurement Management

Process name	Project management process group
Plan Procurement Management	Planning
Conduct Procurements	Executing
Control Procurements	Monitoring and Controlling

The Project Procurement Management Knowledge Area includes the processes involved with purchasing goods or services from vendors, contractors, suppliers, and others outside the project team. As such, these processes involve negotiating and managing contracts and other procurement vehicles and managing changes to the contract or work order. When discussing the Project Procurement Management processes, it's assumed that the discussion is taking place from your perspective as a buyer, whereas sellers are external to the project team. Interestingly, the seller might manage their work as a project, particularly when the work is performed on contract, and you as the buyer become a key stakeholder in their project. In a predictive approach, planning your procurements will happen early on in the project and as the work progresses, you'll execute the procurement and monitor the contractor to ensure you're getting the goods or services that were outlined in the procurement document. If you are hiring outside contractors to help your team produce the project and you're using an agile approach, the procurement must be completed before the project work can start. This means you'll plan and conduct or execute your contract before starting the project and then employ agile techniques to manage the work of the project.

Project Stakeholder Management

There are four processes in the *Project Stakeholder Management Knowledge Area,* as shown in Table 2.10.

TABLE 2.10 Project Stakeholder Management

Process name	Project management process group
Identify Stakeholders	Initiating
Plan Stakeholder Engagement	Planning
Manage Stakeholder Engagement	Executing
Monitor Stakeholder Engagement	Monitoring and Controlling

The Project Stakeholder Management Knowledge Area is concerned with identifying all the stakeholders associated with the project, both internal and external to the organization. These processes also assess stakeholder needs, expectations, and involvement on the project and seek to keep the lines of communication with stakeholders open and clear. The definition of a successful project is one where the stakeholders are satisfied. These processes ensure that stakeholder expectations are met and that their satisfaction is a successful deliverable of the project. When using a predictive approach to manage the project, you'll go to great lengths to identify, plan, and classify your stakeholders early on in the project. An agile or hybrid approach will also involve identification and classification of stakeholders at the beginning of the project, but managing and monitoring stakeholders will occur in each phase or each iteration.

The remainder of this book will deal with processes and process groups as they occur in order (that is, Initiating, Planning, Executing, Monitoring and Controlling, and Closing), because this is the way you will encounter and manage them during a project. I will also add agile and hybrid methodology information that relates to each of these process groups as we go.

Assessing Project Viability

Most organizations don't have the luxury of performing every project that's proposed. Even consulting organizations that sell their project management services must pick and choose the projects on which they want to work. Selection and evaluation methods help organizations decide among alternative projects and determine the tangible benefits to the company of choosing or not choosing the project.

How projects are assessed will vary depending on the company, the people serving on the selection committee, the criteria used, and the project. Sometimes the criteria for selection methods will be purely financial, sometimes marketing, and sometimes they'll be based on public perception or political perception. In most cases, the decision is based on a combination of all these factors and more.

Most organizations have a formal, or at least semiformal, process for selecting and prioritizing projects. In my organization, a steering committee is responsible for project review, selection, and prioritization. A *steering committee* is a group of folks consisting of executive managers or senior managers who represent each of the functional areas in the organization.

Here's how our process works: The steering committee requests project ideas from the business staff and other subject matter experts prior to the beginning of the fiscal year. These project ideas are submitted in writing and contain a high-level overview of the project goals, a description of the deliverables, the business justification for the project, a desired implementation date, what the organization stands to gain from implementing the project, a list of the functional business areas affected by the project, and (if applicable) a benefit–cost ratio analysis (I'll talk about that in a bit).

A meeting is called to review the projects, and a determination is made on each project about whether it will be included on the upcoming list of projects for the new year. Once the no-go projects have been weeded out, the remaining projects are prioritized according to their importance and benefit to the organization using a scoring model. The projects are documented on an official project list, and progress is reported on the active projects at the regular monthly steering committee meetings.

In theory, it's a great idea. In practice, it works only moderately well. Priorities can and do change throughout the year. New projects come up that weren't originally submitted during the call for projects, and they must be added to the list. Reprioritization begins anew, and resource alignment and assignments are shuffled. But again, I'm getting ahead of myself. Just be aware that organizations usually have a process to recognize and screen project requests, accept or reject those requests based on some selection criteria, and prioritize the projects based on some criteria.

Exam Spotlight

According to PMI®, project selection is outside the scope of the project manager's role. In reality, the project manager often participates in the decision and assists with analysis and selection methods. For the exam, remember that project selection is performed by the project sponsor, customer, or subject matter experts.

I'll discuss several project selection methods next. Selection methods are one way the sponsor or customer may choose between projects. The individual opinion, and power, of stakeholders on the selection committee also plays a part in what projects the organization chooses to perform. Don't underestimate the importance of the authority, political standing, and individual aspirations of selection committee members. Those stakeholders who happen to carry a lot of weight in company circles, so to speak, are likely to get their projects approved just because they are who they are. This is sometimes how project selection works in my organization. How about yours?

Using Project Selection Methods

Project selection methods are concerned with the advantages or merits of the product of the project. In other words, selection methods measure the business value of what the product, service, or result of the project will produce and how it will benefit the organization. Selection methods involve the types of concerns about which executive managers are typically thinking. This includes factors such as market share, financial benefits, return on investment, customer retention, customer loyalty, and public perceptions. Most of these are reflected in the organization's strategic goals. Projects, whether large or small, should always be weighed against the strategic plan. If the project doesn't help the organization reach its goals (increased market share, for example), then the project probably shouldn't be undertaken.

The most common form of project selection methods (aside from the big boss saying "make-go" with this project) are benefit measurement methods. These methods employ various forms of analysis and comparative approaches to make project decisions such as benefit–cost ratio analysis, scoring models, and benefit contribution methods that include various cash flow techniques and economic models. The end result you're looking for with this method is whether the project will produce a sustained benefit realization for the organization. The following methods may also be used and documented in the business case to use as criteria for determining whether the project will provide business value to the organization. Let's examine several of these methods, starting with benefit–cost ratio analysis.

Benefit–Cost Ratio Analysis

One common benefit measurement method is the *benefit–cost ratio analysis (BCR)*. The name of this method implies what it does—it compares the cost to produce the product, service, or result of the project to the benefit (usually financial in the form of savings or revenue generation) that the organization will receive as a result of executing the project. Obviously, a sound project choice is one where the costs to implement or produce the product of the project are less than the financial benefits. How much less is the organization's decision. Some companies are comfortable with a small margin, whereas others are comfortable with a much larger margin between the two figures.

 Benefit–cost ratio analysis is also known as *cost–benefit analysis*. The techniques are the same.

When examining costs for a benefit–cost ratio analysis, include the costs to produce the product or service, the costs to take the product to market, and the ongoing operational support costs. For example, let's say your company is considering writing and marketing a database software product that will allow banks to dissect their customer base, determine which types of customers buy which types of products, and then market more effectively to those customers. You will take into account some of the following costs:

- The costs to develop the software, such as program developer costs, hardware costs, and testing costs

- The costs to maintain the software once it's deployed and keep it up to date with the latest patches, security updates, and more

- Marketing costs for advertising, travel costs to perform demos at potential customer sites, and so on

- Ongoing costs such as having customer support staff available during business hours to assist customers with questions and problems

Let's say the cost to produce this software plus the ongoing support costs $5 million. Initial projections look like the demand for this product is high. Over a three-year period, which is the potential life of the software in its proposed form, projected revenues are $24 million. Taking only the financial information into account, the benefits outweigh the costs of this project. This project should receive a go recommendation.

 Projects of significant cost or complexity usually involve more than one benefit measurement method when go or no-go decisions are being made or one project is being chosen over another. Keep in mind that selection methods can take subjective considerations into account as well—the project is a go because it's the new CEO's pet project; nothing else needs to be said.

Scoring Models

Another project selection technique in the benefit measurement category is a *scoring model*, or *weighted scoring model*. My organization uses weighted scoring models not only to choose between projects but also as a method to choose between competing bids on outsourced projects.

Weighted scoring models are quite simple. The project selection committee decides on the criteria that will be used on the scoring model—for example, profit potential, marketability of the product or service, or ability of the company to quickly and easily produce the product or service. Each of these criteria is assigned a weight depending on its importance to the project committee. More important criteria should carry a higher weight than less important criteria.

Then each project is rated on a scale from 1 to 5 (or some such assignment), with the higher number being the more desirable outcome to the company and the lower number having the opposite effect. This rating is then multiplied by the weight of the criteria factor and added to other weighted criteria scores for a total weighted score. Table 2.11 shows an example that brings this together.

In this example, Project A is the obvious choice.

TABLE 2.11   Weighted scoring model

Criteria	Weight	Project A score*	Project A totals	Project B score*	Project B totals	Project C score*	Project C totals
Profit potential	5	5	25	5	25	3	15
Marketability	3	4	12	3	9	4	12
Ease to produce/ support	1	4	4	3	3	2	2
Weighted score	—	—	41	—	37	—	29

*5 = highest

Cash Flow Analysis Techniques

The remaining benefit measurement methods involve a variety of cash flow analysis techniques, including payback period, discounted cash flows, net present value, and internal rate of return. We'll look at each of these techniques individually, and I'll provide you with a crash course on their meanings and calculations.

PAYBACK PERIOD

The *payback period (PBP)* is the length of time it takes the company to recoup the initial costs of producing the product, service, or result of the project. This method compares the initial investment to the cash inflows expected over the life of the product, service, or result. For example, say the initial investment on a project is $200,000, with expected cash inflows of $25,000 per quarter every quarter for the first two years and $50,000 per quarter from then on. The payback period is two years and can be calculated as follows:

> Initial investment = $200,000
>
> Cash inflows = $25,000 * 4 (quarters in a year) = $100,000 per year total inflow
>
> Initial investment ($200,000) – year 1 inflows ($100,000) = $100,000 remaining balance
>
> Year 1 inflows remaining balance – year 2 inflows = $0
>
> Total cash flow year 1 and year 2 = $200,000
>
> The payback is reached in two years.

The fact that inflows are $50,000 per quarter starting in year 3 makes no difference because payback is reached in two years.

The payback period is the least precise of all the cash flow calculations. That's because the payback period does not consider the value of the cash inflows made in later years, commonly called the *time value of money*. For example, if you have a project with a five-year payback period, the cash inflows in year 5 are worth less than they are if you received them today. The next section will explain this idea more fully.

DISCOUNTED CASH FLOWS

As I just stated, money received in the future is worth less than money received today. The reason for that is the time value of money. If I borrowed $2,000 from you today and promised to pay it back in three years, you would expect me to pay interest in addition to the original amount borrowed. If you were a family member or a close friend, maybe you wouldn't, but ordinarily this is the way it works. You would have had the use of the $2,000 had you not lent it to me. If you had invested the money (does this bring back memories of your mom telling you to save your money?), you'd receive a return on it. Therefore, the future value of the $2,000 you lent me today is $2,315.25 in three years from now at 5 percent interest per year. Here's the formula for future value calculations:

$$FV = PV(1+i)^n$$

In English, this formula says the future value (FV) of the investment equals the present value (PV) times (1 plus the interest rate) raised to the value of the number of time periods (n) the interest is paid. Let's plug in the numbers:

$$FV = 2,000(1.05)^3$$
$$FV = 2,000(1.157625)$$
$$FV = \$2,315.25$$

The *discounted cash flow* technique compares the value of the future cash flows of the project to today's dollars. To calculate discounted cash flows, you need to know the value of the investment in today's terms, or the PV. PV is calculated as follows:

$$PV = FV / (1+i)^n$$

This is the reverse of the FV formula we talked about earlier. So, if you ask the question, "What is $2,315.25 in three years from now worth today given a 5 percent interest rate?" you'd use the preceding formula. Let's try it:

$$PV = \$2,315.25 / (1+.05)^3$$
$$PV = \$2,315.25 / 1.157625$$
$$PV = \$2,000$$

$2,315.25 in three years from now is worth $2,000 today.

Discounted cash flow is calculated just like this for the projects you're comparing for selection purposes or when considering alternative ways of doing the project. Apply the PV formula to the projects you're considering, and then compare the discounted cash flows of all the projects against each other to make a selection. Here is an example comparison of two projects using this technique:

Project A is expected to make $100,000 in two years.

Project B is expected to make $120,000 in three years.

If the cost of capital is 12 percent, which project should you choose?
Using the PV formula used previously, calculate each project's worth:

The PV of Project A = $79,719

The PV of Project B = $85,414

Project B is the project that will return the highest investment to the company and should be chosen over Project A.

NET PRESENT VALUE

Projects might begin with a company investing some amount of money into the project to complete and accomplish its goals. In return, the company expects to receive revenues, or cash inflows, from the resulting project. *Net present value (NPV)* allows you to calculate an accurate value for the project in today's dollars. The mathematical formula for NPV is complicated, and you do not need to memorize it in that form for the test. However, you do need to know how to calculate NPV for the exam, so I've given you some examples of a less complicated way to perform this calculation in Table 2.12 and Table 2.13 using the formulas you've already seen.

TABLE 2.12   Project A

Year	Inflows	PV
1	10,000	8,929
2	15,000	11,958
3	5,000	3,559
Total	30,000	24,446
Less investment	—	24,000
NPV	—	**446**

TABLE 2.13   Project B

Year	Inflows	PV
1	7,000	6,250
2	13,000	10,364
3	10,000	7,118
Total	30,000	23,732
Less investment	—	24,000
NPV	—	**(268)**

Net present value works like discounted cash flows in that you bring the value of future monies received into today's dollars. With NPV, you evaluate the cash inflows using the discounted cash flow technique applied to each period the inflows are expected instead of in one sum. The total present value of the cash flows is then deducted from your initial investment to determine NPV. NPV assumes that cash inflows are reinvested at the cost of capital.

Here's the rule: If the NPV calculation is greater than 0, accept the project. If the NPV calculation is less than 0, reject the project.

Look at the two project examples in Tables 2.12 and 2.13. Project A and Project B have total cash inflows that are the same at the end of the project, but the amount of inflows at each period differs for each project. We'll stick with a 12 percent cost of capital. Note that the PV calculations were rounded to two decimal places.

Project A has an NPV greater than 0 and should be accepted. Project B has an NPV less than 0 and should be rejected. When you get a positive value for NPV, it means that the project will earn a return at least equal to or greater than the cost of capital.

Another note on NPV calculations: Projects with high returns early in the project are better projects than projects with lower returns early in the project. In the preceding examples, Project A fits this criterion also.

INTERNAL RATE OF RETURN

The *internal rate of return (IRR)* is the most difficult equation to calculate of all the cash flow techniques we've discussed. It is a complicated formula and should be performed on a financial calculator or computer. IRR can be figured manually, but it's a trial-and-error approach to get to the answer.

Technically speaking, IRR is the discount rate when the present value of the cash inflows equals the original investment. Or stated another way, the IRR is the interest rate that makes the NPV of all the cash inflows equal to zero. When choosing between projects or when choosing alternative methods of doing the project, projects with higher IRR values are generally considered better than projects with low IRR values.

Exam Spotlight

For the exam, you need to know three facts concerning IRR:

- IRR is the discount rate when NPV equals 0.

- IRR assumes that cash inflows are reinvested at the IRR value.

- You should choose projects with the highest IRR value.

You can use one, two, or several of the benefit measurement methods alone or in combination to come up with a selection decision. Remember that payback period is the least precise of all the cash flow techniques, NPV is the most conservative cash flow technique, and NPV and IRR will generally bring you to the same accept/reject conclusion.

You can use these methods to evaluate multiple projects or a single project. You might be weighing one project against another or simply considering whether the project you're proposing is worth performing.

 **Real World Scenario**

Fun Days Vacation Resorts

Nick is a project manager for Fun Days Vacation Resorts. He is working on three different project proposals to present to the executive steering committee for review. As part of the information-gathering process, Nick visits the various resorts pretending to be a guest. This gives him a feel for what Fun Days guests experience on their vacations, and it assists him in presenting alternatives.

Nick has prepared the project overviews for three projects and called on the experts in marketing to help him out with the projected revenue figures. He works up the numbers and finds the following:

- Project A: payback period = 5 years; IRR = 4 percent

- Project B: payback period = 3.5 years; IRR = 3 percent

- Project C: payback period = 2 years; IRR = 3 percent

Funding exists for only one of the projects. Nick recommends Project A and predicts this is the project the steering committee will choose since the projects are mutually exclusive.

Nick's turn to present comes up at the steering committee. Let's listen in on the action:

"On top of all the benefits I've just described, Project A provides an IRR of 8 percent, a full 5 percent higher than the other two projects we discussed. I recommend the committee choose Project A."

"Thank you, Nick," Jane says. "Good presentation." Jane is the executive chairperson of the steering committee and has the authority to break ties or make final decisions when the committee can't seem to agree.

"However, here at Fun Days we like to have our fun sooner rather than later." Chuckles ensue from the steering committee. They've all heard this before. "I do agree that a 4 percent IRR is a terrific return, but the payback is just too far out into the future. There are too many risks and unknowns for us to take on a project with a payback period this long. As you know, our industry is directly impacted by the health of the economy. Anything can happen in five years' time. I think we're much better off going with Project C. I recommend we accept Project C. Committee members, do you have anything to add?"

In the next section, we'll look at two important documents that are usually created by the stakeholder requesting the project. They are the business case and the benefits management plan. These documents can also be used to help in the project selection process. For example, the financial analysis will be documented in the business case, along with several other elements we'll discuss shortly, and presented to the selection committee. The project selection activity occurs no matter the life-cycle methodology you'll use to perform the project. Waterfall projects or agile projects both need a selection and approval process before proceeding with the work of the project.

Assessing Project Needs and Creating the Project Charter

Your first stop in the Initiating process group is a process called *Develop Project Charter*. As the name of the process suggests, your purpose is to create a project charter. The purpose for the Initiating group is to authorize a project, or the next phase of a project, to begin. It also gives the project manager the authority to apply resources to the project. These also happen to be the purposes of a project charter: formally authorizing the project to begin and committing resources.

Exam Spotlight

The project charter (which is an output of the Develop Project Charter process) is the written acknowledgment that the project exists. The project charter documents the name of the project manager and gives that person the authority to assign organizational resources to the project. The project is officially authorized when the project charter is signed.

There are two business documents the project sponsor will need to create before creating the project charter, and they both serve as inputs to the Develop Project Charter process. They are the business case and the benefits management plan. These documents will help determine the need that is driving project creation and the business value the project will bring about. The business case documents the need for the project, and the benefits management plan documents how the project will create, augment, or preserve and improve its benefits to the organization. The benefits management plan is iterative and updated throughout the life cycle of the project. The business case and the benefits management plan are directly related to each other. We'll look at each of these documents next.

Business Case

The purpose of a *business case* is to understand the business need for the project and determine whether the investment in the project is worthwhile. The business case also contains the financial analysis we discussed earlier in this chapter (such as NPV, ROI, IRR, PBP, or BCR), and it describes the business need or demand that brought about the project. As a refresher, these business needs or demands include market demand, organizational need, customer request, technological advance, legal requirements, ecological impacts, and social needs.

 According to the *PMBOK® Guide,* the business case is an economic feasibility study. It is used to track progress and compare project results against the success factors identified in the business case.

Performing a needs assessment is a great first step in building the business case. The needs assessment will examine the needs and demands of the business, marketing opportunities, costs, risks, and more, while taking into account the given assumptions and constraints of the proposed products or services. Once the needs assessment is completed, it's easy to build a business case based on the findings of the assessment. Don't forget that the feasibility study is often the first phase in the project and the findings from it are typically summarized in the business case.

Most sound business case documents contain the following elements:

- Description of the project, including the business need, demand, or opportunity that's driving the project. This section may note the needs assessment and feasibility study findings as well. This section should also include a description of the impact to the organization if the project is not undertaken.

- A high-level description of the scope of the project, which might include a list of high-level deliverables and desired outcomes.

- Names of the stakeholders involved with the project.

- Analysis of the problem or opportunity the project presents, including how the project aligns with the organization's strategic vision and goals, the cause of the problem (or opportunity) that's driving the creation of the project, and a gap analysis showing the existing capabilities of the organization versus how the organization will change if the project is implemented.

- A high-level list and description of risks. (This includes risks known at this time. As the project progresses, more risks will be identified and added to the list.)

- Description of the critical success factors.

- Description of the expected results of each alternative solution and the course of action for each.

- Recommended option. (This includes a summary of the analysis performed to get to this solution; a description of the risks, constraints, and assumptions associated with the recommended option; and a high-level plan to implement the solution outlining major milestones, any project dependencies, and roles and responsibilities of stakeholders.)
- Evaluation statement. (This includes how you'll measure the benefits the project will bring to the organization once it's completed.)

Remember that when your selection committee is analyzing and approving the business case (and thus triggering the Develop Project Charter process), they are completing these activities outside of the boundaries of the project. Typically, the project team is not involved at this stage. In other words, the project isn't a project until the project charter is approved, and the project charter cannot be started until the business case is approved.

One of the elements discussed earlier includes an analysis of the problem or opportunity the project presents. According to the *PMBOK® Guide*, there are three categories you can use for your decision analysis:

Required   Must perform or must have to successfully realize the opportunity, solve the problem, or make a decision.

Desired   The preferred solution or option to fulfill the opportunity, solve the problem, or make a decision.

Optional   Alternative solutions or options for solving the problem or taking advantage of the opportunity, or making a decision. Optional choices include the following:

- Do nothing
- Do the minimum needed
- Do more than the minimum

Once the business case is reviewed by the project selection committee, including weighing all the alternatives, they will make one of two decisions: proceed with the project charter, or deny the request. If the project proceeds, it's good practice to review the business case at each phase gate to make sure the project is successfully producing the objectives and benefits outlined in the business case.

Real World Scenario

The Motor Vehicle Registration App

Jason, Sam, and Kate are web programmers working for the Department of Revenue in the State of Bliss. Ron, their manager, approaches them one day with an idea.

"Team, business unit managers think it would be a great idea to offer motor vehicle owners the ability to renew their license plates on their mobile phone. We already offer them the ability to renew on the Internet, thanks to all your efforts on that project last year. It has been a fabulous success. No other state has had the success that Bliss has had with our Internet renewal system.

"Kate, I know you've had previous experience on projects like this one, but I'm not sure about you guys, so this is new territory for us. I'd like to hear what each of you thinks about this project."

Jason speaks up first. "I think it's a great idea. You know me; I'm always up for learning new things, especially when it comes to programming. When can we start?"

Sam echoes Jason's comments.

Kate says, "Jason and Sam are excellent coders and could work on the programming side of things, but I would have to pick up the interface piece on my own. After we're up and running, we could go over how the interfaces work with the mobile app so Jason and Sam could help me support it going forward. I'd really like to take on this project. It would be good for the team and good for the department."

Ron thinks for a minute. "Let's not jump right into this. I know you're anxious to get started, but I think a feasibility study is in order. The senior director of the motor vehicle business unit doesn't know whether this project is cost justified and has some concerns. A feasibility study will tell us the answers to those questions. It should also help us determine whether we're using the right technology to accomplish our goals, and it will outline alternative ways of performing the project that we haven't considered. I don't want Kate going it alone without first examining all the issues and potential impacts to the organization."

Benefits Management Plan

The *benefits management plan* outlines the intended benefits and business value the project will bring about, how the benefits will be measured, and how they will be obtained (this is typically described in financial terms). We will discuss business value in a larger context in depth in Chapter 3.

According to the *PMBOK® Guide,* a project benefit is the result of actions or behaviors, and/or the value of the product, service, or result the project brings to the organization and the project stakeholders.

The benefits management plan, according to the *PMBOK® Guide,* includes the following elements:

- Target benefits, including tangible and intangible value. Financial value is often a target benefit, and it should always be stated using net present value.

- Strategic alignment to the organization's goals and objectives.

- Time frame in which benefits will come to fruition.

- Benefits owner, who will track, monitor, document, and report when or if benefits are realized.

- Metrics to show how benefits will be measured.

- Assumptions about the project or the benefits that are believed to be true.

- Risks associated with the benefits.

As I mentioned previously, the benefits management plan and the business case are tightly linked. The business case is reviewed at each phase gate, and the benefits management plan is updated iteratively throughout the project life cycle. For example, the benefits management plan will use the preferred option from the business case to outline the expected benefits. The benefit–cost ratio analysis documented in the business case can also be used to describe benefit realization.

Exam Spotlight

Remember that the sponsor is accountable for documenting and updating the business case. The project manager is responsible for tailoring the project documents and the project processes in accordance with the size and complexity of the project and for keeping the project artifacts (such as the business case, benefits management plan, project charter, and the project management plan) coordinated and aligned with each other and with the organization's strategic goals.

One of the roles of the project manager is recommending a strategy for executing the work of the project. This may include performing the project with internal resources, contracting out the work of the project, or using both contract resources and internal resources. If you decide to contract out the project, you'll need to consider agreements such as contracts, memorandums of understanding, verbal commitments, and so on. A contract is applicable when you are performing a project for a customer external to the organization or you are hiring an outside firm to perform the project for your organization. Agreements typically document the conditions under which the project will be executed, the time frame, and a description of the work, so it's important to review these before documenting the project charter.

Enterprise Environmental Factors

Enterprise environmental factors (EEF) are elements you'll want to consider when planning your project. EEFs show up as an input to many of the other processes we'll discuss throughout the book and are most notably found in the Planning processes. EEFs refer to the factors outside the control of the project team that have (or might have) significant influence on the success of the project. According to the *PMBOK® Guide*, environmental factors are both internal and external to the organization. External EEFs are external to the organization and outside the control of the project team. For example, government or industry standards may have legal or statutory authority that dictates specific standards in terms of manufacturing or product specifications. Internal EEFs are internal to the organization but outside the control of the project team: for example, the location of office buildings and workers, resource availability, and organizational culture.

EEFs may drive project compliance requirements. For example, your project may involve creating a website that will accept credit card information. The Payment Card Industry (PCI) standards developed by the PCI Security Standards Council outline how to secure and protect the sensitive data you'll be handling when processing credit card information. If you fail to follow these standards, you could be subject to massive fines by the major credit card companies as well as your bank. When creating the project charter, you should consider specific compliance issues, including government or industry standards, legal and regulatory requirements, marketplace conditions, organizational culture and political climate, organizational governance framework, and stakeholders' expectations and risk thresholds. It is the project manager's responsibility to ensure that any compliance requirements associated with the project are followed and met.

Organizational Process Assets

Organizational process assets (OPAs) are the organization's policies, guidelines, processes, procedures, plans, approaches, and standards for conducting work, including project work. Organizational process assets are divided into two categories: 1) processes, policies, and procedures and 2) organizational knowledge bases. Processes, policies, and procedures refer to a wide range of elements that might affect several aspects of the project, such as project management policies, safety policies, performance measurement criteria, templates, financial controls, communication requirements, issue and defect management procedures, change control procedures, risk control procedures, and the procedures used for authorizing work. These are usually established by the PMO or the organization itself and are not updated as the project progresses. Table 2.14 lists the process, policies, and procedures for OPAs by process group, as outlined in the *PMBOK® Guide*.

TABLE 2.14   OPA process and procedures by process group

Initiating and Planning	Executing, Monitoring and Controlling	Closing
Guidelines for tailoring processes	Change control processes	Closure guidelines or requirements
Organizational policies	Traceability matrices	
Product and project life-cycle procedures	Financial controls	
Templates	Issue and defect management procedures	
Preapproved supplier lists	Resource availability and assignment management	
	Communication requirements	
	Work authorization procedures	
	Templates	
	Work instruction procedures	
	Verification and validation processes	

Organizational knowledge bases refer to items such as lessons learned, process measurement databases, project files, and the information the organization has learned on previous projects (including how to store and retrieve that information). For example, previous project risks, performance measurements, earned value data, and schedules for past projects are valuable resources of knowledge for the current project. This information, also known as historical information, falls into the corporate knowledge base category. If you don't capture and store this information, however, it won't be available when you're starting a new project. You'll want to capture and store information such as project financial data (budgets, costs, overruns), historical information, lessons learned, project files, issues and defects, process measurements, and configuration management knowledge.

According to the *PMBOK® Guide,* the following are the types of organizational knowledge repositories you may encounter:

- Configuration management
- Financial data
- Historical information and lessons learned
- Issue and defect management data
- Data repositories
- Project files from other projects

Organizational process assets and historical information should be reviewed and examined when a new project is starting. Historical information can be useful to project managers and to stakeholders. When you're evaluating new projects, historical information about previous projects of a similar nature can be handy in determining whether the new project should be accepted and initiated. Historical information gathered and documented during an active project is used to assist in determining whether the project should proceed to the next phase. Historical information will help you with the project charter, project scope statement, development of the project management plan, the process of defining and estimating activities, and more during the project-planning processes.

Understanding previous projects of a similar nature—their problems, successes, issues, and outcomes—will help you avoid repeating mistakes while reusing successful techniques to accomplish the goals of this project to the satisfaction of the stakeholders. These issues are documented in the lessons learned document. Lessons learned should be updated throughout the life of the project. You will finalize this document in the Closing process and include all the good, and not so good, things you learned about this project along the way. Many of the processes in the project management process groups have organizational process assets as an input, implying that you should review the pertinent organizational assets that apply for the process you're about to start. For example, when performing the Estimate Costs process, you might find it helpful to review the activity estimates and budgets on past projects of similar size and scope before estimating the costs for the activities on the new project.

According to the *PMBOK® Guide,* the OPAs you may need to consider when working in the Develop Project Charter process include organizational policies, procedures, and processes; portfolio, program, and project governance framework; monitoring and reporting methods; templates; and historical information such as the lessons learned repository.

Exam Spotlight

Remember that organizational process assets encompass many elements, including policies, guidelines, standards, historical information, and so on, and that they're divided into two categories: 1) processes, policies, and procedures, and 2) corporate knowledge base. OPAs are internal to the organization. For the exam, make certain you understand what the organizational process assets entail and that you can differentiate them from the enterprise environmental factors input.

Tools and Techniques

Tools and techniques are multifaceted and include elements such as brainstorming, meetings, focus groups, alternatives analysis, and much more. Tools and techniques are used with the inputs of every process to produce the outputs of that process. For example, in the Develop Project Charter process, the business documents, agreements, EEF, and OPA inputs are examined using the tools and techniques of this process (expert judgment, data gathering, interpersonal and team skills, and meetings) to produce the project charter, which is the primary output of this process along with the assumption log. Throughout the book, we'll examine those tools and techniques that directly relate to the People, Process, and Business Environment domains on the exam. Please be certain to refer to Appendix B for a full list of tools and techniques for every process and reference the *PMBOK® Guide* for further information.

Expert judgement is an important tool and technique that relates to the People domain. Let's take a look at its meaning in the context of the project charter.

Expert Judgment

The concept behind *expert judgment* is to rely on individuals, or groups of people, who have training, specialized knowledge, or skills in the areas you're assessing. These folks might be stakeholders, consultants, other experts in the organization, subject matter experts, the PMO, industry experts, or technical or professional organizations. According to the *PMBOK® Guide*, you should use experts who have knowledge or training in the following: organizational strategy, benefits management, technical industry knowledge, estimating techniques for duration and cost, and risk identification. Expert judgment is a tool and technique used in many other processes as well.

In the case of developing a project charter, expert judgment would be helpful in assessing the inputs of this process, the environmental factors, organizational assets, and historical information. For example, as the project manager, you might rely on the expertise of your executive committee to help you understand how the proposed project gels with the strategic plan, or you might ask stakeholders about any compliance issues the project team should be aware of, or you might rely on team members who have participated in similar projects in the past to make recommendations regarding the proposed project.

Expert judgment will also come in handy with your key stakeholders. You'll rely on their expertise and collaborate with them in order to understand and document the needs of the project and the business problem you're trying to solve. They are people who will tell you what benefits the project should produce.

Other Tools and Techniques

Three other tools and techniques in this process also appear in many other processes. The first is the data gathering technique, which includes brainstorming, focus groups, and interviews. Brainstorming is used with groups of people to gather information, whether pertinent to the topic or not, in a short amount of time. Focus groups are sessions conducted with stakeholders to help identify success criteria, risks, and other information. Interviews are typically one-on-one and used to gather information regarding project requirements, the assumptions that stakeholders are making, defining approval criteria, and identifying constraints. I think you can see how these techniques could be useful in defining the business value and goals of the project.

Interpersonal and team skills are a topic we'll discuss throughout the book. This concerns the ability to relate to others in a genuine manner and entails the ability to manage conflict, facilitate, and manage and conduct meetings. Facilitation involves bringing groups of people together to make decisions or find solutions. The facilitator ensures that all parties are heard and that there is agreement at the end of the meeting. Meeting management refers to being prepared for meetings, such as creating an agenda beforehand, making certain the right people attend the right meetings, and keeping the meeting on track. Meeting minutes and action items are recorded during the meeting and sent to the participants after the meeting concludes.

Meetings is the last tool and technique of this process, and they may entail stakeholder meetings, progress meetings, update meetings, budget meetings, and more. Meetings in terms of the project charter refer to identifying high-level requirements, determining summary milestones, defining success criteria, and more. Meetings will occur throughout the remainder of the project, and you will use them as one mechanism to keep stakeholder expectations aligned with the objectives of the project.

All successful project managers are adept at communicating with others and possess excellent interpersonal skills.

Formalizing and Publishing the Project Charter

The approved *project charter* is the official, written acknowledgment and recognition that a project exists. It ties the work of the project with the ongoing operations of the organization. It's usually issued and signed by the project sponsor, and it gives the project manager the authority to assign organizational resources to the project. It is one of two outputs of this process. The other output is the assumptions log. This log is used to document all the assumptions and constraints for the project from this point forward. You might recall that high-level assumptions and constraints were documented in the business case. If they still apply, they should be transferred to the assumptions log.

The charter documents the business need or demand that the project was initiated to address, and it includes a description of the product, service, or result of the project. In a waterfall, or predictive methodology, the project charter is the first official document of the project. Project charters are often used as a means to introduce a project to the organization. Because this document outlines the high-level project description, the business opportunity or need, the business value the project will bring about, and the project's purpose, executive managers can get a first glance at the benefits of the project. Good project charters that are well documented will address many of the questions your stakeholders are likely to have up front.

Project charters may also be the first document created using a hybrid approach to project management. The idea is to document the need for the project, understand the business value it will bring about, and then break down these needs into small units of work. The charter is the big picture idea, and the benefits management plan outlines the value to the organization. You'll refer to both documents throughout the project whether using a predictive or hybrid methodology.

Pulling the Project Charter Together

According to the *PMBOK® Guide*, to create a useful and well-documented project charter, you should include elements such as the following:

- Purpose or justification for the project
- Project objectives that are measurable
- High-level list of requirements
- High-level description of the project with high-level deliverables
- Overall project risks
- Milestone schedule (summary level)
- Preapproved budget
- List of key stakeholders
- Criteria for project approval (success criteria)
- Criteria for project exit
- Name of the project manager and their authority levels
- Name of the sponsor (or authorizer of the project) and their authority levels

Many of these elements are self-explanatory. Let's look at criteria for project approval (success criteria) and criteria for project exit in a bit more depth.

As a refresher, the project is considered successful when it meets the objectives of the project and the stakeholders are satisfied. This might be easy to say, but how do you know specifically that the project is a success? In many organizations in which I've worked, our projects were determined to be successful (or not) using the universal triad of project success measures: on time, on budget, and within scope. But what if my sponsor isn't happy? It doesn't matter if I'm on time or on budget. They won't see the project as a success. What if my customers don't adopt the new product, service, or result of my project? That's not successful. What if my project outcomes cannot be integrated into the existing organizational infrastructure or business environment? Oops, that's a problem. What if my project takes a long time to complete and over time, it no longer aligns with the organizational goals and mission? Again, not good.

In order to ensure you and your stakeholders know what a successful project looks like, you must define and document the *success criteria*. According to the *PMBOK® Guide,* and my own personal experience, there are two questions that must be asked and answered and agreed to (in writing) by the stakeholders:

- What does success look like?
- How will success be measured?

The answers to these are part of the project approval requirements that should be documented in the project charter.

Project exit criteria are different from success criteria. They focus on the circumstances or conditions under which you can cancel or close the project or phase. For example, an exit criterion for the beginning phase of a project might be whether the project charter or contract is documented, signed, and approved. Exit criteria for a working phase in the project might be whether testing of all interfaces is complete and data is being processed correctly. Exit criteria for the project might include whether the application is free from defects or the app is live and available in the app store.

For the exam, the important factors to remember about the project charter are as follows:

- It is a formal record of the project.
- It authorizes the project to begin.
- It authorizes the project manager to assign resources to the project.
- It shows the organization's commitment to the project.
- It authorizes the project manager to plan, execute, and control the project.
- It documents the business need, justification, and impact.
- It describes the customer's requirements.
- It sets stakeholder expectations.
- It ties the project to the strategic goals of the organization.

Once the charter is signed, the project is formally authorized and work can begin. The project charter should always be written and signed before you begin the Planning process group.

Let's take a brief look at the key stakeholders who might be involved with the project charter and the role they'll play in developing it. We'll discuss project stakeholders in more depth in Chapter 3.

Key Stakeholders

A project is successful when it achieves its objectives by producing deliverables that meet the expectations of the *stakeholders*. Stakeholders are those folks (or organizations) with a vested interest in your project. They may be active or passive as far as participation on the project goes, but the one thing they all have in common is that each of them has something to either gain or lose as a result of the project. The *PMBOK® Guide* states that the project charter forms a partnership between the organization (made up of stakeholders) requesting the project and the one performing the project (the project team and project manager).

The project charter will help ensure that stakeholders' expectations are in alignment with the project scope and with the final results of the project. According to the *PMBOK® Guide,* the project charter is a document issued by the person (or organization) who initiated the project or the project sponsor.

Let's take a look at the roles of some of the key stakeholders and how they can help contribute to creating a comprehensive project charter.

Project Manager

The project manager is the person who assumes responsibility for the success of the project. The project manager should be identified as early as possible in the project and ideally should participate in writing the project charter.

The project charter identifies the project manager and describes the authority the project manager has in carrying out the project. The project manager's primary responsibilities are project planning and then executing and managing the work of the project. By aligning the project charter with the other project planning documents created later in the project, the project manager is assured that everyone knows and understands what's expected of them and what constitutes a successful project. This will ensure that the stakeholders understand the purpose for the project and gives the project manager the ability to continually relay the vision for the project to the team and to the stakeholders. Once the project charter is approved, you'll use it as a basis to optimize alignment between the stakeholder needs and the project objectives. Stakeholder memories can become fuzzy over time. It's your responsibility to keep their expectations in line with the goals of the project, and an approved, signed project charter can help you do that.

Project managers are responsible for setting the standards and policies for the projects on which they work. As a project manager, it is your job to establish and communicate the project procedures to the project team and stakeholders. In turn, the project team is responsible for supporting you by performing the work of the project.

Project managers are business-savvy leaders who can relate the benefits of the project to the organization's mission. They will identify activities and tasks, resource requirements, project costs, project requirements, performance measures, and more. Communication and documentation must become the project manager's best friends. Keeping stakeholders, the project sponsor, the project team, and all other interested parties informed is "job one," as the famous car manufacturer's ads say.

Project managers are the ones who recommend the development life-cycle methodology that is best suited for the project. For example, predictive methodologies are a good choice when requirements are well known and the project is considered low risk. Agile methodologies are a good choice when the requirements are uncertain and the project is high risk, or when you know there will be a lot of changes on the project. Hybrid methodologies blend the best of predictive and agile or combine two or more agile approaches that will work well for the project. Project managers will consider the organizational culture, the team members' knowledge of the development life cycle they are recommending, training needs, management buy-in, and other factors that we will explore in more depth throughout the remainder of the book.

Project Sponsor

Have you ever attended a conference or event that was put on by a sponsor? In the information technology field, software development companies often sponsor conferences and seminars. The sponsor pays for the event, the facilities, and the goodies and provides an opportunity for vendors to display their wares. In return, the sponsor comes out looking like a winner. Because it is footing the bill for all this fun, the sponsor gets to call the shots on conference content, and it gets the prime spots for discussing its particular solutions. Last but not least, it usually provides the keynote speaker and gets to present its information to a captive audience.

Project sponsors are similar to this. In their role as project champion, they rally support from stakeholders and the executive management team. They keep the project front and center at the higher levels of the organization and are the spokesperson for the project.

The *project sponsor* is usually an executive in the organization who has the power and authority to make decisions and settle disputes or conflicts regarding the project. The sponsor takes the project into the limelight, so to speak, and gets to call the shots regarding project outcomes. The project sponsor is also the one with the big bucks who provides funds for your project. The project sponsor should be named in the project charter and identified as the final authority and decision-maker for project issues. Ultimately, the project sponsor is responsible for facilitating the project's success.

In a predictive project management environment, sponsors are actively involved in the Initiating and Planning phases of the project and tend to have less involvement during the Execution and Monitoring and Controlling phases. In an adaptive project management environment, this role is actively engaged in the project throughout its life cycle. It's up to the project manager to keep the project sponsor informed of all project activities, project progress, and any conflicts or issues that arise. The sponsor is the one with the authority to resolve conflicts and set priorities when these things can't be dealt with any other way.

The project charter is typically created by the project sponsor who is requesting the project. You'll want to work closely with the sponsor to validate that their project idea aligns with the organizational strategy and with the business value the project is expected to bring about.

Other Stakeholders

Project managers must work with and gain the support of functional managers in the organization in order to complete the project. Functional managers fulfill the administrative duties of the organization, provide and assign staff members to projects, and conduct performance reviews for their staff. Functional managers may include human resource managers, accounting managers, marketing managers, and so on. It's a good idea to identify the functional managers who will be working on project tasks or assigned project responsibilities in the charter.

The project charter should include a list of all key project stakeholders. Stakeholders may be inside or outside the organization. You know they are a stakeholder if they have a vested interest in your project. This may include government or regulatory agencies, customers, business executives, and more. Stakeholder influences aren't explicitly stated as part of this process, but you'll see in Chapter 3 how the influence and power of one stakeholder (let alone two or three stakeholders) can make or break a project.

Exam Spotlight

In some organizations, the project manager might write the project charter. However, if you (as the project manager) are asked to write the charter, remember that your name should not appear as the author. Because the project charter authorizes the project and authorizes you as the project manager, it doesn't make sense for you to write a document authorizing yourself to manage the project. The author of the charter should be an executive manager in your organization with the power and authority to assign resources to this project. This is usually the project sponsor or project initiator.

Project Charter Sign-Off

The project charter isn't complete until you've received sign-off from the project sponsor, senior management, and key stakeholders. Sign-off indicates that the document has been read by those signing it (let's hope so, anyway) and that they agree with its contents and are on board with the project. It also involves the major stakeholders right from the beginning and should win their continued participation in the project going forward. If someone has a problem with any of the elements in the charter, this is the time to speak up.

Prior to publishing the charter, I like to hold a kickoff meeting with the key stakeholders to discuss the charter and then obtain their sign-off. I think it's imperative for you to identify your key stakeholders as soon as possible and involve them in the creation of the project charter. Remember that stakeholder identification is an ongoing activity.

Signing the project charter document is the equivalent of agreeing to and endorsing the project. This doesn't mean the project charter is set in stone, however. Project charters may change throughout the course of the project. As more details are uncovered and outlined and as the Planning processes begin, more project information will come to light. This is part of the iterative process of project management and is to be expected. Keep in mind that iterative in this sense means to revisit and continually discover information; it does not refer to an agile methodology. The charter will occasionally be revised to reflect these new details, project plans will be revised, and project execution will change to incorporate the new information or direction.

The last step in this process is informing the stakeholders of the approved project charter and distributing it to them. You will distribute a copy of the project charter to the key stakeholders, the customer, the management team, and others who might be involved with the project. This can take several forms, including printing copies, sending copies via electronic formats such as the company email system, or sending a link to the document where they can view it on the company's intranet or project document repository (on behalf of information technology professionals everywhere, this is the preferred option).

Exam Spotlight

It's important that stakeholders have access to the project charter so that you can ensure a common understanding of the key deliverables, major milestones, and their roles and responsibilitie in the project, and that they are on board with accomplishing the project objectives.

Maintaining Project Artifacts

You now have your first project artifact, the project charter. Now's a good time to start thinking about how you will distribute, maintain, and store your project documents and artifacts. We'll talk more about document management in Chapter 10, "Sharing Information," when we cover the Manage Project Knowledge process. However, between now and Chapter 10 I'll be introducing a lot of documents, so let's start thinking about some key aspects of maintaining artifacts.

You'll want to ask some basic questions to determine what, when, where, and who will manage this activity. I can answer the who quickly: it's you as the project manager. It's your responsibility to ensure the project artifacts are distributed and stored appropriately, that they are accessible to the stakeholders, and that the documents are kept up to date with versioning control.

Exam Spotlight

The four main components of maintaining project artifacts are ensuring that they are distributed to the appropriate stakeholders, that they are stored appropriately, that they are kept up to date with version control, and that they are easily accessible by the stakeholders.

Determining where the documents will be stored should be decided early in the project. You'll start accumulating documents quickly. So far, we have the business case, the benefits management plan, and the project charter. You'll need a place to keep these for future reference. We're going to come back to the benefits management plan often and revisit the project charter at least a few more times. Most organizations have a content management system for the purpose of storing and sharing documents and this is the logical place to store your project management documents. The system is easily accessible by all stakeholders and may automatically perform version control so that you know your documents are always up to date. The documents can be easily updated in the system and distributed once the update is complete. Rather than attaching a document to an email and sending it to 27 people, thereby clogging up the email server, you can send stakeholders a link to the document in the content management system. On behalf of all information technology professionals everywhere, we kindly ask that you please use a link to distribute documents rather than email. Thank you.

Introducing the Kitchen Heaven Project Case Study

This chapter introduces a case study that we'll follow throughout the remainder of the book. The case study is updated at the end of every chapter. It's designed to show you how a project manager might apply the material covered in the chapter to a real-life project. As happens in real life, not every detail of every process is followed during all projects. Remember that the processes from the *PMBOK® Guide* that I'll cover in the remaining chapters are project management guidelines. You will often combine processes during your projects, which will allow you to perform several steps at once. The case studies will present situations or processes that you might find during your projects and describe how one project manager resolves them.

Project Case Study: New Kitchen Heaven Store

You are a project manager for Kitchen Heaven, a chain of retail stores specializing in kitchen utensils, cookware, dishes, small appliances, and some gourmet foodstuffs, such as bottled sauces and spices. You're fairly new to the position, having been hired to replace a project manager who recently retired.

Kitchen Heaven currently owns 49 stores in 34 states and Canada. The world headquarters for Kitchen Heaven is in Denver, Colorado. Counting full-time and part-time employees, the company employs 2,500 people, 200 of whom work at headquarters.

The company's mission statement reads, "Great gadgets for people interested in great food."

Recently, the vice president of marketing paid you a visit. Dirk Perrier is a well-dressed man with the formal air you would expect a person in his capacity might have. He shakes your hand and gives you a broad, friendly smile.

"We are opening our 50th store! I don't know if you're familiar with our store philosophy, so let me take a moment to explain it. We like to place our stores in upscale neighborhoods and mixed-use communities where people live, work, and shop. We try to make it easy for our customers to shop with us and attend our cooking classes.

"Our next store is going to be right here in our home area—Colorado Springs. Because this is going to be our 50th store, we plan on having a 50th grand-opening celebration, with the kind of surprises and activities you might expect for such a notable opening.

"Our stores generally occupy from 1,500 to 2,500 square feet of retail space, and we typically use local contractors for the build-out. A store build-out usually takes 120 days from the date the property has been procured until the doors open to the public. I can give you our last opening's project plan so you have a feel for what happens. Your job will be to procure the property, negotiate the lease, procure the shelving and associated store furnishings, get a contractor on the job, and prepare the 50th store festivities. My marketing folks will assist you with that last part.

"We're also updating our website at the same time. We focus on our specialty spice mixes, nonperishable gourmet foods, and the Kitchen Heaven line of must-have kitchen tools on the website.

"We're targeting the type of customer who watches the Food Network channel and must have all the gadgets and tools they see the famous chefs using. So, the stores are upbeat and convey a fun, energetic feel, if you will. I want the updated online store to convey the same feeling. I'm envisioning posting short video clips on how to use some of those gourmet spice mixes and gadgets on the website. I'm hoping to engage some celebrity chefs to star in our video clips.

"Our goal for the updated website is to create a robust e-commerce site that will increase sales by reaching a broad base of customers, particularly those customers who are not close to a retail store. We want to make it easy for people to purchase their kitchen essentials.

"You have nine months to complete the project. Any questions?"

You take a deep breath and collect your thoughts. Dirk has just given you a lot of information with hardly a pause between thoughts. A few initial ideas drift through your head while you're reaching for your notebook.

You've been with the company long enough to know that Dirk is high up in the executive ranks and carries the authority and power to make things happen. Therefore, Dirk is the perfect candidate for project sponsor.

You grab your notebook and start documenting some of the things Dirk talked about, clarifying with him as you write:

1. The project objective is twofold, to open a new store in Colorado Springs nine months from today and to offer Kitchen Heaven products on the website.

2. The retail store should be located in a mixed-use community.

3. The retail store will carry the full line of products from utensils to gourmet food items.

4. The grand opening will be a big celebration because this is the 50th store opening.

5. A new Kitchen Heaven website will be launched at the same time as the new retail store opening.

6. The online store will carry a full line of products from utensils to gourmet food items.

7. The website will contain training videos with celebrity guest appearances. There will be a weekly live webcast demonstrating the use of a product or gourmet food item.

8. The business value for this project is to increase sales, increase market share, and broaden the customer base.

You have a question or two for Dirk.

"Is there a special reason we have to open, let's see, nine months from now, which is the end of January?"

He responds, "Yes, we want the retail store and new online store to open before the first week in February. Early February is when the Garden and Home Show conferences across the country kick off. We'll have a trade show booth at the first show in February here in Colorado, and we'll have a booth at six other shows in surrounding states running through April. We know from experience in other areas that our stores generally see a surge in sales during these months as a result of the trade show. It's a great way to get a lot of advertising out there and let folks know we've updated our website. We will also create a digital marketing campaign that will launch in late January."

"Another question, Dirk. Is there a budget set for this project yet?"

"We haven't set a hard figure," Dirk replies. "But again, from past experience we know it takes anywhere from $1.5 to $2 million to open a new store—and we don't want to forget the big bash for the grand opening. The overhaul of the website can be accomplished with in-house resources. They might need a contractor or consultant, so let's set aside $100,000 for that, just in case."

You look over your notes and ask, "What about the budget for the two celebrity chefs to kick off the video content on the site?"

"Oh, yes," Dirk replies. "You'll have to research that. I don't have a good understanding of their fees."

"Thanks, Dirk. I'll get started writing the project charter right away and I'll put your name on the document because you're the project sponsor."

Dirk concludes with "Feel free to come to me with questions or concerns at any time."

One week later.

You review your notes and reread the project charter you've prepared for the Kitchen Heaven store opening and website launch one last time before looking for Dirk. You finally run across Dirk in a hallway near the executive washroom.

"Dirk, I'm glad I caught you. I'd like to go over the project charter with you before the kickoff meeting tomorrow. Do you have a few minutes?"

"Sure," Dirk says to you. "Let's have it."

"The project charter states the purpose of the project, which is to open the new, 50th Kitchen Heaven retail store in Colorado Springs and launch the new website. I also documented some of the high-level requirements, many of which we talked about last time we met. I documented the assumptions and constraints you gave me with the understanding that we'll define these much more closely when I create the scope statement. I've included a section that outlines a preliminary milestone schedule, and I've included some preliminary benefit measurement calculations. Using your estimate of $2.2 million as our initial budget request and based on the projected inflows you gave me last week, I've calculated a payback period of 17 months, with an IRR of 3 percent."

"That's impressive," replies Dirk. "Let's hope those numbers hold true."

"I think they're reliable figures," you say. "I researched our historical data based on recent store openings in similar-sized cities and factored in the economic conditions of the Colorado Springs area. Since they're on a growth pattern, we think the timing is perfect. I also did some research based on our past performance on the website and our competitors' online stores. Since the economic conditions look to be on a growth pattern in online sales, we think the timing is perfect.

"As you know, the project kickoff is scheduled for tomorrow. I'll need you to talk about the project and the goals, talk about the commitment you'll need from the management team to support this project, and introduce me as the project manager. I will forward a copy of the project charter to the meeting attendees as soon as I leave your office so that they can review it before the meeting. I included a list of the assumptions we've made so far as an appendix to the charter. I'll also discuss the development life-cycle methodology we'll use to conduct this project. As a heads up, we're going to use a hybrid approach. The construction portion of the project will use a predictive methodology and the website refresh will use an agile approach. I'll talk more about that at the kickoff. Lastly, I'll need you to ask everyone present to sign a copy of the project charter."

"Sounds like you've covered everything," Dirk says. "I don't anticipate any problems tomorrow, because everyone is looking forward to this new opportunity."

Project Case Study Checklist

- **Project objective:** To open a new retail store in Colorado Springs by January 25 and launch a new website selling products and offering cooking training videos.

- **Business need or demand for project:** Colorado Springs' population is growing and new live, work, shop communities are planned for next year and beyond. Customer requests have increased in recent months asking for a way to purchase products online. Some customers drive an hour or more to get to the store and would like to have the ability to stock up on essentials without having to drive to the store.

- **Project sponsor:** Dirk Perrier, VP of marketing

- **Project selection methods:** Payback period calculated at 17 months and IRR calculated at 3 percent.

- **Created project charter:** Project charter contains the following:

  - High-level overview of project

  - Business value the project will bring about

  - List of measurable project objectives

  - High-level risks

  - Summary milestone schedule with initial completion date of January 25

  - Summary budget to be determined

  - Project manager authority levels

  - Definition of roles of project sponsor and project manager

- **Determined development life-cycle methodology to use for the project:**

  - Hybrid approach

Next steps: Kickoff meeting set up to discuss charter and obtain sign-off.

Understanding How This Applies to Your Next Project

There are as many ways to select and prioritize projects as there are organizations. You might be profit driven, so money will be king. You might have a stakeholder committee that weighs the pros and cons, or you might have an executive director who determines which project is up next. Scoring models and cash flow analysis techniques are useful on the job. Whether your organization uses these methods or others, an organized, consistent way to select and prioritize projects is necessary. I know I could work the next 100 years straight and probably still not get all the projects completed my organization would like to see implemented. What I've found is that the selection method must be fair and reasonable. If your organization uses an arbitrary method—say you like Tara better than Joe, so Tara's projects always end up on the "yes" list—it won't be long before stakeholders demand that another method be devised to select projects that everyone can understand. Whatever method you're using, stick to it consistently.

If you're like me, when I'm faced with a new project I want to get right to the heart of the matter and understand the purpose of the project. It has been my experience in working with project teams that when the team understands the goals of the project and the value it's expected to deliver, the project is more likely to satisfy the stakeholders and be success-ful. I don't have any scientific evidence for this, but when the teams have a clear under-standing of what they're working on and why, they tend to stay more focused and fewer unplanned changes make their way into the project. Don't assume everyone on the project team understands the goal of the project. It's good practice to review the project goal early in the project and again once the work of the project is under way. Reminding the team of the goal helps keep the work on track.

I usually write a project charter for all but the smallest of projects. I believe the most important sections of the charter are the objectives of the project, the high-level require-ments, and the summary-level milestone schedule. It's important that the goal or objective of the project be written down, no matter how small the project, so that the team and the stakeholders know what they're working toward.

Identifying a few of the key stakeholders early in the project is imperative to project suc-cess. I won't begin writing the charter without stakeholder input.

Always, and I mean *always*, get approval and signatures on the project charter. You will use this document as your basis for project planning, so make certain the sponsor, key stakeholders, and the project manager understand the goals of the project the same way.

Summary

This chapter started with a discussion of the 10 Knowledge Areas. The Knowledge Areas bring together processes that have characteristics in common. They also help you under-stand what types of skills and resources are needed to complete the processes within them.

Project selection methods are most often determined using benefit measurement methods. Benefit measurement methods come in the form of benefit–cost ratio (BCR) analyses, scoring models, and economic analyses, which are comparative approaches. Benefit measurement methods also include cash flow analysis such as payback period (PBP), net present value (NPV), discounted cash flow technique, and internal rate of return (IRR).

Analysis of cash flows includes PBP, discounted cash flows, NPV, and IRR. These last three methods are concerned with the time value of money—or, in other words, converting future dollars into today's value. Generally, projects with a shorter PBP are desired over those with longer PBPs. Projects that have an NPV greater than 0 should be accepted. Projects with the highest IRR value are considered a better benefit to the organization than projects with lower IRR values.

The Project Integration Management Knowledge Area is concerned with coordinating all aspects of the project management plan and is highly interactive. Integration of these processes might occur once, or multiple times as needed, or continuously throughout the project. Integration occurs when project planning, executing, monitoring, and change control are repeated continuously while you're working on the project. According to the *PMBOK® Guide,* performing integration is a critical skill that all project managers should possess.

The project manager is responsible for tailoring the project documents and project processes by considering the size and complexity of the project in order to determine the processes that should be performed and the documents that should be created. It also entails keeping the project artifacts (such as the business case, benefits management plan, project charter, and the project management plan) coordinated and aligned with one another and with the organization's strategic goals.

The Develop Project Charter process is the first process in the Initiating process group. The business case is an input to this process that describes the product, service, or result the project was undertaken to complete. The business case should include the business needs of the organization as well as a high-level scope description and should map to the organization's strategic plan. The benefits management plan is also an input to this process. It outlines the intended benefits and business value the project will bring about, how the benefits will be measured, and how they will be obtained. The benefits management plan also assigns owners to the benefits to ensure they come to fruition.

Enterprise environmental factors (EEFs) are factors outside the control of the project team that might have significant influence on the success of the project. EEFs can be internal or external to the organization and may drive the need for regulatory or standards compliance on your project. Organizational process assets (OPAs) refer to policies, guidelines, and procedures for conducting the project work and are internal to the organization.

Expert judgment, data gathering, interpersonal and team skills, and meetings are the tools and techniques of this process. Experts usually have specialized knowledge or skills and can include staff from other departments in the company, external or internal consultants, and members of professional and technical associations or industry groups. Data-gathering techniques include brainstorming, focus groups, and interviews. Interpersonal and team skills include conflict management, facilitation, and meeting management. Meetings are used to define objectives, success criteria, summary milestones, and more.

The outputs of this process include the project charter and the assumptions log. The project charter is the formal recognition that a project, or the next project phase, should begin. The charter authorizes the project to begin, authorizes the project manager to assign resources to the project, documents the business need and justification, describes the customer's requirements, and ties the project to the ongoing work of the organization. The assumptions log contains a list and description of all the assumptions and constraints on the project for the life of the project.

Exam Essentials

Be able to name the 10 Project Management Knowledge Areas.   The 10 Project Management Knowledge Areas are Project Integration Management, Project Scope Management, Project Schedule Management, Project Cost Management, Project Quality Management, Project Resource Management, Project Communications Management, Project Risk Management, Project Procurement Management, and Project Stakeholder Management.

Be able to define project selection methods.   Project selection methods are used prior to the Develop Project Charter process to determine the viability of the project. The most common are the benefit measurement methods, which include comparative approaches and cash flow analysis.

Be able to describe and calculate the payback period.   The payback period is the amount of time it will take the company to recoup its initial investment in the product of the project. It's calculated by adding up the expected cash inflows and comparing them to the initial investment to determine how many periods it takes for the cash inflows to equal the initial investment.

Be able to denote the decision criteria for NPV and IRR.   Projects with an NPV greater than 0 should be accepted, and those with an NPV less than 0 should be rejected. Projects with high IRR values should be accepted over projects with lower IRR values. IRR is the discount rate when NPV is equal to 0, and IRR assumes reinvestment at the IRR rate.

Be able to describe tailoring.   Tailoring means determining the documents and project process needed for the project when considering the size and complexity of the project. This also involves keeping the project artifacts (such as the business case, benefits management plan, project charter, and the project management plan) coordinated and aligned with one another and with the organization's strategic goals.

Be able to describe integration.   Integration involves repeating processes continuously while you're working on the project and coordinating all aspects of the project management plan. Integration is highly interactive. According to the *PMBOK® Guide,* performing integration is a critical skill that all project managers should possess.

Be able to explain integration when using an agile methodology.   Integration is managed and performed by the project manager when using an agile project management methodology, and the planning, control, and delivery of the product is managed by the agile team.

Be able to describe the purpose of the business case.   The purpose of a business case is to understand the business need for the project and determine whether the investment in the project is worthwhile. It is considered an economic feasibility study. It usually includes a benefit–cost ratio analysis and the needs or demands that brought about the project. This is an input to the Develop Project Charter process.

Be able to describe the purpose of the benefits management plan.   The benefits management plan outlines the intended benefit of the project, how those benefits will be measured, and how they will be obtained. Owners are assigned to each benefit to ensure they are achieved. The benefits management plan is monitored throughout the life of the project. This is an input to the Develop Project Charter process.

Be able to describe the importance of the project charter.   The approved project charter is the document that officially recognizes and acknowledges that a project exists. The charter authorizes the project to begin, authorizes the project manager to assign resources to the project, documents the business need and justification, describes the customer's requirements, and ties the project to the ongoing work of the organization.

Review Questions

You can find the answers to the review questions in Appendix A. Be sure to download the Bonus Exams and Bonus Questions so that you'll have a broader exposure and more experience answering questions related to the topics in this chapter.

1. Which of the following statements is not true regarding EEFs and OPAs?

    A. OPAs are external to the organization.

    B. EEFs are outside the control of the project team.

    C. Resource availability is an example of internal EEF.

    D. Change control processes are an example of an OPA.

    E. EEFs may drive compliance requirements on the project.

2. Which of the following is true regarding the project charter? (Choose two.)

    A. The project charter should be issued by a manager external to the project.

    B. The project charter is not created when you are using a hybrid life-cycle methodology.

    C. The project charter should be issued by the project manager.

    D. The project charter should be issued by the project sponsor.

    E. The project charter assists the project manager in setting a clear vision and mission for the project and aligns stakeholder needs and expectations with the project objectives.

3. The Integration Knowledge Area is highly interactive and may involve iteratively performing the Planning, Executing, and Monitoring and Controlling processes. Which of the following is true regarding this Knowledge Area? (Choose two.)

    A. Integration is managed and performed by the agile team when using an adaptive project management methodology.

    B. The project manager determines which project management processes will interact with one another and integrates them appropriately.

    C. Performing integration is a critical skill all project managers should possess.

    D. Integration can be iterative and focuses on ensuring the schedule is defined, communication is planned and managed, and risks are continually identified.

    E. Integration is only used with predictive project management methodologies.

4. You are the project manager for Fun Days Vacation Resorts. Your new project assignment is to head up the Fun Days resort opening in Austin, Texas. First, you estimate the duration of the project management plan activities. When that activity is complete, you start devising the project schedule, and once the work has started, you will follow the process involved with monitoring and controlling deviations from the schedule. Which of the following is true regarding this question? (Choose two.)

    A. This describes the Project Integration Management Knowledge Area.

    B. This describes a predictive project management methodology.

    C. This describes the Project Scope Management Knowledge Area.

    D. This describes the Project Schedule Management Knowledge Area.

    E. This describes an agile project management methodology.

5. Which of the following is not true about the benefits management plan?

    A. It is created early in the project and is reviewed at each phase gate.

    B. It documents the expected benefits the project will bring about and how the benefits will be measured.

    C. It is an important input to the Develop Project Charter process and helps to define a clear vision and mission for the project.

    D. It creates strategic alignment of benefits to the organizational goals.

    E. It assigns benefit owners to monitor benefit realization.

6. According to the *PMBOK® Guide*, all of the following are elements of a business case except for which one?

    A. The findings from the feasibility study are documented in the business case.

    B. It is preceded by a needs assessment, which will assess the needs and demands of the project, costs, risk, and opportunities.

    C. It names the project manager.

    D. It must be approved before proceeding with the project.

7. Your nonprofit organization is preparing to host its first annual 5K run/walk in City Park. You worked on a similar project for the organization two years ago when it cohosted the 10K run through Overland Pass. Which of the organizational process assets might be most helpful to you on your new project?

    A. The organization's marketing plans

    B. Historical information from a previous project

    C. The marketplace and political conditions

    D. The organization's project management information systems

8. When using an agile methodology to manage your project, all of the following Knowledge Areas have processes that are performed before, during, and after an iteration except for which one?

    A. Risk Management

    B. Schedule Management

    C. Procurement Management

    D. Quality Management

9. You need to collaborate with the stakeholders to determine project approval requirements. Which of the following describes project approval requirements?

    A. This describes what constitutes project success and how it will be measured.

    B. This describes the conditions that must be met in order to close the project.

C. This is used in agile project management to determine if an iteration is successful.

D. This is determined by the project manager in the project charter.

10. You are the project manager for the Late Night Smooth Jazz Club chain, with stores in 12 states. Smooth Jazz is considering opening a new club in Arizona or Nevada. You have derived the following information:

Project Arizona: The payback period is 18 months, and the NPV is (250).

Project Nevada: The payback period is 24 months, and the NPV is 300.

Which project would you recommend to the selection committee?

A. Project Arizona, because the payback period is shorter than the payback period for Project Nevada

B. Project Nevada, because its NPV is a positive number

C. Project Arizona, because its NPV is a negative number

D. Project Nevada, because its NPV is a higher number than Project Arizona's NPV

11. You are the project manager for the Late Night Smooth Jazz Club chain, with stores in 12 states. Smooth Jazz is considering opening a new club in Kansas City or Spokane. You have derived the following information:

Project Kansas City: The payback period is 27 months, and the IRR is 6 percent.

Project Spokane: The payback period is 25 months, and the IRR is 5 percent.

Which project should you recommend to the selection committee?

A. Project Spokane, because the payback period is the shortest

B. Project Kansas City, because the payback period is the longest

C. Project Spokane, because the IRR is the lowest

D. Project Kansas City, because the IRR is the highest

12. Which of the following is true regarding NPV?

A. NPV assumes reinvestment at the cost of capital.

B. NPV decisions should be made based on the lowest value for all the selections.

C. NPV assumes reinvestment at the prevailing rate.

D. NPV assumes reinvestment at the NPV rate.

13. You are the project manager for Insomniacs International. Since you don't sleep much, you get a lot of project work done. You're considering recommending a project that costs $575,000; expected inflows are $25,000 per quarter for the first two years and then $75,000 per quarter thereafter. What is the payback period?

A. 40 months

B. 38 months

C. 39 months

D. 41 months

14. Which of the following is true regarding IRR?

    A. IRR assumes reinvestment at the cost of capital.

    B. IRR is not difficult to calculate.

    C. IRR is a constrained optimization method.

    D. IRR is the discount rate when NPV is equal to zero.

15. Which of the following is not true regarding the purpose of a business case?

    A. It is an economic feasibility study that helps determine whether the investment in the project is worthwhile.

    B. It describes the need or demand that brought about the project.

    C. It describes how the benefits of the project will be measured and obtained.

    D. It contains the results of the benefit measurement methods that will assist in project selection.

16. Which of the following is true regarding stakeholders and the project charter? (Choose three.)

    A. The project charter will be used to ensure that stakeholders' expectations are in alignment with the project scope and with the final results of the project.

    B. The results of the needs assessment are documented in the project charter so that stakeholders' expectations are in alignment with the project scope and with the final results of the project.

    C. The project charter documents the stakeholders who will own monitoring and measuring the benefits of the project to ensure they are realized.

    D. The project charter will be aligned with other project planning documents created later in the project, and this will ensure that the stakeholders understand the project objectives.

    E. The Develop Project Charter process resides in the Project Stakeholder Management Knowledge Area process because this Knowledge Area is concerned that stakeholders' expectations are in alignment with the project scope and with the final results of the project.

    F. The project manager is responsible for informing the stakeholders that the project charter has been approved and ensuring that they receive a copy.

17. Your selection committee is debating between two projects. Project A has a payback period of 18 months. Project B has a cost of $125,000, with expected cash inflows of $50,000 the first year and $25,000 per quarter after that. Which project should you recommend?

    A. Either Project A or Project B, because the payback periods are equal

    B. Project A, because Project B's payback period is 21 months

    C. Project A, because Project B's payback period is 24 months

    D. Project A, because Project B's payback period is 20 months

18. Which of the following is true?

    A. Discounted cash flow analysis is the least precise of the cash flow techniques because it does not consider the time value of money.

    B. NPV is the least precise of the cash flow analysis techniques because it assumes reinvestment at the discount rate.

    C. Payback period is the least precise of the cash flow analysis techniques because it does not consider the time value of money.

    D. IRR is the least precise of the cash flow analysis techniques because it assumes reinvestment at the cost of capital.

19. You are a project manager for Zippy Tees. Your selection committee has just chosen a project you recommended for implementation. Your project is to manufacture a line of miniature stuffed bears that will be attached to your company's trendy T-shirts. The bears will be wearing the same T-shirt design as the shirt to which they're attached. Your project sponsor thinks you've impressed the big boss and wants you to skip to the manufacturing process right away. What is your response?

    A. Agree with the project sponsor because that person is your boss and has a lot of authority and power in the company.

    B. Require that a preliminary budget be established and a resource list be put together to alert other managers of the requirements of this project. This should be published and signed by the other managers who are impacted by this project.

    C. Require that a project charter be written and signed off on by all stakeholders before proceeding.

    D. Suggest that a preliminary business case be written to outline the objectives of the project.

20. Which of the following state the four main components of maintaining project artifacts?

    A. Ensuring that documents are stored appropriately

    B. Ensuring that documents are easily accessible to stakeholders

    C. Ensuring that documents are kept up to date with versioning control

    D. Ensuring that documents are printed and stored in a project notebook(s) for future reference

    E. Ensuring that documents and artifacts are distributed appropriately

Chapter 3

Delivering Business Value

THE PMP® EXAM CONTENT FROM THE PEOPLE DOMAIN COVERED IN THIS CHAPTER INCLUDES THE FOLLOWING:

✓ **Task 1.2: Lead a team**

- 1.2.6 Analyze team members' and stakeholders' influence

✓ **Task 1.4: Empower team members and stakeholders**

- 1.4.3 Evaluate demonstration of task accountability

- 1.4.4 Determine and bestow levels of decision-making authority

✓ **Task 1.6: Build a team**

- 1.6.1 Appraise stakeholder skills

✓ **Task 1.9: Collaborate with stakeholders**

- 1.9.1 Evaluate engagement needs for stakeholders

- 1.9.2 Optimize alignment between stakeholder needs, expectations, and project objectives

- 1.9.3 Build trust and influence stakeholders to accomplish project objectives

✓ **Task 1.10: Build shared understanding**

- 1.10.2 Survey all necessary parties to reach consensus

✓ **Task 1.13: Mentor relevant stakeholders**

- 1.13.1 Allocate the time to mentoring

- 1.13.2 Recognize and act on mentoring opportunities

THE PMP® EXAM CONTENT FROM THE PROCESS DOMAIN COVERED IN THIS CHAPTER INCLUDES THE FOLLOWING:

✓ **Task 2.1 Execute the project with the urgency required to deliver business value**

- 2.1.1 Assess opportunities to deliver value incrementally

- 2.1.2 Examine the business value throughout the project

- 2.1.3 Support the team to subdivide project tasks as necessary to find the minimum viable product

✓ **Task 2.2 Manage communications**

- 2.2.1 Analyze communication needs of all stakeholders

- 2.2.2 Determine communications methods, channels, frequency, and level of detail for all stakeholders

✓ **Task 2.4. Engage stakeholders**

- 2.4.1 Analyze stakeholders

- 2.4.2 Categorize stakeholders

- 2.4.3 Engage stakeholders by category

- 2.4.4 Develop, execute, and validate a strategy for stakeholder engagement

✓ **Task 2.13 Determine appropriate project management methodology/methods and practices**

- 2.13.4 Recommend a project methodology/approach

THE PMP® EXAM CONTENT FROM THE BUSINESS ENVIRONMENT DOMAIN COVERED IN THIS CHAPTER INCLUDES THE FOLLOWING:

✓ **Task 3.1 Plan and manage project compliance**

- 3.1.1 Confirm project compliance requirements

✓ **Task 3.2 Evaluate and deliver project benefits and value**

- 3.2.1 Investigate that benefits are identified

- 3.2.2 Document agreement on ownership for ongoing benefit realization

- 3.2.3 Verify measurement system is in place to track benefits

- 3.2.4 Evaluate delivery options to demonstrate value

✓ **Task 3.3 Evaluate and address external business environment changes for impact on scope**

- 3.3.1 Survey changes to external business environment

✓ **Task 3.4 Support organizational change**

- 3.4.1 Assess organizational culture

- 3.4.3 Evaluate impact of the project to the organization and determine required actions

This chapter starts with a description of organizational structures and how they may impact the project and the project manager's role. Understanding organizational structures is important because it will also help you understand the power, influence, and interests of the stakeholders on your project. Stakeholders work within the organizational structure, so knowing the characteristics of that structure will help you identify stakeholders' level of influence.

Next I'll introduce stakeholders and talk about how to identify them and analyze their needs. If you don't have stakeholders, you don't have a project. And stakeholders bring their own set of baggage—I mean *interests*—to the project, so it is critical to know who they are, what they have to offer to the project, what they hope to get out of the project, and the power and influence they wield.

I'll wrap up this chapter with a discussion of business value. I introduced this topic in Chapter 2, "Assessing Project Needs," with the benefits management plan and the project charter. You'll need to know a bit more detail about business value for the exam, so we'll spend some time going more in depth.

This chapter will conclude the foundational information and Initiating process group. In the next chapter, I'll kick off the Planning process group.

We have a lot of ground to cover in this chapter, so let's get to it.

The process names, inputs, tools and techniques, outputs, and descriptions of the project management process groups and related materials and figures in this chapter are based on content from *A Guide to the Project Management Body of Knowledge (PMBOK® Guide), Sixth Edition* (PMI®, 2017). The references to adaptive and hybrid methodologies, related materials, and figures in this chapter are based on content from the *Agile Practice Guide* (PMI®, 2017).

Understanding Organizational Structures

Just as projects are unique, so are the organizations in which they're carried out. Organizations have their own styles, cultures, and ways of communicating that influence how project work is performed and their ability to achieve project success. Because uniqueness abounds in business cultures, you're likely to find any number of organizational structures. Here are some elements that help frame an organizational structure, according to the *PMBOK® Guide*:

- Alignment with organizational objectives
- Skills and special capabilities
- Escalation path
- Authority levels
- Accountability and responsibility levels
- Ability to adapt
- Efficient and effective performance
- Cost
- Locations
- Communications

According to the *PMBOK® Guide,* examples of organizational structures include simple, multidivisional, functional, project oriented, matrix, virtual, hybrid, or PMO. It's helpful to know and understand the organizational structure and the culture of the entity in which you're working. Companies with aggressive cultures that are comfortable in a leading-edge position within their industries are highly likely to take on risky projects. Project managers who are willing to suggest new ideas and projects that have never been undertaken before are likely to receive a warm reception in this kind of environment. Conversely, organizational cultures that are risk-averse and prefer the follow-the-leader position within their industries are highly unlikely to take on risky endeavors. Project managers with risk-seeking, aggressive styles are likely to receive a cool reception in a culture like this. Keep in mind that organizational structures and culture are independent from the development or life-cycle methodology you will use to manage the project (such as predictive, adaptive, or hybrid). However, organizational structures and culture can still influence how you interact with the stakeholders and team members across the organization. For example, perhaps you are using a hybrid approach and you need key resources available almost full time early in the project to define requirements. Once the project work starts, you may need the same or different resources participating in the Kanban board or iterations, depending on which methodology you're using. If you're in a functional organization, you'll need to follow the chain of command to obtain the resources and ensure their availability when you need them. If you work in a matrixed organization with an easy-going culture, the request for resources may be able to be worked out with the individual rather than going through layers of management.

One of the keys to determining the type of organization you work in is measuring how much authority senior management is willing to delegate to project managers. The level of authority the project manager enjoys is denoted by the organizational structure and by the interactions of the project manager with various levels of management. For example, a project manager within a functional organization has little to no formal authority. Their title might not be project manager; instead, they might be called a *project leader*, a *project coordinator*, or perhaps a *project expeditor*. And a project manager who primarily works with operations-level managers will likely have less authority than one who works with middle- or strategic-level managers.

Exam Spotlight

According to the *PMBOK® Guide*, the project managers' authority levels, the availability of resources, control over the project budget, the role of the project manager, and the project team makeup are influenced by the structure of the organization; by the level of interaction the project manager has with strategic-level managers, middle managers, and operations-level managers; and by the project management maturity levels of the organization.

We'll now look at several types of organizations individually to better understand how the project management role works in each one. Functional, matrix, and project-oriented organizational structures are the most common, so we'll focus more heavily on those. I discussed the PMO in Chapter 1, "Building the Foundation," but I'll give you a brief refresher on this structure in this section.

Functional Organizations

One common type of organization is the *functional organization*. Chances are you have worked in this type of organization. This is probably the oldest style of organization and is, therefore, known as the traditional approach to organizing businesses.

Functional organizations are centered on specialties and grouped by function, which is why they're called functional organizations. As an example, the organization might have a human resources department, finance department, marketing department, and so on. The work in these departments is specialized and requires people who have the skill sets and experience in these specialized functions to perform specific duties for the department. Figure 3.1 shows a typical organizational chart for a functional organization.

You can see that this type of organization is set up to be a hierarchy. Staff personnel report to managers, who report to vice presidents, who report to the CEO. In other words, each employee reports to only one manager; ultimately, one person at the top is in charge. Many companies, as well as governmental agencies, are structured in a hierarchical fashion. In organizations like this, be aware of the chain of command. A strict chain of command might exist, and the corporate culture might dictate that you follow it. Roughly translated: *Don't talk to the big boss without first talking to your boss who talks to their boss who talks to the big boss.* Wise project managers should determine whether there is a chain of command, how strictly it's enforced, and how the chain is linked before venturing outside it.

FIGURE 3.1 Functional organizational chart

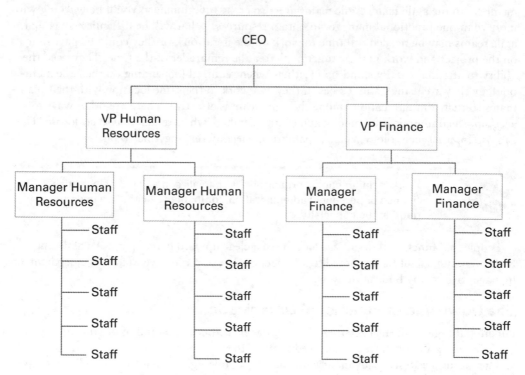

Each department or group in a functional organization is managed independently and has a limited span of control. Marketing doesn't run the finance department or its projects, for example. The marketing department is concerned with its own functions and projects. If it were necessary for the marketing department to get input from the finance department on a project, the marketing team members would follow the chain of command. A marketing manager would speak to a manager in finance to get the needed information and then pass it back down to the project team.

Human Resources in a Functional Organization

Commonalities exist among the personnel assigned to the various departments in a functional organization. In theory, people with similar skills and experiences are easier to manage as a group. Instead of scattering them throughout the organization, it is more efficient to keep them functioning together. Work assignments are easily distributed to those who are best suited for the task when everyone with the same skill works together. Usually, the supervisors and managers of these workers are experienced in the area they supervise and are able to recommend training and career enrichment activities for their employees.

The project manager will need to work with each of the functional managers to obtain resources. Remember that agile teams are self-organized and self-directed. Although that's a good theory, in reality when you're using an agile approach to manage your project and you work in a functional organization, you'll have to get approval to assign these team members to the agile team. Agile teams are also cross-functional, so you'll be working with more than one functional manager to obtain resources. What will be difficult here is that agile teams may be needed full time to work on the iterations, or they could be part-time on the project and work in their functional area the remainder of the time. They need the ability to attend the daily stand-up meetings when required (depending on the agile methodology they are using), and it's best if they can work in the same location with their teammates so that communication among the agile team is effective. There are other ways to accomplish communication among agile team members who are not able to co-locate. They can use Internet technology, video capabilities, teleconferencing, and so on.

 Workers in functional organizations specialize in an area of expertise—finance or personnel administration, for instance—and then become very good at their specialty.

People in a functional organization can see a clear upward career path. An assistant budget analyst might be promoted to a budget analyst and then eventually to a department manager over many budget analysts.

The Downside of Functional Organizations

Functional organizations have their disadvantages. If this is the kind of organization you work in, you probably have experienced some of them.

One of the greatest disadvantages for the project manager is that they have little to no formal authority. This does not mean project managers in functional organizations are doomed to failure. Many projects are undertaken and successfully completed within this type of organization. Good communication and interpersonal and influencing skills on the part of the project manager are required to bring about a successful project under this structure.

In a functional organization, the vice president or senior department manager is usually the one responsible for projects. The project manager will likely not have a lot of authority. In this structure, authority rests with the VP. It's critical for your success to work closely with the VP, get to know them, their management style, and the needs they have in regard to communication and so on. I've seen many a project in this type of organization end up in confusion because the project manager was not close to the VP, and the stakeholders had no issue with going directly to the VP to make demands on the project. If you are communicating and meeting regularly with the VP, you can hopefully avoid this scenario and keep them up to date with the latest status, issues, and decisions along the way.

Managing Projects in a Functional Organization

Projects are typically undertaken in a divided approach in a functional organization. For example, the marketing department will work on its portion of the project and then hand it off to the operations department to complete its part, and so on. The work the marketing department does is considered a marketing project, whereas the work the operations department does is considered an operations project, even though it's all the same project.

Some projects require project team members from different departments to work together at the same time on various aspects of the project. Project team members in this structure will more than likely remain loyal to their functional managers. The functional manager is responsible for their performance reviews, and their career opportunities lie within the functional department—not within the project team. Exhibiting leadership skills by forming a common vision regarding the project and the ability to motivate people to work toward that vision are great skills to exercise in this situation. As previously mentioned, it also doesn't hurt to have the project manager work with the functional manager in contributing to the employees' performance reviews.

Resource Pressures in a Functional Organization

Competition for resources and project priorities can become fierce when multiple projects are undertaken within a functional organization. For example, in my organization, it's common to have competing project requests from three or more departments all vying for the same resources. Thrown into the heap is the requirement to make, for example, mandated tax law changes, which automatically usurps all other priorities. This sometimes causes frustration and political infighting. One department thinks their project is more important than another and will do anything to get that project pushed ahead of the others. Again, it takes great skill and diplomatic abilities to keep projects on track and functioning smoothly.

Project managers have little authority in functional organizations, but with the right skills, they can successfully accomplish many projects. Table 3.1 highlights the advantages and disadvantages of this type of organization.

TABLE 3.1   Functional organizations

Advantages	Disadvantages
This is an enduring organizational structure.	Project managers have little to no formal authority.
There is a clear career path with separation of functions, allowing specialty skills to flourish.	Multiple projects compete for limited resources and priority.
Employees have one supervisor with a clear chain of command.	Project team members are loyal to the functional manager.
Predictive project management methodology works well in this structure; adaptive and hybrid could be challenging.	Organizational structures and culture are independent from the methodology you will use to manage the project. However, adaptive and hybrid methodologies can be challenging in a functional organization because resources are not easily self-organized or self-directed.

Project-Oriented Organizations

Project-oriented organizations are nearly the opposite of functional organizations. The focus of this type of organization is the project itself. The idea behind a project-oriented organization is to develop loyalty to the project, not to a functional manager.

Figure 3.2 shows a typical organizational chart for a project-oriented organization.

FIGURE 3.2   Project-oriented organizational chart

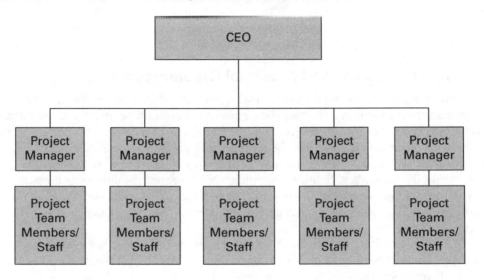

Organizational resources are dedicated to projects and project work in purely project-oriented organizations. Project managers almost always have ultimate authority over the project in this structure and report directly to the CEO. In a purely project-oriented organization, supporting functions such as human resources and accounting might report directly to the project manager as well. Project managers are responsible for making decisions regarding the project and acquiring and assigning resources. They have the authority to choose and assign resources from other areas in the organization or to hire them from outside if needed. For example, if there isn't enough money in the budget to hire additional resources, the project manager will have to come up with alternatives to solve this problem. This is known as a *constraint*. Project managers in all organizational structures are limited by constraints such as scope, schedule, and cost (or budget). Quality is also considered a constraint, and it's generally affected by scope, schedule, and/or cost. We'll talk more about constraints in Chapter 4, "Developing the Project Scope."

You may come across the term *triple constraints* in your studies. The term refers to scope, schedule, and cost constraints. However, please note that the *PMBOK® Guide* also calls constraints *competing demands*, and identifies scope, schedule, cost, risk, resources, and quality as the primary competing demands on projects.

Teams are formed and often *co-located*, which means team members physically work at the same location, which is ideal no matter what type of organizational structure you work in. Having team members co-located is beneficial because communication among the team members is easier; they can physically see the work getting done, and they have easy access to the project manager. When using a predictive development methodology in a project-oriented organization, the project team members typically report to the project manager, not to a functional or departmental manager. When using an adaptive methodology in this organizational structure, teams are self-organized and self-managed and are held accountable by one another rather than a manager. Team members themselves can determine their availability to take on more work or new projects.

One obvious drawback to a project-oriented organization is that project team members might find themselves out of work at the end of the project. A similar situation occurs when you have consultants working on a project. They work on the project until completion and then they are put on the bench, let go at the end of the project, or assigned to another project. Some inefficiency exists in this kind of organization when it comes to resource utilization. If you have a situation where you need a highly specialized skill at certain times throughout the project, the resource you're using to perform this function might be idle during other times in the project.

In summary, you can identify project-oriented organizations in several ways:

- Project managers have high to ultimate authority over the project.

- The focus of the organization is the project.

- The organization's resources are focused on projects and project work.

- Team members are co-located.

- Loyalties are formed to the project, not to a functional manager.

- Project teams are dissolved at the conclusion of the project.

- Project-oriented organizations are ideal for an adaptive project management methodology because teams can easily self-organize and self-manage.

 Real World Scenario

The Project-Oriented Graphic Artist

You've been appointed project manager for your company's website design and implementation. You're working in a project-oriented organization, so you have the authority to acquire and assign resources. You put together your team, including programmers, technical writers, testers, and business analysts. Juliette, a highly qualified graphic arts designer, is also part of your team. Her specialized graphic arts skills are needed only at certain times throughout the project. Juliette has to work in tandem with the programmer and when she has completed the graphic design she's working on for one section of the website, she doesn't have anything else to do until the programmer has the next section of the website ready to go. Depending on how involved the project is and how the work is structured, days or a week or two might pass before Juliette's skills are needed. This is where the inefficiency occurs in a purely project-oriented organization. It's not practical to let her go and then hire her back when she's needed again.

In this situation, you might assign Juliette to other project duties when she's not working on graphic design. Perhaps she can edit the text for the web pages or assist with the design of the upcoming marketing campaign. You might also share her time with another project manager in the organization.

During the Planning process, you will discover the skills and abilities of all your team members so that you can plan their schedules accordingly and eliminate idle time.

Matrix Organizations

Matrix organizations came about to minimize the differences between, and take advantage of, the strengths and weaknesses of functional and project-oriented organizations. The idea at play here is that the best of both organizational structures can be realized by combining them into one. The project objectives are fulfilled, and good project management techniques are utilized while still maintaining a hierarchical structure in the organization.

Employees in a matrix organization often report to one functional manager and to at least one project manager. It's possible that employees could report to multiple project managers if they are working on multiple projects at one time. Functional managers pick up the administrative portion of the duties and assign employees to projects. They also monitor the work of their employees on the various projects. Project managers are responsible for executing the project and giving out work assignments based on project activities. Project managers and functional managers share the responsibility of performance reviews for the employees. Hybrid development approaches are often used in a matrix environment. They allow the functional managers to manage and assign their resources and allow the team members to choose their work items for the upcoming iteration.

 In a nutshell, in a matrix organization, functional managers assign employees to projects, whereas project managers assign tasks associated with the project.

Matrix organizations have unique characteristics. Next, we'll look at how projects are conducted and managed and how project and functional managers share the work in this organizational structure.

Project Focus in a Matrix Organization

Matrix organizations allow project managers to focus on the project and project work just as in a project-oriented organization. The project team is free to focus on the project objectives with minimal distractions from the functional department.

Project managers should take care when working up activity and project estimates for the project in a matrix organization. The estimates should be given to the functional managers for input before publishing. The functional manager is the one in charge of assigning or freeing up resources to work on projects. If the project manager is counting on a certain employee to work on the project at a certain time, the project manager should determine their availability up front with the functional manager. Project estimates might have to be modified if it's discovered that the employee they were counting on is not available when needed.

A hybrid project management methodology could work well in this organizational structure. The organization, the methodology, and the project itself may dictate when to use a predictive methodology and when to use an adaptive project management methodology.

Balance of Power in a Matrix Organization

As we've discussed, a lot of communication and negotiation takes place between the project manager and the functional manager in a matrix organization. This calls for a balance of power between the two, or one will dominate the other.

In a strong matrix organization, the balance of power rests with the project manager. They have the ability to strong-arm the functional managers into giving up their best resources for projects. Sometimes, more resources than necessary are assembled for the project, and then project managers negotiate these resources among themselves, cutting out the functional manager altogether, as you can see in Figure 3.3.

FIGURE 3.3   Strong matrix organizational chart

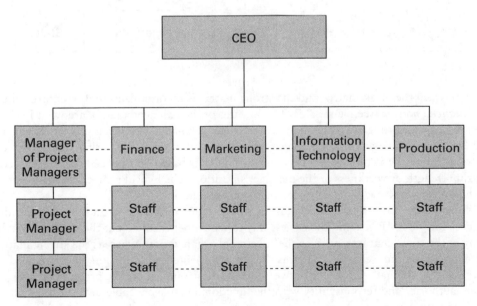

On the other end of the spectrum is the weak matrix (see Figure 3.4). As you would suspect, the functional managers have the majority of the power in this structure. Project managers are simply project coordinators or expeditors with part-time responsibilities on projects in a weak matrix organization. Project managers have limited authority, just as in the functional organization. On the other hand, the functional managers have a lot of authority and make all the work assignments. The project manager simply expedites the project.

FIGURE 3.4   Weak matrix organizational chart

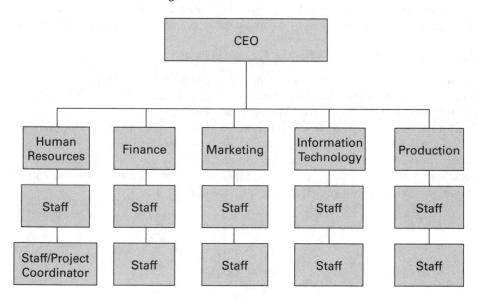

In between the weak matrix and the strong matrix is an organizational structure called the *balanced matrix* (see Figure 3.5). The features of the balanced matrix are what I've been discussing throughout this section. The power is balanced between project managers and functional managers. Each manager has responsibility for his or her part of the project or organization, and employees are assigned to projects based on the needs of the project, not the strength or weakness of the manager's position. The hybrid project management methodology works well in a balanced matrix organization.

Matrix organizations have subtle differences, and it's important to understand their differences for the PMP® exam. The easiest way to remember them is that the weak matrix has many of the same characteristics as the functional organization, whereas the strong matrix has many of the same characteristics as the project-oriented organization. The balanced matrix is exactly that—a balance between weak and strong, where the project manager shares authority and responsibility with the functional manager. Table 3.2 compares all three structures.

TABLE 3.2   Comparing matrix structures

	Weak matrix	**Balanced matrix**	**Strong matrix**
Project manager's title	Project coordinator, project leader, or project expeditor	Project manager	Project manager
Project manager's focus	Split focus between project and functional responsibilities	Projects and project work	Projects and project work
Project manager's power	Minimal authority	Balance of authority and power	Significant authority and power
Project manager's time	Part-time on projects	Full-time on projects	Full-time on projects
Organization style	Most like a functional organization	Blend of weak and strong matrices	Most like a project-oriented organization
Project manager reports to	Functional manager	A functional manager, but shares authority and power	Manager of project managers
Best PM methodology	Predictive or hybrid methodology will likely work best.	All of the methodologies will work here, and hybrid is the most likely choice.	Adaptive works well here.

FIGURE 3.5   Balanced matrix organizational chart

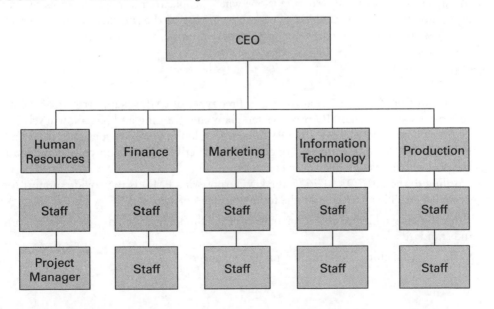

Most organizations today use some combination of the organizational structures described here. They're a *composite* of functional, project-oriented, and matrix structures. It's rare that an organization would be purely functional or purely project oriented. Project-oriented structures can coexist within functional organizations, and most composite organizations are a mix of functional and project-oriented structures.

In the case of a high-profile, critical project, the functional organization might appoint a special project team to work only on that project. The team is structured outside the bounds of the functional organization, and the project manager has ultimate authority for the project. This is a workable project management approach and ensures open communication between the project manager and team members. It is also conducive to an adaptive project management methodology. At the end of the project, the project team is dissolved, and the project members return to their functional areas to resume their usual duties.

Exam Spotlight

Understand the characteristics of each organizational structure and their strengths and weaknesses for the exam.

Other Organizational Structures

A simple, or organic, organizational structure is one that is flexible and tends to be fluid depending on the situation. People usually work in close proximity in this type of structure. A virtual organizational structure is one where people are not physically together at the same place but are connected via a corporate network or the Internet or both. Hybrid structures, as you can imagine, take on aspects of two or more structures.

PMO

You will recall from Chapter 1 that there are three types of PMOs: supportive, controlling, and directive. A supportive PMO provides templates and training and has a low level of control over the project. A controlling PMO concerns conformance to methodologies and frameworks and has a moderate level of control. A directive PMO is controlling and manages the projects. It has a high level of control.

According to the *PMBOK® Guide*, the key purpose of a PMO is to provide support for project managers. This may include the following types of support:

- Providing an established project management methodology, including templates, forms, and standards

- Mentoring, coaching, and training project managers

- Facilitating communication within and across projects
- Managing resources

The PMO usually has responsibility for maintaining and archiving project artifacts for future reference. They establish processes for ensuring the project artifacts are produced in a similar format (in the form of templates), kept up to date, are version controlled, and are distributed to the stakeholders. The PMO ensures that projects are aligned with the strategic objectives of the organization, and it measures the performance of active projects and suggests corrective actions. The PMO evaluates completed projects for their adherence to the project management plan and asks questions like "Did the project meet the time frames established?" and "Did it stay within budget?" and "Was the quality acceptable?" and "Did we bring about business value?"

PMO in an Agile Environment

According to the *Agile Practice Guide* (PMI®, 2017), the PMO's responsibility in an agile environment is to ensure that the right business value is delivered to the right people at the right time. Agile is a collaborative, transparent methodology that is value driven, and as such, the PMO should function in the same manner. In a predictive methodology, the PMO will likely dictate what processes need to be accomplished for the project, and in what order, and also prescribe standardized templates and processes that must be used on the project. In an adaptive or agile methodology, the PMO needs to tailor the processes to the project. This could mean using standardized templates and processes in some cases, or in others, simply acting as a consultant and providing coaching and mentoring to executives and stakeholders regarding the agile process. It's important in an agile environment to solicit and encourage employee engagement. Ask the people who will be working on the project what will work best for them. If the agile team participates in determining the best processes for their project, you'll have much better success at obtaining buy-in and ensuring the agreed-on processes are followed. The key for the PMO is to offer the processes and services that will provide the most value to the project and the organization.

Business partners in the organization are more likely to request PMO services when they perceive that the PMO is providing value to them and the project.

While the *Agile Practice Guide* (PMI®, 2017) notes that an agile PMO must have competencies in several areas in order to serve the needs of projects from different disciplines, I would argue that all PMOs (not just agile) need this expertise. Projects take on many forms and cover as many disciplines as there are industries that practice them—for example, technology, science, healthcare, construction, engineering, transportation, infrastructure, and many more. And generally, within any one industry, you'll almost always have disciplines such as human resource management, finance and accounting, information technology, legal counsel, procurement, general administration, and so on. Project managers, as well as PMOs, should have some understanding of these competencies. Not everyone can know everything, so one project manager may have stronger skills in the finance and accounting area whereas another may have stronger skills in the technology arena, and so on. The PMO will need project managers and business analysts who have a wide range of business expertise and skills.

In keeping with the multidisciplinary theme, the *Agile Practice Guide* (PMI®, 2017) notes that some organizations are transforming their PMOs into agile centers of excellence. The following is a list of some of the ways that PMOs are evolving, along with the services they provide:

Standards   PMOs are adept at developing standards and processes for portfolio, program, and project management. Agile methodologies require tools such as user stories, test cases, and task boards that display workflow. The PMO can assist by standardizing some of these processes and by developing tools, techniques, and templates for agile-based practices.

Training and Mentoring   The PMO can help encourage the adoption of agile methodologies by providing training, mentoring, and coaching to employees and executives on the benefits of agile. Experts in agile technologies within the PMO can become coaches to others and encourage them to transition to agile methodologies and explain the benefits of not only the methodology, but the personal benefit to the employees by improving their skill sets and perhaps obtaining certifications. The PMO should encourage stakeholders and employees to attend agile events and conferences. Speak to your stakeholders in terms of lean thinking principles when explaining the benefits of agile. This includes small and frequent deliveries, frequent reviews, opportunities for immediate feedback, and the ability to perform retrospectives that include discussions on how to improve the product or how to improve the next cycle of work.

Managing Multiple Projects   All PMOs are responsible for managing multiple projects. What's different for an agile PMO is the potential for a significant amount of coordination among and between agile teams and their projects. It's possible for team members to work on more than one agile team (and therefore more than one project) at a time. The PMO will coordinate these resources and efforts. The PMO also helps in determining the right agile process and framework for the project. They also help in assessing project investments and the business value and opportunities they may provide to the organization at the portfolio level.

Agile projects have small, frequent releases that deliver features or functionality at periodic intervals. There may also be a collection of these deliverables, or minimum viable products, that are delivered to the customer as a release. The agile team manages the small, frequent deliverables, and the PMO coordinates the major releases across the program.

Agile is a highly collaborative approach, so good communication from the PMO is required to keep everyone informed.

Organizational Learning   This is another task that all PMOs are responsible for. Think of organizational learning as capturing lessons learned; documenting the agile methodologies used, along with all of the data about those methodologies; documenting whether users' stories were written effectively, how well the team broke down these stories into activities, and estimating what could be completed within an iteration; and so on. This information should be indexed so that it can be easily referenced.

Managing Stakeholders   In an agile methodology, stakeholders are actively engaged in the project and ordinarily interact with the team members on a daily basis. In this context, the agile PMO is responsible for providing training to stakeholders on agile processes, helping them understand the purpose for acceptance testing and how it's performed, coaching them regarding how they can provide constructive feedback to the team to improve both the product and the process, and encouraging and educating them on the importance of subject matter experts participating in the project.

Obtaining and Evaluating Team Leaders   This process is the same in a predictive methodology as in an agile methodology. The PMO recruits, interviews, and hires project managers and evaluates their performance throughout the year. According to the *Agile Practice Guide* (PMI®, 2017), an agile PMO will help develop the interview guidelines that will help ensure potential employees or team members are well versed in agile practices.

Executing Specialized Tasks   This is very similar to the Training and Mentoring topic we covered earlier. The PMO is responsible for providing agile mentors and coaches to the project team and the stakeholders. They also train facilitators to conduct retrospective meetings. A retrospective meeting occurs at the end of the iteration and determines the overall progress of the work, needed changes to schedule or scope, ideas for improvement for the next iteration, and what worked well during this iteration (or not) so that the next iteration, and future iterations, can be improved.

Project-Based Organizations

A *project-based organization (PBO)* is a temporary structure an organization puts into place to perform the work of the project. A PBO can exist in any of the organizational structure types we've just discussed. Most of the work of a PBO is project based and can involve an entire organization or a division within an organization. The important thing to note about PBOs for the exam is that a PBO measures the success of the final product, service, or result of the project by bypassing politics and position and thereby weakening the red tape and hierarchy within the organization. This is because the project team works within the PBO and takes its authority from there.

   Organizations are unique, as are the projects they undertake. Understanding the organizational structure will help you, as a project manager, with the cultural influences and communication avenues that exist in the organization to gain cooperation with stakeholders and project team members and successfully bring your projects to a close.

Exam Spotlight

Table 2-1 in the *PMBOK® Guide* does a great job of laying out the differences in organizational structures, a project manager's role in each, project budgets, and more. Check it out.

Influences of Organizational Structure on Agile Methodologies

The *Agile Practice Guide* (PMI®, 2017) notes that organizational structure is heavily influenced by several factors, which may help or hurt an organization in adapting to agile methodologies, adapting to new information, and recognizing changing market needs. The factors include the following:

- Geography may influence projects due to different locations and cultures, competing interests among the dispersed divisions within the organization, language barriers, and more. Agile approaches can help encourage collaboration when the organization faces geographic challenges.

- Functional organizations are hierarchical, as you learned earlier in this chapter. Due to their chain-of-command structure, it could be difficult to introduce a highly collaborative approach such as agile in this type of organization. There may be resistance, especially from those in positions of authority, so practice some of the mentoring skills we talked about earlier and start educating the stakeholders on the benefits of agile.

- Large project deliverables are not practical when using an agile methodology. Agile is based on small, frequent cycles of delivering minimum viable products, features, or functionality. Consider reducing the size of the deliverables using an agile approach. This will increase the speed with which business value can be realized by the organization.

- Assigning human resources occurs organically on an agile project. Agile team members are self-organized and self-directed. However, the *Agile Practice Guide* (PMI®, 2017) notes that you should consider asking for one resource from each department to work on the project full time until it's completed. Resources should be allocated based on the priority of the project in the organization.

- Procuring project work from external vendors is a common practice. Many organizations rely on a network of vendors to assist in implementing their projects. Remember that the vendors' best interest is the profitability of their own organizations, not yours. That doesn't mean they won't perform good work, but you need to keep this in mind. Another concern with vendor-led projects is that they gain a lot of experience and knowledge while implementing your project (and often learn on the job). This is knowledge and expertise that won't be available to the organization after the vendor has finished the project. This is where agile methodologies can help. Because agile is collaborative and transparent, and cross-functional team members are highly engaged in the project, knowledge transfer will occur as the project work is performed. Additionally, the retrospective meetings at the end of the iteration will help in improving knowledge transfer and capturing lessons learned for future projects.

Identifying Stakeholders

Think of stakeholders and project team members as a highly polished orchestra. Each participant has a part to play. Some play more parts than others, and alas, some don't play their parts as well as others. An integral part of project management is getting to know your stakeholders and the parts they play. You'll remember from Chapter 2 that stakeholders are those people or organizations who have a vested interest in the outcome of the project. They have something to either gain or lose as a result of the project, and they have the ability to influence project results. You should also recall that a project is successful when it achieves its objectives by producing deliverables that meet the expectations of the stakeholders. Stakeholders are critical to the success of your project, so you should spend as much time as needed identifying them, getting to know them, and understanding their needs. I'll talk about the tools and techniques you can use to do so shortly.

Key stakeholders can make or break the success of a project. Even if all the deliverables are met and the objectives are satisfied, if your key stakeholders aren't happy, nobody is happy.

As we discussed in Chapter 2, the project sponsor, generally an executive in the organization with the authority to assign resources and enforce decisions regarding the project, is a stakeholder. The project sponsor serves as the tie-breaking decision-maker and is the final stop on the escalation path. If the project stakeholders cannot make a final decision or come to a conclusion on an issue, the project sponsor gets the final say.

The project sponsor is also the person who approves the project charter and gives the project manager the authority to commence with the project. The customer is a stakeholder, as are contractors and suppliers. The project manager, the project team members, and the managers from other departments, including the operations areas, are stakeholders as well. It's important to identify all the stakeholders in your project up front. Leaving out an important stakeholder or their department's function and not discovering the error until well into the project can be a project killer.

Figure 3.6 shows a sample listing of the kinds of stakeholders involved in a typical project. The org chart view shows a hierarchical view, not an organizational reporting view, of how the stakeholders listed in the chart and in the box below the chart interact with the project manager.

FIGURE 3.6   Project stakeholders

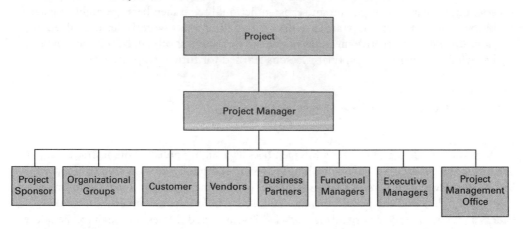

Many times, stakeholders have conflicting interests. It's the project manager's responsibility to understand these conflicts and try to resolve them. It's also the project manager's responsibility to manage stakeholder expectations. Be certain to identify and meet with all key stakeholders early in the project to understand their expectations, needs, and constraints. When in doubt, you should always resolve stakeholder conflicts in favor of the customer. If you can't reach resolution, you'll need to escalate.

Discovering Stakeholders

The *Identify Stakeholders* process involves identifying and documenting all the stakeholders on the project, including their interests, interdependencies, and potential positive or negative impacts on the project. You will document all of this information in the stakeholder register, which is one output of this process. But let's not get ahead of ourselves; it all starts with identifying key stakeholders. This may seem as if it should be fairly easy—however, once you get beyond the obvious stakeholders, the process can become difficult. You may also find that sometimes stakeholders, even key stakeholders, will change throughout the project's life, so you should repeat the Identify Stakeholders process often.

Some of the inputs you should consider in this process include the business case, benefits management plan, and the project charter. Each of these documents contains a list of stakeholders, so you will want to review them when preparing the stakeholder register. In addition, some of the key project management plans, such as the communications management plan, agreements (or contracts), and requirements documentation, will contain references to stakeholders. Historical information, lessons learned, and the stakeholder register from past projects of similar size and complexity will also help in identifying stakeholders. These are all part of the organizational process assets input for this process. We'll look at the enterprise environmental factors (EEFs) input to this process in the next section because they directly relate to the power and influence the stakeholder possesses.

Documents are one place to start to identify stakeholders. Another is by using one of the tools and techniques of this process called *brainstorming*. This technique is a great way to start the process of both identifying and analyzing stakeholders. Gather a few of the key experts, team members, and known stakeholders that you identified from the project charter and business documents and conduct an old-fashioned brainstorming session to determine who else may have an interest in the project. Start by asking who is missing; perhaps go through the org chart department by department to identify more and ask which business processes could be impacted as a result of the project. Don't forget that stakeholders can be internal or external to the organization, so remember to ask about both in the brainstorming session.

Brainwriting is a type of brainstorming you could use where participants are given the question or brainstorming topic ahead of time, giving them time to think about their answers. You'll hold a meeting within a week or so of distributing the questions and host the brainstorming session. A brainwriting meeting can be face to face, or you could use technology such as a video- or web-conferencing tool to hold the meeting.

It's better to start by identifying everyone you can think of and then eliminating them later, rather than missing them early on and having to add them later. Start with a simple list of names, and build on this list by adding interests and influence as you proceed through this process.

I can't say this enough: don't forget important stakeholders. That could be a project killer. Leaving out an important stakeholder and their requirements, or not considering a stakeholder whose business processes are impacted as result of your project, could spell disaster.

Exam Spotlight

Stakeholder identification should occur as early as possible in the project and continue throughout its life. Likewise, the stakeholder analysis and strategy should be reviewed periodically throughout the project and updated as needed. Remember to manage stakeholder satisfaction just as you would any key deliverable on the project. Identifying and analyzing their needs throughout the project will help make certain you are managing their expectations, and thus their satisfaction.

Stakeholder Analysis

According to the *PMBOK® Guide,* stakeholder analysis involves using qualitative and quantitative data to analyze which stakeholders' interests should be considered throughout the project. Those with the most influence will also have the most impact, so you need to know who they are. During stakeholder analysis, you'll want to identify the influences stakeholders have on the project and understand their expectations, needs, and desires.

Stakeholder analysis is performed using tools and techniques such as expert judgment, data gathering, data analysis, data representation, and meetings. Expert judgment in this process has a specific emphasis on involving experts who have knowledge of the culture and politics of the organization, the customers of the organization, specific industry experience (including understanding the particular deliverables of this project), and an understanding of team member skills and capabilities. Use the brainstorming technique with these experts to derive as much information as you can about the stakeholders involved on the project and what these experts know of their behaviors and influence from experience on past projects. Discuss the culture of the organization and how stakeholders adapt to that culture. For example, perhaps you work for an innovative company that is flexible and willing to take risks. Do the stakeholders on your project exhibit this same type of attitude, or do they hold back a bit? Write that down.

My favorite technique is interviewing the stakeholders. Ask them directly what their interests are on the project. Here is a list of some questions you might include in your interviewing process:

- What is your interest in the project?
- What are your expectations regarding the project outcomes?
- What do you anticipate will be your level of involvement with the project?
- What is your knowledge level of the project and of any skills or information needed to produce a successful project?
- What contributions are you anticipating making to the project?
- How will the project impact your organization both positively in the form of benefits or negatively in terms of consequences or impacts? What are those benefits and/or impacts?
- Are there other stakeholders who have expectations of this project that conflict with yours?

The information you're gathering by asking these questions is known as *stakeholder analysis.* In the last section, you identified the stakeholders and created a list of their names. Now, you'll want to add their interests and needs to this list and indicate, based on your interviews, the potential impact their needs or interests may have on the project. I recommend using a simple high-medium-low indicator for potential impacts. You could ask the stakeholder what level of impact they think these needs or desires have on the project as part of your interviewing process. This will give you an idea of the importance of their expectations. As the project manager, you will have to sift through this information and make judgment calls based on the overall objectives of the project. Not every stakeholders' interests will carry the highest impact. Table 3.3 shows what we have collected to date as an example.

TABLE 3.3 Stakeholder interest

Stakeholder name	Needs and interests	Potential impact
Ryan, project sponsor	Need to replace an aging system with a more modern system that's easier to use	High
	Need better reports to inform decision-making	Med
	Need to lower cost of operations	High
Kay, key stakeholder in operations	Need to improve productivity of workers; too much time spent on entering data	High
	Need to improve efficiency of workers; too much time spent looking for key information	High
Joe, key stakeholder in information technology	Need to improve security of data	High
	Need to lower support costs of the system	Med

Be warned that stakeholders are mostly concerned about their own interests and what they (or their organizations) have to gain or lose from the project. In all fairness, we all fall into the stakeholder category, so we're all guilty of focusing on those issues that impact us most.

Categorizing Stakeholders

Now that we know our stakeholders and their interests, we need to categorize them in terms of the power and influence they have in the organization. Influence and power are indicators of how much one stakeholder's opinion or decision will have on others in the organization and how it will affect your project. For example, in every organization I've ever worked in, the chief financial officer (CFO) wields a lot of power and control. (You know the old saying, "She who owns the gold makes the rules.") If the CFO says no, it will likely have a ripple effect throughout the organization and your project. The CFO by virtue of her position and standing in the organization has the authority to say no. Because of this standing, she also has influence on most company decisions. This is what is meant by power and influence.

This is one of the things you probably can't or shouldn't ask the stakeholders directly about. Some stakeholders have inflated opinions of their influence in the organization, and others may think they have little influence when in reality they wield a lot of influence. Deriving this information may require a bit of undercover work on your part as project manager. If you've worked in the organization for any length of time, you probably have a good feel for the influence certain stakeholders may possess. If you've identified stakeholders on your project whom you don't know well, you might want to conduct a brainstorming session with a few key, trusted individuals who can help you identify the level of influence the stakeholders have in the organization. I advise that you don't make this information available to everyone on the project. This is information you need to know as project manager but could potentially damage relationships if it were published.

You must consider certain factors when determining and categorizing power and influence, and they are found in the enterprise environmental input to this process. According to the *PMBOK® Guide*, the EEFs include company culture, organizational structure, governmental or industry standards, global trends, and geographic locations of resources and facilities. The organizational structure will help you understand who has influence and power based on their position and where they reside in the organization, much like the CFO example I explained previously. The CFO likely reports directly to the CEO in a functional organization and the reporting structure alone provides power to this position. Other "C" level positions or vice presidents may have similar power and influence when reporting directly to the CEO. Geographic location can dictate power and influence. The vice president over the largest geographic region or largest sales region, for example, will likely have more power and influence than VPs over smaller regions. There will always be one or two stakeholders who seem to have more power and influence than others, and sometimes for no discernible reason. One of the ways to determine power and influence is by using the brainstorming or interviewing techniques we discussed earlier with your trusted confidants who have previous experience with the stakeholders. These confidants will use their expert judgment in considering organizational culture and hierarchy, and then determine the power and interest levels of your stakeholders. Next you'll plot that information into the models I'll discuss next to document power and influence.

The following models are part of the Data Representation tool and technique of this process that display the categories you'll derive for the stakeholders. According to the *PMBOK® Guide*, the following methods are used for prioritizing and categorizing stakeholders.

Power/Interest Grid, Power/Influence Grid, or Impact/Influence Grid

Each of these grids use two factors to identify the power, authority, interest, or influence level the stakeholder has on the project. For example, the power/interest grid plots power on one axis of the graph and interest on the other. They are most useful on small projects where it's easier to identify the relationship between the stakeholder and the project. Figure 3.7 shows a sample power/interest grid.

FIGURE 3.7   Example of a power/interest grid

$$FV = PV(1+i)^n$$

Stakeholder Cube

A stakeholder cube uses the elements described in the power/interest grid along with other elements, which could include attitudes, influence-level designations, power-level designations, or any combination of elements associated with the power, influence, and abilities of the stakeholders to impact, change, or influence the project. The cube is displayed as a 3D model and allows the project manager and team members to identify stakeholders, understand how their characteristics can impact the project, and better interact with them according to their needs.

Salience Model

This model uses three categories to analyze stakeholders: power, urgency, and legitimacy. Power refers to their ability to influence the project outcomes as we've discussed throughout this section. *Urgency* may seem strange here, but it refers to stakeholders who may have a significant investment or interest in the outcomes of the project (meaning they will require a lot of attention and likely need a lot of handholding throughout the project), how attentive you'll need to be, and whether their participation or needs are time-sensitive and/or time-bound. Legitimacy means they have a legitimate need to participate in or receive the benefits from the project. This model is very helpful when you have a large number of stakeholders and/or a significant amount of interaction among the large community of stakeholders. The Salience model is displayed as a Venn diagram. Each element (power, urgency, legitimacy) is represented in a circle, and the intersection of the three circles shows the strongest influences of the stakeholder at a glance.

This model was developed by Ronald K. Mitchell, Bradley R. Agle, and Donna J. Wood and presented in an article published in *The Academy of Management Review* published in 1997. You can reference the article at www.jstor.org/ stable/259247?seq=22#metadata_info_tab_contents.

Stakeholders who only exhibit one of these three categories (power, legitimacy, or urgency) will not likely have a big impact on the project and your engagement with them will be limited. If stakeholders have overlap in two of the areas, power and legitimacy for example, or all three, they are more likely to exert influence on the project and you may need to devote extra care and attention in ensuring their needs are met.

Mitchell, Agle, and Wood identified three qualitative classes of stakeholders: latent, expectant, and demanding. Latent stakeholders are those who possess only one of the three attributes: power, legitimacy, or urgency. As I stated earlier, they will not likely have a big impact on your project and your engagement with them will be limited. There are three types of latent stakeholders, each associated with one of the classifications. Latent stakeholders who are classified with the power attribute are known as dormant stakeholders. Dormant stakeholders may have power in the organization, but since they do not have an urgent need or a legitimate need to participate in the project (nor are they impacted by it), their power is not likely a threat. The same idea holds true for those stakeholders who are classified in the legitimacy category. They have a legitimate need to participate in or receive the benefits of the project, but they have no power or urgency. The latent legitimate stakeholders are known as discretionary stakeholders. The last type of latent stakeholder is demanding, and you guessed it, they are classified in the urgency category. Again, they don't have power in the organization and no legitimacy in terms of the project—but beware; these stakeholders are like the squeaky wheel. They will pester you repeatedly and could make you think you need to focus attention on them and their needs, but the squeaky wheel syndrome is likely just noise. Without the other two attributes, there isn't much to worry about from the demanding stakeholder.

Expectant stakeholders is the second type of classification. They are stakeholders who possess two of the attributes in the model. Like latent stakeholders, there are three classifications of stakeholders in the expectant category. First is the dominant stakeholder. This includes stakeholders who are classified as having both power and legitimacy. These stakeholders will have an impact on your project, and since they have a legitimate interest in the project and the power to influence others, you'll want to have a close relationship with them. Dependent stakeholders are classified with both the legitimacy and urgency categories. Because they don't have power, they may have to rely on dominant stakeholders to help influence key issues or make decisions on the project that will impact their organization. Last is dangerous stakeholders who possess both the power and urgency attributes. This means they can become not just a squeaky wheel, but one with power who may be able to divert or coerce project resources away from key tasks and have them address their individual needs alone. This can have disastrous impacts on the project, so you'll want to keep a close watch on these stakeholders.

Demanding stakeholders is the third classification of stakeholders. They possess all three attributes: power, legitimacy, and urgency. You'll recall that expectant stakeholders possess two of the three attributes in the Salience model. Adding in the third attribute makes them demanding stakeholders. If any of the expectant stakeholders (dominant, dependent, or dangerous) acquire the third attribute, they become demanding stakeholders. For example, when the dominant stakeholder who is classified with both the power and legitimacy attributes obtains the urgency attribute, their classification changes to a demanding stakeholder.

Table 3.4 shows a recap of the classifications and subclassifications of stakeholders in the Salience model.

Figure 3.8 shows the Salience model along with the intersections of where stakeholders with the classifications listed in the table fall on the diagram.

It's possible for stakeholders to move between all of these categories and classifications during the project. Analysis should be performed throughout the project to reassess and reclassify stakeholders as their needs change and/or as stakeholders are added or dropped from the project.

TABLE 3.4 Salience model categories and classifications

Category	Classification	Subclass	Level of involvement with project
Power	Latent	Dormant	Low
Legitimacy	Latent	Discretionary	Low
Urgency	Latent	Demanding	Low
Power + Legitimacy	Expectant	Dominant	Med or high
Power + Urgency	Expectant	Dependent	Med or high
Legitimacy + Urgency	Expectant	Dangerous	Low; they need the help of dominant or dependent stakeholder
Power + Legitimacy + Urgency	Demanding	N/A	High

FIGURE 3.8 Salience diagram

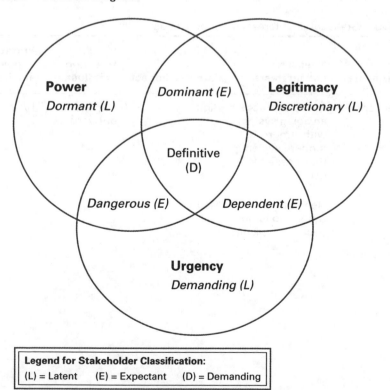

Legend for Stakeholder Classification:
(L) = Latent   (E) = Expectant   (D) = Demanding

Directions of Influence

Just as the title sounds, this classification model describes the direction of influence a stakeholder might have inside and outside the organization, including up, down, sideways, and outward. In my experience, you should pay close attention to those with upward influence. If the project hits some rocky bumps or you find additional funding is needed, these stakeholders will go right to the top of the organization to assure the executives that the project is under control but has hit a snag and needs some help. They'll also be your biggest champions in carrying the great news upward so that you and your team get the recognition they deserve. Be certain to prioritize stakeholders throughout the project, including their direction of influence. Stakeholders have a tendency to come and go on the project, and their relationships within the organization and with one another are often dynamic.

Prioritization

This is another method to categorize your stakeholders in terms of influence levels, interest, impact, and so on. Prioritization comes in handy on complex projects, or projects where you may have dozens of stakeholders and you need to keep track of those with the most power and influence. You could prioritize stakeholders according to rank—for example, the CEO is number one—or you could prioritize their power and interest using high-medium-low indicators as I've done in Table 3.5. I used the table we started earlier and added the stakeholders' direction of influence and prioritization of stakeholders according to power and interest.

TABLE 3.5   Stakeholder interest and prioritization

Stakeholder name	Needs and interests	Potential impact	Direction of influence	Prioritization of power and interest
Ryan, project sponsor	Need to replace an aging system with a more modern system that's easier to use	High	Up, down, outward	H/H
	Need better reports to inform decision-making	Med		

Stakeholder name	Needs and interests	Potential impact	Direction of influence	Prioritization of power and interest
	Need to lower cost of operations	High		
Kay, key stakeholder in operations	Need to improve productivity of workers; too much time spent on entering data	High	Up, sideways, down	M/H
	Need to improve efficiency of workers; too much time spent looking for key information	High		
Joe, key stakeholder in information technology	Need to improve security of data	High	Up, sideways, down	L/H
	Need to lower support costs of the system	Med		

Don't skip this step. Analyzing your stakeholders for power and influence is critical to the success of the project. You need to know that if the CFO says no, it could be impossible to change their mind. If you know that they are close with one of the vice presidents, you may be able to enlist the vice president in helping to change the CFO's mind. You won't know or understand this level of power and interest without performing analysis of the stakeholders. I know it's not nice to play favorites, but when it comes to stakeholders, you as the project manager need to understand who has power and authority, who can influence project outcomes, and which ones you should pay particular attention to so that your project progresses smoothly. Each of these tools and techniques will help you identify and categorize your stakeholders so that you aren't caught off guard.

Remember that the definition of a successful project is one that accomplishes the goals of the project and meets stakeholders' expectations. Understand and document those expectations and analyze the stakeholders regarding their power and influence on the project, and you're off to a good start.

Stakeholder Register

The primary output of this process is the stakeholder register. The stakeholder register contains information about the stakeholder, including their contact information. Remember that you can use the stakeholder register template (part of the OPA inputs to this process) if your PMO or organization has one. The stakeholder register should contain at least the following details, according to the *PMBOK® Guide*:

Identifying Information   This includes items such as contact information, department, role in the project, and so on.

Assessment Information   This includes the information regarding influence, expectations, key requirements, and when the stakeholder involvement is most critical. Not all stakeholders are required for the entire life of the project. Be certain to note when the stakeholders' participation is needed on the project. This should include the phase of the project in which the stakeholder will have the most influence or impact.

Stakeholder Classification   This includes the classification elements we discussed in the "Categorizing Stakeholders" section, including the influence, power, impact, direction of influence, and whether they are internal stakeholders or hold a position outside the organization.

It's important to update the stakeholder register on a routine basis so that new stakeholders are added and those who have finished their roles on the project are noted as completed.

Project documents are usually easily accessible by the project team and stakeholders. Use caution when documenting sensitive information regarding a stakeholder and your strategy for dealing with that stakeholder because that information could become public knowledge and damage your relationship with that stakeholder.

You can easily use Table 3.5 as the basis for your stakeholder register. You'll need to add contact information to this table and the other elements discussed that are required according to the *PMBOK® Guide*.

Some other outputs of note in the Identify Stakeholders process are change requests, which generally occur later in the project once the work of the project has started. However, stakeholders could change, new stakeholders might be added, and/or information you've gathered about the stakeholders could change. Any of these situations could generate change requests, so keep an eye on those stakeholders as the project progresses.

Project management plan updates are an output you will see frequently. Remember that many project documents evolve iteratively throughout the project (or throughout certain project process groups) and may require updates as a result of changing stakeholders. In particular, the requirements management plan, the communications management plan, the risk management plan, and the stakeholder engagement plan will almost always need updates when stakeholder changes or additions occur.

Project documents updates is another output you will see often. The documents you will need to update for this process include the assumption log, the issue log, and the risk register. These documents are frequently updated throughout the life of the project, so keep them handy.

> Stakeholder identification is not a onetime process. You should continue to ask fellow team members and current stakeholders if other stakeholders should be included in the project. And you should continually assess stakeholder needs and power and interests.

Stakeholders on an Agile Project

Agile methodologies have many benefits, and one of the greatest is active stakeholder involvement in all aspects of the project. Agile is a flexible methodology that is highly interactive. It encourages a great deal of communication among the team. It keeps the stakeholders and project team members engaged and utilizes a facilitated process so that the team can interact on a daily basis. Scrum is one of the most well-known agile processes, so we'll look at the stakeholders on this type of project first.

Scrum teams are self-directed, self-managed, adaptable, and highly aware of customer needs. The Scrum team consists of several members: team facilitator, product owner, stakeholders, and cross-functional team members. Here is a brief description of each:

Scrum Master   The Scrum master is also known as project manager, project team lead, team coach, or team facilitator. The Scrum master coordinates the work of the sprint. They also run interference between the team and distractions that might keep them from the work at hand. Scrum masters are facilitators and help educate others in the agile process. They typically do not perform development tasks, but they assist the product owner in maintaining the backlog, prioritizing work, and defining when the work is done. The Scrum master is a facilitator and not a manager. Project team members do not report to the Scrum master.

> In practice you'll often find that projects using Scrum do not have a project manager and the Scrum master takes on many of the functions of the project manager. If you are using a hybrid approach, there could be both a Scrum master and a project manager assigned to the project. Keep in mind for the exam that the *Agile Practice Guide* (PMI®, 2017) states that all projects (including agile) must have project managers.

Product Owner   The *product owner* represents the stakeholders and is the liaison between the stakeholders and the Scrum master. Product owners speak on behalf of the business unit, customer, or the end user of the product and are considered the *voice of the customer.* There should be only one product owner on the team. Communicating with the stakeholders is a critical responsibility of product owners. They communicate progress and milestones achieved. They determine project scope and schedule, and they request the funding needed to complete the work of the project. They manage and prioritize the backlog, which contains the user stories the Scrum team will work on in each sprint.

Stakeholders   Stakeholders are people with a vested interest in the project or the outcomes of the project. They interface with the product owner, who informs them of work progress. It's the product owner's responsibility to keep the stakeholders informed.

Team Members   Team members are responsible for completing backlog items. They sign up for tasks associated with user stories that have been chosen for the sprint, based on the priority of the work and their skill sets. They establish estimates for the work and take on enough tasks to fill the iteration period. Agile teams are self-directed, self-organized, and self-managed.

Table 3.6 shows the various titles that team members might hold in the predictive and adaptive methodologies.

TABLE 3.6   Roles in adaptive methodologies

Role	Waterfall	Scrum	Kanban	Lean	Extreme Programming	Six Sigma
Project sponsor or key project liaison	Project sponsor (usually an executive in the organization)	Product owner (also known as voice of the customer)	Service request manager	Project sponsor (usually a manager, senior manager, or executive in the organization)	Customer	Six Sigma deployment leader, Six Sigma champion

Role	Waterfall	Scrum	Kanban	Lean	Extreme Programming	Six Sigma
Project team members	Various cross-functional titles depending on the project (HR specialist, developer, marketing manager, accounts payable supervisor, etc.)	Scrum team; this team consists of cross-functional members	No set roles or team member names; the team consists of cross-functional team members	Team member	Developer or programmer	Usually Six Sigma–certified team members; also includes cross-functional team members
Project manager	Project manager	Scrum master or team facilitator	Flow master or service delivery manager	Sensei	Manager or tracker; may also be a coach	Six Sigma Black Belt or Six Sigma Master Black Belt

Six Sigma Adaptive Methodology

Six Sigma is concerned with continuous improvement. It's discipline oriented and focuses on developing and practicing disciplines that improve and maintain quality, improve the product, or improve the process. It can also be used as a problem-solving technique.

Six Sigma relies on a methodology called DMAIC. These are phases within the methodology: define, measure, analyze, improve, and control. There is some correlation between these phases and the project management process groups. Initiating and the DMAIC design phase are similar in that the project charter is completed and stakeholders are identified. The measure and analyze phases are similar to Monitoring and Controlling processes where work performance is measured and analyzed. The improve and control phases are also similar to Monitoring and Controlling where changes or corrective actions are identified and implemented. All of the DMAIC phases also interact with the Executing process group. The design phase, for example, outlines the work, the resources needed, and the key milestones, all of which are worked on during the Executing process group. The phases are iterative, much like the project process groups. For example, as you analyze and improve, it may require modifying the work of the project, which means the Executing processes, and perhaps the design and measure phase will be repeated.

Now that we have documented the project charter (in Chapter 2) and identified the stakeholders, let's take a look at how these processes, and future processes, might be completed when using a Six Sigma approach.

Define   This phase is where the project is defined. In project management terms, it would concern defining the project charter, defining key milestones, identifying the project sponsor, and more. Some of the activities that occur in this phase are:

- Identify project goals and objectives
- Identify business value the project will bring about
- Create project charter
- Identify stakeholders
- Identify project sponsor
- Define project resources
- Create process flowcharts
- Create project plan
- Identify key milestones

Measure   This phase is where elements of the project are measured. This may include the processes, the work product the project has produced, the result the project has produced, and more. This phase measures and quantifies problems so that they can be analyzed and corrected in future phases. Some of the activities that occur in this phase are:

- Collect data
- Define benchmarks
- Define measurement (including units of measure)
- Define defects
- Validate the measurement system
- Create or modify process flowcharts

Analyze   This phase concerns researching and determining the root cause of defects. The work of the project is analyzed to determine whether it meets quality standards (or other standards) using several tools and techniques. Some of the activities and tools and techniques used in this phase are:

- Define process and/or product expectations
- Determine sources of variance
- Identify root causes

- Histogram
- 5 Whys
- Pareto chart
- Run chart
- Statistical analysis
- Cause and effect diagrams (Fishbone diagram)
- Scatter charts

Improve   This phase involves improving the process and/or product by eradicating the defects. Some of the activities and tools and techniques used in this phase are:

- Determine possible solutions
- Assess and evaluate possible solutions
- Define tolerance levels
- Brainstorming
- Design of experiments
- Simulation
- Mistake-proofing

Control   This phase concerns putting solutions into place that will control future performance. Some of the activities and tools and techniques used in this phase are:

- Define the monitoring and control system
- Validate the monitoring and control system
- Validate benefits of the improvements (cost savings, profit, etc.)
- Create procedures
- Create standards
- Transfer product or process to operations owner
- Close project
- Close out project documentation
- Communicate project closeout
- Control charts
- Control plans
- Cost savings calculations

You might want to bookmark this section, as I'll refer to the DMAIC in future chapters. I'll relate the project process we are discussing to the DMAIC phase. So far, we have completed the Develop Project Charter process and the Identify Stakeholders process. Both of these processes are completed in the design phase of the DMAIC.

Delivering Business Value

I mentioned earlier that projects can initiate change in an organization—sometimes significant change—by moving the business from one state to another. Think about the humble beginnings of Amazon. Jeff Bezos started out with a simple idea to sell books via the Internet. Keep in mind that the Internet in those days was in its toddler phase and no one was sure how this thing called the World Wide Web would mature. Fast-forward a couple of decades and Amazon has grown to become one of the largest online retailers in the world. You can buy almost anything from Amazon. Along the way, other services and products were introduced including two-day free shipping, video streaming, cloud computing, the Kindle and the Echo, and I'm only scratching the surface. Amazon not only moved their own organization from one business state to another (that is, selling books to selling almost everything), they also changed the way consumers interact with retailers when buying and selling goods, watching movies and video content, storing data, and more.

Business value is a loosely defined term that refers to those values that will lead to short- and long-term benefits for the organization. Benefits to the organization usually translate to benefits to the customer as well. Amazon likely determined to increase business value by increasing revenues (which led to selling more products) and in turn, increased value to the customer because they made it easy to do business with Amazon.

Exam Spotlight

Business value refers to those values that will lead to short- and long-term benefits for the organization.

Business value can mean different things to different organizations. Business value may be tangible or intangible. We talked about the tangible benefits of a project, such as return on investment, payback period, net present value, and more, in Chapter 2. In terms of project delivery, business value (also known as benefits realization) may involve increasing revenues, increasing net profits, expanding into new markets, improving efficiencies, producing goodwill, increasing customer satisfaction, and more. Business value for your project must be defined by the project sponsor and key stakeholders. Defining and analyzing the benefits of the project is a key activity that you as project manager are responsible for accomplishing. The first step is to identify the benefits. It may start with questions the sponsor or key stakeholders begin asking, such as "How can we improve efficiencies and workflows (thus saving time)?" or "How can we increase profits?," which could also be asked as "How can we save money?" or "How can we increase revenues?"

Interviews with the stakeholders and brainstorming sessions are good techniques you can use to help derive business value. The following is a list of potential examples to help get you started. Both tangible and intangible benefits are listed here.

Improve Market Value   This may involve increasing market share, reaching new customers, and providing new or improved services or functionality that your competitors don't provide.

Increase Revenue   This may involve offering new goods and services, decreasing time to market, reducing costs through continuous improvement activities, and more. This category could include any number of examples.

Produce Goodwill   Perhaps your company is involved in social awareness and improving the lives of those who are less fortunate. For example, TOMS shoes use a portion of the proceeds from every pair of shoes sold to give shoes, safe water, and other services to underprivileged communities. As a bonus, producing goodwill may improve brand recognition as well, as discussed next.

Improve Brand Recognition   Brand recognition aids consumers in choosing goods and services over and above all the choices out there. Walk into most grocery stores in North America and you are likely to find at least 20 brands of toothpaste to choose from. Brand recognition involves marketing or generating social awareness (such as TOMS) to draw attention to your company or product.

Improve Employee Engagement   Engaged employees are typically happy employees. Happy employees typically produce good work. Examples here may include bonus or incentive programs, rewards and recognition programs, communication venues, and so on.

Improve Efficiency   This may involve automating tasks to increase accuracy and speed of completion, reducing time spent on processing customer orders, making processes or tasks easier to complete, improving systems by adding new features and functionality that improve workflows, and so on.

Increase Innovation   This is a wide-open topic that may cover almost anything, including using the Internet of Things, creating new infrastructure (or improving existing infrastructure) to accommodate future changes in technology, creating a ride-share program, creating electronic health records, and more.

Increase Shareholder Equity   This is the mantra of students everywhere seeking a business degree: An organization's first job is to increase shareholder wealth. This could be accomplished by performing any of the business value activities in this list (and others not listed here).

Improve Partner Engagement   Many organizations rely on partners and suppliers for components for their products, to deliver their products to consumers, to provide subject matter expertise in specialized fields, and so on.

Improve Customer Satisfaction   This may involve improving response times at the call center, decreasing processing time, addressing customer questions and complaints more thoroughly on the first call, and more.

Adhere to Regulatory or Compliance Standards   This may include implementing changes to accommodate new laws, cybersecurity standards, healthcare industry standards, privacy standards, and many more.

Determining business value is not a process within the process groups like the Develop Project Charter or the Identify Stakeholder process. Determining business value is something you will define at the beginning of the project, and you'll recall from Chapter 2 that both the benefits management plan and the project charter state the business value and benefits the project will bring about. The benefits management plan is more detailed than the project charter and includes specific benefits such as those listed earlier, along with how they will be tracked and measured; an acknowledgment when the benefits are realized; and a benefits owner who is responsible for watching over and reporting on their benefits. You'll want to survey and interview all necessary parties to reach consensus on business value. Keep in mind that business value, as well as project goals, should align with the organization's mission and overall goals. It is the project manager's responsibility to understand and articulate how the project goals and business value equate to the organization's overall mission.

Defining and assessing business value is not a onetime activity. You and the stakeholders need to continually assess business value throughout the project to make sure that you are realizing the benefits the project set out to produce. Doing so requires close collaboration with stakeholders and obtaining their buy-in on benefits definitions and measurements.

Business Value Network

Most businesses cannot perform all aspects of their business on their own. They rely on a network of suppliers and contractors to provide value, in the form of goods and/or services, to the organization. For example, a local restaurant relies on third-party suppliers to provide them with fresh produce for their daily menu offerings. Employees are also a part of this network and, arguably, the most important members.

The idea is that all the members of the *value network* work together to bring business value to their customer, their organization, the end-user customer, shareholders, and more. There may be multiple members of this value network such as employees, customers, stakeholder groups, suppliers, controllers, and watchdogs.

Continuing with the Amazon example, here is what I imagine some of their value network members might be. I'm sure you can probably think of many more.

- Employees
- Shareholders
- Shipping suppliers
- Delivery services

- Suppliers for their "Amazon Basics" line
- Internet of Things innovators
- Whole Foods
- Warehousing
- TV and movie production

Assessing Business Value

According to PMI®, delivering business value is one of the most important aspects of project management. Defining business value is the responsibility of the project sponsor and key stakeholders, but delivering on that value is your responsibility. The project sponsor must clearly articulate the business value so that it's easily understood by you and the project team members. For example, maybe your project involves replacing a bridge that is outdated and no longer safe. The business value is ensuring the safety of the traveling public. Therefore, the project was created to replace a failing bridge and make traveling from the points on either end safe and reliable. The next step is determining how to measure the benefits and value.

One critical aspect of delivering business value involves executing the project with the urgency required to deliver value. Delivering value based on urgency is your responsibility as project manager. In the previous example, replacing a failing bridge likely has a good deal of urgency surrounding it—especially if it's a main thoroughfare for commuters who rely on this bridge to get to work and back every day. Urgency can be subjective, so you'll need to work in collaboration with your project sponsor and key stakeholders to determine and understand the urgency of your project.

NOTE   Urgency is often subjective. Be certain you understand the urgency that the business value needs to bring about and strive to meet the goals of the project in the time frame needed to realize business value.

It's important to know how business value will be achieved. You can do this by creating business value measures, or *key performance indicators (KPIs)*, to measure value. KPIs are metrics used to determine whether the organization, or project in this case, is achieving its goals. You should work with your project sponsor to determine and document these measures in the benefits management plan. Monetary and financial KPIs are straightforward and probably the easiest way to measure value. If the project's business value is to increase revenues, it's easy enough to compare last year's revenues to this year's revenues to determine whether the launch of the new project produced the business value. It may not be as easy to measure a value like goodwill. Perhaps in this case you could consider the organization's social media presence. For example, as people hear your story, they will want to follow the organization, so they will subscribe to social channels and visit the organization's website. Thus, an increase in subscribers may indicate that goodwill has been realized. KPIs for a customer satisfaction goal could be derived from surveys distributed to customers after their interactions with the company.

 Business value measures may also be referred to as key performance indicators, or KPIs. KPIs are typically numeric measures that let your organization know whether they are achieving their goals. For example, a business value stated as "increase sales" may have a KPI metric that states the amount by which sales will increase and over what period of time.

One of the ways to help your sponsor articulate business value measures is by asking the question, "What outcome do you hope to realize once this project is implemented?" From there, continue to ask the "who, what, when, where, why" type questions to determine the measures you'll use to monitor whether business value has been achieved. Here is a list of more example business value measures (these may also apply to KPIs) to help get the discussion started. Your project may involve one or more of these measures. Keep in mind that there could be hundreds of ways to measure value depending on the nature of your project.

- Increase revenues
- Increase profits
- Increase sales
- Improve employee retention
- Improve the quality of your product
- Avoid penalties for compliance violations
- Improve vendor partnerships
- Reduce expenses
- Improve customer satisfaction
- Streamline supply chain management
- Increase search engine results
- Increase social media followers
- Decrease the number of call center interactions
- Streamline operations
- Decrease workers' compensation claims

There are innumerable ways to measure benefits. Make certain you work with the benefit owners to determine how their benefits will be measured and then document the benefits and their measures in the benefits management plan. Don't forget to include some of the value network members, like those described in the "Business Value Network" section of this chapter, when determining benefits and their measurements. You, as the project manager, will work together and collaborate with and across all of these stakeholders to define and promote business value and produce the project results. It's also important that you understand the business value measures and how they will be evaluated. If you don't, you will have a difficult time delivering a successful project. It's your job to champion the business values at all times to the stakeholders and project team, and you won't be able to do that if you don't understand the values and how they will be measured and produced.

Exam Spotlight

The project manager is responsible for delivering the business value by accomplishing the project.

Delivering Business Value Incrementally

It is the project manager's responsibility to report on the progress of the project. One of the ways to accomplish this is by providing updated status on the business value measurements and document the completion of deliverables. The benefits owners should provide you with updates for the status report so that you can include benefits realization and KPI measures in the report.

In some cases, the final business value measurement can't be made until the end of the project. This is typical in a project managed using a waterfall methodology. Although you may be able to measure some deliverables that are completed along the way, you won't know until the end of the project whether it's successful because deliverables may not be completed in logical order, and the dependencies among them sometimes can't be measured until they are all completed.

In an agile methodology approach, small increments of work are produced within defined time periods or iterations. As the project progresses, increments of work are completed and examined by the product owner. Changes can be made quickly, problems are discovered early on, and adaptations can be made in future iterations. At the completion of an iteration, you have a functioning deliverable (or a functioning portion of a deliverable) that is easily measured. Your product owner (who is also the benefits owner in this case) can see the business value (or the absence of it) at the end of the iteration and determine whether the benefit measures have been achieved.

Exam Spotlight

The project manager is responsible for assessing opportunities and project management methodologies that will deliver value incrementally.

At the beginning of the project, you will assess the methodology you'll use to accomplish the work of the project. Whenever possible, use a methodology such as an adaptive or a hybrid approach that will allow you to continuously deliver value over time, incrementally. Using this method, you can assess whether the project is on track, and if not, use corrective actions right then and there to get it back on track. It's much easier to determine project status if you can break the project into small increments of work that are easily measured.

 Real World Scenario

The New Website Project

You've just been charged with managing a project to create a new expense report submission process on the intranet site for your organization. At the beginning of the project, you don't know a lot of detail other than the high-level purpose of the project. The business value is to improve efficiency by automating expense reports. Today they are completed on paper and are prone to error. Most employees don't know or read the travel policy guidelines and their expense reports are filled out incorrectly. The accounts payable team sends back roughly 65 percent of expense reports due to errors, causing a significant amount of rework.

One of your next steps is to discover the elements that should be included on the website. For example, you might need to make the travel policy, travel request forms, expense reimbursement forms, and so on available electronically.

You're using an agile approach to perform this project and have documented all the user stories. You will progressively elaborate requirements for these user stories as they are pulled into each iteration. For example, the agile team will need to ensure expense reimbursement forms calculate per diem based on a 24-hour period and automatically sum expenses by category.

Because these deliverables are unique and can be tested independently, you decide to deliver this project in an incremental fashion and produce business value with each iteration.

The accompanying case study shows an example of how a project can deliver value incrementally. For example, once the travel policy is posted your stakeholder can verify that the policy is available in the right place and easily accessible. There isn't an exact measure here that can be applied, but you can measure whether it's been uploaded and is accessible. Thus, you've delivered an incremental business value because now all employees have access to the travel policy without needing to keep a paper copy handy. Yes, this business value is minor, but it is value and your goal as project manager is to deliver value. The idea here is to deliver business value in iterations or stages and to work in short time periods so that problems are easily discovered and addressed before more work progresses. This strategy also allows for changes along the way.

If your project requires a more predictive methodology, you can still plan incremental deliverables. Perhaps consider dividing the project into phases, where each phase delivers a portion of the business value the overall project was created to bring about. In either methodology, producing phase deliverables or incremental deliverables, the stakeholders are allowed to determine whether the results have value, and whether to continue working in the same manner or reexamine the project scope.

Examining Business Value

The measures you defined to monitor business value creation will be assessed, monitored, and reported on throughout the life of the project. These measures need to be realistic, measurable, and validated in a way to know they are being met. You should document and report on business value measures in the project status report, and this item should be a standing agenda item at project meetings. You will use the benefits management plan, along with the project scope, schedule, and cost baselines, and other project management documents, as the basis for measuring project performance and business value creation. You might also use tools and techniques such as cost and schedule forecasts, the status of deliverables, the status of change requests, costs incurred to date, variance analysis, and more. We'll examine each of these tools and techniques throughout the book, but for now, let's take a quick look at variance analysis. *Variance analysis* measures and compares the work that was planned and documented in the project management documents to the work that was completed. When the performance measures or business value are not as expected, you will document the variance and determine what actions to take to get the project back in alignment with the project plan. In an agile approach, you can take action during each iteration and/or make corrections in the next iteration. In a more traditional project management methodology such as waterfall, variance analysis may not happen until the end of a project or phase, or when a major deliverable is completed. This could take weeks or months, or more, from the start of the project and cause serious delays if changes or corrections are needed to get the project back on track.

As I mentioned, if a variance is discovered, you will need to take *corrective action* to get the project back on track. Corrective action can take many forms, such as adding more resources to the project, increasing the budget, or changing a workflow process. Ultimately, the corrective action should align the anticipated outcomes with the project management plan.

The project manager is responsible for examining the business value measures and comparing them to the project management documents to ensure the goals of the project are met and the business value is realized. This activity is continual and ongoing. Failure to measure and monitor the value you're delivering (or not) to your stakeholders can be a career-limiting move.

Examining business value also concerns managing change requests. I will discuss the change control process in depth in Chapter 11, "Measuring and Controlling Project Performance." In terms of examining the business value, you should know that unending changes can kill a project. You must manage changes according to the change management plan, and all change requests should be evaluated against the original scope, timeline, and budget for the project. Be certain to have a change management plan and process in place and require all stakeholders to follow the process. One technique that will help manage change requests is to review the project goals and business value it's expected to bring about at the beginning of each change meeting.

If you are embarking on a project that is not well defined and anticipate changes along the way, you should consider using an agile project management approach. This approach allows you to deliver incremental business value in small units of time, but it also allows for changes and modifications along the way that won't necessarily disrupt the overall goal of the project.

Subdividing Project Tasks

You will need to support your team members and help them to subdivide project tasks to find the minimal viable product that can be delivered within an iteration. This approach, as I've just hinted at, is typically applied using an agile project management approach. The *minimum viable product* involves breaking down tasks so that you can produce tangible components that have enough features and functionality to allow the customer to examine value and provide feedback to the team. These results should be produced quickly, with as little effort as possible. This might take the form of a prototype or mock-up. It's a test of sorts to see what the customer wants and whether the prototype produces positive results and realizes business value. It isn't a full version of the final end product, but it does represent the end product. It gives the organization, and the project team, data and information about what the customer likes, what works, and what doesn't. The minimum viable product is used to validate the business value and validate the idea for the project. The goal is to release a product or result quickly so that you can validate whether the product or result has value, determine whether the results meet the needs of your customers, and adapt the product based on feedback as you learn more about the customer and the product.

As we looked at earlier in this chapter, the product owner represents the stakeholders and is the liaison between the stakeholders and the project team. When using a Scrum methodology, you will produce a minimum viable product at the end of every iteration and the feedback received from the product owner will be used to influence future iterations. An example may look like this: The product owner will sort the items on their wish list of requirements (known as user stories) in priority order. The team members will break these down into quick activities that can be completed, tested, and shown to the customer in a short amount of time. The idea is that the product owner or customer can "see" the deliverable and realize the value it brings to the project in a short time frame. Adjustments can be made quickly if business value is not realized or problems are spotted.

You now have a solid foundation for where we're headed in the remainder of this book—that is, toward a customer-driven, pull-based project management approach.

Project Case Study: New Kitchen Heaven Retail Store

You've reviewed the project charter with Dirk and a few key stakeholders. Now you have a question or two for Dirk. You knock on his door, and he invites you in.

"How can I help you?" he asks.

"I know that we've documented the benefits management plan and project charter," you begin. "I'd like to spend some more time defining the benefits and the KPIs we'll use to measure them. This will take some involvement on your part, and I'll need to interview more key stakeholders. I want to make sure I get the right people involved in the meetings. Who are the key stakeholders you recommend I speak with?"

"I can think of a few people right off that you don't want to miss. There's Jake Peterson over in facilities. He's in charge of store furnishings, shelving, things like that—any supplies for the stores that aren't retail products. His team also manages the warehouse that we'll be shipping the online orders from. He supervised our last eight store openings and did a terrific job.

"You should also talk to Jill Overstreet, the vice president in charge of retail products. She can help with the initial store stocking, and once the store is open, her group will take over the ongoing operations. She also heads up order fulfillment for our website. All the district managers report to Jill."

"Good—anyone else you can think of?"

"Yes, Ricardo in information technology is a key stakeholder. We'll need to build the infrastructure for the retail store and support this new website endeavor. You should probably talk with Terri, the digital marketing manager on my team, as well."

"Thank you. I'll get with them right away."

After meeting with each of the stakeholders, you devise a stakeholder register with power and influence levels. You also ask if any key stakeholders are missing on the project at this point. The register looks like this:

Stakeholder register

Name	Contact	Needs/interests	Potential impact	Power/influence
Dirk, project sponsor	x2310, dirk@KH.com	New retail store in nine months; increase revenues	High	H/H

Name	Contact	Needs/interests	Potential impact	Power/influence
Jake, facilities	x6174 jake@KH.com	Warehouse build-out; update inventory system	Med	M/H
Jill, sales	x5438 jill@KH.com	Retail and online sales	High	H/H
Ricardo, IT	x6286 ricardo@KH.com	Provide infra-structure for the new store and website	High	M/M
Kate, CFO	x1615 kate@KH.com	Manage and provide funding for the project	High	H/H

In addition to creating the stakeholder register, you interview each of these stakeholders and update the benefits management plan to reflect the business value each of them expects to realize on this project, as well as the KPIs that will be used to measure value, and you've assigned each of them as benefits owners. For example, Jill in sales will measure business value in terms of increasing sales. Her goal is to increase overall sales for the company by 3 percent in the first six months after the new store opening and the website is live. She also wants website sales to account for only 15 percent of total sales at the end of one year after go-live.

You've decided to use a hybrid project management approach to deliver business value incrementally for this project. You want to use the rigor of the Planning processes to document the project objectives, deliverables, and work effort required for the new retail store, and you hope to accomplish the actual work by using an agile methodology. You will ask Dirk for a project manager to manage the website update project, and this project manager will report to you. They will use an agile approach in managing their project.

Project Case Study Checklist

- **Project objective:** To open a new retail store and website by January 25 next year.
- **Business need or demand for project:** Colorado Springs has new mixed-use communities planned to open within the next year and beyond. Customer requests have increased in recent months asking for more products online and cooking tutorials.
- **Project sponsor:** Dirk Perrier, VP of marketing.
- **Business value:** Increase sales, revenues, and market share (increase our customer base). Make shopping with Kitchen Heaven easier.

- **Incremental business value:** You will use a hybrid approach to manage this project. This will ensure incremental business value and the ability to make modifications quickly while receiving feedback for future iterations.

- **Understand the urgency of delivering business value:** You understand that this store and updated website must be open and available by January 25 in order to align with the Home and Garden show events across the United States.

- **Stakeholder needs analysis:** Stakeholders were interviewed and their needs, wants, and expectations recorded and a power/interest grid was created.

- **Stakeholder register created:** Includes the stakeholder names, contact information, assessment information, and stakeholder classification.

- **Organizational structure:** Functional organization with a separate project-oriented department.

Understanding How This Applies to Your Next Project

In your day-to-day work environment, it probably doesn't matter much if you're working in a functional or a strong matrix organization. More important are your communication, conflict management, and negotiation and influencing skills. Good communication is the hallmark of a successful project no matter what type of organizational structure you work in. You'll find leaders and you'll find people who have the title of leader in any structure you work in. Again, the organizational structure itself isn't as important as knowing who the real leaders and influencers are in the organization. The stakeholders with the most power and influence are the people you'll lean on to help with difficult project decisions and hurdles.

Identifying the stakeholders early in the project is imperative to project success. It's also important to understand the role stakeholders will play in the project as well as their power, influence, and interest level. If the CFO has ultimate influence over which projects go forward, you'll want the CFO's buy-in as soon as possible. Involvement in writing the charter, creating the benefits management plan, defining business value, and developing the future Planning process outputs helps ensure their buy-in. At a minimum, it helps reduce surprises midway through the project.

Most projects I've worked on involved more than one stakeholder, and stakeholders often have conflicting interests. On your next project, find out what those stakeholder interests are. Resolving conflicts is easier at the beginning of the project than at the end. You'll likely need both negotiating and influencing skills.

One of the most important concepts to understand from this chapter is that projects bring about business value. Business value can mean different things to different organizations. For example, I've spent the last 20 or so years of my career working in government organizations. Business value for us can mean cost savings, efficiency in conducting business with our constituents, or providing safe and reliable services for our citizens.

Delivering business value is one of the primary functions of the project manager. You can't do this alone—it takes a team of people to make this happen—but you're the one who will champion the business value and ensure the project stays on track to deliver the end product or result of the project.

Delivering business value incrementally is another goal for the project manager. This can be accomplished by breaking the project into phases when using a waterfall methodology to manage the project, or by determining the minimum viable product of each iteration when using an agile methodology. Agile is a methodology that makes it easier to break down tasks into incremental deliverables. The project manager works with the team to subdivide tasks and determine the minimum viable product for each iteration.

Summary

Organizational structures come in a variety of forms, such as functional, project oriented, and matrix. Functional organizations are traditional, with hierarchical reporting structures. Project managers have little to no authority in this kind of organization. Project-oriented organizations are structured around project work, and personnel report to project managers. Project managers have full authority in this organizational structure. Matrix organizations are a combination of the functional and project oriented. A project manager's authority varies depending on the structure of the matrix, be it a weak matrix, a balanced matrix, or a strong matrix.

A project manager's authority level, ability to manage the project budget, and availability of staff are also influenced by the project manager's interactions within the hierarchy of management (strategic, middle, and operations management) and the project management maturity level of the organization. These are all influenced by the organizational structure.

Stakeholders are those people or organizations that have a vested interest in the outcome of the project. Stakeholders include people such as the project sponsor, the customer, key management personnel, operations managers, the project manager, contractors, and suppliers. Projects are considered complete when the project meets or exceeds the expectations of the stakeholders. It's critical to identify, assess, and classify your stakeholders early in the project so that you are aware of the power and influence they may wield on the project.

The power/interest grid and Salience model are good data representation tools you can use to help determine and plot power, influence, and interests for the stakeholders. The Salience model includes three categories of stakeholders: power, legitimacy, and urgency. There are classifications of stakeholders within each of these categories: latent, expectant, and demanding. Each of these classifications contains subclassifications depending on the combination of power, legitimacy, and urgency the stakeholder possesses.

The titles of project stakeholders and participants will vary depending on the methodology you're using to manage the project. For example, the Scrum master helps the team manage and organize work. The product owner is the voice of the customer and serves as the liaison between the Scrum team and the stakeholders. The sensei is similar to a project manager on a Lean project.

Next, I described business value. Business value refers to those values that will lead to short- and long-term benefits for the organization. Business value can mean different things to different organizations. Business value in terms of project delivery may involve increasing revenues, improving efficiencies, producing goodwill, and more. Some examples of business value are increasing market share, increasing revenue, reducing costs, producing goodwill, improving brand recognition, increasing employee engagement, and improving partner engagement.

Business value is documented in the benefits management plan and the project charter. The benefits management plan includes the definition of benefits, the KPIs you'll use to measure achievement of the benefits, and the benefit owners who will monitor and report on benefits realization. The project charter is the document that officially recognizes and acknowledges that a project exists. The charter documents the business need and justification, documents the business value, describes the customer's requirements, and ties the project to the ongoing work of the organization.

It's important to know how business value will be achieved. You can do this by creating measures, or KPIs, to measure value. KPIs are metrics used to determine whether the organization, or project in this case, is achieving its goals. You should work with your project sponsor to determine and document these measures. Business measures may be tangible or intangible. It's best to quantify intangible benefits whenever possible with the use of surveys or other measures.

Exam Essentials

Be able to differentiate the organizational structures and the project manager's authority in each.   Organizations are usually structured in some combination of the following: functional, project-oriented, and matrix (including weak matrix, balanced matrix, and strong matrix). Project managers have the most authority in a project-oriented organization and the least amount of authority in a functional organization.

Understand the Identify Stakeholders process.   The purpose of this process is to identify the project stakeholders, assess their influence and level of involvement, and record stakeholder information in the stakeholder register.

Understand stakeholder analysis.   Stakeholder analysis is performed using qualitative and quantitative data. Some of the tools and techniques used to accomplish this are expert judgment, data gathering, data analysis, data representation, and meetings.

Understand the ways to categorize and display stakeholder analysis.   Categorizing and displaying stakeholder analysis can take several forms including: power/interest grid, power/influence grid, impact/influence grid, stakeholder cube, the Salience model, directions of influence, and prioritization.

Be able to name the categories, qualitative classes, and subclasses of the Salience model.   The categories are power, legitimacy, and urgency. The qualitative classes are latent, expectant, and demanding. The subclasses within latent are dormant, discretionary, and demanding. The subclasses within expectant are dominant, dependent, and dangerous. Demanding stakeholders have the attributes of all three categories: power, legitimacy, and urgency.

Be able to name the key components of the stakeholder register.   One component is the identifying information such as name, title, and contact information. Another component is assessment information, including needs and interests. Last is stakeholder classification, in terms of power and influence.

Be able to name the key roles on a Scrum team.   The key roles are the product owner (also known as the voice of the customer), the Scrum team (these are the team members who make up the team), and the Scrum master and/or team facilitator.

Be able to describe business value and examine this throughout the project.   Business value refers to those values that will lead to short- and long-term benefits for the organization. The project manager should champion the business value vision and examine and measure business value throughout the project.

Understand the opportunities to deliver value incrementally.   It is the responsibility of the project manager to deliver business value. The project manager should help the team define the opportunities to deliver value incrementally.

Support the team to subdivide tasks and find the minimum viable product.   The minimum viable product involves breaking down tasks into tangible components that have enough features and functionality to allow the customer to examine value and provide feedback to the team.

Review Questions

You can find the answers to the review questions in Appendix A. Be sure to download the Bonus Exams and Bonus Questions so that you'll have a broader exposure and more experience answering questions related to the topics in this chapter.

1. A project is considered successful when _____.

   A. the product of the project has been manufactured.

   B. the project sponsor announces the completion of the project.

   C. the product of the project is turned over to the operations area to handle the ongoing aspects of the project.

   D. the project achieves its objectives and meets the expectations of the stakeholders.

2. The VP of customer service has expressed concern over a project in which you're involved. His specific concern is that if the project is implemented as planned, he'll have to purchase additional equipment to staff his customer service center. The cost was not taken into consideration in the project budget. The project sponsor insists that the project must go forward as originally planned or the customer will suffer. Which of the following are true? (Choose two.)

   A. The VP of customer service is correct. Since the cost was not taken into account at the beginning of the project, the project should not go forward as planned. Project initiation should be revisited to examine the project plan and determine how changes can be made to accommodate customer service.

   B. The conflict should be resolved in favor of the customer.

   C. Stakeholder identification should occur as early in the project as possible.

   D. The conflict should be resolved in favor of the VP of customer service.

3. The amount of authority a project manager possesses can be related to which of the following? (Choose three.)

   A. The organizational structure

   B. The power and influence key stakeholders possess on the project

   C. The interaction with various levels of management

   D. The project management maturity level of the organization

   E. The project management methodology you'll use to manage the project

4. Which of the following are true regarding functional organizations? (Choose three.)

   A. All employees report to one manager and have a clear chain of command.

   B. Organizational structures and culture are independent from the project management methodology you'll use to manage the project.

   C. The organization is focused on projects and project work.

   D. Teams are co-located.

   E. Using adaptive and hybrid methodologies in this structure is possible but can be challenging.

5. You have completed the Develop Project Charter and Identify Stakeholder processes. What phase does this relate to in the Six Sigma DMAIC methodology?

  A. Analyze

  B. Control

  C. Measure

  D. Define

6. It is important for a project manager to identify stakeholders and understand their needs, interests, and influence. All of the following are true regarding stakeholders in this context except which one?

  A. Organizational structures and hierarchy, culture, geographic locations, global trends, and locations of resources and facilities can impact stakeholder power, influence, and interest on the project.

  B. Stakeholders often have conflicting interests and the project manager needs to understand the conflicts in order to resolve them and manage stakeholder expectations.

  C. Stakeholders may have the power and influence to cause issues on the project. You'll need to identify and assess them so that you know their needs and can address issues early in a way that keeps the stakeholders engaged.

  D. The Identify Stakeholder process should be repeated often throughout the project.

  E. All of the above.

7. You have performed variance analysis and discovered that you need to take action to get the project back on track. Which of the following are true? (Choose two.)

  A. Corrections and changes due to variance analysis findings can be made quickly using an agile project management methodology.

  B. Variance analysis compares the work performed to the project management baseline documents such as schedule, cost, and scope.

  C. The project sponsor is responsible for ensuring corrective actions are taken to get the project back on track no matter which project management methodology you are using.

  D. Change requests may bring about corrective actions.

8. The VP of customer service has expressed concern over a project you are managing. Her specific concern is that if the project is implemented as planned, she'll be missing an important feature enabling a call-back function where customers will not have to wait online but will receive a call back when they are next up in the queue. This functionality was not taken into consideration in the project plan, and this new wrinkle will delay the project by two months. The business value for this project is improving customer satisfaction. Customer surveys show that satisfaction levels are low when it comes to wait times and overall call resolution time. The CEO was expecting this project to go live early in the fiscal year so that customer service scores would show an improvement, thereby increasing satisfaction along with making the board happy. Which of the following is true regarding this scenario?

  A. This project was brought about due to customer requests.

  B. Customer satisfaction surveys could be used to determine if the business value was met.

   C. The project manager must champion the business value for this project.

   D. The project manager should consider delivering business value incrementally by using an agile approach.

   E. All of the above.

9. Match the following categories associated with analyzing stakeholders in the Salience model with their classification.

Category	Classification
A. Legitimacy + urgency	1. Dangerous
B. Power	2. Dependent
C. Power + urgency	3. Dominant
D. Urgency	4. Dormant
E. Legitimacy	5. Discretionary
F. Power + legitimacy	6. Demanding

10. Which term refers to those values that will lead to short- and long-term benefits for the organization?

   A. Business process improvement

   B. Business value

   C. Continuous improvement

   D. Value network

11. You have identified, obtained contact information for, assessed, classified, and prioritized your stakeholders. Which of the following is true? (Choose three.)

   A. You have documented this information in the stakeholder register.

   B. You should use caution when making some of this information available to all project stakeholders.

   C. You have used the power/interest grid to obtain this information.

   D. You've analyzed influence, power, interest, impact, directions of influence, and power, urgency, and legitimacy as part of the tools and techniques of the Identify Stakeholder process.

12. Match the following roles in the agile methodologies with their titles.

Role	Description
A. Product owner	1. The title similar to a project manager in the Lean methodology
B. Sensei	2. The title similar to a project manager in the Kanban methodology

Role	Description
C. Flow master	3. A Six Sigma title
D. Team facilitator	4. Also known as the Scrum master. Coordinates the work of the sprint
E. Black belt	5. The liaison between the Scrum master and the stakeholders in a Scrum methodology. Also known as the voice of the customer

13. All of the following are examples of business value except which one?

   A. Create a new car model that's never been offered before.

   B. Implement a new accounting system.

   C. Create goodwill.

   D. Improve employee engagement.

   E. A and B

14. The VP of Human Resources has requested a new timekeeping project for their machine shop workers. The business value for this project is improving payroll accuracy. The workers are threatening a strike because of the antiquated system in place now. The current system has ancient time clocks that rarely work, causing the workers to revert to paper timecards. This creates multiple errors in overtime pay and shift differential pay. Which of the following is true regarding this scenario? (Choose three.)

   A. The project sponsor is the one responsible for delivering the business value of this project.

   B. Urgency is objective and is defined by the project sponsor.

   C. The urgency to deliver this project is the potential of an impending strike.

   D. The goals of the project must be met within the time frame outlined in the project plan in order to realize business value.

   E. The project manager should examine business value creation and compare this to the project management documents, such as the project charter, to ensure that the goals of the project are met and the business value is realized.

15. You are using an agile approach to deliver business value incrementally. What other elements of business value should you be focused on? (Choose two)

   A. Examining business value throughout the project

   B. Reporting on business value only at the end of the project

   C. Accepting suggestions about delivering incremental business value and ensuring that business value is achieved when the project is complete

   D. Discussing business value with the project sponsor, recording it in the project charter, and expecting project team members and stakeholders to refer to this periodically

16. How will you know that business value is being achieved?

    A. By observing the work product

    B. By asking the project sponsor

    C. By measuring business value using KPIs

    D. By requiring the project team to report on business value

17. Your project requires contract resources with specific subject matter expertise to deliver business value. All of the following are true except which one?

    A. The contractor is a member of the business value network.

    B. The business value network is made up of external resources such as contractors, subject matter experts, suppliers, and delivery services.

    C. Members of the value network work together to bring about business value to their customer, their organization, the end-user customer, and stakeholders.

    D. It is the project manager's responsibility to ensure that the contractor understands the business value the project was created to bring about.

18. You are using an agile project management methodology to deliver business value. During each iteration, you and the team members are breaking down tasks into tangible components that have enough features and functionality to allow the customer to examine value and provide feedback to the team. Which of the following are true regarding this question when using a Scrum methodology? (Choose two.)

    A. The product owner will determine whether business value has been achieved.

    B. The project team will manage and prioritize the product backlog and choose the user stories for the upcoming iteration that can be broken down to the minimum viable product.

    C. The Scrum master will assist the team in breaking down the user stories and assign each team member tasks for the upcoming iteration.

    D. You are creating the minimum viable product.

19. Which of the following is true regarding assisting the team in subdividing tasks into the minimum viable product?

    A. Each iteration will produce enough features to examine business value.

    B. Each iteration will provide an opportunity for feedback on future iterations.

    C. The minimum viable product allows the customer to see or experience the business value that was created.

    D. All the options are correct.

20. All of the following are true regarding supporting the team in subdividing tasks into the minimum viable product except for which one?

    A. Business value is delivered incrementally.

    B. Adjustments or corrections can only be made at the end of the iteration.

    C. This process is used in an agile project management methodology, allowing small, tangible results to be delivered at the end of each iteration.

    D. Tasks can be completed quickly, tested, and shown to the customer in a short period of time.

Chapter 4

Developing the Project Scope

THE PMP® EXAM CONTENT FROM THE PEOPLE DOMAIN COVERED IN THIS CHAPTER INCLUDES THE FOLLOWING:

✓ **Task 1.2 Lead a team**

- 1.2.1 Set a clear vision and mission

✓ **Task 1.4 Empower team members and stakeholders**

- 1.4.2 Support team task accountability
- 1.4.4 Determine and bestow level(s) of decision-making authority

✓ **Task 1.6 Build a team**

- 1.6.1 Appraise stakeholder skills
- 1.6.2 Deduce project resource requirements

✓ **Task 1.9 Collaborate with stakeholders**

- 1.9.1 Evaluate engagement needs for stakeholders
- 1.9.2 Optimize alignment between stakeholder needs, expectations, and project objectives
- 1.9.3 Build trust and influence stakeholders to accomplish project objectives

✓ **Task 1.10 Build shared understanding**

- 1.10.2 Survey all necessary parties to reach consensus

THE PMP® EXAM CONTENT FROM THE PROCESS DOMAIN COVERED IN THIS CHAPTER INCLUDES THE FOLLOWING:

✓ **Task 2.1 Execute project with urgency required to deliver business value**

- 2.1.1 Assess opportunities to deliver value incrementally
- 2.1.2 Examine the business value throughout the project
- 2.1.3 Support the team to subdivide project tasks as necessary to find the minimum viable product

✓ **Task 2.8 Plan and manage scope**

- 2.8.1 Determine and prioritize requirements
- 2.8.2 Break down scope (e.g., WBS, backlog)
- 2.8.3 Monitor and validate scope

✓ **Task 2.13 Determine appropriate project methodology/methods and practices**

- 2.13.1 Assess project needs, complexity, and magnitude
- 2.13.2 Recommend project execution strategy (e.g., contracting, finance)
- 2.13.4 Recommend a project methodology/approach

THE PMP® EXAM CONTENT FROM THE BUSINESS ENVIRONMENT DOMAIN COVERED IN THIS CHAPTER INCLUDES THE FOLLOWING:

✓ **Task 3.1 Plan and manage project compliance**

- 3.1.1 Confirm project compliance requirements (e.g., security, health and safety, regulatory compliance)
- 3.1.2 Classify compliance categories
- 3.1.4 Use methods to support compliance
- 3.1.5 Analyze the consequences of noncompliance

✓ **Task 3.2 Evaluate and deliver project benefits and value**

- 3.2.1 Investigate that benefits are identified
- 3.2.2 Document agreement on ownership for ongoing benefit realization
- 3.2.3 Verify measurement system is in place to track benefits
- 3.2.4 Evaluate delivery options to demonstrate value

✓ **3.4 Support organizational change**

- 3.4.1 Assess organizational culture
- 3.4.3 Evaluate impact of the project to the organization and determine required actions

Great job! You've successfully completed the project's Initiating processes and published the project charter and the stakeholder register. The project is officially under way. Stakeholders have been identified and informed of the project, you have management buy-in on the project, the project manager has been assigned, and the project objectives, description, and business value have been identified. A solid foundation for the planning process is in place.

In this chapter, we will begin the Planning processes for the project. In fact, I will continue discussing the Planning processes through Chapter 8, "Planning and Procuring Resources." Planning is a significant activity in any project and, if done correctly, will go a long way toward ensuring project success. Keep in mind that planning is an important part of any project no matter what development methodology you're using. You might make the mistake of thinking that planning isn't necessary for agile projects, but that is not a wise assumption. I will discuss how most of the Planning processes work in an agile process as we progress through the next several chapters.

This chapter begins with the Develop Project Management Plan process. This process will describe the overall approach you'll use to manage the project. The result of this process is the project management plan document that describes how you'll execute, monitor, and control the project outcomes as the project progresses and how you'll close out the project once it concludes.

Next we'll cover the Plan Scope Management process. Here you'll learn how to document a plan that outlines how to define, validate, and control the project scope.

The Collect Requirements process is next. During this process, quantified requirements are gathered and documented to ensure that stakeholder needs are met and expectations are managed. This requires close collaboration with the stakeholders and their buy-in.

During the Define Scope process, you'll use the project charter and the requirements documentation—plus some other inputs—and then apply the tools and techniques of this process to come up with the project scope statement. I'll talk in depth about project objectives, requirements, and other elements of writing the project scope statement, which is an output of this process.

Once you have the deliverables and requirements well defined, you'll begin the process of breaking down the work of the project via a work breakdown structure (WBS). You'll accomplish this task in the Create WBS process. The WBS defines the scope of the project and breaks the work down into components that can be scheduled and estimated as well as easily monitored and controlled.

We have a lot to cover in this chapter, so let's get started.

The process names, inputs, tools and techniques, outputs, and descriptions of the project management process groups and related materials and figures in this chapter are based on content from *A Guide to the Project Management Body of Knowledge (PMBOK® Guide), Sixth Edition* (PMI®, 2017). The references to adaptive and hybrid methodologies, related materials, and figures in this chapter are based on content from the *Agile Practice Guide* (PMI®, 2017).

Developing the Project Management Plan

The first process in the Planning process group is the *Develop Project Management Plan* process. It's first for good reasons. This process is part of the Project Integration Management Knowledge Area and is concerned with defining, preparing, coordinating, and then integrating all the various subsidiary project plans into an overall project management plan. Remember that you will tailor this process (and all processes) according to the size and complexity of your project. When tailoring, you should take into account the project life cycle (predictive or adaptive), development life cycle (predictive, adaptive, hybrid), management approaches, knowledge management, change, governance, lessons learned, and benefits.

If you haven't already done so, be sure to conduct a project kickoff meeting with the stakeholders and project team members. At this meeting, you'll review the project charter, explain the business value the project is intended to bring about, and identify key stakeholders and their roles on the project. The kickoff meeting does the following:

- Communicates the start of the project
- Informs stakeholders of the project objectives
- Informs stakeholders of the business value the project will achieve
- Ensures key milestones are identified and a common understanding is established among stakeholders
- Ensures key deliverables are identified and a common understanding is established among stakeholders
- Establishes stakeholder engagement and commitment

Before beginning the processes we'll discuss in this chapter, you need stakeholder buy-in and commitment to the project to successfully document requirements, key milestones, and deliverables. Holding a kickoff meeting will help ensure this happens.

Exam Spotlight

The *PMBOK® Guide* notes that kickoff meetings may occur at two different points in the project, depending on its size. A small project generally involves a small project team that is actively involved in the Planning processes, so it makes sense to hold the kickoff meeting now. For large projects, a kickoff meeting is typically held in the Executing process group because the project manager and a few key team members will perform most of the planning before assigning other resources.

This process involves defining and documenting the processes you're going to use to manage this project. For example, let's say you and the project team have determined you will use project management processes involving costs, human resources, risks, and a project schedule. (Warning: This is a demonstration only—don't try this at home. In reality, professionals perform many more processes than this on a typical project.) Each particular process might have a management plan that describes it. For instance, a cost management plan (an example of a subsidiary plan) would describe how costs will be managed and controlled and how changes to costs will be approved and managed throughout the project. The Develop Project Management Plan process brings all these subsidiary plans together, along with the outputs of the Planning group processes, into one document called the *project management plan.*

 In Chapter 2, "Assessing Project Needs," I talked about *tailoring*—determining which processes within each process group are appropriate for the project on which you're working. Tailoring is also used in the Develop Project Management Plan process because it's here you'll determine what processes to use to best manage the project.

To create and document the plan, you need to gather some inputs and build on the information you've already collected. The inputs you'll want to focus on include the project charter, outputs from other processes, EEFs, and OPAs.

I've covered the project charter previously. Remember that it describes the objectives of the project and includes a list of high-level requirements needed to satisfy stakeholder expectations. It's an important input because you will use the information contained in it to help determine exactly which project management processes to use on the project. Let's take a closer look at the importance of the other inputs to this process.

 I will not name or describe every input, tool and technique, or output of every process. I will focus on the key elements needed for each process. Be sure to check out Appendix B for all the inputs, tools and techniques, and outputs for every process.

Outputs from Other Processes   The project management processes include all the individual processes that make up the process groups we're talking about throughout this book. The Initiating group, for example, has two processes, Planning has a zillion (okay, not that many, but it seems like it), and so on. The outputs from other processes you use on the project become inputs to the Develop Project Management Plan process. For example, the cost management plan we talked about in the introduction is an input. Any processes you use that produce a baseline (such as schedule or cost) or a subsidiary management plan (such as a risk management plan, communications management plan, and so on) are included as inputs to this process.

I'll talk more about baselines and the makeup of individual subsidiary management plans as we discuss the various Planning processes, starting with this chapter and continuing through Chapter 7, "Identifying Project Risks."

Enterprise Environmental Factors   You've seen enterprise environmental factors before. Some of the key elements of the environmental factors you should consider when choosing the processes to perform for this project are standards and regulations (both industry and governmental), legal requirements, project management expertise in the industry or focus area of the project, company culture and organizational structure, governance framework used within the organization, and infrastructure.

Pay particular attention to compliance and regulatory issues when creating all the subsidiary plans that will make up the project management plan. Many industries by the nature of their business are required to adhere to certain regulations and standards—for example, the healthcare industry, pharmaceutical companies, the banking industry, and the construction industry. These standards ensure many things, including safety, environmental protection, health protections, security, products that meet quality standards, food labels that are accurate, products that are reliable and work as advertised, buildings that are safe, and much more.

It's the project manager's responsibility to ensure that any compliance and regulatory requirements are documented in the requirements document and/or the project scope statement. You, as the project manager, need to ask your stakeholders about compliance issues and ensure that they are included as part of the project management plan.

We've discussed organizational structure in Chapter 3, "Delivering Business Value." Remember that organizational structure (such as functional, matrix, projectized) may impact culture and could become a project constraint. But company culture also has its own nuances. You can think of company culture as the character or personality of the organization. Cultures are unique to each organization just as personalities are unique to people. Culture will manifest in the actions and behaviors of the employees of the organization. For example, some organizations follow strict chain-of-command–style communications, which means you wouldn't dream of jumping over your boss's head and communicating to the next level up. Some organizations might be team focused and value a collective contribution versus organizations that value independent contributions. If a team member breaks protocol and attempts to promote their own self-interest rather than the interest of the team in a team-oriented culture, they may find that their teammates avoid them in the lunchroom and "forget" to include them in important meetings. One organization may value leading-edge innovation and risk taking, whereas another may value using their tried-and-true practices and taking only the risks necessary to complete the project. There are as many cultures as there are organizations. Large organizations may have several cultures within the organization that are based on a line of business (such as the operations unit or the accounting unit), geographic or regional locations, product lines, and more. It's important to understand the culture because it can impact the way stakeholders participate on the project and their level of influence on the project. It will also have an impact on team members. For example, if you value being an independent contributor more than being a team contributor, working in an organization that values team collaboration may not feel right to you and you may not produce your best work..

Organizational governance framework (part of the EEFs for this process) include rules, policies and procedures, systems, relationships, and norms. These help to frame the culture and behavior of the organization, which will in turn influence the project, the risk tolerance of the organization, and how people perform. It's important to note that *management elements* are influenced according to the organizational structure (i.e., functional, matrix, or project-oriented) and the organizational governance framework. Example management elements include how work is distributed, authority levels of workers, disciplinary actions, chain of command, fair treatment of employees, fair payment for work performed, communication channels, safety of staff members, morale, and more.

Organizational Process Assets   Some of the critical elements of the organizational process assets input you should consider when choosing the processes to perform for this project are policies, procedures, and standards used in the organization; project management plan template; change control procedures; monitoring and reporting criteria and methods; historical information; and project information from previous projects that are similar to the current project.

As I talked about in Chapter 2, the processes you choose to perform for the project will be based on the complexity of the project, the project scope, and the type of industry in which you work. Your organization's standards, guidelines, and policies or the project management office (PMO) might also dictate the types of processes you'll use for the project. You should also consider whether your organization has existing change control processes in place, templates that you're required to use, or financial controls and processes. Historical information and past project files are useful in helping you decide which processes to use for this project.

Project Complexity

I've mentioned complexity a few times and you'll hear it again throughout the book. Defining a complex project is a bit subjective, but the *PMBOK® Guide* states that complexity usually means that the project has multiple parts, that there are a number of dynamic interactions or connections between the parts, and that there is a synergy of behaviors that come about as a result of these interactions. Complexity comes about because of the organization's system behavior, human behavior, and ambiguity. System behavior refers to the segments and systems within the organization and their interdependencies. How the organization relates to these interdependencies can be simple or complicated, which in turn impacts the complexity of the project. Human behavior refers to the interactions among diverse team members, departments, and individuals.

Ambiguity exists in all organizations, some more pronounced than others. And what does ambiguity mean? Well, I'm not really sure … but I think maybe it has to do with … I'm kidding with you here to show ambiguity in action. Ambiguity means confusion or uncertainty, or not understanding. I've often worked in organizations where the procurement process or the hiring process is somewhat ambiguous. On the surface they appear straightforward, but once you begin the process, the rules seem to change at every turn. This impacts the project because there isn't a clear-cut process for procuring resources and that could cause risk and/or delays.

Data Gathering and Interpersonal Skills

There are four tools and techniques of the Develop Project Management Plan process that you'll use to produce the project management plan. They are expert judgment, data gathering, interpersonal and team skills, and meetings. According to the *PMBOK® Guide*, you'll need to collaborate with stakeholders and team members who possess the following types of expert judgment:

- Tailoring techniques
- Understanding technical and management details that need to be included in the project management plan
- Determining which tools and techniques to use
- Determining resources and assessing skill levels needed for project work
- Determining and defining the amount of configuration management to apply on the project
- Identifying which project documents require formal change control processes
- Prioritizing project work

Some data gathering techniques you might use in the process are brainstorming, checklists, focus groups, and interviews. We looked at both brainstorming and interview techniques in the previous chapter. They come in handy here too in collaborating with stakeholders to determine the technical and management details, tools and techniques, processes, and more that should be considered in defining the project management plan.

Interpersonal and team skills are the same as those used in the Develop Project Charter process: conflict management, facilitation, and meeting management. Meetings in this process are used to help determine the project approach (tailoring the processes to the project), how to execute the work of the project so that its objectives are met, and how the project will be monitored and controlled. A kickoff meeting is an example of a meeting that might occur during this process. According to the *PMBOK® Guide,* a kickoff meeting might be held for small projects just after the creation of the project charter is complete and stakeholders have been identified (which signals the end of Initiating processes and the beginning of the Planning processes). That means for small projects, the team would be involved with the Develop Project Management Plan process because it's the first process in the Planning process group. Kickoff meetings for large projects usually occur as soon as the Executing process group begins (and Planning is complete). Planning processes on large projects are completed mostly by the project management team. Team members are brought onto the project when work is ready to begin, kicking off the Executing processes, and a large kickoff meeting is held to explain the objectives of the project, to gain buy-in from the project team members, and to explain the roles and responsibilities of all stakeholders and team members.

Exam Spotlight

For the exam, remember that small project kickoff meetings occur when Planning begins. Large project kickoff meetings occur when Executing begins.

Documenting the Project Management Plan

The purpose of most processes is, of course, to produce an output. An output is usually a report or document or some type of deliverable. These often become inputs to other processes. In this case, you end up with a document—the project management plan—that describes, integrates, and coordinates baselines and subsidiary plans for the processes you've determined to use for the project. The project management plan can be detailed, or it can be a high-level summary based on the needs of the project.

According to the *PMBOK® Guide,* the project management plan defines how the project is executed, how it's monitored, how it's controlled, and how it's closed. It is progressively elaborated over the life of the project. It also documents the outputs of the Planning group processes, which I'll cover over the next several chapters. The subsidiary management plans that are associated with the processes you'll be using for this project should be documented in the project management plan. Each of these subsidiary management plans might contain the same elements that the overall project management plan does, but they're specifically related to the topic at hand. For example, the cost management plan should define how changes to cost estimates will be reflected in the project budget and how changes or variances with a significant impact should be communicated to the project sponsor and stakeholders. The schedule management plan describes how changes to the schedule will be managed, monitored, and controlled.

The subsidiary management plans might be detailed or simply a synopsis, depending on the needs of the project. I've listed the subsidiary plans along with a brief description next. I will cover each of these plans in more detail throughout the remainder of this book. According to the *PMBOK® Guide*, the subsidiary management plans that make up the project management plan are as follows:

Scope Management Plan   Describes the process for defining, maintaining, and managing project scope; facilitates creating the work breakdown structure (WBS); describes how the product or service of the project is validated and accepted; and documents how changes to the scope will be handled. The project scope management plan defines, maintains, and manages the scope of the project.

Requirements Management Plan   Describes how requirements will be analyzed, documented, traced, reported, and managed throughout the project.

Schedule Management Plan   Describes how the project schedule will be developed and controlled and how changes will be incorporated into the project schedule.

Cost Management Plan   Describes how costs will be managed and controlled and how changes to costs will be approved and managed.

Quality Management Plan   Describes how the organization's quality policy will be implemented. It should address and describe quality control procedures and measures, quality assurance procedures and measures, and continuous process improvement.

Communications Management Plan   Describes the communication needs of the stakeholders, including timing, frequency, and methods of communication.

Risk Management Plan   Describes how risks will be managed and controlled during the project. This should include risk management methodology; roles and responsibilities; definitions of probability and impact; when risk management will be performed; and the categories of risk, risk tolerances, and reporting and tracking formats.

Procurement Management Plan   Describes how the procurement processes will be managed throughout the project. This might include elements such as type of contract, procurement documents, and lead times for purchases.

Resource Management Plan   Documents the roles and responsibilities for project team members, their reporting relationships, and how resources (materials, equipment, and human resources) will be managed.

Stakeholder Engagement Plan   Documents the strategies to use to encourage stakeholder participation, the analysis of their needs and interests and impacts, and the process regarding project decision-making.

The project management plan is not limited to the subsidiary plans listed here. For example, even though the *PMBOK® Guide* does not mention the change management plan in the list of subsidiary plans, you should develop and include a change management plan with your project management plan. The change management plan describes how you will document, address, track, and control changes. You might include other plans and documentation that help describe how the project will be executed or monitored and controlled. Perhaps you're working on a project that requires precise calculations and exact adherence to regulatory requirements. You could include a plan that describes these calculations, how they'll be monitored and measured, and the processes you'll use to make changes or corrections.

The project management plan also includes project baselines, such as these:

- Schedule baseline
- Cost baseline
- Scope baseline

Together, the schedule, cost, and scope baseline make up what's called the *performance measurement baseline*.

Exam Spotlight

Understand that the purpose of the project management plan is to define how the project is executed, monitored and controlled, and closed as well as to document the processes you'll use during the project. It is made up of the subsidiary management plans and project baselines.

The project management plan consists of all the subsidiary plans and all of the project baselines and other information. More specifically, the project management plan should include or discuss the following components:

- Processes you'll use to perform each phase of the project and their level of implementation, the tools and techniques you'll use from these processes, and the interactions and dependencies among the processes
- The life-cycle methodology you'll use for the project and for each phase of the project if applicable
- The development approach you'll use, such as waterfall, agile, or hybrid
- When management reviews should be established (for example at certain times during the project or at completion of milestones), so that the stakeholders can review project progress and ensure it's meeting the objectives and/or take action if it is not on track as planned

- Change management plan describing methods for monitoring and controlling change
- Configuration management plan describing elements of the product, service, or result of the project
- Methods for determining and maintaining the validity of performance baselines, especially scope, schedule, and cost

Once the project management plan is complete and approved, it should be baselined. Most importantly, the scope, schedule, and cost management plans must be baselined so that changes can be closely managed and so that you have a point of reference against which to determine and measure project progress for the remainder of the project. As the project progresses and more and more processes are performed, the subsidiary plans and the project management plan itself might change. After the project management plan is baselined, you will manage these changes using the Perform Integrated Change Control process. All changes should be reviewed following the processes outlined in the change control plan, and the project management plan should be updated to reflect the approved changes.

Exam Spotlight

In practice, you'll find that you'll prepare the project management plan after you've progressed through several of the other Planning processes. It's difficult to finalize some of these subsidiary plans without thinking through or sometimes performing the processes they're associated with first. However, for the exam, remember that Develop Project Management Plan is the first process in the Planning group, and it should be performed first. Updates to the project management plan can and should occur as subsidiary plans are created or changed. However, once the project management plan is baselined, it can be changed only by using the Perform Integrated Change Control process.

Documenting the Project Management Plan Using a Predictive Methodology

The project management plan is a document you'll use throughout the remainder of the project. The project management plan is particularly helpful when using a waterfall or hybrid methodology. It's your road map for the project, and every future decision, work effort, change request, and corrective action will be weighed against this plan. What's important to keep in mind is that collectively, the project management plan (made up of all the subsidiary plans I described earlier in this section) will help you fulfill several objectives outlined in the People domain of the *PMP® Examination Content Outline*, published in June 2019 by PMI®. Creating the plan is a collaborative effort between the project manager, project team, and key stakeholders. Collaborating and then documenting the project management plan help to fulfill these objectives:

- Manage customer expectations and direct the work of the project to achieve the project goals by documenting the goals and ensuring that the project management plan is available to all project participants.

- Optimize alignment between stakeholder needs, expectations, and project objectives by making certain key stakeholders contribute to the plan and that their needs and expectations are documented, and by encouraging them to sign off on the plan, thereby gaining their commitment to the project's success.

- Survey all necessary stakeholders to reach consensus on the goals of the project, the subsidiary plans, the project management plan, and the contents of the plan by including them in the development of the project charter, project scope statement, and other project management plans.

- Conduct benefit analysis with relevant stakeholders to validate project alignment with organizational strategy and with expected business value by including them in the development of the project charter, project scope statement, and other project management plans.

When using an adaptive methodology to manage the project, you will define the work as the project progresses. You won't have a detailed roadmap to rely on as you have in a predictive approach. I'll discuss the agile methodology in more depth when we look at the project scope statement and when we begin collecting requirements for your project later in this chapter.

Plan Scope Management

Plan Scope Management is the first process in the Project Scope Management Knowledge Area. Remember that the purpose of the Project Scope Management Knowledge Area is to describe and control what is and what is *not* considered work of the project. *Scope* is collectively the product, service, or result of the project and the deliverables the project intends to produce. *Deliverables* are measurable outcomes, measurable results, or specific items that must be produced or performed to consider the project or project phase completed.

The primary purpose of this process is twofold: to write the scope management plan and to develop the requirements management plan. The scope management plan is a planning tool that documents how the project team will go about defining project scope, how changes to scope will be maintained and controlled, and how project scope will be validated. The requirements management plan addresses analyzing, documenting, and managing the project requirements. The scope management plan and requirements management plan help set a clear vision and mission for the project and are used to plan and manage the scope of the project. I'll cover the requirements management plan in detail in the section "Documenting the Scope Management Plan" later in this chapter.

Exam Spotlight

According to the *PMBOK® Guide*, the activities and processes in the Planning process group involve defining, assessing, and reviewing detailed project requirements based on the project charter to establish the project deliverables. You will perform these processes with your stakeholders using requirements-gathering techniques. Reviewing lessons learned from previous projects of similar size and scope and reviewing the constraints and assumptions of this project are keys to documenting a detailed list of project deliverables.

For now, you'll concentrate on some of the key inputs to this process, which will help you get started documenting the decisions about the Plan Scope Management process for your project. You've seen the inputs to the Plan Scope Management process before. The key inputs are the project charter, project life cycle description, quality management plan, development approach, EEFs, and OPAs.

Even though I've talked about these before, you'll want to look at specific elements of some of these inputs closely. Defining project scope, and managing that scope as you progress through the project, has a direct relationship to the success of the project. It's difficult to document how you'll define project scope if you don't first understand the purpose behind the project; the product, service, or result you're trying to produce; the business value the project will generate; and the environmental factors and organizational process assets under which you're working. The process of how you'll go about defining and managing scope is what the scope management plan is about.

The project charter describes the purpose and high-level requirements of the project. Analyzing the information in this document will help you determine the appropriate tools and methodologies to use (as well as which processes to perform) for the project. The idea here is that you want to understand the size and complexity of the project so that you don't spend more time than necessary documenting how you're going to go about defining and managing scope.

The project management plan is helpful in developing scope, and special consideration should be given to the quality management plan. This plan documents how the quality policy, methodologies, and standards for the organization, and thus your project, should be implemented. The project life cycle description describes the phases of the project, so this will help clarify the scope boundaries for your project as well. Also note that the development approach, be it predictive (or waterfall), adaptive (agile), or hybrid, will also dictate how scope will be defined. In a waterfall approach, for example, the entire scope statement and all elements of scope will be developed before the work begins; in an agile approach, scope is developed iteratively throughout the project.

One of the environmental factors that might influence the way scope is managed is personnel administration, or more specifically, the people resources involved on the project. Their skills, knowledge, and ability to communicate and escalate issues appropriately might influence the way project scope is managed.

Other environmental factors that might affect scope management are the organization's culture, infrastructure, and market conditions. We've talked about these previously.

Organizational process assets typically include policies and procedures, whether formal or informal. Your organization's policies or the policies and guidelines of your industry might have an effect on scope management, so make certain you're familiar with them. For example, personnel policies might dictate how team members should interact or relate with one another. Let's say you have a team member who has a close relationship with one of the stakeholders. And imagine that the stakeholder wants a change to the project. The stakeholder and team member grab a cup of coffee together at the corner deli, and the next thing you know, the team member is incorporating the change into the project. Scope can't be managed efficiently when it's being changed without the knowledge of the project manager or project team. Personnel policies may help you curb situations like this.

And don't forget historical information and lessons learned. You can review the scope management plan from previous projects of similar size and scope to help you craft the one for this project.

Real World Scenario

Scope Management Plan Requirements

Phil Reid is a gifted engineer. He works as an accident reconstructionist and has a 90 percent success rate at assisting his clients (who are attorneys) in winning court cases. Phil can intuitively and scientifically determine whether the scene of an accident is real or is insurance fraud. Most cases Phil works on are managed as projects because each accident is unique, the cause of each accident is unique, and each investigation has a definite beginning and ending. The attorneys that Phil's organization works with want the final results of the investigation delivered in different formats. Although Phil is exceptionally good at determining the forensic evidence needed to prove or disprove how the accident occurred, he is not at all gifted in oral communication skills. As a result, the scope management plan requires the client to define how the outcome of the investigation should be presented and whether the engineer might be required to testify regarding the results of the investigation. That way, Phil's organization can plan in advance how to use his talents on the project and assign a resource to work with him who has the communication skills and ability to testify if that's required.

Alternatives Analysis

Plan Scope Management has three tools and techniques: expert judgment, data analysis in the form of alternatives analysis, and meetings. The experts you should rely on here include key stakeholders, industry experts, team members with specialized training, and other project managers with previous experience on projects similar to yours. Meetings are a tool and technique in many of the planning processes and are the primary technique you'll use to collaborate with stakeholders, gather their input, and gain consensus and buy-in.

Alternatives analysis is a technique used for discovering different methods or ways of accomplishing the work of the project, or for identifying requirements or project scope. This tool and technique is useful in defining scope and helps the team think outside the box. I like to use a simple, one-word question to help drive alternative analysis: "Why?" From there, you can ask follow-on questions that will lead you to alternative approaches, outcomes, or results. Continue until you have viable alternatives to consider. This strategy is also helpful when determining requirements, defining project scope, and validating and controlling scope later in the project.

Brainstorming might be used to help you discover alternative ways of achieving one of the project objectives. Perhaps the project's budget doesn't allow for a portion of the project that the stakeholders think needs to be included. Brainstorming might uncover an alternative that would meet the stakeholders' need in a different way while still accomplishing the goal.

Pairwise comparisons is another alternatives generation technique that is much like it sounds in that you compare any number of items against one another to determine which option is preferred. This technique typically uses quantitative measures to determine the preferred option.

Lateral thinking is a form of alternatives generation that can be used to help define scope, requirements, or objectives. Edward de Bono created this term and has done extensive research and writing on the topic of lateral thinking. The simplest definition is that it's thinking outside the box. Lateral thinking is a process of breaking apart the problem (or in our case the deliverables and requirements), looking at them from angles other than their obvious presentation, and encouraging team members to come up with ways to solve problems or look at requirements that are not obvious.

Outside the Box

Lateral thinking is a way of reasoning and thinking about problems from perspectives other than the obvious. It challenges our perceptions and assumptions. Consider these two examples of lateral thinking that I crafted based on some puzzles I found at this website: www.folj.com/lateral. Use your favorite search engine and run a query on **lateral thinking puzzles** to find many more examples.

Question:

How could your pet poodle fall from the window of an 18-story building and live?

Answer:

The question asks how your pet could fall from an 18-story building and live; however, the question doesn't state that your pet fell from the 18th floor. Your pet poodle fell from the basement-level window of an 18-story building.

Question:

> Eight chocolates are arranged in an antique candy dish. Eight people each take one chocolate. There is one chocolate remaining in the dish. How can that be?

Answer:

> If there are eight chocolates in an antique dish, how can the last person take the last chocolate yet one remains in the dish? Well, the last person to take a chocolate took the dish as well—therefore, the last chocolate remained in the dish.

Remember these examples the next time you're defining requirements or looking for alternative answers to a problem.

Documenting the Scope Management Plan

The first output of the Plan Scope Management process is the *scope management plan*. This plan describes how the project team will go about defining project scope, validating the work of the project, and managing and controlling scope. According to the *PMBOK® Guide*, the scope management plan should contain the following:

- The process you'll use to prepare the project scope statement. The project scope statement (which I'll define later in this chapter) contains a detailed description of what the deliverables and requirements of the project are and is based on the information contained in the preliminary scope.

- A process for creating, maintaining, and approving the WBS. The WBS further defines the work of the project (as defined in the project scope statement) by breaking down the deliverables into smaller pieces of work.

- A process describing the approval process for the scope baseline and how the scope baseline will be maintained throughout the project.

- A process that defines how the deliverables will be obtained and formally accepted.

Exam Spotlight

The scope management plan is a planning tool that documents how the project team will go about defining project scope, how the work breakdown structure will be developed, how approval of the scope baseline will be obtained, and how the deliverables will be obtained and accepted. The scope management plan is based on the approved project scope. And don't forget, the scope management plan is a subsidiary of the project management plan.

Documenting the Requirements Management Plan

The second output of this process is the *requirements management plan*. This plan is similar to the scope management plan but focuses on the project and product requirements. This plan details how to analyze, document, and manage the project requirements throughout all phases of the project. The development approach you use for the project, such as a waterfall, an agile, or a hybrid approach, will influence how you will manage the project requirements. For example, a predictive or waterfall approach would dictate that all the requirements for each phase be completed before beginning the work of the phase and most certainly before moving on to the next phase of the project. An agile or a hybrid approach might mean that the requirements are not fully defined before the work begins and are progressively elaborated as the project (or phase) progresses.

There are several components of a sound requirements management plan. According to the *PMBOK® Guide*, you should include the following factors in the plan (and you are always free to add more than those noted here):

- How planning, tracking, and reporting of requirements activities will occur
- How changes to the requirements will be requested, tracked, and analyzed along with other configuration management activities
- How requirements will be prioritized
- What metrics will be used to trace product requirements
- What requirements attributes will be documented in the traceability matrix (the last output of this process)
- What elements to include on the traceability matrix

Exam Spotlight

The scope management plan and requirements management plan help set a clear vision and mission for the project and are used to plan and manage the scope of the project.

Collecting Requirements

Now we are getting into the meat of the Planning processes. In the *Collect Requirements* process, we define what the final product or service of the project will look like. You will recall that scope management describes what is and what is not included in the project scope. In this case, we're starting off by defining what *is* included in the work of the project.

Exam Spotlight

This process will gather all the requirements desired for the end product, service, or result of the project. However, not all of the requirements gathered during this process will be included in the end result of the project. The Define Scope process is where you will determine which of the requirements gathered here will be included in the final project. We will look at Define Scope later in this chapter.

Deliverables are measurable outcomes, measurable results, or specific items that must be produced or performed to consider the project or project phase completed. *Requirements* describe the characteristics of the deliverables. They might also describe functionality that a deliverable must have or specific conditions a deliverable must meet to satisfy the objective of the project. Requirements are typically conditions that must be met or criteria that the product or service of the project must possess to satisfy the objectives of the project. Requirements quantify and prioritize the wants, needs, and expectations of the project sponsor and stakeholders. According to the *PMBOK® Guide* (and lots of personal experience), you must be able to measure, trace, and test requirements. It's important that they're complete and accepted by your project sponsor and key stakeholders.

Requirements can take many forms. According to the *PMBOK® Guide*, requirements can be classified into the categories listed here, which may also assist you in elaborating them further:

- Business
- Stakeholder
- Solution
  - Functional
  - Nonfunctional
- Transition and readiness
- Project
- Quality

The primary purpose of the Collect Requirements process is to define and document the project sponsor's, the customer's, and the stakeholder's expectations and needs for meeting the project objectives. In my experience, understanding, documenting, and agreeing on requirements are factors critical to project success. This also requires actively engaged stakeholders who are able to define their needs so that the project team can decompose these needs into the product and project requirements. Recording the requirements and attaining stakeholder approval of the requirements will help you define and manage their expectations throughout the project.

Exam Spotlight

Requirements must be documented, analyzed, and quantified in enough detail that they can be measured once the work of the project begins. Requirements become the basis for developing the WBS and are essential in estimating costs, developing the project schedule, and quality planning.

Requirements gathering and documentation is primarily the responsibility of a business analyst. Project managers may perform this function on very small projects, but trends in project management show these activities are typically performed by the business analyst role. According to the *PMBOK® Guide* and PMI®'s publication titled *Business Analysis for Practitioners: A Practical Guide, business analysis* concerns applying knowledge, skills, and tools and techniques to determine business needs; recommending workable solutions for those needs; documenting and managing stakeholder requirements; encouraging stakeholders to be forthcoming with requirements; and facilitating the implementation of these requirements into the final product, service, or result of the project. *Business analysts* are the people who apply the knowledge and skills to obtain and document the stakeholder requirements.

Business analysts are often the first people assigned to a project, and they are usually engaged in the business case or feasibility phase before the project even gets to the Initiating processes. If business analysts are assigned to the project, all the processes associated with gathering, tracking, recording, analyzing, controlling requirements, and evaluating solutions are the responsibility of that role. The business analyst will follow the project all the way through to implementation and closeout.

When a business analyst is working on a project, they work in coordination and close collaboration with the project manager. The project manager in this scenario is responsible for overseeing that the requirements gathering and tracking activities are being conducted and performed according to the project schedule and budget. They also ensure that the requirements process and business analysts are adding value to the process. It's important that the business analyst and project manager understand their roles on the project and don't try to do each other's jobs. When they are both clear that the business analyst will manage the requirements processes, the project is likely to proceed smoothly and successfully.

Gathering Documents for the Collect Requirements Process

The key inputs of the Collect Requirements process you should consider when creating the requirements documentation are the project charter, project management plan, assumption log, agreements, business documents, lessons-learned register, stakeholder register, EEFs, and OPAs.

The business documents you might consider include the feasibility study or the business case. Agreements (typically contract type documents) contain scope and requirements information for the project, so if you are using vendors to work on your project, or parts of your project, be certain to review the contract scope.

Some of the EEFs you'll focus on in this process are the same as the Plan Scope Management process: culture, infrastructure, personnel administration, and marketplace conditions. OPAs for this process include policies and procedures, and historical information and lessons learned. Historical information and lessons learned are inputs to many Planning processes for good reason. You should review previous projects that are similar in size and scope to your current project so that you can learn what worked well for the project and what didn't. You may also find project documents you can repurpose for the current project.

Let's take a little time to understand the assumptions and constraints that are documented in the assumption log in more detail before we move on to the tools and techniques for this process.

Project Assumptions and the Assumption Log

You've probably heard the old saying about the word *assume*, something about what it makes out of "u" and "me." In the case of project management, however, throw this old saying out the window, because it's not true.

Assumptions, for the purposes of project management, are things you believe to be true. For example, if you're working on a large construction project, you might make assumptions about the availability of materials. You might assume that concrete, lumber, drywall, and so on are widely available and reasonably priced. You might also assume that finding contract labor is either easy or difficult, depending on the economic times and the availability of labor in your locale. Each project will have its own set of assumptions, and the assumptions should be identified, documented, and updated throughout the project.

It's essential to understand and document the assumptions you're making, and the assumptions your stakeholders are making, about the project. Did I mention you should write these down? Record them all in the assumption log. It's also important to find out as many of the assumptions as you can up front. Projects can fail, sometimes after lots of progress has been made, because an important assumption was forgotten or the assumption was incorrect. Defining new assumptions and refining old ones through the course of the project are forms of progressive elaboration. You can use techniques such as brainstorming or lateral thinking to help discover and document both assumptions and constraints (which we'll cover next).

Let's say you make plans to meet your friend for lunch at 11:30 on Friday at your favorite spot. When Friday rolls around, you assume they're going to show up, barring any catastrophes between the office and the restaurant. Project assumptions work the same way. For planning purposes, you presume the event or thing you've made the assumptions about is true, real, or certain. You might assume that key resources will be available when needed on the project. Document that assumption. If Sandy is the one and only resource who can perform a specific task at a certain point in the project, document your assumption that Sandy will be available and run it by her manager. If Sandy happens to be on a plane for Helsinki at the time you thought she was going to be working on the project, you could have a real problem on your hands.

Other assumptions could be factors, such as vendor delivery times, product availability, contractor availability, the accuracy of the project plan, the assumption that key project members will perform adequately, contract signing dates, project start dates, and project phase start dates. This is not an exhaustive list, but it should get you thinking in the right direction. As you interview your stakeholders, ask them about their assumptions and add them to your list. All of the assumptions you're uncovering could lead to discovering requirements that weren't previously thought about. Use brainstorming exercises with your team and other project participants to come up with additional assumptions.

Think about some of the factors you usually take for granted when you're trying to identify assumptions. Many times they're the elements everyone expects will be available or will behave in a specific way. Think about factors such as key team members' availability, access to information, access to equipment, management support, and vendor reliability. From there, determine what requirements are needed to support these assumptions.

Try to validate your assumptions whenever possible. When discussing assumptions with vendors, make them put those assumptions in writing. In fact, if the services or goods you're expecting to be delivered by your suppliers are critical to the project, include a clause in the contract to create a contingency plan in case your suppliers fail to perform. For example, if you're expecting 200 computers to be delivered, configured, and installed by a certain date, require the vendor to pay the cost of rental equipment in the event the vendor can't deliver on the promised due date.

Remember, when assumptions are incorrect or not documented in the assumption log, it could cause problems partway through the project and might even be a project killer.

Project Constraints

Constraints are anything that either restricts the actions of the project team or dictates the actions of the project team. Constraints put you in a box. (I hope you're not claustrophobic.) As a project manager, you have to manage the project constraints, a task that sometimes requires creativity.

In my organization—and I'm sure the same is true in yours—we have far more project requests than we have resources to work on them. In this case, resources are a constraint. You'll find that a similar phenomenon occurs on individual projects as well. Almost every project you'll encounter must work within the triple constraint combination of scope, time, and cost. The quality of the project (or the outcome of the project) is affected by how well these three constraints are managed. Usually, one or two constraints apply (and sometimes all three), which restricts the actions of the project team. You might work on projects for which you have an almost unlimited budget (don't we wish!), but time is the limitation.

For example, if the president mandated that NASA put an astronaut on Mars by the end of 2030, you'd have a time-constrained project on your hands.

Other projects might present the opposite scenario. You have all the time you need to complete the project, but the budget is fixed. Still other projects might incorporate two or more of the project constraints. Government agencies are notorious for starting projects that have at least two and sometimes all three constraints. For example, new tax laws are passed requiring updates to the systems that calculate and track this tax. Typically, a due date is given when the tax law takes effect, and the organization responsible is required to implement the changes with no additions to budget or staff. In other words, they are told to use existing resources to accomplish the objectives of the project, and the specific requirements, or scope, of the project are such that they cannot be changed to try to meet the time deadline.

As a project manager, one of your biggest jobs is to balance the project constraints while meeting the expectations of your stakeholders. On most projects, you will always have to balance the triple constraints and/or quality in addition to these. For example, if your primary constraint is completing the project by the end of the year, you will need to balance the other constraints: scope and cost (and/or quality). As the saying goes, "I can give you good, cheap, or on time—pick two."

Constraints can take on many forms and aren't limited to time, cost, and scope. Anything that impedes your project team's ability to perform the work of the project or specifically dictates the way the project should be performed is considered a constraint. Constraints can come from inside or outside the project or organization. Let's say to fulfill some of the deliverables of your project you'll have to purchase a large amount of materials and equipment. Procurement processes may be so cumbersome that ordering supplies for a project adds months to the project schedule. The procurement process itself becomes a constraint because of the methods and procedures you're required to use to get the materials.

You're likely to encounter the following constraints on your future projects:

Time Constraints   As I said, time can be a project constraint. This usually comes in the form of an enforced deadline, commonly known as the "make it happen now" scenario. If you are in charge of the company's holiday bash scheduled for December 10, your project is time constrained. Once the invitations are out and the hall has been rented, you can't move the date. All activities on this project are driven by the due date.

Budget Constraints   Budgets, or cost, are another element of the classic triple constraint. Budgets limit the project team's ability to obtain resources and potentially limit the scope of the project. For example, a desired requirement cannot be part of this project because the budget doesn't support it.

Scope Constraints   Scope is the third element of the original triple constraints. Scope defines the deliverables of the project, and you may have situations where scope is predefined by your project sponsor. Alternatively, sometimes budget constraints will impact the scope of the project and require you to cut back on the deliverables originally planned.

Quality Constraints   Quality constraints typically are restricted by the specifications of the product or service. Most of the time, if quality is a constraint, one of the other constraints—time or budget—has to give. You can't produce high quality on a restricted budget and within a restricted time schedule. Of course, there are exceptions—but only in the movies.

Schedule Constraints   Schedule constraints can cause interesting dilemmas for the project manager. For example, say you're the project manager in charge of building a new football stadium in your city. The construction of the stadium will require the use of cranes—and crane operators—at certain times during the project. If crane operators are not available when your schedule calls for them, you'll have to make schedule adjustments so that the crane operators can come in at the right time.

Resource Constraints   Resources could be a constraint from a few different perspectives, including availability of key resources both internally and in the marketplace, skill levels, and personality. You may also have availability issues or quality problems with nonhuman resources, like materials and goods. Human resources constraints are something I deal with on every project.

Technology Constraints   Technology is marvelous. In fact, how did humans survive prior to the invention of computers and cell phones? However, it can also be a project constraint. For example, your project might require the use of new technology that is still so new it hasn't been released on a wide-scale basis or hasn't been adequately tested to determine stability in production. One impact might be that the project will take an additional six months until the new technology is ready and tested.

Directive Constraints   Directives from management can be constraints as well. Your department might have specific policies that management requires for the type of work you're about to undertake. This might add time to the project, so you must consider those policies when identifying project constraints. When you're performing work on contract, the provisions of the contract can be constraints.

Constraints, particularly the classic triple constraints, can be used to help drive out the objectives and requirements of the project. If it's difficult to discern which constraint is the primary constraint, ask the project sponsor something like this: "Ms. Sponsor, if you could have only one of these two alternatives, which would you choose? The project is delivered on the date you've stated, or we don't spend one penny more than the approved budget." If Ms. Sponsor replies with the date response, you know your primary constraint is time. If push comes to shove during the project, the budget might have to give because time cannot.

You'll want to understand what the primary constraint is on the project. If you assume the primary constraint is budget when in actuality the primary constraint is time, in the immortal words of two-year-olds worldwide, "Uh-oh." Understanding the constraints and which one carries the most importance will help you later in the project Planning process group with details such as scope planning, scheduling, estimating, and project management plan development. That's assuming your project makes it to the Planning process group.

Which brings me back to the assumption log. Constraints are documented right along with assumptions in the assumption log. Did I mention the importance of writing these down and reviewing them periodically throughout the project?

Gathering and Documenting Requirements

Your communication skills are about to come in handy. Gathering and documenting requirements is not a task for the faint of heart. Because defining and producing requirements are so critical to the success of the project, I recommend using team members with excellent communication skills to perform this task. If they have the ability to read minds, all the better. Stakeholders almost always know what they want the end product to look like but often have difficulty articulating their needs. An expert communicator can read between the lines and ask probing questions that will draw the information out of the stakeholder. Even better, if you are lucky enough to have a business analyst assigned to your project, they are usually experts at asking the right questions and taking high-level concepts from stakeholders and breaking them down into specific requirements.

There are several tools and techniques you'll use to gather and document requirements, and we'll cover each of them next. They include the following:

- Expert judgment
- Data gathering (brainstorming, interviews, focus groups, questionnaires and surveys, benchmarking)
- Data analysis (document analysis)
- Decision-making (voting, multicriteria decision analysis, autocratic decision-making)
- Data representation (affinity diagrams, mind mapping)
- Interpersonal and team skills (nominal group, observation/conversation, facilitation)
- Context diagram
- Prototypes

Expert Judgment

Business process owners are those people who are experts in their particular area of the business. They are invaluable resources to the project manager and in gathering requirements for the project. They are usually the mid-level managers and line managers who still have their fingers in the day-to-day portion of the business. For example, it takes many experts in various areas to produce and market a great bottle of beer. Machinists regulate and keep the stainless steel and copper drums in top working order. Chemists check and adjust the secret formulas brewing in the vats daily. Graphic artists must develop colorful and interesting labels and ads to attract the attention of those thirsty patrons. Of course, those great TV commercials advertising the tasty brew are produced by yet another set of business experts, and managers at all levels review and approve the work. These are the kinds of people you'll interview and ask to assist you in identifying requirements.

Data Gathering Tools and Techniques

There are several data gathering tools and techniques in this process you can use to help identify the requirements of the project. Some of these tools and techniques will be used in other Planning processes as well. The following data gathering tools and techniques are used for the Collect Requirements process:

Brainstorming   This involves getting a group of experts together to help identify potential requirements. A facilitator usually conducts these sessions and all ideas (there are no bad ideas in brainstorming) are shared, collected, prioritized, and then determined to be valid requirements or not.

Interviews   Interviews are typically one-on-one conversations with stakeholders. Interviews can be formal or informal and generally consist of questions prepared ahead of time. The advantages to this tool are that subject matter experts and experienced project participants can impart a lot of information in a short amount of time and typically have a good understanding of the features and functions needed from the project deliverables. You should record the responses during the interviews, and don't be afraid to ask spontaneous questions as they occur to you during the interview.

Focus Groups   Focus groups are usually conducted by a trained moderator. The key to this tool lies in picking the subject matter experts and stakeholders to participate in the focus group.

Questionnaires and Surveys   The questionnaires and surveys in this case would be written with a focus on gathering requirements. This data-gathering technique is useful when teams are not colocated or you need responses quickly.

Benchmarking   This technique is used in the Quality processes as well and involves comparing measurements against standards to determine performance. It can also compare processes, operations, management practices, and so on against other departments, organizations, or industries to help refine and promote best practices or come up with ideas to improve current practices.

Data Analysis

Data analysis for this process involves reviewing documents with the purpose of identifying requirements or information associated with potential requirements. The documents might include agreements, business plans, business processes, business rules, process flows, issue logs, regulatory documents, marketing information, and more.

Decision-Making

The purpose of decision-making in this process is to ultimately come to conclusions regarding the requirements needed to fulfill the project objectives. There are two options you can use for this tool and technique: voting and multicriteria decision analysis. Voting can be a straightforward technique where stakeholders simply vote yes or no on including the requirements. The *PMBOK® Guide* outlines several voting methods that are a bit more complex than a simple yes/no vote. Stakeholders should agree beforehand which of the following methods they will use for voting. The methods are as follows:

Unanimity   This decision is unanimous in that all stakeholders agree.

Majority   This occurs when the majority of stakeholders agree on the decision.

Plurality   This involves counting the excess of votes where the largest segment of a group decides but is not necessarily a majority. This is typically used when two or more options are being voted on.

Autocratic   This involves one person making the decision on behalf of the group. Multicriteria decision analysis is another method of making a decision where a matrix is used to analyze criteria identified ahead of time (such as risk levels valuation).

Data Representation

This tool and technique is concerned with how you will present the information you're gathering and documenting. In this process, affinity diagrams and mind-mapping methods are helpful for visually displaying the interplay of requirements.

Affinity diagrams are what you might end up with after a brainstorming session. These diagrams group like ideas and solutions so that it's easier to analyze them. Mind mapping is another data representation technique in which participants first use brainstorming techniques to record their ideas. Whiteboards and flip charts are great tools to use with this process. The facilitator uses the whiteboard to map ideas and, using a mind-mapping layout, group similar topics together. Several mind-mapping software packages are available on the market that can greatly assist with this process. Mind mapping allows the participants to gain an understanding of common ideas and themes, create new ideas, and understand differences.

Interpersonal and Team Skills

Several techniques can be used to assist in defining requirements within this heading, including the nominal group technique, observation and conversation, and facilitation.

The nominal group technique can be used for several Planning processes, and we'll look at this one more than once. For this process, understand that this technique enhances brainstorming and involves a facilitator and a voting process.

Observation and conversation is a way to elicit requirements from stakeholders by observing their actions or speaking with them one on one. Observation is also known as *job shadowing* because you sit beside the stakeholder and physically watch each step they take in performing their task or function. This helps you, or the business analyst, to define the requirements in terms the project team will understand. Sometimes, stakeholders have difficulty explaining their requirements in an understandable manner, so actually observing them performing their tasks is a great way to obtain firsthand knowledge of their requirements. An equally effective method is for the business analyst or project manager to perform the task or function themselves so that they see what the stakeholder experiences (this is known as a *participant observer*). This method may also reveal elements of the process that are not apparent when discussing the process verbally with a stakeholder.

Facilitation may involve workshops or interactive participant meetings where requirements are identified. Facilitated sessions are usually performed by someone who does not have a vested interest in the outcome so that they can remain objective and keep everyone focused on the objectives of the session. It builds camaraderie and trust and can often turn into lively, spirited discussions where everyone is engaged and consensus is easier to obtain. According to the *PMBOK® Guide*, facilitation sessions might include joint application design (JAD) and development sessions, quality function deployment (QFD), or user stories. JAD is used primarily on technical projects where the program developers sit with the business process experts in a facilitated session and discuss requirements and the software development process at the same time. QFD sessions involve identifying essential characteristics of the product being developed. This focuses on gathering information about the customer needs. This method is used primarily in the manufacturing industry. We'll look at user stories next.

Exam Spotlight

The primary difference between focus groups (discussed earlier in this section) and facilitated workshops is that focus groups are gatherings of prequalified subject matter experts and stakeholders, and facilitated workshops consist of cross-functional stakeholders who can define cross-functional requirements. Differences among stakeholders can be resolved more quickly and consensus is more easily attained in a facilitated workshop environment.

User Stories

When you're using an agile methodology to manage your project, the requirements of the project will be defined and elaborated during each iteration or workflow stage, rather than in a detailed requirements document. In an agile methodology, you'll document what's known as user stories. *User stories* are similar to requirements, and they are written from the perspective of the customer. User stories document the person or people who will benefit from this requirement. This isn't usually a named person but rather a role, responsibility, or department. The "who" of the user stories are also known as actors. The user story also contains a brief description of the features or functionality that are desired by the actor. Last but not least, the user story describes why this requirement is needed by the actor and why it's valuable to them. This could be described in terms of business value as well. User stories should contain acceptance criteria that define how you will know that value has been achieved. We'll talk about acceptance criteria later in this chapter.

Context Diagram

A *context diagram* visually shows the business process or system, where or how actors interact with the system at differing points in the process, and how the interactions between them take place.

Prototypes

Prototyping is a technique that involves constructing a working model or mock-up of the final product with which the participants can experiment. The prototype does not usually contain all the functionality the end product does, but it gives participants enough information that they can provide feedback regarding the mock-up. Storyboarding is a prototyping technique used in the film industry, instructional design projects, and technology and software development projects. In a technology project, for example, the storyboard may walk through the steps involved when a user opens an app and clicks the various buttons or menu options. You can experience a simulated mock-up of what end users will see when they interact with the app by storyboarding all the screens or options that will be presented to them. This is an iterative process where participants experiment and provide feedback and the prototype is revised and the cycle starts again. Prototypes are often used in iteration-based agile methodologies.

Finalizing Requirements

Now that you've employed the tools and techniques of this process to gather requirements, you'll want to record them in a requirements document. (Remember that you'll use a requirements document in a predictive or hybrid methodology and user stories in an adaptive methodology.) Stakeholders sometimes have short memories, particularly on long-term projects, so documenting requirements and obtaining their approval is essential for project success. You will use the requirements documentation throughout the project to manage stakeholder and customer expectations. This is a lot easier to accomplish when they've agreed to the requirements ahead of time and you have their approval documented.

I've already mentioned the first output of this process, which is requirements documentation. The other output of this process is the requirements traceability matrix. I'll describe each in detail next.

Requirements Documentation

As I mentioned earlier, requirements quantify and prioritize the wants, needs, and expectations of the project sponsor and stakeholders to achieve the project objectives. Requirements typically start out high level and are progressively elaborated as the project progresses. You must be able to track, measure, test, and trace the requirements of the project. It is relatively easy to test or measure the requirements in an agile methodology because you'll know at the end of the work period if the product, prototype, or deliverable meets the criteria defined in the user story.

In a predictive methodology, you never want to find yourself at the end of the project (or phase) and discover you have no way to validate the requirements. If you can't measure or test whether the requirements satisfy the business need of the project, the definition of success is left to the subjective opinions of the stakeholders and team members.

You've worked hard to gather and define requirements, and you don't want all that effort going to waste. This output involves recording the requirements in a requirements document. The *PMBOK® Guide* does not dictate the format of this document and acknowledges that it can be formal with lots of detail or a simple list categorized by stakeholder and priority. However, it does state that the requirements can be classified into the following categories:

- Business requirements, such as the business needs or opportunities to be realized and the reasons why the project was started

- Objectives of the project and the business objectives the project hopes to fulfill

> The first two items in this list—business need for the project and objectives—are also needed for the traceability matrix.

- Stakeholder requirements describing stakeholder needs.
- Solution requirements, including both functional and nonfunctional requirements:
  - Functional requirements describe how the product will perform.
  - Nonfunctional requirements describe characteristics needed for the requirement to function, such as security needs, performance, and reliability.

> *Functional requirements* is a term used often in software development. It typically describes a behavior, such as calculations or processes that should occur once data is entered. In non-software terms, *functional requirements* might describe specifications, quantities, colors, and more. *Nonfunctional requirements* refers to elements that are related to the product but don't describe the product directly. In the case of a software product, this could be a security requirement or performance criteria.

- Quality requirements describing the criteria needed to ensure the deliverable or requirement was fulfilled successfully.
- Transition and readiness requirements that are needed to ensure successful transition to the end or future state. Training and data conversion are examples of transition and readiness requirements.
- Project requirements, which might include milestone completion dates, constraints, and contract commitments.

One of the most important elements of the requirements document that isn't in the preceding list is the signatures of the key stakeholders indicating their acceptance of the requirements. They will also sign the scope statement, which we'll talk about in the section "Defining Scope" later in this chapter.

Requirements Traceability Matrix

The last output of the Collect Requirements process is the requirements traceability matrix. The idea behind the traceability matrix is to document where the requirement originated, document what the requirement will be traced to, link each requirement to a testing strategy, and then follow it through to delivery or completion. Table 4.1 shows a sample traceability matrix with several attributes that identify the requirement.

TABLE 4.1   Requirements traceability matrix

Unique ID	Description of requirement	Source	Priority	Test scenario	Owner	Status and date
001	Requirement one	Project objective	B	User acceptance	HR specialist	Approved 3/15

Each requirement should have its own unique identifier. You could devise a numbering system that defines both the category of the requirement and a unique, ascending number— for example, HR (for human resources) 001—or a simple numbering system, as shown in this example, may suffice. The description should be brief but have enough information to easily identify the requirement. The source column refers to where the requirement originated. Requirements may come from many sources, including project objectives, business needs, product design, the work breakdown structure, and deliverables.

Priority refers to the importance of the requirement. You can use any prioritization process, like a simple numbering system or an alpha system, as the example shows. For example, perhaps an "A" is essential to project success and a "B" is highly desirable. The description and definition of the priority system should be included in the requirements management plan.

The test scenario in this example is where you record how the requirement will be tested (or during which project phase) and the owner of the test item, who will also decide whether the test scenario passes or fails. Status may capture information that refers to whether the requirement was approved; if it was added, deferred, or canceled; and so on.

The traceability matrix may also include the WBS deliverable the requirement is associated with, how it ties to the product design, and/or the product development.

> **Exam Spotlight**
>
> According to the *PMBOK® Guide*, the requirements traceability matrix helps ensure that business value is realized when the project is complete because each requirement is linked to a business and project objective.

The next process in the Planning process group is the Define Scope process. We'll look at that next.

Defining Scope

Now that you've documented the project requirements, you're ready to further define the needs of the project in the *Define Scope* process. Scope can refer to product scope (the features and characteristics that describe the product, service, or result of the project) or project scope (the project management work). We'll look at both product and project scope in this part of the chapter. The project scope statement (an output of this process) is a detailed description of the deliverables of the project and the work needed to produce them. This process is progressively elaborated as more detail becomes known.

> **Exam Spotlight**
>
> You'll want to pay particular attention to the accuracy and completeness of this process. Defining project scope is critical to the success of the project because it spells out exactly what the product or service of the project looks like. Conversely, poor scope definition might lead to cost increases, rework, schedule delays, and poor morale.

First, you'll examine the inputs and tools and techniques of this process.

The key inputs to the Define Scope process are the project charter, assumption log, requirements documentation, and risk register. We'll also look at some elements of the enterprise environmental factors and organization process assets. We've discussed many of these inputs previously, so I'll point out some of the key elements you'll want to consider when creating the project scope statement.

Some of the important elements from the project charter that you'll want to consider when writing the project scope statement are the project description, project objectives, the characteristics of the product of the project, and the process for approving the project. *Objectives* are quantifiable criteria used to measure project success. They describe "what" you're trying to do, accomplish, or produce. Quantifiable criteria should always include schedule, cost, and quality measures, and may also include other measures that are applicable to your project, such as business measures or quality targets. These objectives will be broken down shortly into deliverables that will describe the objectives outlined in the charter.

The project documents that are useful for this process include the assumption log, requirements documentation, and risk register. We'll look at the assumption log in more detail in the next section. You may recall that this is an output of the Develop Project Charter process. We covered the requirements documentation in the previous section. The risk register is a list of all the risks associated with the project, along with information such as the probability of the risk occurring and the impact it will have if it does occur. We will discuss the risk processes in detail in a later chapter.

During the Define Scope process, you will determine which requirements will be included in the final product, service, or result of the project. They will be progressively elaborated and then documented in detail in the scope statement. The idea here is that you know some information when the charter is being written and more information comes to light when you work with the stakeholders to document the requirements of the product of the project. Finally, those requirements are decided on, further elaborated, and documented (again) in much more detail in the scope statement. Keep in mind that not all the requirements that you gathered and documented in the Collect Requirements process may be included in the scope statement. You will reexamine those requirements here and determine which ones are keepers and which ones are not.

Some of the key EEFs for this process include the organizational culture, its infrastructure, personnel administration practices, and the marketplace conditions. The OPAs include historical information from previous projects, policies and procedures, templates that may exist to write the scope statement, and lessons learned from previous projects.

Exam Spotlight

If the project charter is missing or was not created, you'll need to develop the information normally found in the charter (or obtain it from other sources) to use as the foundation for creating the project scope statement.

The tools and techniques you'll use in this process to produce the project scope statement include expert judgment, alternatives analysis, multicriteria decision analysis, facilitation, and product analysis.

You learned about expert judgment previously and about interpersonal and team skills in the form of facilitation and facilitated workshops earlier in this chapter. We also covered alternatives analysis and decision-making earlier in this chapter, so let's move on to product analysis.

Product analysis goes hand in hand with the product scope description and, therefore, is most useful when the project's end result is a product. Product analysis is a method for converting the product description and project objectives into deliverables and requirements. According to the *PMBOK® Guide*, product analysis might include performing value analysis, product breakdown, requirements analysis, systems-engineering techniques, systems analysis, and value-engineering techniques to further define the product or service.

Exam Spotlight

It's beyond the scope of this book to go into the various analysis techniques used in product analysis. For exam purposes, remember that product analysis is a tool and technique of the Define Scope process and memorize the list of analysis techniques that might be performed in this process.

Writing the Project Scope Statement

The purpose of the *project scope statement* is to document the project objectives, deliverables, and the work required to produce the deliverables so that it can be used to direct the project team's work and as a basis for future project decisions. The scope statement is an agreement between the project management team and the project customer that states precisely what the work of the project will produce. Simply put, the scope statement tells everyone concerned with the project exactly what they're going to get when the work is finished.

Exam Spotlight

Understand that the purpose of the scope statement, according to the *PMBOK® Guide*, is to provide all the stakeholders with a foundational understanding of the project and product scope. It describes the project deliverables in detail and sets a clear vision for the project. Also remember that the scope statement defines and progressively elaborates the work of the project. It guides the work of the project team during the Executing process, and all change requests will be evaluated against the scope statement. If the change request is outside the bounds of the project scope as documented in the project scope statement, it should be denied.

Since the project scope statement serves as a baseline for the project, if questions arise or changes are proposed later in the project, they can be compared to what's documented in the scope statement. Making change decisions is easier when the original deliverables and requirements are well documented. You'll also know what is out of scope for the project simply because the work isn't documented in the scope statement (or conversely, deliverables or other elements are documented and noted as being specifically out of scope). The criteria outlined in the scope statement will also be used to determine whether the project was completed successfully. I hope you're already seeing the importance of documenting project scope.

Understanding the Scope Statement Components

According to the *PMBOK® Guide*, the project scope statement should include all of the following:

- Product scope description
- Acceptance criteria
- Project deliverables
- Project exclusions

We'll look at each of these next. Keep in mind that if the details surrounding these are spelled out in other documents, you don't have to reenter all the information in the scope statement. Simply reference the other document in the scope statement so that readers know where to find it.

Exam Spotlight

You may be thinking that the project scope statement has some of the same information as the project charter. Keep in mind that the project charter is a high-level description of the project (and it names the project manager), whereas the project scope statement is a detailed description of the project and product scope and describes how the final product or result of the project will be accepted, along with the other items mentioned earlier.

Product Scope Description　The *product scope description* describes the characteristics of the product, service, or result of the project. If the product scope description is contained in the project charter, you can reference the project charter in the project scope statement, or you can copy and paste the information from the project charter into the scope statement. It won't hurt anything to have it in both places and will make reading the scope statement easier.

Acceptance Criteria   *Acceptance criteria* include the process and criteria that will
be used to determine whether the deliverables and the final product, service, or results
of the project are acceptable and satisfactory. Acceptance criteria help you describe
project success because it defines the specifications the deliverables must meet to be
acceptable to the stakeholder. Acceptance criteria might include any number of elements, such as quality criteria, fitness for use, and performance criteria. This component should also describe the process stakeholders will use to indicate their acceptance
of the deliverables. Acceptance criteria should be defined for user stories as well. The
difference is that acceptance criteria will be tested or verified for each user story at the
end of the iteration, rather than weeks or months later in a predictive methodology.

Project Deliverables   As I stated earlier in this chapter, deliverables are measurable
outcomes, measurable results, or specific items that must be produced or performed
to consider the project or project phase completed. Deliverables should be specific and
verifiable. For example, one of your deliverables might include widgets with a 3-inch
diameter that will in turn be assembled into the final product. This deliverable, a
3-inch-diameter widget, is specific and measurable. However, if the deliverable was
not documented or not communicated to the manager or vendor responsible for manufacturing the widgets, there could be a disaster waiting to happen. If they deliver
2-inch widgets instead of the required 3-inch version, it would throw the entire project
off schedule or perhaps cause the project to fail. This could be a career-limiting move
for the project manager because it's the project manager's responsibility to document
deliverables and monitor the progress of those deliverables throughout the project.
Most projects have multiple deliverables. As in this example, if you are assembling
a new product with many parts, each of the parts might be considered independent
deliverables.

A project deliverable is typically a unique and verifiable product or result or a service
that's performed. The product or service must be produced or performed to consider
the project complete. The deliverables might also include supplementary outcomes such
as documentation or project management reports.

Don't forget compliance and regulatory standards when documenting your deliverables. If your industry standards require an independent audit to ensure compliance,
for example, the audit needs to be one of the deliverables on your project. Also keep
in mind that some compliance or regulatory requirements may be ongoing and extend
into the operations of the organization once the project is completed. For example,
audits may be required on an annual basis. You'll have the audit performed as part of

the project work, but the operations unit will need to continue this activity for the life of the product.

Project Exclusions   *Project exclusions* are, as you'd guess, anything that isn't included as a deliverable or work of the project. You'll want to note the project exclusions in the project scope statement so that you can continue to manage stakeholder expectations throughout the project.

The bottom line is this: No matter how well you apply your project skills, if the wrong deliverables are produced or the project is managed to the wrong objectives, you will have an unsuccessful project on your hands.

Critical Success Factors

Deliverables and requirements are sometimes referred to as *critical success factors*. Critical success factors are those items that must be completed in order for the project to be considered complete. For example, if you're building a bridge, one of the deliverables might be to produce a specific number of trusses that will be used to help support the bridge. Without the trusses, the bridge can't be constructed. The trusses, in this case, are critical success factors. Not all deliverables are critical success factors, but many of them will fall into this category and should be documented as such.

Documenting *All* the Deliverables and Requirements

One of the project manager's primary functions is to accurately document the deliverables and requirements of the project and then manage the project so that they are produced according to the agreed-upon criteria. Deliverables describe the components of the goals and objectives in a quantifiable way. Requirements are the specifications of the deliverables. Remember for the exam that according to the *PMBOK® Guide*, requirements are documented in the Collect Requirements process and this process is performed *before* the Define Scope process. In real practice, it's easier to define the scope and the deliverables first, then define the requirements.

The project manager should use the project charter as a starting point for identifying and progressively elaborating project deliverables, but it's possible that only some of the deliverables will be documented there. Remember that the charter was signed by a manager external to the project, and it was the first take at defining the project objectives and deliverables. As the project manager, it's your job to make certain *all* the deliverables are identified and documented in the project scope statement. That's because the scope statement (not the project charter) serves as the agreement among stakeholders—including the customer of the project—regarding what deliverables will be produced to meet and satisfy the business needs of the project.

Interview the stakeholders, other project managers, project team members, customers, management staff, industry experts, and any other experts who can help you identify all the deliverables of the project. Depending on the size of the project, you might be able to accomplish this in a group setting using simple brainstorming techniques, but large, complex projects might have scope statements for each major deliverable on the project. Remember that the project scope statement is progressively elaborated into finer detail and is used later to help decompose the work of the project into smaller tasks and activities.

Approving and Publishing the Project Scope Statement

Just like the project charter, the project scope statement should be approved, agreed on, published, and distributed to the stakeholders, key management personnel, and project team members. This isn't noted in the *PMBOK® Guide*, but remember that the project scope statement serves as an agreement between the project management team and the project customer that states precisely what the work of the project will produce. You'll want a formal sign-off procedure that's documented as part of the *approval requirements* section of the scope statement. When stakeholders sign off and agree to the scope statement, they're agreeing to the deliverables and requirements of the project. As with the project charter, their agreement and endorsement of the project requirements and deliverables will likely encourage and sustain their participation and cooperation throughout the rest of the project. That doesn't mean they'll agree to everything as the project progresses, but it does mean the stakeholders are informed and will likely remain active project participants.

Remember that the definition of a successful project is one that accomplishes the goals of the project and meets stakeholders' expectations. Understand and document those expectations and you're off to a good start.

The last output of this process is project documents updates. When you're in the midst of defining deliverables, you'll often find that changes to the original project objectives, requirements, or stakeholder register will occur. This may require updates to the stakeholder register, the requirements documentation, the requirements traceability matrix, and the assumption log. Changes to scope may also occur later in the project. When the changes are approved, you'll need to update the scope statement and notify stakeholders that changes have been made.

Managing the Product Backlog

When using a predictive development methodology, the project scope statement is fully defined before the work begins. The agile approach is not concerned with defining scope early in the project because requirements will be discovered and fine-tuned as the project progresses, thus delivering value to the customer incrementally. According to the *Agile Practice Guide* (PMI®, 2017), an agile project has continually evolving requirements, a high degree of uncertainty, and high risk. In an agile project, small increments of work are produced in each iteration that the customer reviews for accuracy and completeness. Prototypes are often used as well. This methodology provides frequent feedback to the team and allows them to make modifications to the work as it progresses, evolving the requirements, so that the deliverables align with customer expectations. This helps eliminate what I call the "surprised stakeholder" effect, which occurs more often than you'd think in a predictive methodology. For example, you go to great effort to define the requirements up front, work tirelessly for months or years on the project, and once it's finally delivered to the customer they tell you, "That's what I said I wanted but it's not what I need." Ugh. Iterating the requirements as you go in an agile method will help prevent this.

Agile methodologies use a *product backlog* to contain the user stories we talked about during the Collect Requirements process. The product backlog is a list of all the deliverables and requirements (user stories) that are needed to complete the project. You need enough user stories in the product backlog to start the work, and new user stories can be defined and added to the backlog as the work progresses. You can use the tools and techniques described in this process to discover and document the user stories.

The product owner (you'll recall this is the liaison with the agile team who speaks on behalf of the stakeholders) is responsible for maintaining and prioritizing the product backlog. The user stories are prioritized so that the most important are at the top of the list (and are usually well defined) and user stories with less importance fall beneath these on the list. User stories should be small, detailed units of work that the team can understand and finish during an iteration. The product owner defines and manages the user stories, but the agile team breaks the user stories down into tasks that can be accomplished during the iteration. The user stories should also contain acceptance criteria so that at the end of the iteration, the team can test or measure whether the deliverable meets the criteria.

Remember that agile teams are cross-functional teams that are self-organized and self-directed and are usually small in number. An agile development team rarely has more than five to nine members, and seven members is ideal. Also keep in mind there should only ever be one product owner on the project.

One more point you should keep in mind when using an agile approach is that if you're using Six Sigma, the requirements and project scope are documented during the define phase of the DMAIC.

🌐 Real World Scenario

Mountain Streams Services

Maria Sanchez is the CEO of Mountain Streams Services. She recently accepted a prestigious industry award on behalf of the company. Maria knows that without the dedication and support of her employees, Mountain Streams Services wouldn't have achieved this great milestone.

Maria wants to host a reception for the employees and their guests in recognition of all their hard work and contributions to the company. Maria has appointed you to arrange the reception.

The reception is scheduled for August 20, and Maria has given you a budget of $125 per person. The company employs 200 people. The reception should be semiformal.

You've documented the deliverables as follows:

- Location selection
- Food and beverage menu
- Invitations
- Entertainment
- Insurance coverage
- Decorations
- Photographer
- Agenda

In addition to the deliverables, you want to go over the following requirements with Maria to be certain you are both in agreement:

- The location should be in the downtown area.
- Employees are encouraged to bring one guest but no children.
- There will be an open bar paid for by Maria.
- The agenda will include a speech by Maria, followed by the distribution of bonus checks to every employee. This is to be kept secret until the reception.
- The decorations should include gold-trimmed fountain pens with the company logo at every place setting for the attendees to keep.

Once you've documented all the particulars, you ask to speak with Maria to go over this project scope statement and get her approval before proceeding with the project.

Creating the Work Breakdown Structure

Have you ever mapped out a family tree? In the *Create WBS* process, you'll construct something like it called a *work breakdown structure (WBS)*. It maps the deliverables of the project with subdeliverables and other components stemming from each major deliverable in a tree or chart format. Simply put, a WBS is a deliverables-oriented hierarchy that defines and organizes all of the work of the project and *only* the work of the project. The items defined on the WBS come from the approved scope statement. Like the scope statement, the WBS serves as a foundational agreement among the stakeholders and project team members regarding project scope.

Exam Spotlight

Subdividing deliverables into smaller components is the purpose of the Create WBS process. The *PMBOK® Guide* calls this decomposition, which is also a tool and technique of this process.

The WBS will be used throughout many of the remaining Planning processes and is an important part of project planning. You'll want to use the WBS for all projects, including predictive, agile, and hybrid methodologies. As you probably have concluded, everything you've done so far builds on the previous step. The project charter, requirements documentation, and project scope statement outline the project objectives, requirements, and deliverables. Now you'll use that comprehensive list of requirements and deliverables to build the framework of the WBS.

 I can't stress enough the importance of the work you've done up to this point. Your WBS will be only as accurate as your list of requirements and deliverables. The deliverables will become the groupings that will form the higher levels of the WBS from which activities will be derived later in the Planning processes.

The WBS should detail the full scope of work needed to complete the project. This breakdown will smooth the way for estimating project cost and time, scheduling resources, and determining quality controls later in the Planning processes. Project progress will be based on the estimates and measurements assigned to the WBS segments. So, again, accuracy and completeness are required when composing your WBS.

You'll use some of the documents we've already created to assist in defining the WBS. Remember that the scope management plan describes how you will create the WBS from the elements listed in the scope statement, and it also outlines how to obtain approval for and maintain the WBS. The approved project scope statement is used to define and organize the work of the project into the WBS. Make certain you're using the most current version of the scope statement. Also note that the WBS, just like the project scope statement, contains the work of the project and only the work of the project.

Decomposing the Deliverables

Decomposition is the primary tool and technique you'll use to create the WBS. It involves breaking down the deliverables into smaller, more manageable components of work. The idea here is to break down the deliverables to a point where you can easily plan, execute, monitor and control, and close out the project deliverables. Decomposition typically pertains to breaking deliverables down into smaller deliverables, or component deliverables, where each level of the WBS (or each level of decomposition) is a more detailed definition of the level above it.

This breaking down or decomposing process will accomplish several tasks for you, one of which is improving estimates. It's easier to estimate the costs, time, and resources needed for individual work components than it is to estimate them for a whole body of work or deliverable. Using smaller components also makes it easier to assign performance measures and controls. These give you a baseline to compare against throughout the project or phase. Finally, assigning resources and responsibility for the components of work makes better sense because several resources with different skills might be needed to complete one deliverable. Breaking them down ensures that an assignment, and the responsibility for that assignment, goes to the proper parties.

According to the *PMBOK® Guide*, decomposition is a five-step process:

1. **Identify the deliverables and work.** This step involves identifying all the major project deliverables and related work. You can use the expert judgment technique to analyze the project scope statement and identify the major deliverables.

2. **Organize the WBS.** This step involves organizing the work of the project and determining the WBS structure.

3. **Decompose the WBS components into lower-level components.** WBS components, like the deliverables and requirements, should be defined in tangible, verifiable terms so that performance and successful completion (or delivery) are easily measured and verified. Each component must clearly describe the product, service, or result in verifiable terms, and it must be assigned to a unit in the organization that will take responsibility for completing the work and making certain of its accuracy.

4. **Assign identification codes.** This step is a process where you assign identification codes or numbers to each of the WBS components.

5. **Verify the WBS.** This involves examining the decomposition to determine whether all the components are clear and complete. You'll also determine whether each component listed is absolutely necessary to fulfill the requirements of the deliverable, and verify that the decomposition is sufficient to describe the work.

Of course, you won't perform this process alone. That's where the expert judgment tool and technique comes into play. You'll work with others such as team members, stakeholders, other experts with specific training or industry knowledge, those who have worked on similar projects in the past, and industry aids such as templates.

You can now plug the components you and the team have identified into the WBS. This all sounds like a lot of work. I won't kid you—it is, but it's essential to project success. If you don't perform the WBS process adequately and accurately, you might end up setting yourself up for a failed project at worst or for lots of project changes, delayed schedules, and increased costs at best—not to mention all those team members who'll throw up their hands when you return to them for the third or fourth time to ask that they redo work they've already completed. I know you won't let this happen, so let's move on to constructing the WBS.

The Create WBS process has several outputs, one of which is the WBS. You'll look at the specifics of how to create the WBS next.

Constructing the WBS

There is no "right" way to construct a WBS. In practice, the chart structure is used quite often. (This structure resembles an organization chart with different levels of detail.) But a WBS could be composed in outline form as well. The choice is yours. You'll look at a couple of ways shortly, along with some figures that depict the different levels of a WBS.

According to the *PMBOK® Guide*, you can organize the WBS in several ways:

Major Deliverables and Subcomponents   The major deliverables of the project are used as the first level of decomposition in this structure. If you're opening a new store, for example, the deliverables might include determining location, store build-out, furnishings, product, and so on. I'll talk about subcomponents next.

Subcomponents That May Be Executed Outside the Project Team   Another way to organize the work is by subcomponents or subprojects within the project. Perhaps you're expanding an existing highway and several subcomponents are involved. Some of your first levels of decomposition might include these subcomponents: demolition, design, bridgework, and paving. Each of the project managers will develop a WBS for their subcomponent that details the work required for that deliverable. When subcomponent work is involved, often the subcomponent work is contracted out. In this example, if you contracted out the paving deliverable, this subcomponent requires its own WBS, which the seller (the paving subcontractor) is responsible for creating as part of the contract and contract work.

Project Phases   Many projects are structured or organized by project phases. For example, let's say you work in the construction industry. The project phases used in your industry might include project initiation, planning, designing, building, inspection, and turnover. A feasibility study might be a deliverable under the project initiation phase, blueprints might be a deliverable under the planning phase, and so on. Each phase listed here would be the first level of decomposition (that is, the first level of the WBS); their deliverables would be the next level, and so on.

We'll take a look at some example WBS structures next.

Understanding the Various WBS Levels

Although the project manager is free to determine the number of levels in the WBS based on the complexity of the project, all WBS structures start with the project itself. Some WBS structures show the project as level one. Others show the level under the project, or the first level of decomposition, as level one. The *PMBOK® Guide* notes that level one is the project level, so I'll follow that example here.

The first level of decomposition might be the deliverables, phases, or subprojects, as I talked about earlier. (Remember that the first level of decomposition is actually the second level of the WBS because the project level is the first level.) This applies to agile and hybrid projects as well, where the second level of the WBS might be a phase or a release. The levels that follow show increasingly more detail and might include more deliverables followed by requirements. Each of these breakouts is called a *level* in the WBS. The lowest level of any WBS is called the *work package level*. The goal is to construct the WBS to the work package level, where you can easily and reliably estimate cost and schedule dates. For example, the work package level for an agile or hybrid project is the user story. This is a unit of work that describes the functionality or features the user story will produce. User stories are easy to estimate and schedule and are broken down further into tasks during the planning meeting. Keep in mind that not all of the deliverables may need the same amount of decomposition. Also realize that performing too much decomposition can be as unproductive as not decomposing enough. My rule of thumb is, when the work outlined in the work package level is easily understood by team members and can be completed within a reasonable length of time, you've decomposed enough. If you decompose to the point where work packages are describing individual activities associated with higher-level elements, you've gone too far.

Exam Spotlight

Remember that each descending level of the WBS is a more detailed description of the project deliverables than the level above it. Each component of the WBS should be defined clearly and completely and should describe how the work of the project will be performed and controlled. Collectively, all the levels of the WBS roll up to the top so that all the work of the project is captured (and no additional work is added). According to the *PMBOK® Guide*, this is known as the 100 percent rule.

There is some controversy among project managers over whether activities should be listed on the WBS. In practice, I often include activities on my work breakdown structure for *small* projects only because it facilitates other Planning processes later. In this case, the activities are the work package level. However, you should realize that large, complex projects do not include activities on the WBS. For the exam, remember that you will decompose activities during the Define Activities process (which I'll talk about in Chapter 5, "Creating the Project Schedule") and that activities are *not* part of the WBS.

The easiest way to describe the steps for creating a WBS is with an example. Let's suppose you work for a software company that publishes children's games. You're the project manager for the new Billy Bob's Bassoon game, which teaches children about music, musical rhythm, and beginning sight reading. The first box on the WBS is the project name; it appears at the top of the WBS, as shown in Figure 4.1, and is defined as WBS level one.

FIGURE 4.1 WBS levels one and two

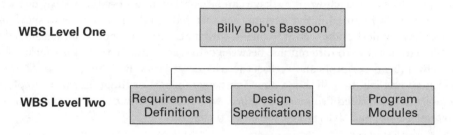

The next level is the first level of decomposition and should describe the major deliverables for the project. In this example, some of the deliverables might be requirements definition, design specifications, and programming. This isn't an exhaustive list of deliverables; in practice, you would go on to place all of your major deliverables into the WBS as level two content.

Level three content might be the component deliverables that are further broken out from the major deliverables of level two, or it might be the products or results that contribute to the deliverable. The Billy Bob's Bassoon example shows further deliverables as level three content. See Figure 4.2 for an illustration of the WBS so far.

FIGURE 4.2 WBS levels one, two, and three

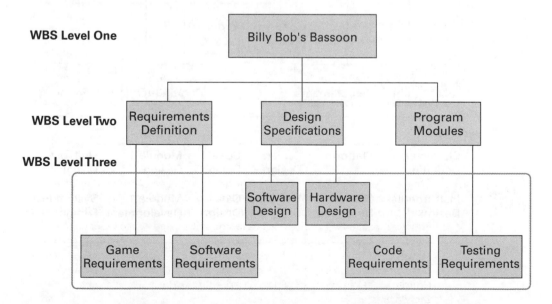

Large, complex projects are often composed of several subprojects that collectively make up the main project. The WBS for a project such as the Billy Bob's Bassoon game would show the subprojects as level one detail. These subprojects' major deliverables would then be listed as level two content, perhaps more deliverables as level three, and so on.

The goal here is to eventually break the work out to the point where the responsibility and accountability for each work package can be assigned to an organizational unit or a team of people. In Figure 4.3, I've decomposed this WBS to the fourth level to show an even finer level of deliverable detail. Remember that activities are not usually included in the WBS. An easy way to differentiate between deliverables and activities is to use nouns as the deliverable descriptors and verbs as activity descriptors. Reaching way back to my grade-school English, I recall that a noun is a person, place, or thing. In this example, the deliverables are described using nouns. When we get to the activity list, you might use verbs like *define*, *design*, and *determine* to describe them.

FIGURE 4.3 Figure WBS levels one, two, three, and four

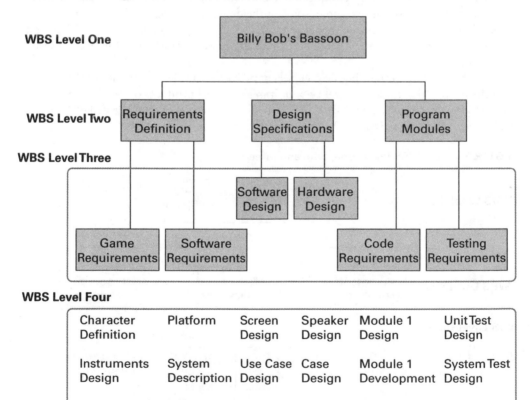

You can see from these illustrations how a poorly defined scope or inadequate list of deliverables will lead to a poorly constructed WBS. Not only will this make the WBS look sickly, but the project itself will suffer and might even succumb to the dreaded premature project demise. The final cost of the project will be higher than estimated, and lots of rework (translation: late nights and weekends) will be needed to account for the missing work not listed on the WBS. You can construct a good WBS and maintain a healthy project by taking the time to document all the deliverables during the Define Scope process.

WBS Templates

Work breakdown structures can be constructed using WBS templates or the WBS from a similar, completed project. Although every project is unique, many companies and industries perform the same kind of projects repeatedly. The deliverables are similar from project to project, and they generally follow a consistent path. WBS templates can be used in a case like this as a tool to simplify the WBS process, saving the project manager time.

Don't get too carried away when creating a WBS. The object is to define the work of the project so you can easily plan, manage, monitor, and control the work. But you don't want to take this too far. If you decompose the work to the point that you're showing every minute detail, you've ventured into inefficiency and will find it more difficult to plan and manage. In addition, you're potentially stifling the creativity of the people working on the project because everything is so narrowly defined.

Sometimes, particularly when working on large projects that consist of several subprojects, some of the subprojects might not be scheduled until a future date. Obviously, it makes sense to develop the WBS in detail at that future date when the deliverables and subprojects are better known and more details are available. This technique is called *rolling wave planning*. The idea behind this technique is that you elaborate the work of the project to the level of detail you know about at the time. If a subproject or deliverable is scheduled sometime in the future, the only component that might appear on the WBS today is the subproject itself. As you get closer to the subproject, you'll elaborate the details of the subproject and record them on the WBS.

Exam Spotlight

Understand that rolling wave planning is a process of elaborating deliverables, project phases, or subprojects in the WBS to differing levels of decomposition depending on the expected date of the work. Work in the near term is described in more detail than work to be performed in the future.

Understanding the Unique WBS Identifiers

Each work package in the WBS is assigned a unique identifier known as a code of accounts identifier. This is typically a number, or a combination of letters and numbers, that's used to track the costs, schedules, and scope associated with that work package. Code of accounts identifiers can also be assigned to other unique components of the WBS, but for study purposes, we'll focus on the work package level.

A *control account* consists of two or more work packages. And remember that a work package can be assigned to only one control account. A control account is used to summarize cost, schedule, and scope for all the work package levels assigned to it. Think of them as a checkpoint to monitor performance for the deliverable. A control account is typically inserted between a deliverable and the work packages needed to fulfill that deliverable. Let's look at an example.

The unique identifiers for the requirements definition branch of the WBS might look something like this:

10-1 Billy Bob's Bassoon

10-2 Requirements Definition

10-3 Game Requirements

     10-3-1 Character Definition

     10-3-2 Instruments Design

10-4 Software Requirements

     10-4-1 Platform

     10-4-2 System Description

In this example, we have four work packages with code of accounts identifiers as follows: 10-3-1, 10-3-2, 10-4-1, and 10-4-2. We will assign them all to one control account so that we can monitor earned value performance measures as well as costs, scope, and schedule for this component of the project. The control account could be inserted after the 10-2 Requirements Definition identifier because it will incorporate all the work packages needed for this deliverable. Most projects will usually have more than one control account.

Defining Work Packages

Work packages are the units of work that can be assigned to one person, or a team of people, with clear accountability and responsibility for completing the assignment, and they can be monitored and controlled throughout the project. Assignments are easily made at the work package level but can be made at any level in the WBS. The work package level is where time estimates, cost estimates, and resource estimates are determined. On an agile or hybrid project, user stories are the work packages.

As mentioned earlier, the project manager is free to determine the number of levels in the WBS based on the complexity of the project. You need to include enough levels to accurately estimate project time and costs but not so many levels that it's difficult to distinguish between the components. Regardless of the number of levels in a WBS, the lowest level in a WBS is called the *work package level*. The work package level for an agile or hybrid project is the user story.

Work package levels on large projects can represent subprojects that are further decomposed into their own work breakdown structures. They might also consist of project work that will be completed by a vendor, another organization, or another department in your organization. If you're giving project work to another department in your organization, you'll assign the work packages to individual managers, who will in turn break them down into activities during the Define Activities process later in the Planning process group.

Work packages might be assigned to vendors or others external to the organization. For example, perhaps one of the deliverables in your project is special packaging and a vendor is responsible for completing this work. The vendor will likely treat this deliverable as a project within its own organization and construct its own WBS with several levels of decomposition. However, for your project, it's a deliverable listed at the work package level of the WBS.

This might seem self-evident, but work packages not shown on the WBS are not included in the project. The same holds true for the project scope statement—if the deliverable isn't noted there, it isn't part of the project.

 Real World Scenario

The Lincoln Street Office Building

Flagship International has just purchased a new building to house its growing staff. The folks at Flagship consider themselves very lucky to have won the bid on the property located in a prime section of the downtown area. The building is a historic building and is in need of some repairs and upgrades to make it suitable for office space. Constructing the renovations will require special handling and care, as outlined in the *Historical Society Building Revisions Guide*.

Alfredo Martini is the project manager assigned to the renovation project. Alfredo has already determined the deliverables for this project. In so doing, he has discovered that he will not be able to manage all the work himself. He will need several subproject managers working on individual deliverables, all reporting to him. Alfredo calls a meeting with the other project managers to develop the WBS. Let's eavesdrop on the meeting.

"As you all know, we're planning to move into the Lincoln Street building by November 1. There is quite a bit of work to do between now and then, and I'm enlisting each of you to manage a segment of this project. Take a look at this WBS."

Here's how a portion of the WBS that Alfredo constructed looks:

Lincoln Street Building Renovation

2.0 Facility Safety

2.1 Sprinkler System

2.2 Elevators

2.3 Emergency Evacuation Plans

3.0 Asbestos Abatement

3.1 Inspection and Identification

3.2 Plans for Removal

4.0 Office Space

4.1 Building Floor Plans

4.2 Plans for Office Space Allocation

4.3 Plans for Break Room Facilities

4.4 Plans for Employee Workout Room

Alfredo continues, "I'm going to manage the Facility Safety project. Adrian, I'd like you to take the Asbestos Abatement project, and Orlando, you're responsible for the Office Space project."

"Alfredo," Adrian says, "asbestos abatement is going to take contractors and specialized equipment. We don't have staff to do these tasks."

"I understand. You'll need to take charge of securing the contractor to handle this. Your responsibility will be to manage the contractor and keep them on schedule," Alfredo answers. "Also do not forget that you need to follow the policies and procedures as outlined in the *Historical Society* renovation guide. I think I remember seeing some guidelines regarding asbestos abatement."

Orlando reminds Alfredo that he has missed a deliverable on the WBS. "Part of the Office Space project needs to include the network communications and telecommunications equipment rooms. I don't see that on here."

"Good point, Orlando," Alfredo says. "The level two and level three elements of this WBS are not complete. Each of you has been assigned to the subproject level, level one. Your first assignment is to meet back here in two weeks with a WBS for your subproject. I'd like to see some ideas about the staff assignments you'd make at the work package level and how long you think these components will take. We'll refine those after we meet next."

Backlog

Once the work package level, or user stories, are defined on an agile project, they are recoded in the product backlog. Earlier in this chapter, I introduced the product backlog. As you'll recall, this contains a list, in priority order, of all the user stories (deliverables and requirements) that the product owner defined at the beginning of the project. The *backlog* consists of the user stories that the product owner chooses from the product backlog, usually the next user story (or stories) in priority order, that will be worked on during the iteration. In other words, at the beginning of each iteration, the next user story (or stories) is pulled from the product backlog into the iteration backlog and the features of that user story are produced and delivered to the product owner at the end of the iteration. (You'll recall from Chapter 3 that Scrum is a pull-based system and that a sprint, also known as an iteration, is a time-bound period, generally two to four weeks in length.)

The user stories are the "what" you're going to deliver during that iteration. They are typically a deliverable, or a functioning portion of a deliverable. They may also include fixes to the work of previous iterations, new features, and more. After identifying the user stories, the agile team will break them down into individual tasks. The tasks, when completed, will fulfill the deliverable. This is a form of decomposition because they are breaking user stories (deliverables) into smaller units of work. They won't follow the five steps outlined in the predictive methodology described earlier in this section because they won't construct a WBS. They will break down the work into tasks and then work on the tasks. The agile team members choose the tasks they want to work on during the iteration and are responsible for completing their tasks so that the deliverable is completed by the end of the iteration. You'll recall that each user story should have acceptance criteria that you can use to test and measure whether the deliverable is what's expected and delivers value. The product owner and agile team members may work together to define the acceptance criteria. This will be examined at the end of the iteration, thereby validating the work.

In the Six Sigma methodology, the WBS is constructed during the design phase of the DMAIC. Design involves defining the stakeholders, creating the project charter, defining resources, developing milestones, and breaking down the work of the project into the WBS.

Finalizing the WBS

Remember how I said that everything you've done so far has built upon itself? This is an important concept you should know for the scope baseline, the primary output of this process. The *scope baseline* for the project is the approved project scope statement, the WBS, the work package, the planning package, and the WBS dictionary. (You'll recall that the project scope statement serves as a baseline for future project decisions because it helps the team determine whether requested changes are inside or outside of scope.) In other words, these documents together describe in detail all the work of the project. From these documents, you'll develop schedules, assign resources, and monitor and control the work of the project according to what's described here. We've already talked about the project scope statement and the WBS, but let's look at the planning package and the WBS dictionary.

Exam Spotlight

The scope baseline is used for predictive life-cycle methodologies and is defined as the approved version of the detailed project scope statement, the WBS, and the WBS dictionary. Understand this concept for the exam. This process does not name the WBS as an output; the WBS is included as part of the scope baseline output.

Planning Package

A planning package is situated below the control account on the WBS and above the work package. It details information about the work required but does not track schedule activities. Let's revisit our example from earlier in this chapter.

   10-1 Billy Bob's Bassoon

   10-2 Requirements Definition

      10-2-1 Control Account for Requirements Definition

   10-3 Game Requirements

      10-3-1 Character Definition

      10-3-2 Instruments Design

   10-4 Software Requirements

      10-4-1 Platform

      10-4-2 System Description

We could insert a Planning Package directly after the Control Account, or we could insert a Planning Package directly after 10-3 Game Requirements and one directly after 10-4 Software Requirements. Keep in mind that this example WBS is abbreviated. An actual WBS for this project, and the levels pertaining to this deliverable, would contain many more lines and many other work packages than what's shown in this example.

Planning packages are part of the control account and are always inserted below them. A control account might contain one planning package or multiple planning packages.

WBS Dictionary

The *WBS dictionary*, an element of the scope baseline, is where work component descriptions listed on the WBS are documented. According to the *PMBOK® Guide*, the WBS dictionary should include the following elements for each component of the WBS:

- Code of accounts identifier
- Description of the work of the component
- Organization responsible for completing the component
- List of schedule milestones
- Scheduled activities associated with the schedule milestones
- Required resources
- Cost estimates
- Quality requirements
- Criteria for acceptance
- Technical references
- Contract or agreements information
- Constraints and assumptions

Let's look at an example of what some of the elements of a WBS dictionary entry might look like. You'll use the work package level called Inspection and Identification defined in the sidebar "The Lincoln Street Office Building" earlier. The WBS dictionary entry for this might look like the following:

3.1 Inspection and Identification

Description of work—Inspect the building for asbestos, and identify all areas where it's found. Update the plan for removal (WBS 2.2) with each location identified.

Responsible organization—Adrian in facilities will hire and oversee a contractor to perform this work.

Schedule milestones—Inspection and identification to start after contractor is identified and hired (no later than July 1). Work should be completed no later than September 15.

Contract information—Two contractors have been identified as qualified and experienced in this type of work. Contract process should close no later than June 12.

If the WBS and the WBS dictionary are constructed well, you've given yourself a huge helping hand with the remaining Planning processes. The completion of many of the remaining processes depends on the project scope statement and WBS being accurate and complete. You'll use the work packages created here to further elaborate the work into activities. From there, you can estimate costs, develop schedules, and so on. The WBS is an essential tool for project planning, so keep it handy.

You can see from the examples you've walked through in this chapter how new deliverables or requirements might surface as a result of working on the WBS. These requested changes should be reviewed and either approved or denied using your change control processes. For example, if you identified a new requirement during the WBS process, you'll need a change request to update the project scope statement and/or the requirements documentation to add the new requirement. The assumption log will likely have to be updated as a result of this process also, but it will not require a change approval. It's a running list of assumptions and constraints and can be modified at any time. Any changes that occur that include new or modified deliverables or requirements need the appropriate documents updated to reflect the approved changes.

I recommend reviewing *Practice Standard for Work Breakdown Structures, Third Edition*, published by PMI® to see examples of industry-specific WBS templates, or you can modify the templates for particular business areas that may not be represented.

Project Case Study: New Kitchen Heaven Retail Store

The project charter kickoff meeting was held and well attended. You're ready to start gathering requirements and writing the project scope statement.

You review your notes and reread the first draft of the project scope statement you've prepared for the Kitchen Heaven store and website before looking for Dirk. After your meetings with the stakeholders, you were better able to refine the project objectives and deliverables.

"Dirk, I'm glad I caught you. I'd like to go over the project scope statement with you before I give it to the stakeholders. Do you have a few minutes?"

"Sure," Dirk says. "Let's have it."

"The project objective is to open the 50th Kitchen Heaven retail store by January 25 and launch the updated website. When I met with Jake, he confirmed it takes 120 days to do the store build-out. That includes having the shelves set up and in place, ready to stock with inventory."

Dirk asks whether Jake told you about his store location idea.

"Yes, Jake gave me the name of the leasing agent, and I've left her a voicemail. The sooner we can get that lease signed, the better. It takes Jake 120 days to do the build-out, and Jill said she needs two weeks' lead time to order the initial inventory and stock the shelves. That puts us pretty close to our January 25th deadline, counting the time to get the lease papers signed."

"Sounds good so far," Dirk replies. "What else?"

You continue, "I've included an updated description of the products and services the new store will offer, based on the documentation that was written from the last store opening. Jill reviewed the updates to the description, so we should be in the clear there. The store will include some new lines that we've decided to take on—cookware from famous chefs, that kind of thing.

"Jake has already made contact with a general contractor in Colorado Springs, and he is ready to roll once we've signed the lease.

"As you know, I've asked Ruth to manage the website launch. I'm considering this a sub-project. She's also documented the requirements and deliverables for the launch. She met with Ricardo and Jill to understand their needs and expectations. I will incorporate the project documents she creates with mine so that I have oversight over all aspects of the project.

"One more thing, Dirk. Since we're including the big bash at the grand opening as part of the deliverables, I talked to your folks in marketing to get ideas. They are thinking we should have giveaways as door prizes and that we will want the food catered. They also thought having live cooking demonstrations with local chefs would be a good attraction."

"Sounds like you're on the right track. So, what's next?" Dirk asks.

"Once you approve the scope statement, I'd like to send a copy to the stakeholders. My next step is to break down the deliverables and requirements I've documented here into the WBS so we can get rolling on the work of the project."

Project Case Study Checklist

The main topics discussed in the case study are as follows:

Constraints: January 25th date to coincide with Home and Garden show

Assumptions: These are the assumptions:

- A store build-out usually takes 120 days.

- Jill Overstreet will help with the initial store stocking.

- Jake Peterson will provide supplies for the stores that aren't retail products, such as store furnishings and shelving, and can help with the store build-out as well.

- The budget for the project will be between $2 million and $2.5 million.

The project scope statement includes the following:

Project objectives: Open 50th store by January 25 in Colorado Springs. Launch the updated website on January 25.

Project deliverables:

- Build out storefront, including shelving.

- Retail product line will be delivered two weeks prior to grand opening.

- Have grand opening party with cooking demos.

- Begin digital marketing campaign by January 7 to advertise the new store and the updated website.

Project requirements:

- Sign lease within 14 days.

- Offer a new line of cookware.

- Have classroom space in back of store for cooking demos and classes.

- Provide celebrity chef videos on the website.

    Constraints: January 25 date will coincide with Home and Garden Show.

    Fund limitations: Spend no more than $2.5 million on the project.

    Assumptions: (These are the same as listed earlier.) Decomposed deliverables into a WBS.

The WBS includes the following:

- Level one is the project.

- Level two is the retail store opening and the website launch. Both are listed as subprojects.

- Level three is deliverables for each of the subprojects.

- Last level of WBS is the work package level, where time and cost estimates can be defined in the next process.

Understanding How This Applies to Your Next Project

In this chapter, you dealt with the realities of life on the job. The reality is, many project managers I know are managing several projects at once as opposed to one large project. Although every concept presented in this chapter is a sound one, it's important to note that you have to balance the amount of effort you'll put into project management processes against the size and complexity of the project.

As a manager who prides herself and her team on excellent customer service, I have once or twice gotten my team into precarious situations because I was so focused on helping the customer that I hurt my team and our department in the process. If you're wondering how that happened, it was because we didn't take the time to document the scope of the project and the final acceptance criteria. In one case, in the interest of getting the project completed quickly because of our customer's own internal deadlines, we decided the project was straightforward enough that we didn't need to document deliverables. The customer promised to work side by side with us as we produced the work of the project. Unfortunately, that wasn't the case, and we didn't meet the expectations of our customer. Further, after we did implement the project (two months behind schedule), we went through another six weeks of "fixes" because of the miscommunication between the customer and the project team on what constituted some of the features of the final product. There's always a great reason for cutting corners—but they almost always come back to haunt you. My advice is to always create a scope statement and a requirements document and get stakeholder signatures on both. (In practice, small projects can include both the deliverables and requirements within the scope statement.)

Decomposing the deliverables is the first step toward determining resource requirements and estimates. A WBS is always a good idea, no matter the size of the project. I have to admit I have cheated a time or two on small projects and used the project schedule as the WBS. In all fairness, that worked out fine when the team was small and no more than three or four people were working on the project. If you get many more than four people on the project team, it can be a little cumbersome to track deliverables with a schedule only. The WBS is the perfect tool to use to assign names to work packages, and it's the foundation for determining estimates for the work of the project.

The five-step process outlined by the *PMBOK® Guide* works very well. Starting with the 50,000-foot view, the team determines the major deliverables of the project. From there, the deliverables are decomposed into ever smaller units of work. The trick here is to break the work down into measurable units so that you can verify the status of the work and the completion and acceptance of the work when you're finished. If you have "fuzzy" WBS levels or work packages, you won't be able to determine status accurately. In the IT field, we have a saying about the status of projects: "It's 90 percent complete." The problem is it always seems that the last 10 percent takes twice as long to complete as the first 90 did. If you've taken the time to document a WBS, you'll have a much better idea of what that 90 percent comprises. The last step is the verification step where you determine whether everything you've identified in the WBS is absolutely necessary to fulfill the work of the project and whether it's decomposed enough to adequately describe the work. It has been my experience that documenting the WBS will save you time later in the Planning processes, particularly when developing the project schedule and determining the project budget.

I believe the most important idea to take from this chapter is a simple one: always document the requirements, and always get sign-off from the stakeholders.

Summary

This chapter started you on the road to project planning via the Develop Project Management Plan process, the Plan Scope Management process, the Collect Requirements process, the Define Scope process, and the Create WBS process. We covered a lot of material in this chapter. Everything you've learned so far becomes the foundation for further project planning.

The output of the Develop Project Management Plan process is the project management plan, which is concerned with defining, coordinating, and integrating all the ancillary project plans and baselines. The purpose of this plan is to define how the project is executed, how it's monitored and controlled, and how it's closed.

The primary output of the Plan Scope Management process is the scope management plan. This plan is an element of the project management plan that describes how the project team will go about defining project scope, validating the work of the project, and managing and controlling scope.

The Collect Requirements process involves gathering and documenting the requirements of the project. It's important that requirements be measurable, traceable, testable, and so on. Measurement criteria for project requirements are agreed on by the stakeholders and project manager. Additionally, requirements should be tracked in a traceability matrix that documents where they originated, the results of the tests, the priority of the requirement, and more.

The product backlog is used in an adaptive methodology. The product backlog is a list of all the deliverables and requirements (user stories) that are needed to complete the product of the project. You'll document the user stories at the beginning of the project, before the work starts, and then compile them in the product backlog. User stories will be chosen at the beginning of each iteration, and the team will break down the user stories into tasks to complete the deliverable (or functioning portion of a deliverable) by the end of the iteration.

Constraints restrict or dictate the actions of the project team. Constraints usually involve time, cost, and scope but can also include schedules, technology, quality, resources, risk, and more. Assumptions are things believed to be true. You'll want to document project assumptions and validate them as the project progresses. Both assumptions and constraints are documented in the assumption log.

The project scope statement is produced during the Define Scope process. It describes the project deliverables. The project scope statement serves as an agreement between the project management team and the project customer that states precisely what the work of the project will produce. The scope statement, along with the WBS, work packages, planning packages, and the WBS dictionary, forms the scope baseline that you'll use to weigh future project decisions, most particularly change requests. The scope statement contains a list of project deliverables that will be used in future Planning processes. The project scope statement contains many elements, including product scope description, product acceptance criteria, deliverables, and exclusions from scope.

A WBS is a deliverable-oriented hierarchy of project work. The highest levels of the WBS are described using nouns, and the lowest levels are described with verbs. Each element in the WBS has its own set of objectives and deliverables that must be met to fulfill the deliverables of the next highest level and ultimately the project itself. In this way, the WBS validates the completeness of the work.

The lowest level of the WBS is known as the work package level. This allows the project manager to determine cost estimates, time estimates, resource assignments, and quality controls. The work package level for an agile or hybrid project is the user story.

Exam Essentials

Be able to state the purpose of the Develop Project Management Plan process.   It defines, coordinates, and integrates all subsidiary project plans.

Be able to describe the purpose of the scope management plan.   The scope management plan has a direct influence on the project's success and describes the process for determining project scope, facilitates creating the WBS, describes how the product or service of the project is validated and accepted, and documents how changes to scope will be handled. The scope management plan is a subsidiary plan of the project management plan. The scope management plan defines, maintains, and manages the scope of the project.

Understand the purpose of the project scope management plan and requirements management plan.   The scope management plan and requirements management plan help set a clear vision and mission for the project and are used to plan and manage the scope of the project.

Understand the purpose of the project scope statement.   The project scope statement serves as an agreement between the project management team and the project customer that states precisely what the work of the project will produce. The project objectives and deliverables and their quantifiable criteria are documented in the scope statement and are used by the project manager and the stakeholders to determine whether the project was completed successfully. It also serves as a basis for future project decisions.

Be able to describe the product backlog.   The product backlog is a list of all the user stories (deliverables and requirements) that are needed to complete the project. You'll document the user stories at the beginning of the project, before the work starts, and then compile them in the product backlog.

Be able to describe the backlog.   The backlog consists of the user stories that will be worked on during the iteration. The agile team breaks down the user stories into tasks, and each team member chooses the tasks they want to work on during the iteration.

Be able to define project constraints and assumptions.   Project constraints limit the options of the project team and restrict their actions. Sometimes constraints dictate actions. Time, budget, and scope are the most common constraints. Assumptions are conditions that are presumed to be true or real. They are both documented in the assumption log.

Be able to define a WBS and its components.   The WBS is a deliverable-oriented hierarchy. It uses the deliverables from the project scope statement or similar documents and decomposes them into logical, manageable units of work. The first level of decomposition is the major deliverable level or subproject level, the second level of decomposition is a further elaboration of the deliverables, and so on. The lowest level of any WBS is called a work package. User stories are the work package level for an agile or hybrid project.

Understand the difference between code of accounts identifiers and control accounts.   Code of accounts identifiers are assigned to all elements of a WBS. These are a unique numbering system that ties the work elements to the code of accounts. A control account consists of two or more work packages (each with their own code of accounts identifier) and/or one or more planning packages. A control account is inserted between a major deliverable and the work package, and it tracks cost, schedule, and scope and is also used to perform earned value measurements. A planning package is inserted between the control account and the work package level and does not include schedule activities.

Be able to name the components of the scope baseline.   The scope baseline consists of the approved project scope statement, WBS, work packages, planning packages, and the WBS dictionary.

Review Questions

You can find the answers to the review questions in Appendix A. Be sure to download the Bonus Exams and Bonus Questions so that you'll have a broader exposure and more experience answering questions related to the topics in this chapter.

1. You are a project manager for Laredo Pioneer's Traveling Rodeo Show. You're heading up a small project to promote a new line of souvenirs to be sold at the shows. You know that the purpose of the project scope management plan and requirements management plan provide which of the following? (Choose three.)

   A. They help set a clear vision and mission for the project.

   B. They are used to further elaborate the project charter, which contains a list of high-level requirements needed to satisfy stakeholder expectations before starting the project management plan.

   C. They serve as an agreement between the project management team and the project customer to document and agree on the work of the project and what it will produce.

   D. They are both used to plan and manage the scope of the project.

2. You are a project manager for Laredo Pioneer's Traveling Rodeo Show. You're heading up a small project to promote a new line of souvenirs to be sold at the shows. You've held the kickoff meeting and are ready to begin creating the project management plan. You know that a kickoff meeting is important for which of the following reasons? (Choose three.)

   A. It ensures that compliance and regulatory requirements are documented and discussed and will be included in the project scope statement.

   B. It communicates the start of the project and informs stakeholders of the project objectives and the business value the project will bring about.

   C. It ensures that a common understanding is established among stakeholders because the key milestones and key deliverables are identified.

   D. It ensures that policies, procedures, and standards used in the organization have been considered and discussed.

   E. It helps establish stakeholder engagement and commitment.

3. You are a project manager responsible for the construction of a new office complex. You are taking over for a project manager who recently left the company. The prior project manager completed the project scope statement and scope management plan for this project. In your interviews with some key team members, you conclude which of the following? (Choose two.)

   A. The project scope statement assesses the stability of the project scope and outlines how scope will be verified and used to control changes. The team members know that project scope is measured against the product requirements and that the scope management plan is based on the approved project scope.

   B. The scope management plan describes how project scope will be managed and controlled and how the WBS will be created and defined. They know that product scope is measured against the product requirements and that the scope management plan is based on the approved project scope.

C. The scope management plan is deliverables-oriented and includes cost estimates and stakeholder needs and expectations. They understand that project scope is measured against the project management plan and that the scope management plan is based on the approved project charter.

D. The project scope statement describes how the high-level deliverables and requirements will be defined and verified. They understand that product scope is measured against the project management plan and that the scope management plan is based on the approved project charter.

E. The project scope management plan defines, maintains, and manages the scope of the project.

4. Unanimity, majority, plurality, and autocratic are four examples of which of the following techniques and what they are used for? (Choose two.)

A. They are used to help stakeholders come to conclusions and agreement on the requirements needed to fulfill the project objectives.

B. Interviews, which is a tool and technique of the Define Scope process.

C. Facilitated workshops technique, which is a tool and technique of the Define Scope process.

D. Decision-making techniques, which is a tool and technique of the Collect Requirements process.

E. They are used to help stakeholders visualize decisions using tools such as mind mapping or affinity diagrams.

5. You are working on a project where the requirements and scope are well defined, but you know there will be changes when performing the work of the project. Your stakeholders will need to provide continuous feedback during the development stages of this project. Which of the following are true when determining the deliverables for the project? (Choose three.)

A. You should use a predictive approach to manage this project. You will document a project scope statement because it describes how the team will define and develop the work breakdown structure.

B. You should use a hybrid approach for this project. Because the requirements and deliverables are well defined, you may choose to document a project scope statement that further elaborates the deliverables of the project and serves as a basis for future project decisions.

C. You should document the project scope statement so there is agreement between the project management team and the project customer and everyone knows what the work of the project will produce.

    D. You may choose to document the project scope statement because it assesses the reliability of the project scope and describes the process for verifying and accepting completed deliverables.

    E. You should use a hybrid approach for this project because the stakeholders would like to provide continuous feedback on the deliverables once the work of the project starts. User stories will be used to fulfill the deliverables of the project and will be pulled to the backlog.

6. You are a project manager for an agricultural supply company. You have interviewed stakeholders and gathered their project requirements in the Collect Requirements process. Which of the following is true regarding the process to which this question refers?

    A. The requirements document lists the requirements and describes how they will be analyzed, documented, and managed throughout the project.

    B. Requirements documentation consists of formal, complex documents that include elements such as the business need of the project, functional requirements, nonfunctional requirements, impacts to others inside and outside the organization, and requirements assumptions and constraints.

    C. The requirements documentation details the work required to create the deliverables of the project, including deliverables description, product acceptance criteria, exclusions from requirements, and requirements assumptions and constraints.

    D. The requirements traceability matrix ties requirements to project objectives, business needs, WBS deliverables, product design, test strategies, and high-level requirements and traces them through to project completion.

7. Which of the following makes up the scope baseline when using a predictive life-cycle approach?

    A. The approved project scope statement

    B. The approved scope management plan

    C. The approved WBS

    D. The approved WBS dictionary

    E. A, B, C, D

    F. A, C, D

8. Which of the following statements is true regarding brainstorming and lateral thinking? (Choose two.)

    A. They are techniques that can be used in several Planning processes to help determine project scope, requirements, assumptions, constraints, and more.

    B. Lateral thinking is a form of alternatives generation that's considered thinking outside the box.

    C. They are decision-making techniques used to help stakeholders form consensus.

    D. Brainstorming is a technique involving plurality decision-making.

9. Your company, Kick That Ball Sports, has appointed you as project manager for its new Cricket product line introduction. This is a national effort, and all the retail stores across the country need to have the new products on the shelves before the media advertising blitz begins. The product line involves three new products, two of which will be introduced together and a third one that will follow within two years. You are ready to create the WBS. All of the following are true except for which one?

   A. The WBS may be structured using each product as a level one entry.

   B. The WBS should be elaborated to a level where costs and schedule are easily estimated. This is known as the work package level.

   C. Rolling wave refers to how all levels of the WBS collectively roll up to reflect the work of the project and only the work of the project.

   D. Each level of the WBS represents verifiable products or results.

10. You are a project manager for Giraffe Enterprises. You've recently taken over for a project manager who lied about his PMI® certification and was subsequently fired. Unfortunately, he did a poor job of defining the project scope. The project scope statement is the last document the previous project manager created. You will need to create the remaining subsidiary project management plans and documents. All of the following could happen if you don't correct this except for which one?

   A. The stakeholders will require overtime from the project team to keep the project on schedule.

   B. The poor scope definition will adversely affect the creation of the work breakdown structure, and costs may increase.

   C. The project scope statement is used to document the process for defining, maintaining, and managing project scope.

   D. The project costs could increase, there might be rework, and schedule delays might result.

11. You are the project manager for Lucky Stars nightclubs. They specialize in live country and western band performances. Your newest project is in the Planning process group. You are working on the WBS. The finance manager has given you a numbering system to assign to the WBS. Which of the following is true?

   A. The numbering system is a unique identifier known as the WBS dictionary, which is used to assign quality control codes to the individual work elements.

   B. The numbering system is a unique identifier known as the WBS dictionary, which is used to track the descriptions of individual work elements.

   C. The numbering system is a unique identifier known as the control account, which is used to track time and resource assignments for individual work elements.

   D. The numbering system is a unique identifier known as the code of accounts identifier, which is used to track the costs of the WBS elements.

12. You are a project manager working on a large, complex project. You've constructed the WBS for this project, and all of the work package levels are subprojects of this project. You've requested that the subproject managers report to you in three weeks with their individual WBSs constructed. Which of the following statements are correct? (Choose three.)

   A. The work package level is decomposed to create the activity list.

   B. The work package level is the lowest level in the WBS and is known as a user story on an agile project.

   C. The work package level facilitates resource assignments.

   D. The work package level facilitates cost and time estimates.

13. You are a project manager working on a new software product your company plans to market to businesses. The project sponsor told you that the project must be completed by September 1. The company plans to demo the new software product at a trade show in late September and, therefore, needs the project completed in time for the trade show. However, the sponsor has also told you that the budget is fixed at $85,000, and it would take an act of Congress to get it increased. You must complete the project within the given time frame and budget. Which of the following is the primary constraint for this project?

   A. Budget

   B. Scope

   C. Time

   D. Quality

14. Which of the following statements about decomposition is true?

   A. Decomposition is a five-step process used to break down the work of the project.

   B. Decomposition requires expert judgment along with close analysis of the project scope statement.

   C. Decomposition is a tool and technique used to create a WBS and subdivide the major deliverables into smaller components until the work package level is reached.

   D. Decomposition is used in the agile methodology to create users stories at the work package level. They are documented in the product backlog and pulled from here into the iteration backlog at the beginning of each iteration.

   E. All of the above.

15. You are documenting acceptance criteria for your deliverables. Which of the following are true? (Choose two.)

   A. They are documented in the project scope management plan in a predictive methodology.

   B. They are not used in an adaptive methodology because the work is defined at the beginning of the iteration and verified at the end of the iteration.

    C. They are used in a predictive methodology to test or measure whether the deliverables of the project are acceptable and satisfactory when the deliverables are completed.

    D. They are documented in user stories in an adaptive methodology to test or measure whether the deliverables of the project are acceptable and satisfactory at the end of the iteration.

16. You are a project manager for a documentary film company. The company president wants to produce a new documentary on the efforts of heroic rescue teams and get it on air as soon as possible. She's looking to you to make this documentary the best that has ever been produced in the history of this company. She guarantees you free rein to use whatever resources you need to get this project done quickly. However, the best photographer in the company is currently working on another assignment. Which of the following is true?

    A. The primary constraint is time because the president wants the film done quickly. She told you to get it to air as soon as possible.

    B. Resources are the primary constraint. Even though the president has given you free rein on resource use, you assume she didn't mean those actively assigned to projects.

    C. The schedule is the primary constraint. Even though the president has given you free rein on resource use, you assume she didn't mean those actively assigned to projects. The photographer won't be finished for another three weeks on his current assignment, so schedule adjustments will have to be made.

    D. The primary constraint is quality because the president wants this to be the best film ever produced by this company. She's given you free rein to use whatever resources are needed to get the job done.

17. Your project depends on a key deliverable from a vendor you've used several times before with great success. You're counting on the delivery to arrive on June 1. This is an example of a/an _____.

    A. Constraint

    B. Objective

    C. Assumption

    D. Requirement

18. You are creating your project management plan. The project you are working on involves accepting credit cards on your organization's website. This will require yearly audits to ensure you are following best practices in developing your website and not exposing sensitive customer information. Which of the following EEFs should you consider given this scenario? (Choose two.)

    A. Technology issue

    B. Regulatory standard

    C. Compliance issue

    D. Constraint

19. Your company provides answering services for several major catalog retailers. The number of calls coming into the service center per month has continued to increase over the past 18 months. The phone system is approaching the maximum load limits and needs to be upgraded. You've been assigned to head up the upgrade project. Based on the company's experience with the vendor who worked on the last phone upgrade project, you're confident they'll be able to assist you with this project as well. Which of the following is true?

   A. You've made an assumption about vendor availability and expertise. The project came about because of a business need.

   B. Vendor availability and expertise are constraints. The project came about because of a business need.

   C. You've made an assumption about vendor availability and expertise. The project came about because of a market demand.

   D. Vendor availability and expertise are constraints. The project came about because of a market demand.

20. Which of the following is not a major step of decomposition?

   A. Identify major deliverables.

   B. Identify resources.

   C. Identify components.

   D. Verify correctness of decomposition.

Chapter

5

Creating the Project Schedule

THE PMP® EXAM CONTENT FROM THE PEOPLE DOMAIN COVERED IN THIS CHAPTER INCLUDES THE FOLLOWING:

✓ **Task 1.2 Lead a team**

- 1.2.1 Set a clear vision and mission

✓ **Task 1.6 Build a team**

- 1.6.2 Deduce project resource requirements

THE PMP® EXAM CONTENT FROM THE PROCESS DOMAIN COVERED IN THIS CHAPTER INCLUDES THE FOLLOWING:

✓ **Task 2.1 Execute project with urgency required to deliver business value**

- 2.1.3 Support the team to subdivide project tasks as necessary to find the minimum viable product

✓ **Task 2.6 Plan and manage schedule**

- 2.6.1 Estimate project tasks (milestones, dependencies, story points)

- 2.6.2 Utilize benchmarks and historical data

- 2.6.3 Prepare schedule based on methodology

- 2.6.5 Modify schedule, as needed, based on methodology

✓ **Task 2.12 Manage project artifacts**

- 2.12.2 Validate that the project information is kept up to date and accessible to all stakeholders

✓ **Task 2.13 Determine appropriate project methodology/methods and practices**

- 2.13.1 Assess project needs, complexity, and magnitude

- 2.13.2 Recommend project execution strategy

- 2.13.14 Recommend a project methodology/approach

THE PMP® EXAM CONTENT FROM THE BUSINESS ENVIRONMENT DOMAIN COVERED IN THIS CHAPTER INCLUDES THE FOLLOWING:

✓ **Task 3.4 Support organizational change**

- 3.4.3 Evaluate impact of the project to the organization and determine required actions

The Planning process group has more processes than any other process group. As a result, a lot of time and effort goes into the Planning processes of any project. On some projects, you might spend almost as much time planning the project as you do executing and controlling it. This isn't a bad thing. The better planning you do up front, the more likely you'll have a successful project. Planning is essential on all projects no matter the development methodology you're using. We'll talk about many aspects of agile planning in this chapter.

This is another fun-filled, action-packed chapter. We'll start off in a predictive methodology approach by defining the schedule management plan, and then move on to define the activities that become the work of the project. The WBS will come in handy here, so keep it close. Then we'll sequence the activities in their proper order, estimate the resources we'll need to complete the work, and estimate how long each activity will take. Last but not least, we'll develop the project schedule.

Everything you've done up to this point and the processes we'll discuss in this chapter will help you create an accurate project schedule. You'll use these documents (along with several other documents you've created along the way) throughout the Executing and Monitoring and Controlling processes to help measure the progress of the project.

Along the way, I'll describe how these processes work in an adaptive project management approach. Remember that project work in an agile methodology is performed in small increments of time called sprints, iterations, or releases. You may not develop a full-blown project schedule with this approach as you would when using a predictive methodology. I'll talk about how work is scheduled in an agile methodology, and we'll look at the burndown chart, a Kanban board, and a Scrum board. Let's get going.

The process names, inputs, tools and techniques, outputs, and descriptions of the project management process groups and related materials and figures in this chapter are based on content from *A Guide to the Project Management Body of Knowledge (PMBOK® Guide), Sixth Edition* (PMI®, 2017). The references to adaptive and hybrid methodologies, related materials, and figures in this chapter are based on content from the *Agile Practice Guide* (PMI®, 2017).

Creating the Schedule Management Plan

The *Plan Schedule Management* process describes how the project schedule will be developed, executed, and controlled as well as how changes will be incorporated into the project schedule. According to *A Guide to the Project Management Body of Knowledge (PMBOK® Guide), Sixth Edition*, the primary benefit of this process is that it helps guide and direct how the project schedule will be managed. The only output of this process is the *schedule management plan*. Let's first look at the inputs and the tools and techniques of this process and spend most of our time examining the schedule management plan itself.

The inputs of this process include the project charter, project management plan (scope management plan, development approach), enterprise environmental factors, and organizational process assets.

Enterprise environmental factors include the culture of the organization and availability of resources as well as the scheduling software, the guidelines for the tailoring approach you'll use, and estimating of data that might be found in commercial databases. The organizational process assets that are important to this process are templates, change control processes, historical information, control tools for managing schedules, and reporting tools.

We have seen all of the tools and techniques before. They are expert judgment, data analysis (alternatives analysis), and meetings.

The key to this process, as I stated earlier, is the schedule management plan, which is an element of the project management plan. It is the only output of this process, and it serves to describe how the project schedule will be developed, monitored, controlled, and changed. According to the *PMBOK® Guide*, several elements make up the schedule management plan. Do note that there are elements in this plan that relate to an adaptive development cycle. This implies that you should create a schedule management plan whether using a predictive, hybrid, or adaptive approach. In practice, I've never developed a schedule management plan for an agile project, but be aware for the exam that PMI® thinks it's a good idea. Be certain to review all the elements in this plan. I have highlighted some of the most important elements here:

Schedule Model Development   This refers to the methodologies and tools you'll use to develop the schedule (for example, Oracle Primavera or Microsoft Project), along with the data they contain. It may also describe how to maintain the schedule model along with the reporting formats, links to the WBS, and guidelines for releases and iterations if you are using an agile approach.

Iteration Definition   The agile development approach is sometimes performed using iterations. Iterations are time bound, and the schedule management plan should document the expected length of an iteration.

Accuracy Levels   This element describes the rounding you'll use when deriving activity duration estimates. For example, you might round to the nearest week, day, or hour depending on the complexity of the project.

Units of Measure   This element also concerns activity duration estimates as well as schedule activities. This describes what measure you'll use when developing the schedule, such as hours, days, weeks, or some other measure.

Control Thresholds   Control thresholds refer to the level of variance the schedule can experience before you take action. Again, depending on the complexity of the project, this might be a generous amount of time or a very limited amount of time. You can express thresholds in terms of hours or days (as an example, a slippage of greater than three days requires action) or, most typically, as a percentage of time.

Performance Measurement Rules   This refers to where and what types of measures you'll use to verify schedule performance such as schedule variance and schedule performance index, or other earned value management techniques. This could include designating levels on the WBS and/or determining what type of earned value measurement technique you'll use.

Defining Activities

Now you're off and running toward the development of your project schedule. To develop the schedule, you first need to define the activities, sequence them in the right order, estimate resources, and estimate the time it will take to complete the tasks. I'll cover the Define Activities process here and the Sequence Activities process next, and I'll pick up with the estimating processes in the next section.

 Define Activities and Sequence Activities are separate processes, each with its own inputs, tools and techniques, and outputs. In practice, especially for small to medium-sized projects, you can combine the Create WBS process we talked about in Chapter 4, "Developing the Project Scope," with these processes and complete them all at once.

The *Define Activities* process is a further breakdown of the work package elements of the WBS. It documents the specific activities needed to fulfill the deliverables detailed on the WBS and the project scope statement. Like the work package level of the WBS, activities can be easily assigned, estimated, scheduled, and controlled. The Define Activities process might be performed by the project manager, or it might be performed when the WBS is broken down to the subproject level.

The following are the key inputs that you should examine when starting the Define Activities process: schedule management plan, scope baseline, organizational culture, templates, and lessons learned. The tools and techniques that will help you in the Define Activities process include expert judgment, decomposition, rolling wave planning, and meetings.

We covered most of these techniques in the previous chapter. Decomposition in this process involves breaking the work packages into smaller, more manageable units of work called *activities*. These are not deliverables but the individual units of work that must be completed to fulfill the work packages and the deliverables listed in the WBS. Activities will help in later Planning processes to define estimates and create the project schedule. Activity lists (which are one of the outputs of this process) from prior projects can be used as templates in this process. Rolling wave planning involves planning near-term work in more detail than future-term work. This is a form of progressive elaboration. Expert judgment relies on project team members with prior experience developing project scope statements and WBSs. When you're using an adaptive methodology, you'll look for experts who have worked on projects using user stories and backlogs and who understand how to break down a user story into activities.

Exam Spotlight

The purpose of the Define Activities process is to decompose the work packages into schedule activities, where the basis for estimating, scheduling, executing, and monitoring and controlling the work of the project is easily supported and accomplished. In an adaptive methodology, this involves breaking user stories down into tasks or activities. The team members will choose the tasks they want to work on during the iteration and estimate the time needed to complete the task.

Creating the Activity List

The key outputs you'll produce in the Define Activities process are the activity list, activity attributes, and milestone list. Let's look at each next.

Activity List

The *activity list* should contain a list of all the schedule activities that will be performed for the project, with a scope of work description of each activity and an identifier (such as a code or number) so that team members understand what the work is and how it is to be completed. The schedule activities are individual elements of the project schedule, and the activity list document is part of the project documents. To keep your sanity, and that of your team members, be certain to enter the activity names onto the schedule the same way they appear on the activity list.

Activity Attributes

Activity attributes describe the characteristics of the activities and are an extension of the activity list. Activity attributes will change over the life of the project as more information is known. In the early stages of the project, activity attributes might include the activity ID, the WBS identification code it's associated with, and the activity name. As you progress through the project and complete other Planning processes, you might add predecessor and successor activities, logical relationships, leads and lags, resource requirements, and constraints and assumptions associated with the activity. According to the *PMBOK® Guide*, activity attributes might help you understand where the work needs to take place for that activity, the project calendar assignment for the activity, and the type of effort needed to complete the work. Schedule development relies on activity attributes to help decide how to order and prioritize activities. We'll cover these topics throughout the remainder of this chapter.

The activity attributes are used as input to several processes, including the Develop Schedule process that we'll talk about in the section "Developing the Project Schedule" later in this chapter.

In practice, I like to tie the activity list to the WBS. Remember from Chapter 4 that each WBS element has a unique identifier, just like the activities in the activity list. When recording the identifier code for the activity list, I'll use a system whereby the first three or four digits represent the WBS element the activity is tied to and the remaining digits refer to the activity itself.

Milestone List

Milestones are typically major accomplishments of the project and mark the completion of major deliverables or some other key event in the project. For example, approval and sign-off on project deliverables might be considered milestones. Other examples might be the completion of a prototype, system testing, and contract approval. The milestone list records these accomplishments and documents whether the milestone is mandatory or optional. The milestone list is part of the project documentation and is also used to help develop the project schedule.

Breaking Down User Stories

As you've seen so far, there are instances where the *PMBOK® Guide* states that comments about the agile approach should be included in some of the Planning documents we've discussed. The *PMBOK® Guide* is heavily focused on process and tools, and this works well when using a predictive methodology to manage a project. When using an adaptive methodology to manage the project, there is minimal focus on process and maximum focus on delivering value to the customer. According to the *Agile Manifesto*, these four values form the foundation for any agile project management approach:

- Value individuals and interactions over process and tools.
- Value working software over comprehensive documentation.
- Value customer collaboration over contract negotiation.
- Value responding to change over following a plan.

When using an agile methodology, it's more important to deliver value to the customer and not worry too much about the processes used to get there. There is little focus on delivering documentation or following a plan. It's about providing a workable deliverable (or portion of a deliverable) during each iteration, obtaining feedback, and making changes as needed in the next iteration. At the end of each iteration, results are inspected and reviewed and then the cycle repeats, always focusing on the goal of producing value during the iteration.

The *Agile Manifesto* also outlines 12 principles for agile methodologies. Having a customer focus and embracing change sit at the core of these principles. The following is a summarized list of the 12 principles:

- Customer satisfaction is achieved by means of frequent and continuous delivery.
- Accept changes to requirements at all stages of the project.
- Provide frequent delivery of working software.
- Perform daily collaboration between the business personnel and agile teams throughout the project.
- Build the project around motivated team members, trust the team to do the job, and provide them with the support they need.
- Hold face-to-face conversations.
- The primary measure of progress is working code.
- Development should be sustainable and all cross-functional team members should be able to maintain a consistent, steady pace of work throughout the project.
- Create good design and have a continuous focus on technical excellence.
- Value simplicity. The team tackles only the work needed to accomplish the objective in front of them.
- Agile teams are self-organized and self-directed.
- Hold retrospective meetings and take the time to discuss how to be more effective. Adjust behaviors to achieve effective outcomes.

The Define Activities process in an agile approach will be performed by breaking down the user stories into activities. This happens at the beginning of the iteration. An iteration is a short, time-bound period of work, and it kicks off with an *iteration planning meeting*.

During the meeting, the product owner identifies the user stories that should be worked on during the upcoming iteration, pulls them from the product backlog, and places them into the iteration backlog. Team members then choose which user stories to work on during the iteration and break them down into activities or tasks. They also estimate the time it will take to complete the activities and determine how much of the work can be accomplished during the iteration. Because the agile team is self-directed, they have the ability to use the tools and techniques that work best for them to break down tasks and determine estimates. They are the experts in the work and therefore have the freedom to choose their tasks and the tools and techniques they will use to perform the tasks. The agile team will not likely go to the rigor of documenting the tasks in an activity list and ascribing attributes to them. Shhh, please don't tell PMI® I said that.

The *PMBOK® Guide* emphasizes the step-by-step process approach, completing all of the processes within a process group one after the other. This approach applies to projects managed using a predictive methodology, and perhaps in some cases a hybrid methodology. Agile is intended to deliver results quickly and, in practice, doesn't follow or emphasize the rigor that the *PMBOK® Guide* brings to a project.

Exam Spotlight

The product owner prioritizes the backlog items based on business need, risk, and value to the organization. The agile team members break down the user stories into manageable portions of work that can be completed in an iteration.

Understanding the Sequence Activities Process

Now that you've identified the schedule activities, you need to sequence them in a logical order and find out whether dependencies exist among the activities. The interactivity of logical relationships must be sequenced correctly in order to facilitate the development of a realistic, achievable project schedule.

Consider a classic example. Let's say you're going to paint your house but unfortunately, it's fallen into a little disrepair. The old paint is peeling and chipping and will need to be scraped before a coat of primer can be sprayed on the house. After the primer dries, the painting can commence. In this example, the primer activity depends on the scraping. You can't—okay, you *shouldn't*—prime the house before scraping off the peeling paint. The painting activity depends on the primer activity in the same way. You really shouldn't start painting until the primer has dried.

During *Sequence Activities*, you will use a host of inputs and tools and techniques to produce the primary output, which is the project schedule network diagrams. You've already seen all the inputs to this process. They are the project management plan (schedule management plan, scope baseline), project documents (activity list, activity attributes, milestone list, assumption log), enterprise environmental factors, and organizational process assets.

One of the enterprise environmental factors considered in this process that you should know about for the exam is the work authorization system. A work authorization system is a formal, documented procedure that describes how to authorize work and how to begin that work in the correct sequence and at the right time. This system describes the steps needed to issue a work authorization, the documents needed, the method or system you'll use to record and track the authorization information, and the approval levels required to authorize the work. You should understand the complexity of the project and balance the cost of instituting a work authorization system against the benefit you'll receive from it. This might be overkill on small projects, and verbal instructions might work just as well. For the purposes of the exam, understand that project work is assigned and committed via a work authorization system and is used throughout the Executing processes.

Work is usually authorized using a form that describes the task, the responsible party, anticipated start and end dates, special instructions, and whatever else is particular to the activity or project. Depending on the organizational structure, the work is generally assigned and authorized by either the project manager or the functional manager. Remember that the system includes documentation on who has the authority to issue work authorization orders.

We'll look at several new tools and techniques next.

Precedence Diagramming and Leads and Lags

Sequence Activities has four tools and techniques, all of which are new to you:

- Precedence diagramming method (PDM)
- Dependency determination and integration
- Leads and lags
- Project management information system

I'll switch the order of these and cover dependency determination and integration first. In practice, you'll define dependencies either before or while you're using the PDM to draw your schedule network diagram. To make sure you're on the same page with the *PMBOK*® *Guide* terminology regarding dependencies, I'll cover them first and then move on to the other tools and techniques.

Dependency Determination and Integration

Dependencies are relationships between the activities in which one activity is dependent on another to complete an action, or perhaps an activity is dependent on another to start an action before it can proceed. Dependency determination is a matter of identifying where those dependencies exist. For example, thinking back to the house-painting example, you couldn't paint until the scraping and priming activities were completed. You'll want to know about four types of dependencies for the exam:

- Mandatory dependencies
- Discretionary dependencies
- External dependencies
- Internal dependencies

As you've probably guessed, the *PMBOK® Guide* defines dependencies differently depending on their characteristics:

Mandatory Dependencies   *Mandatory dependencies*, also known as *hard logic* or *hard dependencies*, are defined by the type of work being performed. The scraping, primer, and painting sequence is an example of mandatory dependencies. The nature of the work itself dictates the order in which the activities should be performed. An activity with physical limitations is a telltale sign that you have a mandatory dependency on your hands. The project team defines and determines mandatory dependencies, and they do this when they are performing the Sequence Activities process.

Discretionary Dependencies   *Discretionary dependencies* are defined by the project team. Discretionary dependencies are also known as *preferred logic*, *soft logic*, or *preferential logic*. These are usually process- or procedure-driven or "best practice" techniques based on past experience. For example, both past experience and best practices on house-painting projects have shown that all trim work should be hand-painted whereas the bulk of the main painting work should be done with a sprayer. These activities could be performed at the same time (therefore, they are not mandatory dependencies). One side of the house might be spray-painted while the other side is having the trim work done. Sometimes, depending on the activity, it is likely more efficient and less risky to perform the activities in sequence.

Exam Spotlight

Discretionary dependencies have a tendency to create arbitrary total float values that will limit your options when scheduling activities that have this type of dependency. If you are fast-tracking to compress your schedule, you should consider changing or removing these dependencies.

External Dependencies   *External dependencies* are, well, external to the project. This might seem obvious, but the *PMBOK® Guide* points out that even though the dependency is external to the project (and, therefore, a non-project activity), it impacts project activities. For example, perhaps your project is researching and marketing a new drug. The FDA must approve the drug before your company can market it. This is not a project activity, but the project cannot move forward until approval occurs. That means FDA approval is an external dependency.

Internal Dependencies   *Internal dependencies*, another somewhat obvious dependency, are internal to the project or the organization. They may, however, still be outside of your control. For example, perhaps before implementing a new time-tracking system in your maintenance shop, the operations department has decided to study the business rules regarding time tracking. Examining and updating the business rules and processes needs to be completed before the time-tracking system can be installed and your project can proceed.

Keep in mind that these dependencies could coexist. For example, you may have a mandatory external dependency and a discretionary internal dependency, and so on.

Once you've identified the dependencies and assembled all the other inputs for the Sequence Activities process, you'll take this information and produce a diagram—or schematic display—of the project activities. The project schedule network diagram shows the dependencies—or logical relationships—that exist among the activities. You can use one of the other tools and techniques of this process to produce this output. You'll now examine each in detail.

Precedence Diagramming Method

The *precedence diagramming method (PDM)* is what most project management software programs use to sequence activities into a schedule model. Precedence diagrams use boxes or rectangles (called *nodes*) to represent the activities. The nodes are connected with arrows showing the dependencies between the activities. This method is also called *activity on node (AON)*.

The minimum information that should be displayed on the node is the activity name, but you might put as much information about the activity on the node as you'd like. Sometimes the nodes are displayed with activity name, activity number, start and stop dates, due dates, slack time, and so on. Figure 5.1 shows a PDM—or AON—of the house-painting example.

Exam Spotlight

For the exam, remember that the PDM uses only one time estimate to determine duration.

FIGURE 5.1   Example of a PDM or AON

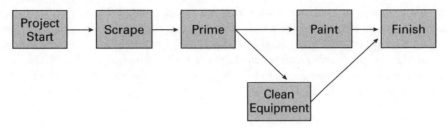

The PDM is further defined by four types of *logical relationships*. The terms *dependencies* and *precedence relationships* also are used to describe these relationships. You might already be familiar with these if you've used Microsoft Project or similar project management software. The four dependencies, or logical relationships, are as follows:

Finish-to-Start (FS)   The finish-to-start relationship is the most frequently used relationship. This relationship says that the predecessor—or *from* activity—must finish before the successor—or *to* activity—can start. In PDM diagrams, this is the most often used logical relationship.

Start-to-Finish (SF)   The start-to-finish relationship says that the predecessor activity must start before the successor activity can finish. This logical relationship is seldom used.

Finish-to-Finish (FF)   The finish-to-finish relationship says that the predecessor activity must finish before the successor activity finishes.

Start-to-Start (SS)   I think you're getting the hang of this. The start-to-start relationship says that the predecessor activity must start before the successive activity can start.

Exam Spotlight

For the exam, know that finish-to-start is the most commonly used dependency in the PDM method and that start-to-finish is rarely used. Also remember that according to the *PMBOK® Guide*, each activity on the network diagram is connected by at least one predecessor and one successor activity *except* the first and last activity.

Keep these logical relationships (or dependencies) in mind when constructing your project schedule network diagram.

Applying Leads and Lags

Leads and lags should be considered when determining dependencies. *Lags* occur when time elapses between two activities, which delay *successor activities* (those that follow a predecessor activity) from starting, and as a result, time is added either to the start date or to the finish date of the activity you're scheduling. *Leads*, conversely, speed up the successor activities, and as a result, time needs to be subtracted from the start date or the finish date of the activity you're scheduling.

Exam Spotlight

Leads and lags speed up or delay successor activities but should not replace schedule logic.

Let's revisit the house-painting example to put all this in perspective. In order to paint, you first need to scrape the peeling paint and then prime. However, you can't begin painting until the primer has dried, so you shouldn't schedule priming for Monday and painting for Tuesday if you need the primer to dry on Tuesday. Therefore, the priming activity generates the need for lag time at the end of the activity to account for the drying time needed before you can start painting.

Lead time works just the opposite. Suppose, for this example, you could start priming before the scraping is finished. Maybe certain areas on the house don't require scraping, so you don't need to wait until the scraping activity finishes to begin the priming activity. In this example, lead time is subtracted from the beginning of the priming activity so that this activity begins prior to the previous activity finishing.

You might also use schedule network templates in this process. These are not a named tool and technique of this process but may come in handy on your next project. Schedule network templates are like the templates I've talked about in previous processes. Perhaps the project you're working on is similar to a project that has been completed in the past. You can use a previous project schedule network diagram as a template for the current project. Or you might be working on a project with several deliverables that are almost identical to projects you've performed in the past or the deliverables on the existing project are fairly similar; in that case, you can use the old schedule network diagrams, or even the same schedule network diagrams, as templates for the project. Templates can be used for certain portions of the project schedule or for the entire project. If you are using templates for portions of the project schedule, they are known as subnetwork templates or fragment network templates.

Project Management Information System

The *project management information system (PMIS)* is an automated system used to document the project management plan and component plans, to facilitate the feedback process, to revise the documents, to create reports, and to develop the schedule. It incorporates the configuration management system and the change control system, both of which I'll cover in Chapter 11, "Measuring and Controlling Project Performance." Later in the project, the PMIS can be used to control changes to any of the plans. When you're thinking about the PMIS as an input (that is, as part of the enterprise environmental factors), think of it as a collection and distribution point for information as well as an easy way to revise and update documents. When you're thinking about the PMIS as a tool and technique, think of it just that way—as a tool to facilitate the automation, collection, and distribution of data and to help monitor processes such as scheduling, resource leveling, budgeting, and web interfaces.

Project Schedule Network Diagrams

There are only two outputs of the Sequence Activities process: project schedule network diagrams and project documents updates (activity attributes, activity list, assumption log, milestone list). I've just spent a good deal of time describing the project schedule network diagrams you can construct using the PDM technique. You typically generate project schedule network diagrams on a computer. Like the WBS, these diagrams are visual representations of the work of the project and might contain all the project details, or they might contain only summary-level details, depending on the complexity of the project. Summary-level activities are a collection of related activities, also known as *hammocks*. Think of hammocks as a group of related activities rolled up into a summary heading that describes the activities likely to be contained in that grouping.

Keep in mind that the construction of these project schedule network diagrams might bring activities to light that you missed when defining your activity list, or it might make you break an activity down into two activities in places where you thought one activity might work. If this is the case, you will need to update the activity list and the activity attributes. The other project document update that may be required as a result of this process is an update to the risk register. After the activities are sequenced, the next steps involve estimating the resources and estimating the durations of the activities so that they can be plugged into the project schedule. We'll look at these topics in the next sections of this chapter.

I'll discuss the process of sequencing activities for adaptive methodologies in the section titled "Using a Kanban Board and Scrum Board" later in this chapter. This activity is combined with creating a schedule in an agile approach, and you'll understand the process better by looking at it as one activity.

Estimating Activity Resources

All projects, from the smallest to the largest, require resources. The term *resources*, in this case, does not mean just people; it means all the physical resources required to complete the project. According to the *PMBOK® Guide*, this includes people, equipment, supplies, materials, software, hardware—the list goes on depending on the project on which you're working. The *Estimate Activity Resources* process is concerned with determining the types of resources needed (both human and materials) and in what quantities for each schedule activity within a work package.

The *PMBOK® Guide* notes that Estimate Activity Resources should be closely coordinated with the Estimate Costs process (I'll talk about Estimate Costs in Chapter 6, "Developing the Project Budget and Engaging Stakeholders"). That's because resources—whether people or material or both—are typically the largest expense you'll have on any project. Identifying the resources becomes a critical component of the project planning process so that estimates—and ultimately the project budget—can be accurately derived.

The Estimate Activity Resources process has several inputs you'll consider when performing the estimating process, most of which you already know. One input you'll want to refer to is the resource management plan, which documents the roles and responsibilities for project team members, their reporting relationships, and how resources (materials, equipment, and human resources) will be managed throughout the project. Three new inputs that you haven't seen before are resource calendars, risk register, and cost estimates (all part of the project documents input). I'll cover each of these next.

Resource calendars are an output of the Acquire Project Team process and are updated as a result of performing other processes. Both of these processes are performed during the Executing process group, so you may find this input perplexing here. However, you may have some resource availability information (resource calendars) on a preliminary basis during the Estimate Activity Resources process, and you'll further define the resource calendar when resources are assigned to the project later in the Executing processes. In practice, you may find that you perform the Acquire Project Team process during the later stages of the Planning portion of the project rather than in the Executing process as outlined in the *PMBOK® Guide*.

The *resource calendars* input describes the time frames in which resources (both human and material) are available. They outline workdays, normal business hours, shift-work hours if they apply, and days off such as holidays or non-work days like weekends. They look at a particular resource or groups of resources and their skills, abilities, quantity, and availability. Perhaps your project calls for a marketing resource and the person assigned to the marketing activities is on an extended vacation in October. The resource calendar would show this person's vacation schedule. Resource calendars also examine the quantity, capability, and availability of equipment and material resources that have a potential to impact the project schedule. For example, suppose your project calls for a hydraulic drill and your organization owns only one. The resource calendar will tell you whether it's scheduled for another job at the same time it's needed for your project.

The risk register is an output of the Identify Risks process. It is a list of identified risks and their potential responses. Activity cost estimates are an output of the Estimate Costs process and are the costs that are associated with completing the activity. We will look at both of these in more depth in future chapters.

How to Estimate Activity Resources

Your goal with the Estimate Activity Resources process is to determine the activity resource requirements, including quantity and availability. This is an important function for the project manager since you'll need this information to construct a project schedule, and more importantly, to deliver a successful project. Identifying and estimating the resources needed for the project should occur early in the project so that you can secure the resources you need and ensure availability.

This process has seven tools and techniques to help accomplish this output: expert judgment, bottom-up estimating, analogous estimating, parametric estimating, data analysis (alternatives analysis), project management information systems, and meetings. We'll cover all the estimating techniques in the Estimate Activity Durations process in the next section following this process. Estimating resources and estimating activity durations use similar estimating techniques. You already know what expert judgment and meetings entail, so let's look at the nuances of the alternatives analysis and PMIS for this process:

Alternative Analysis   Alternative analysis is used when thinking about the methods you might use to accomplish the activities your resources have been assigned. Many times, you can accomplish an activity in more than one way, and alternative analysis helps you decide among the possibilities. For example, a subcompact car drives on the same roads a sports car travels. The sports car has a lot more features than the subcompact, it's faster, it's probably more comfortable, and it has a visual appeal that the subcompact doesn't. The sports car might be the valid resource choice for the project, but you should consider all the alternatives. The same idea applies to human resources in that you might apply senior-level resources versus junior-level resources, or you could add resources to speed up the schedule. You may also use make-rent-or-buy analysis when determining alternative resources.

Project Management Information System   Project management software can help plan, organize, and estimate resource needs and document their availability. It might also help you produce resource breakdown structures, resource rates, resource calendars, and availability.

Documenting Resource Requirements

The purpose of the Estimate Activity Resources process is to develop the resource require-ments output. This output describes the types of resources and the quantity needed for each activity associated with a work package. You should prepare a narrative description for this output that describes how you determined the estimate, including the information you used to form your estimate and the assumptions you made about the resources and their avail-ability. You could do this by preparing a table similar to Table 5.1. Feel free to add other information you may find useful to this example.

TABLE 5.1   Resource requirements

Resource type	Description	Quantity	Estimating method	Activity/work package
C++ developer	Senior level program devel-opers	2	Data analysis	Write the algorithm for predicting arrival times for on-demand services
Virtual server	Needed to host the new appli-cation	1	Analogous estimating	Set up the infrastructure

Work package estimates are derived by taking a cumulative total of all the schedule activities within the work package.

You'll use the resource requirements in the next process (Estimate Activity Durations) to determine the length of time each activity will take to complete. That, of course, depends on the quantity and skill level of the resources assigned, which is the reason you estimate resources before you try to determine duration.

The other outputs of this process are the basis of estimates, the resource breakdown structure, and project documents updates (activity attributes, assumption log, lessons-learned register).

The basis of estimates output contains all the supporting detail regarding the esti-mates. This should include assumptions about the estimates, the methods used to derive the estimate, constraints, what resources you used to help develop the estimates whether human or database related, a confidence level of the estimate, and any risks associated with the estimate that may change the estimate as the project progresses.

The *resource breakdown structure (RBS)* is much like an organizational breakdown structure, but the RBS lists the resources by category and type. You may have several cat-egories of resources, including labor, hardware, equipment, supplies, and so on. *Type* describes the types of resources needed, such as skill levels or quality grades of the material.

The project documents updates portion of this output refers to updating the activity attributes, assumption log, and the lessons-learned register with changes to any of the ele-ments you've recorded here.

Estimating Resources in an Adaptive Methodology

Planning for resources when using an adaptive methodology can be challenging. According to the *Agile Practice Guide* (PMI®, 2017), you should have agreements worked out with your vendor partners that allow for fast fulfillment of your orders (or resource needs) and use Lean methods to eliminate waste in your processes. This strategy will allow you to control costs and meet the schedule timeline. Estimating resources in an agile methodology can be conducted using some of the same techniques you'll use in a predictive methodology. Agile teams consist of cross-functional team members, and you'll want to ensure their availability by checking the resource calendars. You'll need to identify the skills needed to produce the work of the project. Perhaps your project needs a web designer, a web developer, an interface developer, a database developer, and a graphic designer. This will be the agile team makeup for the project from a skills perspective. These same team members could be working on other agile or predictive projects. So, you will need to know their availability and capacity for work.

If you're working in a functional organization, you may have to coordinate the use of resources for your project with the functional managers. For example, software development projects will require technical skills and a product owner who can speak on behalf of the business. Most of the technical resources will likely come from the same department, or maybe even the same team. The team members are responsible for breaking down user stories into tasks and determining how much work they can take on during an iteration. They will self-organize by choosing the user stories they have the skills to complete. For example, the developer will choose development activities and the database administrator will choose database-related activities.

You should consider other resources that may be needed throughout the project when using an agile approach. For example, technology projects often require hardware purchases, licenses for software components they're using or will incorporate into the final product, and so on. The team facilitator will ensure the resources are identified and procured at the right time so that they are available when the team is ready.

You can see how these "Activity" processes have built on one another. First you defined the activities, then you determined dependencies and sequenced them in the correct order, and next you determined what types and quantities of resources are required to complete the activities. Now you're ready to begin estimating the duration of these activities so that you can plug them into the project schedule.

Estimating Activity Durations

The *Estimate Activity Durations* process attempts to estimate the work effort, resources, and number of work periods needed to complete each activity. The *duration estimates* are the primary output of this process. These are quantifiable estimates expressed as the number of work periods needed to complete a schedule activity. Work periods are usually expressed in hours or days. However, larger projects might state duration in weeks or months. Work periods are the activity duration estimates, and they become inputs to the Develop Schedule process.

When estimating activity duration, be certain to include all the time that will elapse from the beginning of the activity until the work is completed. Let's consider the earlier example of the house-painting project. You estimate that it will take three days, including drying time, to prime the house. Now, let's say priming is scheduled to begin on Saturday but your crew doesn't work on Sunday. The activity duration in this case is four days, which includes the three days to prime and dry plus the Sunday the crew doesn't work. Most project management software programs will handle this kind of situation automatically once you've keyed in the project calendar and work periods.

Progressive elaboration comes into play during this process also. Estimates typically start at a fairly high level, and as more details are known about the deliverables and their associated activities, the estimates become more accurate. You should rely on those folks who have the most knowledge of the activities you're trying to estimate to help you with this process.

According to the *PMBOK® Guide,* there are a few factors you should consider when determining duration estimates. The first consideration is that some activities will dictate a certain duration. For example, new server hardware may require a burn-in period of a certain number of hours or days. No amount of effort can alter the duration of this activity.

You may think that adding more resources to the activity will help accomplish the work faster. In some cases, adding too many resources will result in them stepping on one another or actually taking more time to complete the activity because too many people are involved. More resources on an activity is not necessarily a good thing and could manifest in what's known as the *law of diminishing returns.* At some point, additional resources will have the effect of degrading the duration of the activity rather than improving it. When a team member tells you, "It will take me more time to show the new person how to perform this activity than it will for me to do it myself," you may have an example of the law of diminishing returns. Another way to think about it is this saying, "Nine women cannot have a baby in one month." Simply put, there is a point where the energy expended does not produce enough benefit to warrant the level of effort you're putting into the activity.

Advances in technology should be taken into consideration when estimating resources. New technologies may reduce the duration of an activity or require an adjustment in the types or amounts of resources needed.

Lastly, you should consider the motivation of the team members working on these activities. Procrastination may come into play and endanger the timeline when team members wait until the day before the due date to perform the work and meet the deadline. This is also known as *student syndrome* because many students are notorious for waiting before the night before a paper is due to start writing it. Or the opposite effect may occur where they slowly work toward the due date, dragging out the work so that they don't have to take on more tasks (also known as *Parkinson's law.*)

Project Calendars and Other Considerations

The inputs to this process include the project management plan (schedule management plan, scope baseline), project documents (activity attributes, activity list, assumption log, lessons-learned register, milestone list, project team assignments, resource breakdown structure, resource calendars, resource requirements, risk register), enterprise environmental factors, and organizational process assets.

A few of the important elements regarding these inputs apply here as you've seen in past processes: databases, productivity metrics, historical information regarding durations on similar projects, project calendars, scheduling methodology, and lessons learned. The *project calendars* (which list company holidays, shift schedules, and so on) are considered a part of the organizational process assets, and activity resource requirements are especially useful during this process.

Estimating Techniques

There are new tools and techniques we'll look at in the Estimate Activity Durations process that will help in determining the duration estimates, the primary output of this process. The tools and techniques are

- Expert judgment
- Analogous estimating
- Parametric estimating
- Three-point estimating
- Bottom-up estimating
- Data analysis (alternatives analysis, reserve analysis)
- Decision-making
- Meetings

We'll take a look at each of these tools and techniques next, and how they will be used in helping to determine duration estimates, with the exception of meetings, which we discussed earlier.

Expert Judgment

The experts in this process are the staff members who will perform the activities. Because they are doing the work, they are more likely to come up with an accurate estimate. Typically, these team members apply expert judgment because of their experience with similar activities on past projects. You should be careful with these estimates, though, because they are subject to bias and aren't based on any scientific means. Your experts should consider that resource levels, resource productivity, resource capability, risks, and other factors can impact estimates. It's good practice to combine expert judgment with historical information and to use as many experts as you can.

Analogous Estimating

Analogous estimating, also called *top-down estimating*, is a form of expert judgment. With this technique, you will use the actual duration of a similar activity completed on a previous project to determine the duration of the current activity—provided the information was documented and stored with the project information on the previous project. This technique is most useful when the previous activities you're comparing are similar to the activity you're estimating and don't just appear to be similar. You want the folks who are working on the estimate to have experience with these activities so that they can provide reasonable estimates. This technique is especially helpful when detailed information about the project is not available, such as in the early phases of the project.

Top-down estimating techniques are also used to estimate total project duration, particularly when you have a limited amount of information about the project. The best way to think about top-down techniques is to look at the estimate as a whole. Think about being on a mountaintop where you can see the whole picture as one rather than all the individual items that make up the picture.

For instance, let's return to the house-painting example. You would compare a previous house-painting project to the current house-painting project if the houses are of similar size and the paint you're using is the same quality. You can use the first house-painting project to estimate the project duration for the second house-painting project because of the similarities in the project.

Top-down techniques are useful when you're early in the project Planning processes and are just beginning to flesh out all the details of the project. Sometimes during the project selection process, the selection committee might want an idea of the project's duration. You can derive a project estimate at this stage by using top-down techniques.

Exam Spotlight

The *PMBOK® Guide* states that analogous estimating is a gross value estimating technique. It also notes that you can use analogous estimating to determine overall project duration and cost estimates for the entire project (or phases of the project). For the exam, remember that the analogous technique is typically less time-consuming and less costly than other estimating techniques but it's also less accurate.

Parametric Estimating

Parametric estimating is a quantitatively based estimating method that multiplies the quantity of work by the rate or uses an algorithm in conjunction with historical data to determine cost, budget, or duration estimates. The best way to describe it is with an example. Suppose you are working on a companywide network upgrade project. This requires you to run new cable to the switches on every floor in the building. To come up with an estimate, you can use parametric estimates to determine activity duration estimates by taking a known element—in this case, the amount of cable needed—and multiplying it by the amount of time it takes to install a unit of cable. In other words, suppose you have 10,000 meters of new cable to run. You know from past experience it takes one hour to install 100 meters. Using this measurement, you can determine an estimate for this activity of 100 hours to run the new cable. Therefore, the cable activity duration estimate is 100 hours.

Exam Spotlight

The *PMBOK® Guide* states that you can also use parametric estimating to determine time estimates for the entire project or portions of the project when you are using this technique in conjunction with other estimating techniques. For the exam, remember that a statistical relationship exists between historical data and other variables (as explained in the cable example earlier) when using parametric estimates and that this technique can be highly accurate if the data you are using is reliable.

Three-Point Estimating

Three-point estimating, as you can probably guess, uses three estimates that, when averaged, come up with a final estimate. The three estimates you'll use in this technique are the most likely estimate (tM), an optimistic estimate (tO), and a pessimistic estimate (tP). The most likely estimate assumes there are no disasters and the activity can be completed as planned. The optimistic estimate is the fastest time frame in which your resource can complete the activity. The pessimistic estimate assumes the worst happens, and it takes much longer than planned to get the activity completed. You'll want to rely on experienced folks to give you these estimates. Then you can choose to use one of two formulas to calculate the expected duration estimate (tE). The first formula, called the *triangular distribution*, consists of summing the optimistic (tO), the pessimistic (tP), and the most likely (tM) estimates and then dividing that sum by 3. The formula looks like this:

$$tE = (tO + tM + tP) / 3$$

Use the triangular distribution when you don't have enough historical data to assist with the estimates or you're using expert judgment to derive estimates.

The second formula is called a *beta distribution*, which is taken from the program evaluation and review technique (PERT) that we will review in depth in the Develop Schedule process later in this process. The formula for beta distribution, or PERT, looks like this:

$$tE = (tO + 4tM + tP) / 6$$

The beta distribution method is preferred when you have a good set of historical data and samples to base the estimates on.

Bottom-Up Estimating

Bottom-up estimating is performed by obtaining individual estimates for each project activity and then adding them all to arrive at a total estimate for the work package. Bottom-up estimating is a good technique to use when you aren't confident about the type or quantity of resources you'll need for the project. This is an accurate means of estimating, provided the estimates at the schedule activity level are accurate. However, it takes a considerable amount of time to perform bottom-up estimating because every activity must be assessed and estimated accurately to be included in the bottom-up calculation. The smaller and more detailed the activity, the greater the accuracy and cost of this technique.

Data Analysis—Reserve Analysis

Contingency reserves—also called *buffers* or *time reserves* in the *PMBOK® Guide*—means a portion of time (or money when you're estimating budgets) that is added to the schedule to account for risk or uncertainty. You might choose to add a percentage of time or a set number of work periods to the activity or the overall schedule or both. Contingency reserves are calculated for known risks that have documented contingency or mitigation response plans to deal with the risk event, should it occur, but you don't necessarily know how much time it will take to implement the mitigation plan and potentially perform rework. For example, you know it will take 100 hours to run new cable based on the quantitative estimate you came up with earlier. You also know that sometimes you hit problem areas when running the cable. To make sure you don't impact the project schedule, you build in a reserve time of 10 percent of your original estimate to account for the problems you might encounter. This brings your activity duration estimate to 110 hours for this activity. Contingency reserves can be and should be modified as the project progresses. As you use the time, or find you don't need the time, you will modify the reserve amounts.

Management reserves are a type of reserve used for unknown events. Since they are unknown, you have not identified them as risks. Management reserves are for that funny feeling you have that something could come up that you haven't thought about during the Planning process. Management reserves set aside periods of time for this unknown work but are not included in the schedule baseline. Keep in mind that this is not time that is available to throw in extra deliverables that didn't make it into the scope statement. Management reserves must be used for project work that is within scope. If you do use management reserves during the project, you must change the schedule baseline to reflect the time used.

> **Exam Spotlight**
>
> Contingency reserves are included in the schedule baseline; management reserves are not. Contingency reserves are for potential work identified during the Risk Planning processes. Management reserves are for unknown circumstances that have not been previously identified but require work that is within the scope of the project. Management reserves that are used on the project require a change to the schedule baseline.

Decision-Making Techniques

Decision-making techniques include the voting techniques we discussed in Chapter 3, "Delivering Business Value," as well as brainstorming, and the Delphi or nominal group techniques. These techniques get your team members involved and will help improve the accuracy of your estimates. Brainstorming is an age-old technique where all participants have an opportunity to speak up. No idea is a bad idea with this technique, and it's essential that the facilitator not allow participants to get into judging contests or debates on the merits of the ideas proposed during the brainstorming session.

The Delphi technique is similar to brainstorming in that you involve subject matter experts in determining estimates. Their experiences with the organization and on similar past projects will help improve the accuracy of the estimates. Because you have them involved in the process and they know that the estimates derived from this exercise will be attached to the project schedule, they are likely to provide more accurate estimates and work hard to meet or beat them.

Another voting technique noted in the *PMBOK® Guide* is called the *fist to five* technique, which is commonly used on projects using an agile methodology. This technique is simple in that team members hold up a fist if they are not in agreement or will not support the idea proposed (or activity duration in this case), or they can hold up five fingers if they are in full support of the idea, or any number of fingers between a fist and five. If someone holds up fewer than three fingers, they are called on to discuss why they are opposed to the idea and the issues they are concerned about if this idea is adopted. This technique is sometimes known as *fist of five*.

Duration Estimates

Everything I've discussed to this point has brought you to the primary output of this process: the duration estimates. You use the inputs and tools and techniques to establish these estimates. As mentioned earlier, activity duration estimates are estimates of the required work periods needed to complete the activity. This is a quantitative measure usually expressed in hours, days, weeks, or months.

One factor to note about your final estimates is that they should contain a range of possible results. In the cable-running example, you would state the activity duration estimates as "100 hours ± 10 hours" to show that the actual duration will take at least 90 hours and might go as long as 110 hours—or you could use percentages to express this range.

The other outputs of Estimate Activity Durations are the basis of estimates and project documents updates. The information that may need to be revisited and updated as a result of this process includes the activity attributes and the assumptions you made regarding resource availability and skill levels, as well as the lessons-learned register.

Now that you have all the activity information in hand, along with a host of other inputs, you're ready to develop the project schedule.

Exam Spotlight

Remember that you perform the activity processes in this order: Define Activities, Sequence Activities, Estimate Activity Resources, and Estimate Activity Durations. Develop Schedule comes after you've completed all of these processes.

 ### Real World Scenario

Desert State University (DSU)

DSU has hired a contract agency to create its new registration website. The website will allow students in good academic standing to register for classes over the Internet. You have been appointed as the project manager for the DSU side of this project. You'll be working with Henry Lu from Websites International to complete this project.

Henry has given you an activity list and asked for time estimates that he can plug into the project plan.

Your first stop is Mike Walters's desk. He's the expert on the existing registration system and he'll be writing the interface programs to accept registration data from the new website. Mike will also create the download that the Internet program will use to verify a student's academic standing. Mike has created other programs just like this in the past. His expertise and judgment are very reliable.

The next stop is Kate Langdon. She's the new team leader of the testing group. Kate has been with DSU for only one month. Since she has no experience working with DSU data and staff members, she tells you she'll get back to you within a week with estimates for the testing activities. She plans to read through the project binders of some similar projects and base her estimates against the historical information on similar projects. She'll run the estimates by her lead tester before giving them to you.

You've asked both of your resources to provide you with three-point estimates. Mike's estimates are an example of using the tool and technique of expert judgment to derive activity duration estimates. The estimates expected from Kate will be derived using historical information (implied by the research she's going to do into past similar projects) and expert judgment because she's involving her lead tester to verify the estimates.

Estimating Activity Durations Using Adaptive Methodologies

In Chapter 4, we talked about collecting requirements and documenting project scope, both of which are needed to determine activities and then estimate their durations. As we've discussed, in an adaptive methodology those processes are completed by documenting user stories. This activity is conducted at the beginning of the project when the product owner identifies all the user stories needed to produce the outcome of the project. Don't forget that in an adaptive approach, they can add user stories later in the project. However, the majority are defined at the beginning of the project.

When using an iteration-based agile approach, the team facilitator will work with the team members to examine all the user stories and estimate how many iterations will be required to complete all of the user stories (in other words, all the work of the project). This estimate can be expressed in hours or team days. Team days are the total cumulative hours all the team members make up together. For example, if you have five team members working full time on the project, one team day consists of 40 hours (8 hours a day × 5 team members). You can use any or all of the estimating techniques discussed in this chapter to estimate the task durations. Expert judgment is used frequently to estimate agile tasks because agile teams are small, and the resources providing the estimates are also performing the work of the project. After estimating the tasks and their approximate durations, the Scrum master will roll this up into a total number of iterations needed to complete the project. Remember that iterations are typically two to four weeks in length, and agile is flexible. There will be changes and new requirements introduced as the project progresses, so the number of iterations may change.

Let's say our project estimate shows that 12 iterations are needed to complete the work. The Scrum master may use what's called a *burndown chart* to track this work. A burndown chart shows the remaining time and work effort for the iteration. It displays the number of iterations on the horizontal axis and the backlog items (expressed in hours of effort) on the vertical axis. At the end of each day, team members update their estimates for the remaining amount of work, which then updates the burndown chart. As the project progresses and iterations are completed, the amount of work will decrease. Team members and stakeholders can visually see the amount of work remaining in the iteration. Figure 5.2 shows an example burndown chart.

Another way to estimate task durations in agile is by identifying story points. *Story points* are defined by the project team members themselves. They are a unit of measure agreed on by the team and are used to estimate the amount of work to complete a user story. Typically, the story point is a comparative approach based on the complexity of the tasks as they relate to one another, and on team members' past experience working on similar tasks. However, story points can be anything the team agrees on as a measurement. You could use a chart similar to Figure 5.2 to show story points and the amount of work remaining on the project. Story points in this case could be plotted on the y-axis in Figure 5.2 instead of task hours. Burndown charts that use story points are called *product burndown charts* or *release burndown charts*.

FIGURE 5.2 Burndown chart

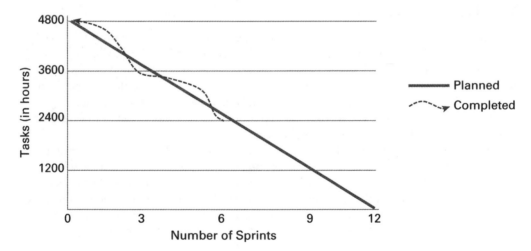

Another estimating concept you should know for the exam is *velocity*. Velocity is generally used in Scrum to determine how long it will take to complete the backlog. Velocity reflects the speed at which the team is working. You can use velocity to estimate the duration of the iteration or the duration of the project. For example, let's say the team has worked on a similar project in the past. They know that their average velocity on that project was 20 story points per iteration. If your project has 100 total story points, it will take 5 iterations to complete the project (100 story points ÷ 20 story points per iteration). If your iterations are 2 weeks in duration, the total duration of the project will be 10 weeks (2 week iteration × 5 total iterations = 10 weeks). Velocity estimates will improve as iterations are completed and the team improves their estimates. You can use this same concept with user stories instead of velocity. For example, let's say your team can complete one user story in two days. If you have 50 user stories total, it will take 100 days to complete all the user stories (50 user stories × 2 days each = 100 days).

Developing the Project Schedule

The *Develop Schedule* process is the heart of the Planning process group. This is where you lay out the schedule for your project activities, determine their start and finish dates, and finalize activity sequences and durations. Develop Schedule, along with Estimate Activity Resources and Estimate Activity Durations, is repeated several times before you come up with the project schedule. Most project management software programs today can automatically build a schedule for you once you've entered the needed information for the activities. The project schedule, once it's approved, serves as the *schedule baseline* for the project that you can track against in later processes.

Remember that you cannot perform Develop Schedule until you have completed at least the following processes in the Planning group (some of these can be performed at the same time for smaller, less complex projects): Collect Requirements, Define Scope, Create WBS, Define Activities, Sequence Activities, Estimate Activity Resources, Estimate Activity Durations, and Plan Resource Management. In practice, it's also beneficial to perform Identify Risks, Perform Qualitative Risk Analysis, Perform Quantitative Risk Analysis, Plan Risk Responses, and Plan Procurement Management prior to developing the schedule.

There is a lot of material to cover in this process, so grab a cup of coffee, tea, or a soda now. I'll start with the inputs to the Develop Schedule process and then follow up with an in-depth discussion of the tools and techniques of the process. These techniques will help you get to the primary outputs of this process: the project schedule and the schedule baseline.

Gather Documents to Assist in Developing the Schedule

There are several inputs you'll need to gather for this process. The more important of these inputs include the schedule management plan, scope baseline, basis of estimates, durations estimates, milestone list, project schedule network diagrams, resource calendars, and team assignments. This is not an exhaustive list, so refer to Appendix B if you're curious about the other inputs.

You can see how important it is to perform all the Planning processes accurately because the information you derive from almost every process in the Planning group is used somewhere else in another Planning process, many of them here. Your project schedule will reflect the information you know at this point in time. If you have incorrectly estimated activity durations or didn't identify the right dependencies, for example, the inputs to this process will be distorted and your project schedule will not be correct. It's definitely worth the investment of time to correctly plan your project and come up with accurate outputs for each of the Planning processes.

As with several other processes, you should pay particular attention to constraints and assumptions when creating the project schedule. Constraints are with you throughout the life of the project. The most important constraints to consider in the Develop Schedule process are time constraints, and they fall into two categories: imposed dates and key events/ major milestones. Imposed dates restrict the start or finish date of activities. The two most common constraints, *start no earlier than* and *finish no later than*, are used by most computerized project management software programs. Let's look once again at the house-painting example. The painting activity cannot start until the primer has dried. If the primer takes 24 hours to dry and is scheduled to be completed on Wednesday, this implies that the painting activity can *start no earlier than* Thursday. This is an example of an imposed date.

Key events or milestones refer to the completion of specific deliverables by a specific date. Stakeholders, customers, or management staff might request that certain deliverables be completed or delivered by specific dates. Once you've agreed to those dates (even if the agreement is only verbal), it's often cast in stone and difficult to change. These dates, therefore, become constraints.

 Be careful of the delivery dates you commit to your stakeholders or customers. You might think you're simply discussing the matter or throwing out ideas, whereas the stakeholder might take what you've said as fact. Once the stakeholder believes the deliverable or activity will be completed by a specific date, there's almost no convincing them that the date needs changing.

Developing the Project Schedule

The primary outputs of Develop Schedule are the schedule baseline and the project schedule. The schedule baseline is the approved version of the project schedule. You can employ several tools and techniques to produce these outputs. The tools and techniques you choose depend on the complexity of the project. For the exam, however, you'll need to know them all.

Develop Schedule has eight tools and techniques:

- Schedule network analysis
- Critical path method
- Resource optimization
- Data analysis (what-if scenarios, simulation)

- Leads and lags
- Schedule compression
- Project management information system
- Agile release planning

A lot of information is packed into some of these tools and techniques, and you should dedicate study time to each of them for the exam. We'll look at each one next.

Schedule Network Analysis

Schedule network analysis produces the project schedule model. It involves calculating early and late start dates and early and late finish dates for project activities (as does the critical path method). It uses a schedule model and other analytical techniques such as critical path and critical chain methods, what-if analysis, and resource leveling (all of which are other tools and techniques in this process) to help calculate these dates and create the schedule. These calculations are performed without taking resource limitations into consideration, so the dates you end up with are theoretical. At this point, you're attempting to establish the time periods within which the activities can be scheduled. Resource limitations and other constraints will be taken into consideration when you get to the outputs of this process.

Critical Path Method

The *critical path method (CPM)* is a schedule network analysis technique that estimates the minimum project duration. It determines the amount of float, or schedule flexibility, for each of the network paths by calculating the earliest start date, earliest finish date, latest start date, and latest finish date for each activity (without taking resource availability into account). This is a schedule network analysis technique that relies on sequential networks (one activity occurs before the next, a series of activities occurring concurrently is completed before the next series of activities begins, and so on) and on a single duration estimate for each activity. The precedence diagramming method (PDM) can be used to perform CPM. Keep in mind that CPM is a method to determine schedule durations without regard to resource availability.

The *critical path (CP)* is generally the longest full path on the project. Any project activity with a float time that equals 0 or with negative float is considered a critical path task. The critical path can change under a few conditions. When activities with float time use up all their float, they can become critical path tasks. Or you might have a milestone midway through the project with a *finish no later than* constraint that can change the critical path if it isn't met.

Float time is also called *slack time*, and you'll see these terms used interchangeably. There are two types of float: total float and free float. *Total float (TF)* is the amount of time you can delay the earliest start of a task without delaying the ending of the project. *Free float (FF)* is the amount of time you can delay the start of a task without delaying the earliest start of a successor task.

In the following section, you'll calculate the CP for a sample project, and I'll illustrate how you derive all the dates, the CP, and the float times.

Gathering Activity and Dependency Information

Let's say you are the project manager for a new software project. Your team will be developing a custom application that manages, tracks, and analyzes charitable contributions to a variety of organizations managed by your parent company. You need to devise a software system that tracks all the information related to the contributions, the donors, and the receivers and that also supplies the management team with reports that will help them make good business decisions. For purposes of illustration, I'm showing only a limited portion of the tasks that you would have on a project like this.

You'll start this example by plugging information from the processes you've already completed into a table (a complete example is shown later in Table 5.2 in the section "Calculating the Critical Path"). The list of activities comes from the Define Activities process. The durations for each activity are listed in the Duration column and were derived during the Estimate Activity Durations process. The duration times are listed in days.

The Dependency column lists the activities that require a previous activity to finish before the current activity can start. You're using only finish-to-start relationships. For example, you'll see that activity 2 and activity 4 each depend on activity 1 to finish before they can begin. The dependency information came from the Sequence Activities process. Now, you'll proceed to calculating the dates.

Calculating the Forward and Backward Pass

Project Deliverables is the first activity and, obviously, where the project starts. This activity begins on April 1. Project Deliverables has a 12-day duration. So, take April 1 and add 12 days to this to come up with an early finish date of April 12. Watch out, because you need to count day 1, or April 1, as a full workday. The simplest way to do this calculation is to take the early start date, add the duration, and subtract 1. Therefore, the early finish date for the first activity is April 12. By the way, we are ignoring weekends and holidays for this example. Activity 2 depends on activity 1, so it cannot start until activity 1 has finished. Its earliest start date is April 13 because activity 1 finished at the end of the previous day. Add the duration to this date minus 1 to come up with the finish date.

You'll notice that since activity 4 depends on activity 1 finishing, its earliest start date is also April 13. Continue to calculate the remaining early start and early finish dates in the same manner. This calculation is called a *forward pass*.

To calculate the latest start and latest finish dates, you begin with the last activity. The latest finish for activity 9 is July 10. Since the duration is only one day, July 10 is also the latest start date. You know that activity 8 must finish before activity 9 can begin, so activity 8's latest finish date, July 9, is one day prior to activity 9's latest start date, July 10. Subtract the duration of activity 8 (three days) from July 9 and add one day to get the latest start date of July 7. You're performing the opposite calculation that you did for the forward pass. This calculation is called a *backward pass*, as you might have guessed. Continue calculating the latest start and latest finish through activity 4.

Activity 3 adds a new twist. Here's how it works: Activity 7 cannot begin until activity 3 and activity 6 are completed. No other activity depends on the completion of activity 3. If activity 7's latest start date is June 29, activity 3's latest finish date must be June 28. June 28 minus eight days plus one gives you a latest start date of June 21. Activity 3 depends on activity 2, so activity 2 must be completed prior to beginning activity 3. Calculate these dates just as you did for activities 9 through 4.

Activity 1 still remains. Activity 4 cannot start until activity 1 is completed. If activity 4's latest start date is April 13, the latest finish date for activity 1 must be April 12. Subtract the duration of activity 1, and add 1 to come up with a latest start date of April 1. Alternatively, you can calculate the forward pass and backward pass by saying the first task starts on day 0 and then adding the duration to this. For example, activity number 1's earliest start date is April 1, which is day 0. Add 12 days to day 0, and you come up with an earliest finish date of April 12.

You determine the calculation for float/slack time by subtracting the earliest start date from the latest start date. If the float time equals 0, the activity is on the critical path.

Calculating the Critical Path

To determine the CP duration of the project, add the duration of every activity with zero float. You should come up with 101 days because you're adding the duration for all activities except for activity 2 and activity 3. A critical path task is any task that cannot be changed without impacting the project end date. By definition, these are all tasks with zero float.

Another way to determine the critical path is by looking at the network diagram. If the duration is included with the information on the node or if start and end dates are given, you simply calculate the duration and then add the duration of the longest path in the diagram to determine the CP. However, this method is not as accurate as what's shown in Table 5.2.

Figure 5.3 shows the same project in diagram form. The duration is printed in the top-right corner of each node. Add the duration of each path to determine which one is the critical path.

Remember that CP is usually the path with the longest duration. In Figure 5.3, path 1-2-3-7-8-9 equals 34 days. Path 1-4-5-6-7-8-9 equals 101 days; therefore, this path is the critical path.

Calculating Expected Value Using PERT

Program evaluation and review technique (PERT) is a method that the U.S. Navy developed in the 1950s. The Navy was working on one of the most complex engineering projects in history at the time—the Polaris Missile Program—and needed a way to manage the project and forecast the project schedule with a high degree of reliability. PERT was developed to do just that.

TABLE 5.2   CPM calculation

Activity number	Activity description	Dependency	Duration	Early start	Early finish	Late start	Late finish	Float/slack
1	Project Deliverables	—	12	4/1	4/12	4/1	4/12	0
2	Procure Hardware	1	2	4/13	4/14	6/19	6/20	67
3	Test Hardware	2	8	4/15	4/22	6/21	6/28	67
4	Procure Software Tools	1	10	4/13	4/22	4/13	4/22	0
5	Write Code	4	45	4/23	6/6	4/23	6/6	0
6	Test and Debug	5	22	6/7	6/28	6/7	6/28	0
7	Install	3, 6	8	6/29	7/6	6/29	7/6	0
8	Training	7	3	7/7	7/9	7/7	7/9	0
9	Acceptance	8	1	7/10	7/10	7/10	7/10	0

FIGURE 5.3   Critical path diagram

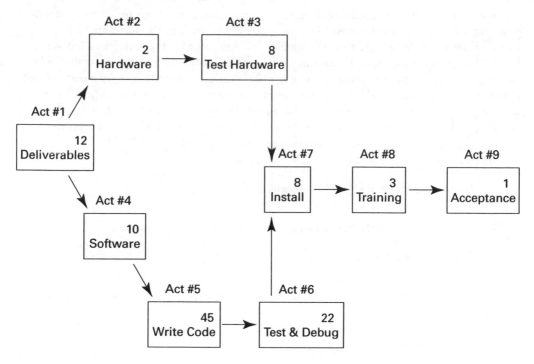

PERT and CPM are similar techniques. The difference is that CPM uses the most likely duration to determine project duration, whereas PERT uses what's called *expected value* (or the weighted average). Expected value is calculated using the three-point estimates for activity duration (I talked about three-point estimates earlier in this chapter) and then finding the weighted average of those estimates (I'll talk about weighted average in the next section, "Calculating Expected Value"). If you take this one step further and determine the standard deviation of each activity, you can assign a confidence factor to your project estimates. Without getting too heavily involved in the mathematics of probability, understand that for data that fits a bell curve—which is what you're about to calculate with the PERT technique—the following is true:

- Work will finish within plus or minus three standard deviations 99.73 percent of the time.

- Work will finish within plus or minus two standard deviations 95.45 percent of the time.

- Work will finish within plus or minus one standard deviation 68.27 percent of the time.

Calculating Expected Value

The three-point estimates used to calculate expected value are the optimistic estimate, the pessimistic estimate, and the most likely estimate. Going back to the software example, let's find out what these three time estimates might look like for the activity called Write Code. You get these estimates by asking the lead programmer, or key team member, to estimate the optimistic, pessimistic, and most likely duration for the activity based on past experience. Other historical information could be used to determine these estimates as well. Say in this case that you're given 38 days for the optimistic time, 57 days for the pessimistic, and 45 days for the most likely. (Forty-five days was derived from the Estimate Activity Durations process and is the estimate you used to calculate CPM.)

The formula to calculate expected value is as follows:

$$\left(\text{optimistic} + \text{pessimistic} + \left(4 \times \text{mostlikely} \right) \right) / 6$$

The expected value for the Write Code activity is as follows:

$$\left(38 + 57 + \left(4 \times 45 \right) \right) / 6 = 45.83$$

The formula for standard deviation, which helps you determine confidence level, is as follows:

$$\left(\text{pessimistic} - \text{optimistic} \right) / 6$$

The standard deviation for your activity is as follows:

$$\left(57 - 38 \right) / 6 = 3.17$$

You could say the following, given the information you now have:

- There is a 68.27 percent chance that the Write Code activity will be completed in 42.66 days to 49 days.
- There is a 95.45 percent chance that the Write Code activity will be completed in 39.49 days to 52.17 days.

You calculated the range of dates for the 68.27 percent chance by adding and subtracting one standard deviation, 3.17, from the expected value, 45.83. You calculated the 95.45 percent chance by multiplying the standard deviation times 2, which equals 6.34, and adding and subtracting that result from the expected value to come up with the least number of days and the most number of days it will take to finish the activity. Generally speaking, two standard deviations, or 95.45 percent, is a close enough estimate for most purposes.

Determining Date Ranges for Project Duration

Let's bring your table of activities back and plug in the expected values and the standard deviation for each (see Table 5.3).

The higher the standard deviation is for an activity, the higher the risk. Because standard deviation measures the difference between the pessimistic and the optimistic times, a greater spread between the two, which results in a higher number, indicates a greater risk. Conversely, a low standard deviation means less risk.

Now let's look at the total project duration using PERT and the standard deviation to determine a range of dates for project duration. You should add only the tasks that are on the critical path. Remember from the CPM example that activities 2 and 3 are not on the critical path, so their expected value and standard deviation calculations have been left blank in this table. When you add all the remaining tasks, the total expected value duration is 102.99 days, or 103 days rounded to the nearest day.

Your next logical conclusion might be to add the Standard Deviation column to get the standard deviation for the project. Unfortunately, you cannot add the standard deviations because you will come out with a number that is much too high. Totaling the standard deviations assumes that all the tasks will run over schedule, and that's not likely. It is likely that a few tasks will run over but not every one of them. So now you're probably wondering how to calculate the magic number.

You might have noticed an extra column at the right called SD Squared (or variance). This is the standard deviation squared—or for those of you with math phobias out there, the standard deviation multiplied by itself.

Once you have calculated the standard deviation squared for each activity, add the squares, for a total of 14.98. There's one more step, and you're done. Take the square root of 14.98 (you'll need a calculator) to come up with 3.87. This is the standard deviation you will use to determine your range of projected completion dates. Here's a recap of these last few calculations:

Total expected value = 103.00

Sum of SD Squared = 14.98

Square root of SD Squared = 3.87

TABLE 5.3 PERT calculation

Activity number	Activity description	Optimistic	Pessimistic	Most likely	Expected value	Standard deviation (SD)	SD squared
1	Project Deliver-ables	10	14	12	12.00	0.67	0.45
2	Procure Hardware	—	—	—	—	—	—
3	Test Hardware	—	—	—	—	—	—
4	Procure Software Tools	8	14	10	10.33	1.00	1.0
5	Write Code	38	57	45	45.83	3.17	10.05
6	Test and Debug	20	30	22	23.00	1.67	2.79
7	Install	5	10	8	7.83	0.83	0.69
8	Training	3	3	3	3.00	0	0
9	Acceptance	1	1	1	1.00	0	0
Totals for CP tasks					102.99		14.98

You can now make the following predictions regarding your project:

- There is a 68.27 percent chance that the project will be completed in 99.13 days to 106.87 days.
- There is a 95.45 percent chance that the project will be completed in 95.26 days to 110.74 days.

Exam Spotlight

For the exam, I recommend that you know that one standard deviation gives you a 68 percent (rounded) probability and two standard deviations give you a 95 percent (rounded) probability. Also, know how to calculate the range of project duration dates based on the expected value and standard deviation calculation. You probably don't need to memorize how to calculate the standard deviation because most of the questions give you this information. You should, however, memorize the PERT formula and know how it works. It wouldn't hurt to memorize the standard deviation formula as well—you never know what might show up on the exam.

PERT is not used often today. When it is, it's used for very large, highly complex projects. However, PERT is a useful technique to determine project duration when your activity durations are uncertain. It's also useful for calculating the duration for individual tasks in your schedule that might be complex or risky. You might decide to use PERT for a handful of the activities (those with the highest amount of risk, for example) and use other techniques to determine duration for the remaining activities.

Critical Chain Method

Although this is not an official tool and technique of this process, you should know the critical chain method for the exam. The *critical chain method* is a schedule network analysis technique that will modify the project schedule by accounting for limited or restricted resources, or for unforeseen project issues, by adding buffers to any schedule path. First, construct the project schedule network diagram using the critical path method. You will apply the duration estimates, dependencies, and constraints and then enter resource availability. Buffers, called *feeding buffers*, are added at this time as well. The idea behind feeding buffers is similar to that of contingencies. Adding buffer activities (which are essentially nonwork activities) to the schedule gives you a cushion of time that protects the critical path and thus the overall project schedule from slipping. Feeding buffers are added to noncritical chain-dependent tasks that feed into the critical chain. Project buffers are a type of buffer that is added at the end of the critical chain. After adding these buffer activities, you should schedule your critical path tasks at their latest start and finish dates.

Once this modified schedule is calculated, you'll often find that it changes the critical path. The new critical path showing the resource restrictions and feeding buffers is called the *critical chain*.

Critical chain uses both deterministic (step-by-step) and probabilistic approaches. A few steps are involved in the critical chain process:

- Construct the schedule network diagram using activity duration estimates (you'll use nonconservative estimates in this method).

- Define dependencies

- Define constraints.

- Calculate critical path.

- Enter resource availability into the schedule.

- Recalculate for the critical chain.

The critical chain method typically schedules high-risk tasks early in the project so that problems can be identified and addressed right away. It allows for combining several tasks into one task when one resource is assigned to all the tasks.

Exam Spotlight

CPM manages the total float of schedule network paths, whereas critical chain manages buffer activity durations. Critical chain is built on CPM and protects the schedule from slipping.

Resource Optimization Techniques

Earlier, I said that CPM and PERT do not consider resource availability. Now that you have a schedule of activities and have determined the critical path, it's time to plug in resources for those activities and adjust the schedule or resources according to any resource constraints you discover. Remember that you identified resource estimates during the Estimate Activity Resources process. Now during Develop Schedule, resources are assigned to specific activities. Usually, you'll find that your initial schedule has periods of time with more activities than you have resources to work on them. You will also find that it isn't always possible to assign 100 percent of your team members' time to tasks. Sometimes your schedule will show a team member who is overallocated, meaning that individual is assigned to more work than she can physically perform in the given time period. Other times, the team worker might not be assigned enough work to keep her busy during the time period. This problem is easy to fix. You can assign underallocated resources to multiple tasks to keep her busy. Adjusting the schedule for overallocated resources is a harder problem to fix. We will look at three techniques that optimize resources to prevent overallocation where possible: resource leveling, resource smoothing, and reverse resource allocation scheduling. You should use these techniques with CPM-based schedules.

Resource Leveling

Resource leveling—also called the *resource-based method*—is used when resources are overallocated, when they are only available at certain times, or when they are assigned to more than one activity at a time. In a nutshell, resource leveling attempts to balance out the resource assignments to get tasks completed without overloading the individual. You accomplish this by adjusting the start and finish dates of schedule activities based on the availability of resources. This typically means allocating resources to critical path tasks first, which often changes the critical path and, in turn, the overall project end date.

The project manager can accomplish resource leveling in a couple of other ways as well. You might delay the start of a task to match the availability of a key team member, or you might adjust the resource assignments so that more tasks are given to team members who are underallocated. Generally speaking, resource leveling of overallocated team members extends the project end date. If you're under a date constraint, you'll have to rework the schedule after assigning resources to keep the project on track with the committed completion date. You can accomplish this with resource smoothing, which we'll look at next.

Resource Smoothing

Resource smoothing accommodates resource availability by modifying activities within their float times without changing the critical path or project end date. That means you'll also use this technique when you need to meet specific schedule dates and are concerned about resource availability.

There are several ways you can accomplish this. You can adjust the resource assignments so that more tasks are given to team members who are underallocated. You could also require the resources to work mandatory overtime—that one always goes over well! Perhaps you can split some tasks so that the team member with the pertinent knowledge or skill performs the critical part of the task and the noncritical part of the task is given to a less skilled team member. Other methods might include moving key resources from noncritical tasks and assigning them to critical path tasks or adjusting assignments. Reallocating those team members with slack time to critical path tasks to keep them on schedule is another option. Don't forget—fast tracking is another way to keep the project on schedule.

Reverse Resource Allocation Scheduling

Reverse resource allocation scheduling is a technique used when key resources—like a thermodynamic expert, for example—are required at a specific point in the project and they are the only resource, or resources, available to perform these activities. This technique requires the resources to be scheduled in reverse order (that is, from the end date of the project rather than the beginning) in order to assign this key resource at the correct time.

> **Exam Spotlight**
>
> Resource leveling can cause the original critical path to change and can delay the project's completion date. Resource smoothing modifies activities within their floats without changing the critical path or project end date. It's used when changes to the critical path cannot or should not be made. Reverse resource allocation scheduling is used when specific resources are needed at certain times.

Data Analysis (What-if Scenario Analysis, Simulation)

Network modeling techniques typically include the use of what-if scenario analysis and simulation.

What-if scenario analysis uses different sets of activity assumptions to produce multiple project durations. For example, what would happen if a major deliverable is delayed or the weather prevents you from completing a deliverable on time? What-if analysis asks the question, "What if (fill in the blank) happens on the project?" and attempts to determine the potential positive and/or negative impacts to the project. What-if questions help determine the feasibility of the project schedule under adverse conditions. They are also useful to the project team in preparing risk responses or contingency plans to address the what-if situations. Worst-case what-if scenarios may result in a no-go decision.

Simulation techniques use a range of probable activity durations for each activity (often derived from the three-point estimates), and those ranges are then used to calculate a range of probable duration results for the project itself. Monte Carlo is a simulation technique that runs the possible activity durations and schedule projections many, many times to come up with the schedule projections and their probability, critical path duration estimates, and float time.

> **Exam Spotlight**
>
> For the exam, remember that Monte Carlo is a simulation technique that shows the probability of all the possible project completion dates. Simulation is also used in Six Sigma during the improve phase of the DMAIC methodology.

Leads and Lags

I talked about leads and lags earlier in this chapter. You'll recall that lags delay successor activities and require time added either to the start date or to the finish date of the activity you're scheduling. Leads require time to be subtracted from the start date or the finish date of the activity. Keep in mind that as you go about creating your project schedule, you might need to adjust lead and lag times to come up with a workable schedule.

Schedule Compression

Schedule compression is a form of mathematical analysis that's used to shorten the project schedule duration without changing the project scope. Compression is simply shortening the project schedule to accomplish all the activities sooner than estimated.

Schedule compression might happen when the project end date has been predetermined or if, after performing the CPM or PERT techniques, you discover that the project is going to take longer than the original promised time. In the CPM example, you calculated the end date to be July 10. What if the project was undertaken and a July 2 date was promised? That's when you'll need to employ one or both of the duration compression techniques: crashing and fast tracking.

Crashing

Crashing is a compression technique that looks at cost and schedule trade-offs. Crashing the schedule is accomplished by adding resources—from either inside or outside the organization—to the critical path tasks. It wouldn't help you to add resources to noncritical path tasks; these tasks don't impact the schedule end date anyway because they have float time. Crashing could be accomplished by requiring mandatory overtime for critical path tasks or requiring overnight deliveries of materials rather than relying on standard shipping times. You may find that crashing the schedule can lead to increased risk or increased costs or both.

 Be certain to check the critical path when you've used the crashing technique because crashing might have changed the critical path. Also consider that crashing doesn't always come up with a reasonable result. It often increases the costs of the project as well. The idea with crashing is to try to gain the greatest amount of schedule compression with the least amount of cost.

Fast Tracking

I talked about fast tracking in Chapter 1, "Building the Foundation." *Fast tracking* is performing two tasks or project phases in parallel that were previously scheduled to start sequentially. Fast tracking can occur for the entire duration of the task or phase or for a portion of the task or phase duration. It can increase project risk and might cause the project team to have to rework tasks. Fast tracking will work only for activities that can be overlapped. For example, it is often performed in object-oriented programming. The programmers might begin writing code on several modules at once, out of sequential order and prior to the completion of the design phase. However, if you remember our house-painting example, you couldn't start priming and painting at the same time, so fast tracking isn't a possibility for those activities.

Project Management Information System (PMIS)

Given the examples you've worked through on Develop Schedule and resource leveling, you have probably already concluded how much a scheduling tool, which is part of an overall PMIS, might help you with these processes. They will automate the mathematical calculations (such as forward and backward passes) and perform resource-leveling functions for you. Obviously, you can then print the schedule that has been produced for final approval and ongoing updates. It's common practice to email updated schedules with project notes so that stakeholders know what activities are completed and which ones remain to be done.

It's beyond the scope of this book to go into all the various software programs available to project managers. Suffice it to say that scheduling tools and project management software range from the simple to the complex. The level of sophistication and the types of project management techniques that you're involved with will determine which software product you should choose. Many project managers that I know have had great success with Microsoft Project software and use it exclusively. It contains a robust set of features and reporting tools that will serve most projects well.

Don't forget that you are the project manager, and your good judgment should never be usurped by the recommendation of a software product. Your finely tuned skills and experience will tell you whether relationship issues between team members might cause bigger problems than what the resource-leveling function indicates. Constraints and stakeholder expectations are difficult for a software package to factor in. Rely on your expertise when in doubt. If you don't have the experience yet to make knowledge-based decisions, seek out another project manager or a senior stakeholder, manager, or team member and ask them to confirm whether you're on the right track. Here's a word of caution: Don't become so involved with the software that you're managing the software instead of managing the project. Project management software is a wonderful tool, but it is not a substitute for sound project management practices or experience.

Agile Release Planning

You'll recall that in the agile methodology, deliverables and requirements start out as user stories. User stories are then decomposed into backlog items that the team members work on during each iteration. To accommodate large projects when using the agile methodology, you'll find it helpful to define releases, which are usually a significant feature or portion of functionality that will exist in the final product. The team also determines how many releases it will take to complete the final product, service, or result of the project as well as how many iterations it will take to complete each release. We'll cover this in more detail in the "Using a Kanban Board and Scrum Board" section later in this chapter. It will be easier to understand the concepts when presented together.

Project Schedule and the Schedule Baseline

We've spent a good deal of time getting to the final outcome of this process, which is the project schedule. Along with the schedule, the other key output of this process is the schedule baseline. We'll take a look at both of these next. Take note that the project schedule and schedule baseline will carry forward throughout the remainder of the project.

 Real World Scenario

Sunny Surgeons, Inc.

Kate Newman is a project manager for Sunny Surgeons, Inc., a software company that produces software for handheld devices for the medical profession. The software allows surgeons to keep notes regarding patients, upcoming surgeries, and ideas about new medicines and techniques to research. Kate's latest project is to write an enhanced version of the patient-tracking program with system integration capabilities to a well-known desktop software product used by the medical industry.

The programming department has had some recent turnover. Fortunately, Stephen, the senior programmer who led the development effort on the original version of the patient tracker, still works for Sunny. His expertise with handheld technologies, as well as his knowledge of the desktop software product, makes him an invaluable resource for this project.

Kate discovered a problem during the development of the project schedule. Stephen is overallocated for three key activities. Kate decides to see what his take is on the situation before deciding what to do.

Stephen, the eternal optimist programmer who loves his job and does all but sleep in his office at night, says he can easily complete all the activities and that Kate shouldn't give it a second thought. He also suggests to Kate that Karen Xio, a junior programmer on his team who worked on the last project with him, might be able to handle the noncritical path task on her own, with a little direction from Stephen.

Kate thinks better of the idea of overallocating her key project resource, even if he does think he can do the entire thing single-handedly. She decides to try some resource leveling to see what turns up.

Kate discovers that rearranging the order of activities, along with assigning Karen to handle the noncritical path activity, might be a possible solution. However, this scenario lengthens the project by a total of eight days. Since Kate knows the primary constraint on this project is quality, she's fairly sure she can get a buy-off from the project sponsor and stakeholders on the later schedule date. She can also sell the resource-leveled schedule as a low-risk option as opposed to assigning Stephen to all the activities and keeping the project end date the same. Overallocating resources can cause burnout and stress-related illnesses, which will ultimately have a negative impact on the project schedule.

Project Schedule

The purpose of the Develop Schedule process is to analyze most of the steps we've talked about so far, including sequencing activities, determining their durations, considering schedule constraints, and analyzing resource requirements. One of the primary outputs of this process is the *project schedule*, which presents the start and finish dates for each of the project activities, the duration of activities, dependencies among activities, milestones, and resources in a project schedule model. Determining human resource assignments occurs in the Acquire Project Team process, and depending on the size and complexity of your project or your organization's culture, this process might not be completed yet. If that's the case, the project schedule is considered preliminary until the resources are assigned to the activities.

In *PMBOK® Guide* terms, the project schedule is considered preliminary until resources are assigned. In reality, keep in mind that once you've published the project schedule (even though it's in a preliminary state), some stakeholders might regard it as the actual schedule and expect you to keep to the dates shown. Use caution when publishing a schedule in its preliminary form.

The project schedule should be approved by stakeholders and functional managers, and they should all sign off on it. This assures you that they have read the schedule, understand the dates and resource commitments, and will likely cooperate. You'll also need to obtain confirmation that resources will be available as outlined in the schedule when you're working in a functional organization. The schedule cannot be finalized until you receive approval and commitment for the resource assignments outlined in it.

Once the schedule is approved, it will become your baseline for the remainder of the project. Project progress and task completion will be monitored and tracked against the project schedule to determine whether the project is on course as planned.

Exam Spotlight

For the exam, remember that the project schedule is based on the timeline (derived from the activity estimates we calculated earlier in this chapter), the scope document (to help keep track of major milestones and deliverables), and resource plans. These plans are all used as references when creating the schedule.

According to the *PMBOK® Guide*, the schedule models are called *presentations*. You can present the schedule in a variety of ways, some of which are variations on what you've already seen. Project schedule network diagrams, like the ones discussed earlier, will work as schedule diagrams when you add the start and finish dates to each activity.

These diagrams usually show the activity dependencies and critical path. Figure 5.4 shows a sample portion of a project schedule network diagram highlighting the programming activities.

FIGURE 5.4   Project schedule network diagram with activity dates

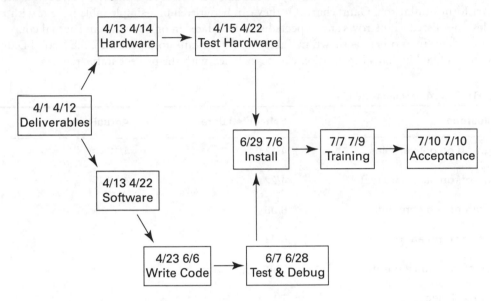

Gantt charts are easy to read and commonly used to display schedule activities. Depending on the software you're using to produce the Gantt chart, it might also show activity sequences, activity start and end dates, resource assignments, activity dependencies, and the critical path. Figure 5.5 is a simple example that plots various activities against time. These activities do not relate to the activities in the tables or other figures shown so far. Gantt charts are also known as *bar charts*.

FIGURE 5.5   Gantt chart

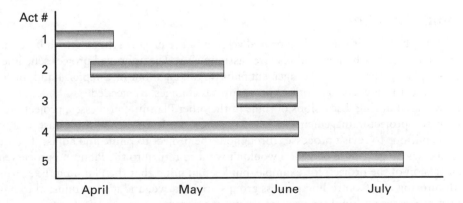

Milestone charts are another way to depict schedule information. Milestones mark the completion of major deliverables or some other key events in the project. For example, approval and sign-off on project deliverables might be considered a milestone. Other examples might be completion of a prototype, system testing, and contract approval.

Milestone charts might show the key events and their start or completion dates in a bar chart form similar to a Gantt chart. Or they can be written in a simple table format with milestones listed in the rows and expected schedule dates in one column and actual completion dates in another, as shown in Table 5.4. As the milestones are met, the Actual Date column is filled in. This information can be included with the project status reports.

TABLE 5.4 Milestone chart

Milestone	Scheduled date	Actual date
Sign-off on deliverables	4/12	4/12
Sign-off on hardware test	4/22	4/25
Programming completed	6/06	
Testing completed	6/28	
Acceptance and sign-off	7/10	
Project closeout	7/10	

Milestone charts are an ideal way to present information to the executive team. They are high level, easy to read, and to the point. Bar charts (or Gantt charts) are great for presenting the schedule to the management level. They have a bit more detail and are a quick way to see at a glance what resources are required when.

Schedule Baseline

The schedule baseline is the final, approved version of the project schedule with baseline start and baseline finish dates and resource assignments. The approved project schedule becomes a part of the project management plan we talked about in Chapter 4 and, once approved, must follow change control procedures if changes are needed.

As I've noted during discussions of some of the other Planning processes, project planning and project management are iterative processes. Rarely is anything cast in stone. You will continue to revisit processes throughout the project to refine and adjust. Eventually, processes do get put to bed. You wouldn't want to return to the Planning process at the conclusion of the project, for example, but keep in mind that the Planning, Executing, and Monitoring and Controlling process groups are iterative, and it's not unusual to have to revisit processes within these process groups as you progress on the project.

In practice, for small to medium-sized projects, you can complete Define Activities, Sequence Activities, Estimate Activity Resources, Estimate Activity Durations, and Develop Schedule at the same time with the aid of a project management software tool. You can produce Gantt charts; you can produce the critical path, resource allocation, and activity dependencies; you can perform what-if analysis; and you can produce various reports after plugging your scheduling information into most project management software tools. Regardless of your methods, be certain to obtain sign-off on the project schedule and provide your stakeholders and project sponsor with regular updates. Make the schedule available to stakeholders by posting it to the project site and be certain to notify them when there are updates. Keep your schedule handy—there will likely be changes and modifications as you go. While you're at it, be certain to save a schedule baseline for comparative purposes. Once you get into the Executing and Monitoring and Controlling processes, you'll be able to compare what you planned to do against what actually happened.

Schedule Data

The schedule data refers to documenting the supporting data for the schedule. The minimum amount of information in this output includes the milestones, schedule activities and activity attributes, and the assumptions and constraints regarding the schedule. You should document any other information that doesn't necessarily fit into the other categories. Always err on the side of too much documentation rather than not enough.

You will have to be the judge of what other information to include here because it will depend on the nature of the project. The *PMBOK® Guide* suggests that you might include schedule contingencies, alternative schedules, cash-flow projections, and resource histograms. Chapter 8, "Planning the Team Structure and Procurement Fundamentals," contains an example of a resource histogram if you want to peek ahead. Resource histograms typically display hours needed on one axis and period of time (days, weeks, months, years) on the other axis. You might also include alternative schedules or contingency schedule reserves in the schedule data section.

As with many of the other processes you've seen in this chapter, creating the project schedule may require updates to the activity resource requirements document, activity attributes, project calendars, resource calendars, change requests, and the risk register.

It's up to the project manager to tailor the scheduling processes for the project, as with all the Planning processes. Consider the life cycle approach you're using for the project (agile, predictive, or hybrid), the resource availability of key team members, the project complexity and/or any uncertainty surrounding the project, and the technology you'll use to construct, store, and display the project schedule model.

Using a Kanban Board and Scrum Board

Adaptive methodologies perform work in short, frequent cycles. Each cycle produces a minimum viable product that the stakeholders review and then provide immediate feedback to the team in order to modify or adjust the requirements to produce business value. In an iteration-based agile approach, user stories are decomposed into activities that are prioritized and worked on during each iteration. This is known as *iterative scheduling.* That is, user stories are pulled from the backlog and placed into the iteration backlog to be worked on during the iteration (a time-bound period of two to four weeks). Iterative scheduling is similar to rolling wave planning, which also elaborates the work of the project to the level of detail you know at the time.

A flow-based agile approach such as Kanban pulls work from the backlog according to the team's capacity to perform the work. This is known as *on-demand scheduling* or *pull-based scheduling.* Work is pulled on demand, as capacity is available.

Kanban is based on Lean-thinking principles and helps you stay organized, keeps team members collaborating, and keeps everyone informed. You'll recall from our previous discussions that Kanban is based on Toyota's just-in-time (JIT) production model and is a pull-based concept where work progresses to the next step only when resources are available. Kanban is used on software development projects, manufacturing projects, construction projects, and others where the work is balanced against available resources or available capacity for work.

Kanban is a visual process where the work is displayed on a board. Each task or work component is represented by what's known as a *card* or a *task.* A card could also be a sticky note. The board has columns that represent stages of work, and the cards are moved from stage to stage as the work progresses. These stages are not time bound; they are capacity bound. You can have any number of columns on a Kanban board, but fewer is better. The idea is that you can visually see the stages of work and where bottlenecks in the process are occurring so you can solve them and ultimately improve the process. Here's a simplified example. Your team has two software architects, five software developers, and one quality assurance (QA) reviewer, as represented in Figure 5.6. All of the cards start in the Backlog column and wait for team members to pull them to the next stage. The architects pull cards from the backlog as they have capacity, but their work capacity is limited to five cards. The developers are also limited to five cards at a time. The QA reviewer has the capacity to review and test only three cards at a time. The maximum number of cards are represented in the column headings in Figure 5.6. In this example, the quality assurance area is the bottleneck since the QA reviewer can't pull work at the same rate at which their teammates can produce and complete their cards. The QA reviewer must wait for a card to be pulled into the Deploy stage before they are able to pull the next card from development into QA.

In Kanban, the topmost card in any column is the most important. Team members are always aware which card to work on next. As the card is pulled from one column to the next, team members might add notes to the card and the card owner may change hands. This is the case in the example I've provided. The developers pull cards from the architect review column and become the new owner of the task. When the Backlog column is empty, and in our example all the cards have moved to the deploy stage, the project is finished.

According to the *PMBOK Guide,* the principles of Kanban are to start where you are, with the current state, and work incrementally toward the future state. Kanban allows for evolutionary change. Kanban encourages all team members to act as leaders. The core properties of Kanban, according to the *PMBOK® Guide,* are the visualization of the work of the project and the stages, or workflow, the work is progressing through. Kanban is capacity based and limits the amount of work in each stage. Feedback is built into the process at every stage, and the team works collaboratively toward continuous improvement.

FIGURE 5.6   Sample Kanban board

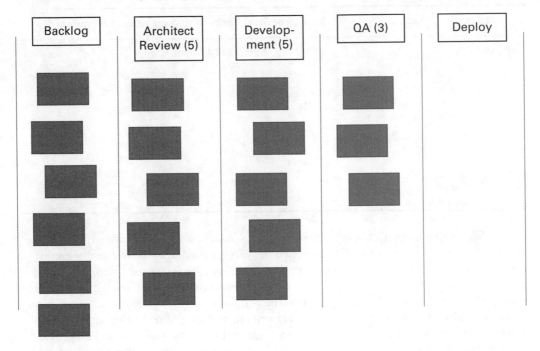

Scrum Board

Decomposing the user stories into activities is a way of delivering a minimum viable product. You'll recall from Chapter 3 that a minimum viable product involves breaking down user stories so that you can produce tangible components that have enough features and functionality to allow the customer to examine value and provide feedback to the team. You want to produce results quickly so that you can validate whether the product or result has value and meets the needs of your customers.

In the Scrum methodology, the project team keeps track of work using a Scrum board or task board. Visually, this board looks almost identical to the Kanban board. The difference between a Scrum board and a Kanban board is that the Scrum board shows the work of the sprint. Therefore, a Scrum board is time bound, not capacity bound. You'll recall that activity duration estimates are decided at the beginning of the sprint where the product owner identifies the user stories to work on during the sprint. Then team members break these down into tasks, or into story points, and estimate how much work they can complete in a sprint. Usually, the user stories are added to the first column of the Scrum board, the tasks are broken down and added to the second column, and the remaining columns show the stages of work during the sprint. This enables all team members and the product owner to see the progress of the sprint at a glance and ensures that important tasks are not forgotten. Table 5.5 shows what a sample Scrum board might look like. Work that isn't completed in a sprint becomes the first priority in the next sprint, unless the product owner says otherwise.

TABLE 5.5   Scrum board

User stories	Tasks	Tasks in progress	Tasks in testing	Tasks completed
User story 1			Task 1 (user story 1)	
		Task 2 (user story 1)		
	Task 3 (user story 1)			
User story 2	Task 1 (user story 2)			

The Scrum master may choose to use either the Scrum board or a burndown chart to show the progress of the sprint. We discussed the various burndown charts earlier in this chapter from the perspective of the entire project. You can also construct a burndown chart for a sprint showing the time on the x-axis (10 workdays for a two-week sprint, for example), and task hours or story points on the y-axis.

Both Kanban and Scrum allow the project team to deliver business value incrementally because work is produced in each stage or each sprint and the product owner can provide real-time feedback on the value provided. These agile methodologies allow for changes, as well as additions to scope. Keep in mind that many software programs are available that help you manage the backlog items, Kanban boards, and Scrum boards in a more automated fashion. Some teams like the sticky-backed notes on a whiteboard that's hung in the work area because they can see the progress of the work at any time, without having to log in to a program.

 When using Six Sigma, the project schedule is produced during the define phase of DMAIC.

Combining Techniques

The *Agile Practice Guide* (PMI®, 2017) recommends using a combination of techniques to manage the multiple, varying sizes of projects that exist in large organizations. For example, they may have a mix of projects ranging from small and simple to large and complex that affect the entire organization. You should consider using a combination of predictive, adaptive, and hybrid to manage these projects rather than declaring that all projects must adhere to one and only one methodology. Remember that you'll need to tailor your approach for each project, and it's helpful if you have established principles and practices that you can apply across the various projects. This may include combining the best of one approach with another and coming up with a methodology that works best for the project and the organization.

Agile Release Planning

Agile release planning is an output of the Develop Schedule process. This is a method you can use to accommodate large projects using an agile development methodology. Agile release planning involves defining releases, which are usually a significant feature or portion of functionality that will exist in the final product. You can think of release planning somewhat like project phases, which are used in a predictive methodology. Release planning starts at a high level with a product vision that drives the road map for the project (think of this as the project scope description), as well as the release plans. Based on this product vision, the team estimates how many releases it will take to complete the final product, service, or result of the project. The product owner will define user stories at the beginning of the project, but they need to define only enough user stories for the first release. User stories may also be created and added at the beginning of each release. The team will assess the user stories at the beginning of each release and determine how many iterations it will take to complete the release.

Let's say we are developing a mobile application that will track the number of business miles driven, track the service record for the vehicle, and calculate reimbursement rates. The product vision is a mobile app to track vehicle mileage and maintenance. The releases are

Release 1 = Business miles driven

Release 2 = Vehicle maintenance tracking

Release 3 = Reimbursement calculations

Release 1 contains four iterations. Iteration 1 within Release 1 contains two user stories, each with their own tasks. Each release has its own release plan with the number of iterations required for that release and the user stories within each iteration. A visual example of Release 1 might look like Figure 5.7.

Applying Process Groups in an Agile Approach

You will find in practice that you may not need to perform all the processes from the Initiating, Planning, Executing, Monitoring and Controlling, and Closing process groups when you're using an agile or hybrid methodology. However, note that the *PMBOK® Guide, Appendix X3* states that the processes within these five process groups will be performed within each iteration if you are using a sequential, iteration-based approach like agile release planning, or Scrum (which uses sprints). The *PMBOK® Guide* notes that this is required in order to manage projects that will incur a high degree of change or where project requirements are uncertain. Remember that you will tailor the processes to meet the needs of the project or organization, and that doing so will help you streamline the processes to fit your project.

FIGURE 5.7 Example product vision release plan and iteration plan

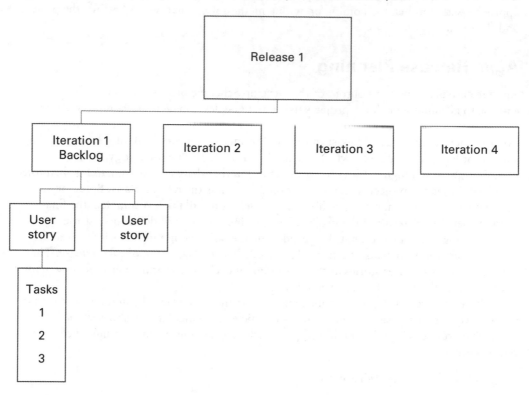

If you are using a continuous or flow-based approach like Kanban, you will perform the processes from the five process groups in a continuous fashion throughout the project. In this way, the processes are adaptive and continuous. Kanban is a continuous, pull-based system that is performed iteratively so that the overhead of performing the process groups is minimized because this occurs on a continual basis, rather than a start-and-stop approach in an iteration-based approach.

And remember that in a predictive or an adaptive approach, you will tailor the project management processes to fit the needs of the project and the organization, which means that some processes may be combined and processes that don't apply to the project may be skipped.

 Real World Scenario

Project Case Study: New Kitchen Heaven Retail Store

You worked with the stakeholders to document the activity list last week. After creating the first draft of the project schedule network diagram, you went back to each of them to ask for time estimates for each of the activities.

Ruth worked with Ricardo to define his requirements and create estimates. They are shown here:

1. Procure a cabling vendor. This takes 30 to 45 days. This activity can be done concurrently with the other activities listed here. Ricardo will perform this activity.

2. Run Ethernet cable throughout the building. This activity depends on the lease being signed and must finish before the build-out can start. The estimated time to complete is 16 hours, which was figured using parametric estimating techniques. Ricardo has one person on staff who can complete this specialized activity. His first available date is September 5.

3. Purchase the router, switch, server, and rack for the equipment room and the four point-of-service terminals. Delivery time is two weeks. Ricardo will perform this activity.

4. Install the router and test the connection. Testing depends on the cable installation. The time estimate to install is eight hours. Ricardo's staff will perform this activity.

5. Install the switch. Based on past experience, the time estimate to install is two days. Ricardo's staff will do this activity.

6. Install the server and test. The testing depends on the cable installation. Based on past experience, the time estimate to install is three days. Ricardo's staff will do this activity.

7. Ruth also worked with the web design team to define the user stories for the new website. Some of the user stories are adding the new store location and phone number to the lookup function, adding the new gourmet products, defining the video training, adding video content, and updating the web pages using the new wireframe. Ricardo has asked his applications programming team to self-organize and perform these activities using a Scrum approach.

You've worked with Jake and Jill and written similar lists with estimates and potential resource assignments for their teams. You begin to align all the activities in sequential order and discover a problem. Jill needs 14 calendar days to hire personnel and stock shelves, meaning that the build-out must be finished by January 11. Build-out takes approximately 120 days and can't start before September 20 because of the contractor's availability. This is a problem because Ricardo's Ethernet cable vendor isn't available until September 5, and he needs two days to complete the cabling. This pushes out the build-out start date by one week, which means the project completion date, or store-opening date, is delayed by one week.

After gathering more information from Ricardo, you head to Dirk's office.

"So, Dirk," you conclude after filling him in on all the details, "we have two options. Hire a contractor to perform the cable run since Ricardo's person isn't available or push the store opening out by one week."

Dirk asks, "How much will the contractor charge to run the cable, and are they available within the time frame you need?"

"Yes, they are available, and I've already requested that Ricardo book the week of August 21 to hold this option open for us. They've quoted a price of $10,000."

"Okay, let's bring in the contractor. At this point, $10,000 isn't going to break the budget. How is that planning coming anyway? Signed a lease yet?"

"Yes, we've signed the lease. Jake has been meeting with Gomez Construction on the build-out. We've used Gomez on three out of the last five new stores and have had good luck with them."

You spend the next couple of days working on the project schedule in Microsoft Project, clarifying tasks and activities with Jake, Ricardo, and Jill. You decide that a Gantt chart will work excellently for reporting status for this project. You stare intensely at the problem you see on the screen. The Grand Opening task is scheduled to occur 13 days later than when you need it! It must happen January 25, not February 9 as the schedule shows. You trace the problem back and see that the Grand Opening task depends on Train Store Personnel, which itself depends on several other tasks, including Hire Store Personnel and Install and Test Hardware. Digging deeper, you see that build-out can't begin until the Ethernet cable is run throughout the building. Ricardo already set up the time with the contractor to run the cable on August 21. This date cannot move, which means build-out cannot start any sooner than August 23, which works with Gomez's availability.

You pick up the phone and dial Jake's number. "Jake," you say into the receiver, "I'm working on the project schedule, and I have some issues with the Gomez activity."

"Shoot," Jake says.

"Gomez Construction can't start work until the Ethernet cable is run. I've already confirmed with Ricardo that there is no negotiation on this. Ricardo is hiring a contractor for this activity, and the earliest they can start is August 21. It takes them two days to run the cable, which puts the start date for build-out at August 23."

"What's the problem with the date?" Jake asks.

"Jill wants to have the build-out finished prior to hiring the store personnel. During the last store opening, those activities overlapped, and she said it was unmanageable. She wants to hire folks and have them stock the shelves in preparation for store opening but doesn't want contractors in there while they're doing it. An August 23 start date for Gomez puts us at a finish date of December 23. Realistically, with folks taking off for the holidays, this date will likely push to the first week in January, which is too late to give Jill time to hire and stock shelves. My question is this: Is 120 days to finish a build-out a firm estimate?"

"Always—I've got this down to a science. Gomez has worked with me on enough of these build-outs that we can come within just a couple of days of this estimate either way," Jake says.

You pick up your schedule detail and continue: "I've scheduled Gomez's resource calendar as you told me originally. Gomez doesn't work Sundays, and neither do we. Their holidays are Labor Day, a couple of days at Thanksgiving, Christmas, and New Year's, but this puts us too far out on the schedule. Our January 25 opening must coincide with the Garden and Home Show dates."

"I can't change the 120 days. Sounds like you have a problem."

"I need to crash the schedule," you say. "What would the chances be of Gomez agreeing to split the build-out tasks? We could hire a second contractor to come in and work alongside Gomez's crew to speed up this task. That would shorten the duration to 100 days, which means we could meet the date."

"Won't happen. I know Gomez. They're a big outfit and have all their own crews. We typically work with them exclusively. If I brought another contractor into the picture, I might have a hard time negotiating any kind of favors with them later if we get into a bind."

"All right," you say. "How about this? I'm making some changes to the resource calendar while we're talking. What if we authorize Gomez's crew to work six 10-hour days, which still leaves them with Sundays off, and we ask them to work on Labor Day and take only one day at Thanksgiving instead of two? We'll also have to ask that they don't take any more than two or three days off at Christmas."

"I think Gomez would go for that. You realize it's going to cost you?"

"Project management is all about trade-offs. We can't move the start date, so chances are the budget might take a hit to accommodate schedule changes or risk. Fortunately, I'm just now wrapping up the final funding requirements, so if you can get me the increased cost from Gomez soon, I'd appreciate it. This change will keep us on track and resolve Jill's issues too."

"I don't think Gomez's crew will mind the overtime during the holiday season. Everyone can use a little extra cash at that time of year, it seems. I'll have the figures for you in a day or two."

Project Case Study Checklist

The main topics discussed in the case study are as follows:

Sequence activities

Estimate activity durations

Estimate activity resources

Create user stories

Create product backlog

Developing project schedule

   Calendars

   Lead and lag times

   Critical path

Duration compression

   Crashing

   Fast tracking

Utilizing project management software

Producing project schedule

   Milestones

   Gantt chart

   Resource leveling

Understanding How This Applies to Your Next Project

The schedule management plan documents how you will define, monitor, control, and change the project schedule. It is the only output of the Plan Schedule Management process. Define Activities and Sequence Activities are the first two processes in the Activity sequence you'll complete on the road to creating your project schedule. You perform Estimate Activity Resources to determine the resource requirements and quantity of resources needed for each schedule activity. For small to medium-sized projects, I've found that you can perform this process at the same time as the Estimate Activity Durations process. If you work in an organization where the same resource pool is used for project after project, you already know the people's skills sets and availability, so you can perform this process at the same time you're creating the project schedule. The same logic holds for projects where the material resources are similar for every project you conduct. If you don't need to perform this process, I recommend that you create a resource calendar at a minimum so that you can note whether team members have extended vacations or family issues that could impact the project schedule. Needless to say, if you're working on a large project or your project teams are new for every project, you should perform the Estimate Activity Resources process rather than combining it with Develop Schedule. It will come in handy later when you're ready to plug names into the activities listed on the project schedule.

Estimate Activity Durations is a process you'll perform for most projects on which you'll work. For larger projects, PERT gives you estimates with a high degree of reliability, which are needed for projects that are critical to the organization, projects that haven't been undertaken before, or projects that involve complex processes or scope. It's easy to create a spreadsheet template to automatically calculate these estimates for you. List your schedule activities in each row, and in the individual columns to the right, record the most likely, pessimistic, and optimistic estimates. The final column can hold the calculation to perform the weighted average of these three estimates, and you can transfer the estimates to your schedule. You can easily add columns to calculate standard deviation as well.

In theory, if you've performed all the Activity processes, the schedule should almost be a no-brainer. You can plug the activity list, resources, estimates, and successor and predecessor tasks into the schedule. From there, you will want to take the next step and determine the critical path. The critical path is, well, critical to your project's success. If you don't know which activities are on the critical path, you won't know what impacts delays or risk events will have on the project. No matter how big or small the project, be sure that you know and understand the critical path activities.

Keep in mind that agile projects will start with user stories in the product backlog. You can use the estimating techniques outlined in this chapter to help determine the number of iterations needed to complete the work of the project as well as an overall project duration.

Summary

Great job! You've made it through the Planning activities associated with the Project Schedule Management Knowledge Area. I covered several processes in this chapter, including Plan Schedule Management, Define Activities, Sequence Activities, Estimate Activity Resources, Estimate Activity Durations, and Develop Schedule.

Define Activities uses the scope baseline (which includes the project scope statement, WBS, and WBS dictionary) to help derive activities. Activities are used to help develop a basis for estimating and scheduling project work during the Planning processes and for executing and monitoring and controlling the work of the project in later processes.

The Sequence Activities process takes the activities and puts them in a logical, sequential order based on dependencies. Dependencies exist when the current activity relies on some action from a predecessor activity or it impacts a successor activity. Four types of dependencies exist: mandatory, discretionary, external, and internal. PDM (also known as AON) is a method for displaying project schedule network diagrams. PDM has four logical relationships, or dependencies: finish-to-start, start-to-finish, finish-to-finish, and start-to-start.

The Estimate Activity Resources process considers all the resources needed and the quantity of resources needed to perform project activities. This information is determined for each activity and is documented in the resource requirements output. Remember that this process belongs to the Project Resource Management Knowledge Area.

Duration estimates are produced as a result of the Estimate Activity Durations process. Activity duration estimates document the number of work periods needed for each activity, including their elapsed time. Analogous estimating—also called top-down estimating or gross value estimating—is one way to determine activity duration estimates. You can also use top-down techniques to estimate project durations and total project costs. Parametric estimating techniques multiply a known element—such as the quantity of materials needed—by the time it takes to install or complete one unit of materials. The result is a total estimate for the activity. Three-point estimates use two formulas to calculate estimates, including triangular distributions (an average estimate based on the most likely estimate, a pessimistic estimate, and an optimistic estimate) and beta distributions (the PERT formula). Reserve analysis takes schedule risk into consideration by adding a percentage of time or another work period to the estimate just in case you run into trouble.

PERT calculates a weighted average estimate for each activity by using the optimistic, pessimistic, and most likely times. It then determines variances, or standard deviations, to come up with a total project duration within a given confidence range. Work will finish within plus or minus one standard deviation 68.27 percent of the time. Work will finish within plus or minus two standard deviations 95.45 percent of the time.

Develop Schedule is the process in which you assign beginning and ending dates to activities and determine their duration. You might use CPM to accomplish this. CPM calculates early start, early finish, late start, and late finish dates. It also determines float time. All tasks with zero float are critical path tasks. The critical path is the longest path of tasks in the project.

Schedules sometimes need to be compressed to meet promised dates or to shorten the schedule times. Crashing looks at cost and schedule trade-offs. Adding resources to critical path tasks or approving over time are two ways to crash the schedule. Fast tracking involves performing tasks (or phases) in parallel that were originally scheduled to start one after the other. Crashing may change the critical path; fast tracking does not. Fast tracking usually increases project risk. You can use Monte Carlo analysis, a what-if scenario technique, in the Develop Schedule process to determine multiple, probable project durations.

Resource leveling is used when resources are overallocated and may create changes to the critical path and project end date. Resource smoothing modifies activities within their floats without changing the critical path or project end date.

The project schedule presents the activities in graphical form through the use of project schedule network diagrams with dates, Gantt charts, milestone charts, and project schedule network diagrams.

When using an agile approach, the user stories are defined at the beginning of the project. Activity estimates can be performed using the techniques outlined in the chapter to determine the number of iterations the project will take. The Scrum team determines velocity by defining story points. The number of total story points is divided by the team's average velocity per iteration to determine the number of iterations needed to complete the work of the project. Agile methodologies use Kanban boards, Scrum boards, and burn-down charts to display the work of the project.

Exam Essentials

Be able to describe the purpose of the Estimate Activity Resources process.   The purpose of Estimate Activity Resources is to determine the types of resources needed (human, equipment, and materials) and in what quantities for each schedule activity within a work package.

Be familiar with the tools and techniques of Estimate Activity Durations.   The tools and techniques of Estimate Activity Durations are expert judgment, analogous estimating, parametric estimating, three-point estimating, bottom-up estimating, data analysis (alternatives analysis, reserve analysis) decision-making, and meetings.

Know the difference between analogous estimating and bottom-up estimating.   Analogous estimating is a top-down technique that uses expert judgment and historical information. Bottom-up estimating performs estimates for each work item and rolls them up to a total.

Be able to calculate the critical path.   The critical path includes the activities with durations that add up to the longest path of the project schedule network diagram. Critical path is calculated using the forward pass, backward pass, and float calculations.

Be able to define a critical path task.   A critical path task is a project activity with zero or negative float.

Be able to describe and calculate PERT duration estimates.   This is a weighted average technique that uses three estimates: optimistic, pessimistic, and most likely. The formula is as follows: (optimistic + pessimistic + (4 × most likely)) / 6.

Be able to describe the difference between resource leveling and resource smoothing. Resource leveling can change the critical path and project end date. Resource smoothing does not change the critical path or project end date.

Be familiar with the duration compression techniques.   The duration compression techniques are crashing and fast tracking.

Be able to describe a critical chain.   The critical chain is the new critical path in a modified schedule that accounts for limited resources and feeding buffers.

Know the key outputs of the Develop Schedule process.   The key outputs are the project schedule and schedule baseline.

Understand the estimating techniques for a project.   Expert judgment is used most frequently on agile projects, but other techniques such as parametric estimating or bottom-up estimating will work.

Understand how to determine activity durations for an agile project.   Activity durations can be calculated using the number of iterations needed to complete the work, using the number of story points needed to complete the work, or using average velocity to determine how many story points can be completed in each iteration.

Be able to describe a Kanban board and a Scrum board and the difference between them.   Kanban boards and Scrum boards are visual displays of the work of the project. Kanban uses cards or tasks, and Scrum uses tasks. Tasks are added to the first column, usually the backlog or user story column, and then broken down into tasks and pulled into the remaining columns as the work finishes. Kanban boards and their work are capacity bound. The work is progressive and continuous. There isn't a start and stop date. Scrum boards are time bound and display the work of the sprint.

Review Questions

You can find the answers to the review questions in Appendix A. Be sure to download the Bonus Exams and Bonus Questions so that you'll have a broader exposure and more experience answering questions related to the topics in this chapter.

1. You are the project manager for Changing Tides video games. You are in the process of breaking the work packages into activities. Which of the following options are true? (Choose two.)

   A. When using an agile methodology, activity lists are created using only the tools and techniques of the Define Activities process.

   B. The iteration planning meeting is the only time that activities are identified and estimated when using an agile methodology.

   C. You can use rolling wave planning, a form of progressive elaboration, to help create activity lists in both predictive and adaptive methodologies.

   D. User stories are chosen by the product owner at the iteration planning meeting and the team members break them down into activities for the iteration.

2. You are the project manager for Changing Tides video games. Your project is a bit ambiguous at first. It's similar in complexity and magnitude to a project that the team members worked on last year. The key stakeholder wants to be actively involved and be able to modify the requirements as the project progresses. The team is ready to define and estimate activities. Which of the following statements are true? (Choose three.)

   A. You should use an agile methodology to deliver business value incrementally and accommodate the need for changes to the requirements.

   B. You should use an agile methodology, and the schedule management plan is not needed with this approach.

   C. When using an agile methodology, there is a maximum focus on value to the customer and minimal focus on process.

   D. Backlog items are prioritized based on business need, risk, and value to the organization. The most important user stories are at the top of the backlog list.

3. Your project's primary constraint is quality. To make certain the project team members don't feel too pressed for time and to avoid schedule risk, you decide to use which of the following activity estimating tools?

   A. Three-point estimates

   B. Analogous estimating

   C. Reserve analysis

   D. Parametric estimating

4. You have been hired as a contract project manager for Grapevine Vineyards. Grapevine wants you to design an Internet wine club for its customers. One of the activities for this project is the installation and testing of several new servers. You know from past experience it takes about 16 hours per server to accomplish this task. Since you're installing 10 new servers, you estimate this activity to take 160 hours. Which of the estimating techniques have you used?

   A. Parametric estimating

   B. Analogous estimating

   C. Bottom-up estimating

   D. Reserve analysis

5. Which of the following statements describe the activity list? (Choose three.)

   A. The activity list is an output of the Define Activities process.

   B. The activity list includes all activities of the project.

   C. The activity list is an extension of and a component of the WBS.

   D. The activity list includes an identifier and description of the activity.

6. You have been hired as a contract project manager for Grapevine Vineyards. Grapevine wants you to design an Internet wine club for its customers. Customers must register before being allowed to order wine over the Internet so that legal age can be established. You know that the module to verify registration must be written and tested using data from Grapevine's existing database. This new module cannot be tested until the data from the existing system is loaded. This is an example of which of the following?

   A. Preferential logic

   B. Soft logic

   C. Discretionary dependency

   D. Hard logic

7. Match the following values from the *Agile Manifesto*.

Table of Agile Manifesto Principles

Values this	Over this
A. Responding to change	1. Contract negotiation
B. Working software	2. Following a plan
C. Customer collaboration	3. Process and tools
D. Individuals and interactions	4. Comprehensive documentation

8. You are working on a project that requires resources with expertise in the areas of hospitality management and entertainment. You have prepared the project schedule and notified the functional managers about the resources you'll need, and when, from their areas. What are the other things you should do?

 A. Make the schedule available to stakeholders by posting it to the project site so that it is accessible to them.

 B. Be certain to notify stakeholders of any schedule updates or changes.

 C. Save a schedule baseline for comparative purposes in case changes are made to the baseline.

 D. All of the above.

9. Which logical relationship does the PDM use most often?

 A. Start-to-finish

 B. Start-to-start

 C. Finish-to-finish

 D. Finish-to-start

10. You are a project manager for Picture Shades, Inc. Your company manufactures window shades that have replicas of Renaissance-era paintings for hotel chains. Picture Shades is taking its product to the home market, and you're managing the new project. It will offer its products at retail stores as well as on its website. You're developing the project schedule for this undertaking and have determined the critical path. Which of the following statements is true?

 A. You calculated the most likely start date and most likely finish date, float time, and weighted average estimates.

 B. You calculated the activity dependency and the optimistic and pessimistic activity duration estimates.

 C. You calculated the early and late start dates, the early and late finish dates, and float times for all activities.

 D. You calculated the optimistic, pessimistic, and most likely duration times and the float times for all activities.

11. You are a project manager for Picture Shades, Inc. Your company manufactures window shades that have replicas of Renaissance-era paintings for hotel chains. Picture Shades is taking its product to the home market, and you're managing the new project. It will offer its products at retail stores as well as on its website. You're developing the project schedule for this undertaking. Looking at the following graph, which path is the critical path?

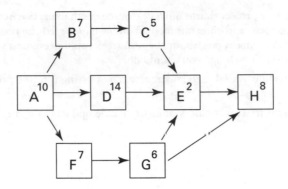

A. A-B-C-E-H

B. A-D-E-H

C. A-F-G-H

D. A-F-G-E-H

12. Use the following graphic to answer this question. If the duration of activity B was changed to 10 days and the duration of activity G was changed to 9 days, which path is the critical path?

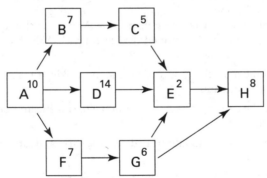

A. A-B-C-E-H

B. A-D-E-H

C. A-F-G-H

D. A-F-G-E-H

13. Which of the following statements is true regarding the critical path?

A. It should not be compressed.

B. It allows for looping and branching.

C. The critical path technique is the same as PERT.

D. It's the duration of all tasks with zero or negative float.

14. You are a project manager for Move It Now trucking company. Your company specializes in moving household goods across the city or across the country. Your project involves upgrading the nationwide computer network for the company. Your lead engineer has given you the following estimates for a critical path activity: 60 days most likely, 72 days pessimistic, 48 days optimistic. What is the weighted average or expected value?

    A. 54

    B. 66

    C. 60

    D. 30

15. You are a project manager for Move It Now trucking company. Your company specializes in moving household goods across the city or across the country. Your project involves upgrading the nationwide computer network for the company. Your lead engineer has given you the following estimates for a critical path activity: 60 days most likely, 72 days pessimistic, 48 days optimistic. What is the standard deviation?

    A. 22

    B. 20

    C. 2

    D. 4

16. If you know the expected value is 500 and the standard deviation is 12, you can say with approximately a 95 percent confidence rating which of the following?

    A. The activity will take from 488 to 512 days.

    B. The activity will take from 464 to 536 days.

    C. The activity will take from 494 to 506 days.

    D. The activity will take from 476 to 524 days.

17. This contains all the supporting detail regarding activity estimates and includes assumptions, the methods used to derive the estimates, constraints, resources used to develop the estimates, confidence levels, and risks.

    A. Contingency estimates

    B. Basis of estimates

    C. Reserve estimates

    D. Buffer estimates

18. You are the project manager working on a research project for a new drug treatment. Your preliminary project schedule runs past the due date for a federal grant application. The manager of the R&D department has agreed to release two resources to work on your project to meet the federal grant application date. This is an example of _____.

    A. Crashing

    B. Fast tracking

    C. Resource leveling

    D. Adjusting the resource calendar

19. You are the project manager for Rivera Gourmet Adventure Vacations. Rivera combines the wonderful tastes of great gourmet food with outdoor adventure activities. Your project involves installing a new human resources software system. Your stakeholders understand this is a large undertaking and that you might experience some schedule slippage. Jason, the database analyst working on this project, is overallocated. He is critical to the success of the project, and you don't want to burn him out by overscheduling him. Which of the following actions should you take?

A. You should use fast tracking to smooth out resource overallocation.

B. You should use crashing to resource level the critical path tasks.

C. You should use resource leveling to balance out resource assignments.

D. You should use resource smoothing to smooth out resource assignments.

20. Your Scrum team's velocity rate is 24 story points per iteration. There are 165 total story points. Which of the following statements are true regarding this question? (Choose two.)

A. Velocity is used to estimate the capacity of the Scrum team.

B. This information can be tracked on a burndown chart.

C. The project needs seven iterations to complete all of the work of the project.

D. A Kanban board can display this information.

Chapter

6

Developing the Project Budget and Engaging Stakeholders

THE PMP® EXAM CONTENT FROM THE PEOPLE DOMAIN COVERED IN THIS CHAPTER INCLUDES THE FOLLOWING:

✓ **Task 1.2 Lead a Team**

- 1.2.6 Analyze team members' and stakeholders' influence

- 1.2.7 Distinguish various options to lead various team members and stakeholders

✓ **Task 1.6 Build a Team**

- 1.6.1 Appraise stakeholder skills

- 1.6.2 Deduce project resource requirements

- 1.6.3 Continuously assess and refresh team skills to meet project needs

✓ **Task 1.9 Collaborate with stakeholders**

- 1.9.1 Evaluate engagement needs for stakeholders

- 1.9.2 Optimize alignment between stakeholder needs, expectations, and project objectives

- 1.9.3 Build trust and influence stakeholders to accomplish project objectives

✓ **Task 1.10 Build shared understanding**

- 1.10.1 Break down situation to identify the root cause of a misunderstanding

- 1.10.2 Survey all necessary parties to reach consensus

- 10.1.4 Investigate potential misunderstandings

✓ **Task 1.13 Mentor relevant stakeholders**

- 1.13.1 Allocate the time to mentoring

- 1.13.2 Recognize and act on mentoring opportunities

THE PMP® EXAM CONTENT FROM THE PROCESS DOMAIN COVERED IN THIS CHAPTER INCLUDES THE FOLLOWING:

✓ **Task 2.1 Execute project with urgency required to deliver business value**

- 2.1.1 Assess opportunities to deliver value incrementally

- 2.1.2 Examine the business value throughout the project

✓ **Task 2.2 Manage communications**

- 2.2.1 Analyze communication needs of all stakeholders

- 2.2.2 Determine communication methods, channels, frequency, and level of detail for all stakeholders

- 2.2.4 Confirm communication is understood and feedback is received

✓ **Task 2.4 Engage stakeholders**

- 2.4.4 Develop, execute, and validate a strategy for stakeholder engagement

✓ **Task 2.5 Plan and manage budget and resources**

- 2.5.1 Estimate budgetary needs based on the scope of the project and lessons learned from past projects

- 2.5.2 Anticipate future budget challenges

- 2.5.4 Plan and manage resources

✓ **Task 2.12 Manage project artifacts**

- 2.12.1 Determine the requirements (what, when, where, who, etc.) for managing the project artifacts

- 2.12.2 Validate that the project information is kept up to date (i.e., version control) and accessible to all stakeholders

✓ **Task 2.13 Determine appropriate project methodology/methods and practices**

- 2.13.1 Assess project needs, complexity, and magnitude

- 2.13.2 Recommend project execution strategy (e.g., contracting, finance)

- 2.13.4 Recommend a project methodology/approach (i.e., predictive, agile, hybrid)

- 2.13.5 Use iterative, incremental practices throughout the project life cycle (e.g., lessons learned, stakeholder engagement, risk)

THE PMP® EXAM CONTENT FROM THE BUSINESS ENVIRONMENT DOMAIN COVERED IN THIS CHAPTER INCLUDES THE FOLLOWING:

✓ **3.2 Evaluate and deliver project benefits and value**

- 3.2.5 Appraise stakeholders of value gain progress

Two of the most important documents you'll prepare for any project are the project schedule and the project budget. You'll use the schedule and budget documents throughout the Executing and Monitoring and Controlling processes to measure progress and determine whether the project is on track. The budget is easier to prepare after the activities have been defined and the resource estimates calculated. So now that we have the schedule in hand, we're going to spend our time in this chapter developing the budget. We'll perform three processes that will lead us to the cost baseline output. The cost baseline is the authorized budget. The processes are Plan Cost Management, Estimate Costs, and Determine Budget. We'll cover several tools and techniques in these processes that you'll want to understand for the exam. Before we get into the details of the Estimate Costs and Determine Budget processes, we'll talk about the cost management plan and its importance in guiding the development of the budget.

Next, we'll talk about the stakeholder engagement plan, which documents stakeholder interests and needs and their potential impacts on the project with a focus on strategizing how best to engage them in the project. We'll wrap up the chapter by discussing how project information is documented and communicated. I've talked a lot about documentation so far, and I will discuss it more in this chapter. Documentation is something you will do throughout the remainder of the project, and the Plan Communications Management process details how to collect information, how to store it, and when and how to distribute it to stakeholders.

The process names, inputs, tools and techniques, and outputs, and the descriptions of the project management process groups and related materials and figures in this chapter, are based on content from *A Guide to the Project Management Body of Knowledge (PMBOK® Guide), Sixth Edition* (PMI®, 2017). The references to adaptive and hybrid methodologies, related materials, and figures in this chapter are based on content from the *Agile Practice Guide* (PMI®, 2017).

Creating the Cost Management Plan

At this point in the project, you have an exhaustive breakdown of project activities, and you have some pretty good duration estimates. Now here's the question that's forever on the mind of the executive management team: how much is this project going to cost? The purpose of the Estimate Costs process is to answer that question.

Every project has a budget, and part of completing a project successfully is completing it within the approved budget. Sometimes project managers are not responsible for the budget on the project. Instead, this function is assigned to a functional manager who is responsible for tracking and reporting all the project costs. Keep in mind that if you don't have responsibility for the project budget, your performance evaluation for the project should not include budget or cost measurements. In my experience, most project managers do have responsibility for the project budget, so don't skip this chapter and think someone else will manage the budget for you. Sorry.

Before diving into the Estimate Costs and Determine Budget process details, you should know that these processes are governed by a cost management plan that is created when you perform the *Plan Cost Management* process. You should know a couple of facts about this process and the cost management plan for the exam.

Performing Plan Cost Management

The purpose of the Plan Cost Management process is to produce the cost management plan. Stakeholders are almost always concerned about costs. And remember that each of them, or the departments they represent, may have a different perspective on the costs of the project, including accounting for costs at different times in the project and in differing ways.

Exam Spotlight

Project costs are always a constraint on any project. Know and understand your stakeholders' view of costs, how they are calculated, and when and how the costs are recognized from an accounting perspective.

Let's take a quick look at the inputs and tools and techniques of this process. This process has four inputs: project charter, project management plan (schedule management plan, risk management plan), enterprise environmental factors, and organizational process assets. The *PMBOK® Guide* doesn't mention the scope statement as an input. In reality, you'll also want to reference this document when developing the cost management plan. Make certain all of your documents and the schedule baseline are accurate and kept up to date with proper version control.

The tools and techniques of the Plan Cost Management process are expert judgment, data analysis, and meetings. Data analysis, in the form of alternatives analysis, is used to choose methods of funding (funding with debt or equity or using funds from the project itself) and whether items should be leased or purchased.

Creating the Cost Management Plan

The only output of this process is the cost management plan. The *cost management plan* establishes the policies and procedures you'll use to plan and execute project costs. It also documents how you will estimate, manage, and control project costs. You will use documents such as the approved project charter, the project scope statement, the project schedule, resource estimates, and the risk management plan to help in developing this plan and managing project costs. Like all the other management plans, the cost management plan is a subsidiary plan of the project management plan.

Exam Spotlight

According to the *PMBOK® Guide,* you need to consider the project scope, schedule, project resources, and the approved project charter when developing the cost management plan. You will use estimating techniques to determine costs and use the estimates to help control project costs.

According to the *PMBOK® Guide,* this plan includes, but is not limited to, the following elements:

Level of Precision and Accuracy   This refers to the precision level you'll use to round activity estimates—hundreds, thousands, and so on. Level of accuracy is based on the scope and complexity of the activities and the project itself. It also describes the acceptable ranges for establishing cost estimates (for example, plus or minus 5 percent).

Units of Measure   This refers to the unit of measure you'll use to estimate resources—for example, hours, days, weeks, or a lump sum amount.

Organizational Procedures Links   In Chapter 4, "Developing the Project Scope," we talked about the WBS and the identifiers associated with each component of the WBS. These identifiers are called the code of accounts. A *control account (CA)* is a point where several factors such as actual cost, schedule, and scope can be used to determine earned value performance measures. The control account is used in the Project Cost Management Knowledge Area to monitor and control project costs. The control account is typically associated with the work package level of the WBS, but control accounts could be established at any level of the WBS. The control account also has a unique identifier that's linked to the organization's accounting system, sometimes known as a *chart of accounts.*

Exam Spotlight

The cost management plan is established using the WBS and its associated control accounts.

Control Thresholds   Most actual project costs do not match the estimates exactly. The control threshold refers to the amount of variance the sponsor or stakeholders are willing to allow before action is required. Document the threshold amount as a percentage of deviation allowed from the cost baseline.

Rules of Performance Measurement   This component of the cost management plan refers to how you will set the earned value management measurements. It's here you'll document where the control accounts exist within the WBS, the earned value management (EVM) techniques you'll use to measure performance, and the equations you'll use for calculating estimate at completion (EAC) forecasts and other measurements.

Reporting Formats   This refers to the types of cost reports you'll produce for this project and how often they'll be created.

Additional Details   This refers to any other information you might want to capture regarding cost activities, including items such as who will perform the cost activities, how a cost is recorded, and a description of the funding alternatives and choices used for the project.

The key to determining accurate cost estimates (and accurate time estimates, as you discovered in Chapter 5, "Creating the Project Schedule") is the WBS. Next, we'll look at how to determine cost estimates for the WBS components.

Estimating Costs

The *Estimate Costs* process develops a cost estimate for the resources (human and material) required for each schedule activity. This includes weighing alternative options and examining risks and trade-offs. Some alternatives you may consider are make-versus-buy, buy-versus-lease, and sharing of resources across either projects or departments.

Let's look at an example of trade-offs. Many times software development projects take on a life of their own. The requested project completion dates are unrealistic; however, the project team commits to completing the project on time and on budget anyway. How do they do this? They do this by cutting things such as design, analysis, and documentation. In the end, the project might get completed on time and on budget, but was it really? For example, the costs associated with the extended support period, because of poor design and lack of documentation and the hours needed by the software programmers to fix the reported bugs, weren't included in the original cost of the project (but they should have been). Therefore, the costs actually exceed what was budgeted. You should examine trade-offs such as these when determining cost estimates.

When you are determining cost estimates, be certain to include all the costs required to complete the work of the project over its entire life cycle. The primary focus for costs on most projects is resources. Resources, both human and materials and supplies, are typically the most expensive elements of any project. You also need to consider costs such as support and maintenance costs, or costs associated with maintaining the final product, service, or result of the project. Be certain to consider the costs of ongoing support or recurring costs such as maintenance, support, and service costs after the project is complete, and include these costs with the original budget request for the project. Considering *all* the costs of the project is imperative if you work in an organization that does budget planning months, if not years, in advance. As in the earlier example, software projects often have warranty periods that guarantee bug fixes or problem resolution within a certain time frame. After the warranty period expires, you move into the yearly maintenance and support period, which comes with a cost. These costs are typically calculated as a percentage of the total software cost. The management team should be made aware that these costs will continue to recur for the length of time the customer owns the software. It's also good practice to understand these costs prior to making go/no-go project decisions.

 Don't confuse pricing with Estimate Costs. If you are working for a company that performs consulting services on contract, for example, the price you will charge for your services is not the same as the costs to perform the project. The costs are centered on the resources needed to produce the product, service, or result of the project. The price your company might charge for the service includes these costs, along with a profit margin.

Many of the inputs of the Estimate Costs process are already familiar to you. We talked about the cost management plan earlier in this chapter. We'll look briefly at the key inputs next so that you can see the primary elements that you should consider when creating the project budget. The key inputs in this process are the scope baseline, project schedule, risk register, some new EEFs, and OPAs. Let's look at each of them next.

Scope Baseline

It's pop quiz time. Do you remember the elements of the scope baseline? They are the project scope statement, the WBS, and the WBS dictionary. You'll want to consider a few key elements from the project scope statement in this process, including key deliverables, constraints, assumptions, and acceptance criteria. I can safely say that every project I've ever worked on had a limited budget, which is a classic example of a project constraint. You should also understand other constraints that have the potential to impact costs, such as required delivery dates or availability of resources. Project assumptions regarding costs might include whether to incorporate indirect costs into the project estimate. We'll talk more about this topic in the section "Establishing the Cost Baseline" later in this chapter.

The WBS, as we've discussed, serves as the basis for estimating costs. It contains all the project deliverables and the control accounts that are typically established at the work package level (but that can be assigned to any level of the WBS). The WBS dictionary describes the deliverables, work components, and other elements of the WBS.

When you're considering deliverables, think about those that may have contractual obligations that should be taken into account when determining cost estimates. Perhaps you have deliverables that have legal or governmental regulations that will require additional expenses to fulfill. Health, safety, security, licenses, performance, intellectual property rights, and environmental factors are some of the other elements of the scope baseline you should keep in mind when estimating costs, according to the *PMBOK® Guide*.

Exam Spotlight

Scope definition is a key component of determining the estimated costs and should be completed as early in the project as possible, because it's easier to influence costs in the beginning phases of the project. But you can't influence costs if you don't understand the project scope.

Project Schedule

We determined the types and quantities of resources we needed in Chapter 5 by using the Estimate Activity Resources (this process is closely associated with Estimate Costs, according to the *PMBOK® Guide*) and Estimate Activity Durations processes. Resource requirements and duration estimates are the key outputs you should consider when estimating costs.

Be aware that duration estimates can affect costs. For example, you must account for costs such as interest charges when you're financing the work of the project. Also consider fluctuations in costs that can occur due to seasonal or holiday demands or collective bargaining agreements. Watch for duration estimates that are calculated for resources who are scheduled to work for a per-unit period of time. These duration estimates can be incorrect (not too far off, you hope) and can end up costing you more. For example, I recently bought a new home requiring a move of about eight miles. The moving company told me that the work was performed on a per-hour basis. The person providing the estimate assured me he had been doing this for over 25 years and that his estimates were typically right on the money. Unfortunately, I swallowed that line and the estimate I was given was wildly incorrect. It took them almost twice the amount of time I was quoted, and you guessed it, the total ended up being almost twice the original estimate.

Risk Register

The risk register is an output of the Identify Risks process that we'll discuss in Chapter 7, "Identifying Project Risks." When developing project cost estimates, you should take into account the cost of implementing risk response plans (identified in the risk register), particularly those with negative impacts on the project. Generally, negative risk events that occur early in the project are more costly than those that occur later in the project. This is because risks that occur early in the project may impact schedule, budget, and/or quality and, depending on their impacts, could actually cause a cancellation of the project. When the project is in the beginning stages, risks are not known or have not been fully identified. If an unknown risk occurs that wasn't identified and does not have a risk response plan, it can devastate a project in the early phases. Risks that occur in the later phases of the project are usually known risks that were identified in the Planning processes and recorded in the risk register. And typically risks with high impacts or high probability have detailed response plans and risk owners who will implement the plans as soon as the risk occurs, thereby reducing the cost of the risk.

As the project progresses, work is completed and it's much less likely that a risk that occurs will have significant costs (because much of the work is already completed).

Enterprise Environmental Factors

According to the *PMBOK® Guide*, the enterprise environmental factors (EEFs) you should consider in this process are market conditions, published commercial information, and exchange rates and inflation. Market conditions help you understand the materials, goods, and services available in the market and what terms and conditions exist to procure those resources. Published commercial information refers to resource cost rates. You can obtain these rates from commercial databases or published seller price lists. Currency exchange rates can work to the organization's benefit ... or not.

Organizational Process Assets

The organizational process assets (OPAs) considered in the process are similar to those we've seen before. Historical information and lessons learned on previous projects of similar scope and complexity can be useful in determining estimates for the current project—particularly if the past projects occurred recently. Your organization's business office or PMO may also have cost-estimating templates you can use to help with this process and/or cost-estimating policies that you should consider. You could use cost-estimating worksheets from past projects as templates for the current project as well.

Estimating Techniques

The Estimate Costs process has eight tools and techniques used to derive estimates:

- Expert judgment
- Analogous estimating
- Parametric estimating
- Bottom-up estimating
- Three-point estimating
- Data analysis (alternatives analysis, reserve analysis, cost of quality)
- Project management information system
- Decision-making (voting)

I covered expert judgment, analogous estimating, parametric estimating, three-point estimating, reserve analysis, and decision-making techniques in previous chapters. The majority of these are also tools and techniques of the Estimate Activity Durations process used to help determine schedule estimates. All of the information we discussed in Chapter 5 applies here as well, except you're using the tools and techniques to derive cost estimates. Three-point estimates (triangular distribution, a simple averaging or weighted averaging, and beta distribution [PERT] formulas) are used in this process when you want to improve your estimates and account for risk and estimation uncertainty. In the case of reserve analysis, you're adding cost reserves (or contingencies), not schedule reserves, during this process. Contingency reserves, like schedule reserves, might be calculated for the whole project, one or more activities, or both, and they may be calculated as a percentage of the cost or as a fixed amount. Contingency reserves are used for known-unknown risks or issues that may impact the project. Known-unknown means the risk is known but the consequences are unknown. And as with the schedule reserve, you should add contingency reserves to the cost baseline but should not add management reserves to the cost baseline. Management reserves are considered a part of the project budget. You could aggregate these cost contingencies and assign them to a schedule activity or a WBS work package level. As more information becomes known further into the project, you may be able to reduce the contingency reserves, turn them back to the organization, or eliminate them altogether.

The project management information system can help you quickly determine estimates given different variables and alternatives. Some systems are quite sophisticated and use simulation and statistical techniques to determine estimates, whereas others are less complex. A simple spreadsheet program can do the trick much of the time. We'll look at the remaining tools and techniques next.

Exam Spotlight

According to the *PMBOK® Guide*, if the Project Cost Management Knowledge Area includes predicting the potential financial performance of the product of the project, or if the project is a capital facilities project, you may also use additional tools and techniques in the process, such as return on investment, discounted cash flow, and payback period. We talked about these techniques in Chapter 2, "Assessing Project Needs."

Bottom-Up Estimating

You learned about this technique in Chapter 5, but there are a few pointers to consider for this process. This technique estimates costs associated with every activity individually and then rolls them up to derive a total project cost estimate. You wouldn't choose this technique to provide a cost estimate for the project during Initiating if one were requested because you don't have enough information at that stage to use it. Instead, use the top-down estimating technique (analogous estimating) when a project cost estimate is needed early in the project selection stage. Bottom-up estimating will generally provide you with the most accurate cost estimates, but it is the most time-consuming estimating technique of all those mentioned here. However, the size and complexity of the project impacts the accuracy you can achieve using this technique.

Cost of Quality

The *cost of quality (COQ)* is the total cost to produce the product or service of the project according to the quality standards. Cost of quality is a topic that we will cover when we look at the quality topics in more depth. For this process, understand that quality is not free and that you will need to include COQ estimates in your final project budget.

Exam Spotlight

For the exam, make sure you're familiar with all the Estimate Costs tools and techniques, their benefits, and under what conditions you use them.

Estimating Costs for an Agile Project

The agile methodology is used when there is a high degree of uncertainty and the project scope and requirements need to evolve as the project progresses. Since you don't know the specifics of the project scope when you start the project, and you know the requirements will change as you produce deliverables, it will be difficult to estimate costs up front. Consider using an analogous method at the beginning of the project to come up with a high-level cost estimate. You could also consider historical information and using subject matter experts to help determine this high-level estimate. If you're hiring contract resources to help with the project, for example, you'll need to know a ballpark figure for the time and amount of resources you'll need. Once you begin each iteration (or each release if you're using an agile release planning method), you can perform detailed cost estimates at the beginning of the release or at the iteration planning meeting.

Exam Spotlight

The *Agile Practice Guide* (PMI®, 2017) notes that when you have a project with uncertain requirements that are constrained by budget, you'll need to adjust the project scope and schedule in order to stay within the budget. When the schedule changes, you'll need to update the budget.

Creating the Cost Estimates

The primary output of the Estimate Costs process is cost estimates. These are quantitative amounts—usually stated in monetary units—that reflect the cost of the resources needed to complete the project activities. The tools and techniques I just described help you derive these estimates. Resources in this case include human resources, material, equipment, services, information technology needs, facilities, leases, rentals, exchange rates (if it applies), financing and/or interest costs (if it applies), and so on, as well as any contingency reserve amounts and inflation factors (if you're using them).

Estimates should be updated throughout the course of the project as more information comes to light. Estimates are performed at a given period of time with a limited amount of information. As more information becomes available, your cost estimates, and therefore your overall project estimate, will become more accurate and should be updated to reflect this new information. According to the *PMBOK® Guide*, the accuracy of estimates during the Initiating phase of a project have a rough order of magnitude (ROM) of –25% to +75%, and as more information becomes available over the course of the project, the definitive estimate moves to a range of –5% to +10%.

The remaining outputs of the Estimate Costs process are the basis of estimates and project documents updates. The basis of estimates is the supporting detail for the activity cost estimates and includes any information that describes how the estimates were developed, what assumptions were made during the Estimate

Costs process, and any other details you think are needed. According to the *PMBOK® Guide*, the basis of estimates should include at least the following:

- A description of how the estimate was developed or the basis for the estimate.
- A description of the assumptions made about the estimates or the method used to determine them.
- A description of the constraints.
- A description of risks regarding cost estimating.
- A range of possible results. You should state the cost estimates within ranges such as $5,000 ± 10%.
- The confidence level regarding the final estimates.

 Real World Scenario

This Older House

Janie is an accomplished project manager. She and her husband recently purchased an 80-year-old home in need of several repairs and modern updates. She decided to put her project management skills to work on the house project. First, they hired a general contractor to oversee all the individual projects needed to bring the house up to date. Janie worked with the general contractor to construct a WBS, and she ended up with 23 work packages. With each work package (or multiple work packages in some cases) assigned to a subcontractor, it was easy to track who was responsible for completing the work and for determining duration estimates. Janie and the general contractor worked together to determine schedule dependencies and make certain the work was performed in the correct order and that each subcontractor knew when their activity was to begin and end.

Some of the cost estimates for certain work packages were easy to determine using the parametric estimating method. Others required expert judgment and the experience of the general contractor (analogous techniques) to determine a cost estimate. Resource rates for laborers for some of the work packages were agreed to when the subcontractors bid on the work. Once Janie had all the cost estimates, she used the bottom-up estimating technique to come up with an overall cost estimate for the project. She added a contingency reserve in addition to the overall estimate for unforeseen risk events.

The last output of this process is project documents updates. Cost variances will occur and estimates will be refined as you get further into your project. As a result, new assumptions or constraints may come to light, so you'll need to update the assumption log. The lessons-learned register should be updated with the techniques you used to develop cost estimates for this project, and the risk register may also require an update after cost estimates are complete.

Estimate Costs uses several techniques to make an accurate assessment of the project costs. In practice, using a combination of techniques is your best bet to come up with the most reliable cost estimates. The cost estimates will become an input to the Determine Budget process, which allows you to establish a baseline for project costs to track against.

Establishing the Cost Baseline

The next process involves determining the authorized cost baseline, which is the primary output of the Determine Budget process. The *Determine Budget* process aggregates the cost estimates of activities and establishes a cost baseline for the project that is used to measure performance of the project throughout the remaining process groups. Only the costs associated with the project become part of the authorized project budget. For example, future period operating costs are not project costs and therefore aren't included in the project budget (but they are included in the cost estimates that are created in the Estimate Cost process we just discussed).

The *cost baseline* is the total expected cost for the project. According to the *PMBOK® Guide*, the cost baseline is a time-phased budget, and it must be approved by the project sponsor (and key stakeholders if appropriate for your project). As discussed earlier, the cost baseline does not include management reserves. And if the cost baseline is not approved, it is not valid. When you're using earned value management techniques to measure project performance, the cost baseline is also known as the performance measurement baseline (PMB). We'll talk about budget at completion and earned value management techniques in detail in Chapter 11, "Measuring and Controlling Project Performance." Remember that costs are tied to the financial system through the chart of accounts—or code of accounts—and are assigned to project activities at the work package level or to control accounts at various points in the WBS. The budget will be used as a plan for allocating costs to project activities.

 As I've discussed with several other processes, in practice you can sometimes perform the Estimate Costs and Determine Budget processes at the same time.

Outputs from other Planning processes, including the Create WBS, Develop Schedule, and Estimate Costs processes, must be completed prior to working on Determine Budget because some of their outputs become the inputs to this process. The inputs for Determine Budget are as follows:

Project Management Plan (Cost Management Plan, Resource Management Plan, and Scope Baseline)   The cost management plan documents how the project costs will be developed, managed, and controlled throughout the project. The resource management plan documents how the resources will be defined, staffed, managed and controlled, and released from the project when their work is completed. The scope baseline includes the project scope statement, the WBS, and the WBS dictionary. The scope statement describes the constraints of the project you should consider when developing the budget. The WBS shows how the project deliverables are related to their components, and the work package level typically contains control account information (although control accounts can be assigned at any level of the WBS).

Project Documents (Basis of Estimates, Cost Estimates, Project Schedule, Risk Register)   The basis of estimates and cost estimates are outputs of the Estimate Costs process. Cost estimates are determined for each activity within a work package and then summed to determine the total estimate for a work package. The basis of estimates contains all the supporting detail regarding the estimates. You should consider assumptions regarding indirect costs and whether they will be included in the project budget. Indirect costs cannot be directly linked to any one project. They are allocated among several projects, usually within the department or division in which the project is being performed. Indirect costs can include items like building leases, management and administrative salaries (those not directly assigned full time to a specific project), and so on. The project schedule contains information that is helpful in developing the budget, such as start and end dates for activities and milestones. Based on the information in the schedule, you can determine budget expenditures for calendar periods.

The risk register contains a list of risks that could occur on the project. Risks with a high impact and/or high probability of occurring will likely have response plans that could add costs to the project, so you should review them before preparing the budget.

Business Documents (Business Case, Benefits Management Plan)   The business case documents the reason the project came about, the critical success factors, and financial factors such as cost–benefit analysis. The benefits management plan outlines the intended benefits and business value the project will bring about, how the benefits will be measured, and how they will be obtained (this is typically described in financial terms).

Agreements   Agreements include cost information for purchased goods or services that you should include in the overall project budget.

Enterprise Environmental Factors   The EEFs you should consider in this process are currency exchange rates and fluctuations in currency rates over time if the project is a multi-year project.

Organizational Process Assets   The organizational process assets that will assist you with the work of this process include cost budgeting tools, the policies and procedures your organization (or PMO) may have regarding budgeting exercises, historical information and lessons learned, and reporting methods.

Techniques for Developing the Project Budget

The Determine Budget process has six tools and techniques, including some (it's more than two) you haven't seen before: expert judgment, cost aggregation, data analysis (reserve analysis), historical information review, funding limit reconciliation, and financing. I've covered expert judgment previously. Let's look at the remaining tools and techniques.

Cost Aggregation   Cost aggregation is the process of tallying the activity cost estimates at the work package level and then totaling the work package levels to higher-level WBS component levels (such as the control accounts). Then all the costs can be aggregated to obtain a total project cost.

Data Analysis (Reserve Analysis)   We talked about reserve analysis in the section "Estimating Techniques" section" earlier in this chapter. Reserve analysis works the same for the Determine Budget process. Contingency costs are considered and included in the aggregation of control accounts for both activity cost estimates and work package estimates. Management reserves are not included as part of the cost baseline (an output of this process) but *should* be included in the project budget. Management reserves are also not considered when calculating earned value measurements.

Reserve analysis should also contain appropriations for risk responses. I'll talk about several categories and tools of reserve analysis in the Plan Risk Responses process in Chapter 7. Additionally, you'll want to set aside money for management reserves for unknown risks. This is for the unforeseen, unplanned risks that might occur. Even with all the time and effort you spend on planning, unexpected issues do crop up. It's better to have the money set aside and not need it than to need it and not have it.

Historical Information Review   Analogous estimates and parametric estimates can be used to help determine total project costs. Remember from Chapter 5 that analogous estimates are a form of expert judgment. Actual costs from previous projects of similar size, scope, and complexity are used to estimate the costs for the current project. This is helpful when detailed information about the project is not available or when it's early in the project phases and not much information is known, such as projects in an adaptive methodology.

Parametric estimates are quantitatively based and multiply the amount of time needed to perform an activity by the resource rate to determine total cost. Quantifiable measures used with the parametric method are easily defined and easily scalable from large to small projects. Analogous and parametric estimating techniques are more accurate when the historical data you're using is accurate.

Funding Limit Reconciliation   Funding limit reconciliation involves reconciling the amount of funds to be spent with the amount of funds budgeted for the project. The organization or the customer sets these limits. Reconciling the project expenses will require adjusting the schedule so that the expenses can be smoothed. You do this by placing imposed date constraints (I talked about these in the Develop Schedule process in Chapter 5) on work packages or other WBS components in the project schedule.

Financing   This refers to obtaining funding for the project and taking into account how that funding will be tracked, reported, and used throughout the project. For example, it's common for state governments to obtain funding for a project from the federal government in the form of grants. Those federal funds come with pretty strict accounting and reporting requirements. If the agency fails to adhere to those requirements, or it spends the money on items other than those intended to fulfill the objectives of the project, they face the loss of future funding and may also have to pay back the funds already granted.

Developing the Cost Baseline

The goal of Determine Budget is to develop a cost baseline for the project that you can use in the Executing and Monitoring and Controlling processes to measure performance. In addition, you'll establish the project funding requirements. The outputs of the Determine Budget process include the cost baseline, project funding requirements, and project documents updates.

You now have all the information you need to create the cost baseline. We've covered the project documents updates in other processes. Let's look at the other two outputs next.

Documenting the Cost Baseline

You develop the cost baseline, the first output of Determine Budget, by aggregating the costs of the WBS work packages, including contingency reserves, into control accounts. All the control accounts are then aggregated, and together they make up the cost baseline. Most projects span some length of time, and most organizations time the release of funding with the project. In other words, you won't get all the funds for the project at the beginning of the project; they'll likely be disbursed over time. The cost baseline provides the basis for measurement, over time, of the expected cash flows (or funding disbursements) against the requirements, including contingency reserves. This is also known as the project's *time-phased budget*.

Management reserves are part of the project budget, but they are *not* included in the cost baseline.

The project budget consists of several components, including:

- Activity costs plus contingency reserves (these are aggregated to the work package level)

- Work package costs plus contingency reserves (these are aggregated to the control accounts)
- Management reserves for the cost baseline

Figure 6.1 depicts all the elements contained in the project budget.

FIGURE 6.1   Project budget

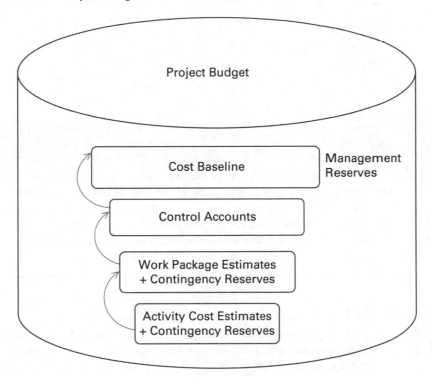

Cost baselines can be displayed graphically, with time increments on one axis and dollars expended on the other axis, as shown in Figure 6.2. The costs shown on this graph are cumulative costs, meaning that what you spent this period is added to what was spent last period and then charted. Many variations of this graph exist showing dollars budgeted against dollars expended to date. Cost budgets can be displayed using this type of graph as well, by plotting the sum of the estimated costs excepted per period.

The cost baseline should contain the costs for all the expected work on the project. You would have identified these costs in the Estimate Costs process. You'll find most projects' largest expense is resource costs (as in labor and materials costs). In the case of projects where you're purchasing the final product, the purchase price is the largest cost.

FIGURE 6.2   Cost baseline

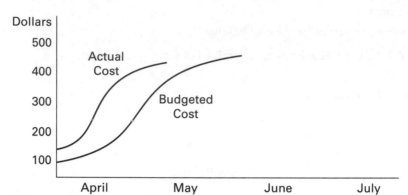

Exam Spotlight

For the exam, remember that cost baselines are displayed as an S curve. The reason for this is that project spending starts out slowly, gradually increases over the project's life until it reaches a peak, and then tapers off again as the project wraps up. Large projects are difficult to graph in this manner because the timescale isn't wide enough to accurately show fluctuations in spending. Also remember that the cost baseline does *not* include management reserves, but the project budget does include them.

    Large projects might have more than one cost baseline. For example, you might be required to track human resources costs, material costs, and contractor costs separately.

You'll revisit the cost baseline when you learn about the Control Costs process and examine different ways to measure costs in Chapter 10, "Sharing Information."

Gathering the Project Funding Requirements

Project funding requirements describe the need for funding over the course of the project, and they are derived from the cost baseline. Funding for a project is not usually released all at once. Some organizations may release funds monthly, quarterly, or in annual increments or other increments that are appropriate for your project.

As I said earlier, spending usually starts out slowly on the project and picks up speed as you progress. Sometimes, the expected cash flows don't match the pace of spending. Project funding requirements account for this by releasing funds in increments based on the cost baseline plus management reserves that may be needed for unanticipated events. Figure 6.3

shows the cost baseline, the funding requirements, and the actual expenses plotted on the S curve. You can see in this figure that actual costs have exceeded the authorized cost baseline. The difference between the funding requirements and the cost baseline at the end of the project is the management reserve.

FIGURE 6.3 Cost baseline, funding requirements, and cash flow

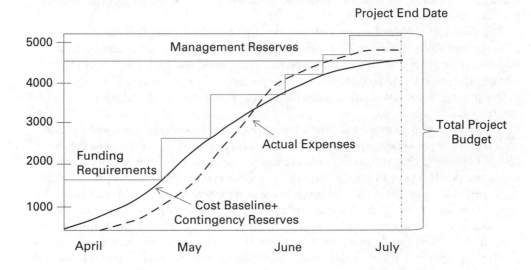

We've now completed two critical components of the project management plan: the project schedule and the project budget. As with all important project documents, you should post it to the project site so that it is accessible by all the stakeholders. As I've mentioned several times throughout the book, make certain to keep this document up to date with any changes as you progress through the project and be certain to apply version controls to the document. Version controls ensures that as changes are made, the documents are updated and named with the latest version, or date, or whatever versioning element you choose to use. When you reference the document, you can be assured that everyone is looking at the correct document because you will point them to the latest version.

Exam Spotlight

The *Agile Practice Guide* (PMI®, 2017) points out that when working on an agile project where cost or budget is the primary constraint, you will need to adjust schedule and scope to accommodate cost.

Understanding Stakeholders

We spent a good deal of time in Chapter 3, "Delivering Business Value," identifying stakeholders, analyzing their needs, categorizing them, and determining their power and influence. All this information was documented in the stakeholder register. After identifying stakeholders, the next step is determining how to effectively engage them on the project and on project decisions.

The *Plan Stakeholder Engagement* process (a part of the Project Stakeholder Management Knowledge Area) focuses on effectively engaging stakeholders by understanding their needs and interests, understanding the good and bad things they might bring to the project, and recognizing how the project will impact them. The only output of this process is the stakeholder engagement plan, which documents all the items I just mentioned.

Stakeholder management is almost a full-time job. Stakeholders are a diverse group with their own personalities, interests, expectations, and more. It is critical to the success of the project that you, as the project manager, understand all the key stakeholders' needs and devise a plan to engage them in the project and manage their expectations. In my experience, the most successful projects are those where the key stakeholders and I have established a strong, professional relationship. That means we trust one another, are able to ask questions, and can discuss constructive alternative ideas without feeling threatened. It also means that I often need to put myself in their shoes and understand how the project impacts them and their departments and devise ways to work together to meet their day-to-day needs while also completing the work of the project.

Stakeholder involvement will vary as the project progresses. I make it a point to meet with the key stakeholders as early in the project as possible. Their involvement early on is important because you'll need them to make decisions about scope, budgets, and timelines and, more important, to help smooth the way and remove obstacles that are impeding progress. As the work gets under way and the project progresses, their involvement may lessen, but it's important to note that their continued engagement, even on a limited level, is critical to the success of the project.

Exam Spotlight

For the exam, remember that stakeholder involvement must occur throughout the life of the project and is a critical success factor for overall project success.

We have seen most of the inputs to this process before with the exception of the communications management plan and the risk management plan. We'll look at the communications management plan later in this chapter and the risk management plan in Chapter 7.

Next we'll take a look at some of the new tools and techniques in this process.

Analyzing Stakeholders

There are several tools and techniques of this process: expert judgment, data gathering (benchmarking), data analysis (assumption and constraint analysis, root cause analysis [RCA]), decision-making (prioritization/ranking), data representation (mind mapping, stakeholder engagement assessment matrix), and meetings.

Expert judgment, as you recall, means meeting with others who know something about the project. In this case, that likely includes senior managers, key stakeholders, other project managers who have worked with your key stakeholders in the past, consultants, industry experts, and more. We've covered benchmarking previously, and assumption and constraint analysis should be self-explanatory.

Root cause analysis in this process refers to understanding your stakeholders at a level that you can ensure they are engaged and involved, and you can choose the optimal strategy to maintain their engagement throughout the project. The 5 Whys questioning technique works well for determining root cause analysis. This technique was originally developed by Toyota and involves asking "why" five times in succession to help identify the root cause of an issue. We'll look at this technique in more depth in Chapter 7.

Decision-making in the form of prioritization and ranking is used to understand stakeholder requirements and rank them according to their needs. You will also use this tool and technique to rank the stakeholders themselves. Typically, the project sponsor holds the number one ranking among stakeholders, followed by those stakeholders with the most influence on the project, and so on in descending order.

Data representation techniques allow you to classify the engagement levels of your stakeholders and monitor and modify them throughout the project. According to the *PMBOK® Guide*, there are five levels of classification of stakeholder engagement:

Unaware   Stakeholders are not engaged in the project.

Resistant   Stakeholders are not supportive of the project and may actively resist engaging.

Neutral   Stakeholders are neither supporting nor resisting the project and may be minimally engaged.

Supportive   Stakeholders have positive expectations of the project and are supportive and engaged.

Leading   Stakeholders are actively engaged in the project and helping to ensure its success.

The purpose behind the classification is to document what levels of engagement your stakeholders are at during each phase of the project. It helps you to plan what levels are needed for coming phases and to create action plans for those stakeholders who are not at the desired level of engagement for the given phase of the project. You can use a couple of methods to determine engagement. Mind mapping diagrams are simple to construct; each branch of the mind map represents a stakeholder and their interest and engagement levels.

You might construct a simple table like the one shown in Table 6.1 to indicate and document the level of engagement of each stakeholder. Or you could hold meetings with groups of stakeholders, or one on one, to determine levels of engagement.

TABLE 6.1   Stakeholder engagement assessment matrix

	Unaware	Resistant	Neutral	Supportive	Leading
Stakeholder A		Existing level		Preferred level	
Stakeholder B			Existing level		Preferred level
Stakeholder C					Existing level

Stakeholder Engagement Plan

The only output of this process is the stakeholder engagement plan. This plan documents the engagement levels of stakeholders and captures the strategies you can use to keep stakeholders engaged throughout the project. (The stakeholder engagement table we talked about in the last section gets documented here.) This is a key document if constructed correctly. It will help you encourage stakeholders to be productive and effective on the project because it will help you identify strategies to keep them engaged based on their interests and classifications. According to the *PMBOK® Guide*, the stakeholder engagement plan is created after the Identify Stakeholders process is completed. The stakeholder engagement plan should be updated throughout the project, and especially when stakeholders are added to or drop off the project. The following elements are among those that might be a part of this plan:

- Relationships between and among stakeholders (who do they influence, and who influences them?)

- Communication requirements for stakeholders engaged in the current phase of the project

- Details regarding the distribution of information, such as the language of the various stakeholders, the format of the information, and the level of detail needed

- Reasons for distributing the information, including how the stakeholders may react and/or the potential for changing their level of engagement

- Timing of the information distributions and their frequency

- The process for updating and modifying the stakeholder engagement plan due to the changing needs of the stakeholders and changes as the project progresses

The stakeholder engagement plan is a subsidiary of the project management plan and assists you in effectively managing stakeholders' expectations and engaging them in project decisions.

Exam Spotlight

For the exam, remember that the stakeholder engagement plan documents the strategies to use to encourage stakeholder participation, the analysis of their needs and interests and impacts, and the process regarding project decision-making. To create this plan, you will work with them to analyze their needs and interests and to understand the potential impact each one of them could have on the project.

Mentoring Stakeholders

One of the key roles the project manager plays on the project is managing stakeholders, managing their expectations, and keeping them engaged. One of the ways to manage these activities is to mentor your key project stakeholders. Many times, stakeholders are not aware of all the processes involved in project management, nor are they familiar with agile methodologies. They also underestimate the amount of time they will need to be involved in the project. They think that a couple of quick meetings will suffice when in reality, you'll need them to make decisions, resolve issues, perform activities, review activities, test, read project documents and provide sign-off, and much more.

This is where mentoring can be helpful. I understand this could seem uncomfortable, especially if you need to mentor an executive or senior member of the organization. But you don't have to tell them overtly you are mentoring them! Look for opportunities that will allow you to coach and mentor, such as inviting them for coffee to discuss the project processes, development life-cycle methodologies, and how critical their expertise and engagement is to the success of the project. Or set up a meeting to walk through an overview of your expectations of their involvement. Take a few minutes at the beginning of every status meeting (or every meeting they attend) to coach the stakeholders on the upcoming processes and the methodology being used to manage the project, and to remind them of the importance of their role. I've often worked with stakeholders who didn't understand the project processes or their part in the processes, or the role of the project manager on the project. Often, stakeholders are not aware of agile methodologies and don't realize how they differ from the predictive methodologies. Taking the time to mentor them on decision-making, why the project processes are important, the advantages of the methodology you'll use to manage the project, and how to work with you to deliver a successful project is time well spent.

Another way to look for opportunities to mentor is to use what's happening in the moment to educate the stakeholders on methodology or process. This may occur during a status meeting, as work is being performed, during a review session, when they are providing feedback, and so on.

Mentoring may also be needed to help your executives or stakeholders understand the agile methodology. According to the *Agile Practice Guide* (PMI®, 2017), you could consider providing (or recommending) training for your executives to educate them in the agile methodologies. The guide also recommends talking to them about agile in terms of short-term cycles of deliverables, frequent reviews, small incremental improvements, and retrospectives (all of which are Lean thinking principles).

You might want to review the resource management plan when thinking about mentoring stakeholders or team members. This plan documents the roles and responsibilities for project team members, their reporting relationships, and how resources (materials, equipment, and human resources) will be managed.

No matter how you choose to mentor the stakeholders, always start out describing their importance to the project. Make them feel appreciated, valued, and needed. Most of the time, making them feel appreciated will be all the opening you need. Ask if they understand everything during your sessions so that you can take the time to clarify if there is a confusing topic or concept. At the end of the session, ask for their commitment. It will take time to mentor stakeholders, so make certain you are allocating time to this important activity.

Engaging Stakeholders in an Adaptive Methodology

As you've learned, one of the key factors in ensuring success when using an adaptive methodology is small, dedicated, cross-functional teams. These cross-functional teams help align with the business units in the organization, thus ensuring the project is creating business value that is meaningful to the organization. Adaptive methodologies focus on eliminating waste, Lean thinking, and pull-based systems. When using this methodology, stakeholders are highly engaged on the project. This engagement is critical to project success because adaptive projects also experience a high degree of change and you'll need the stakeholders right by your side to guide the team, make decisions, and redirect when needed. This tends to lead to higher satisfaction among the stakeholder groups because they are actively engaged; they feel that they are part of the process and that their needs are heard and addressed. Here are some of the benefits noted in the *Agile Practice Guide* (PMI®, 2017) of stakeholder engagement in an adaptive methodology:

▪ Leads to high satisfaction among stakeholders

▪ Reduces risk

▪ Increases trust

- Reduces costs
- Increases the chances for project success
- Allows transparency
- Enables quick discovery and resolution of issues

Stakeholder buy-in is a critical success factor on any project, no matter what methodology you're using to manage the project. You can ensure stakeholder commitment by confirming that executive buy-in is in place, because the executive sponsor will make certain everyone is on board. Executive buy-in for an adaptive methodology is twofold; they need to commit to the project itself and to the agile methodology. According to the *Agile Practice Guide* (PMI®, 2017), if you don't have executive support, you may have difficulty with the teams and their approach to working on the project and getting the team to work in an agile method. If the executive doesn't buy in, why should they? Their old, predictive methods and styles will come into play, and it will be difficult to help them shift focus to a more adaptive approach. Do yourself a favor by ensuring your executive understands the benefits of the agile approach, including receiving frequent deliverables, providing continuous feedback to the team, and having the ability to change and modify requirements as the project progresses.

Communicating the Plan

Communication is critically important on your project, and poor (or limited) communication is a common reason projects struggle to succeed. It is the mechanism you use to convey information to your project team, stakeholders, and contractors. It involves providing project status, alerting others to risks—and well, you get the picture.

A few key points are in order in this section to frame up communications throughout the project. First, communication should be clear, concise, and tailored to the audience so that they can understand the message you are sending. Things like spelling and grammar do make a difference (thank you to language teachers everywhere!). Remember that your team may consist of members from other countries and cultures. For example, English is often the common language used when team members are from multiple countries. If your grammar is awkward and you are not phrasing your requests properly, someone who is reading (or listening) who is not a native English speaker could misinterpret your message. This scenario is equally true for whatever language you're using as the native language for the project, so be clear, concise, and accurate in your use of language.

Another pet peeve—I mean pointer for clear communications—is logical progression and flow. I have in the past worked with team members who are all over the place in describing a problem or issue, and I find I have to ask several clarifying questions to get to the root of the issue. Start at the beginning and progress through the message in a logical manner so that one fact unfolds from another.

Common courtesy applies here as it does to much of life. Be polite, don't interrupt others when they're talking, listen actively and intently (instead of formulating your response in your head before the other person has finished their thought), and follow up with clarifying questions or an indication of some sort that you heard them speak and you understand.

Considerations for tailoring communication and the communication processes include the geographic locations of your team members, stakeholders' relationships to the project (including those external to the organization), the technology you'll use to perform communications, and the language you'll use as the standard for the project.

Communication in agile environments typically occurs more frequently due to the nature of how the agile process works. The product owner is involved in every iteration and typically up to speed on the status of the project and the upcoming work for the next iteration. If your project is conducted in a waterfall or hybrid methodology, practice communicating as though you were using an agile approach. It's better to over-communicate than not to communicate enough—although there is a fine line, which we'll talk about when we discuss the communication processes throughout the remainder of the book.

Agile teams are highly collaborative, which improves communication among the team members, ensures knowledge sharing, and provides team members with the ability to choose the tasks they want to work on during the project.

I've talked a good deal about documentation so far, and this topic will continue to come up throughout the remainder of the book. "Is that documented?" should be an ever-present question on the mind of the project manager. Documentation can save your bacon, so to speak, later in the project. Documentation is only one side of the equation—communication is the other. You and your stakeholders need to know who gets what information and when.

Planning Communications

The *Plan Communications Management* process involves analyzing and determining the communication needs of the stakeholders, defining the types of information needed, the format for communicating the information, how often it's distributed, the level of detail that's needed, and who prepares it. All of this is documented in the communications management plan, which is an output of this process.

Pop quiz: Do you remember where else the communications management plan belongs? I'll give you the answer later in the section "Documenting the Communications Management Plan."

The project management plan is a key input to this process. You'll recall that the project management plan defines how the subsidiary plans will be defined and integrated into the overall project management plan. As such, it's rich with constraints and assumptions that you should review as they pertain to stakeholder communication needs.

The *PMBOK® Guide* notes that all the elements described in the enterprise environmental factors and in the organizational process assets influence this process. Some of the important enterprise environmental factors to take note of include personnel administration, stakeholder risk thresholds, established communication channels, geographic locations, and organization culture. Take special note of the lessons learned and historical information elements of the organizational process assets input. Information you learn as you're progressing through the project is documented as lessons learned. This information is helpful for future projects of similar scope and complexity. Historical information is also useful to review when starting the Plan Communications Management process. Either of these documents might contain information about communication decisions on past projects and their results. Why reinvent the wheel? If something didn't work well on a past project, you'd want to know that before implementing that procedure on this project, so review past project documentation.

Exam Spotlight

The *PMBOK® Guide* notes that there is a difference between effective and efficient communication. *Effective communication* refers to providing the information in the right format for the intended audience at the right time. *Efficient communication* refers to providing the appropriate information at the right time—that is, only the information that's needed at the time.

Determining Communication Needs

The Plan Communications Management process concerns defining and documenting the types of information you're going to deliver, the format it will take, to whom it will be delivered, and when. The process consists of eight tools and techniques to help determine these elements: expert judgment, communication requirements analysis, communication technology, communication models, communication methods, interpersonal team skills (communication styles assessment, political awareness, cultural awareness), data representation (stakeholder engagement assessment matrix), and meetings. Expert judgment in this process involves people who have special skills or knowledge of the political climate of the organization, its culture and customers, change management practices, communications technologies within the organization, industry knowledge regarding the project outcomes, and other stakeholders. Data representation refers to the stakeholder engagement assessment matrix that we discussed earlier in this chapter. Meetings, the last tool and technique of this process, are used to help gather information to determine the communication needs of the stakeholders and other information for the communications management plan. We'll look at the remainder of the tools and techniques for this process next.

Communications Requirements Analysis

Communications requirements analysis involves analyzing and determining the communication needs of the project stakeholders. According to the *PMBOK® Guide*, you can examine several sources of information to help determine these needs, including the following:

- Company and departmental organizational charts
- Stakeholder responsibility relationships
- Other departments and business units involved with the project
- The number of resources involved with the project and where they're located in relation to project activities
- The number of communication channels
- Internal needs that the organization may need to know about the project
- Development approach used on the project (agile, waterfall, or hybrid)
- Legal requirements
- External needs that organizations such as the media, government, or industry groups might have that require communication updates
- Stakeholder information, including communication needs, engagement levels, and more. (This was documented in the stakeholder register, an output of Identify Stakeholders, and the stakeholder engagement plan, the output of the Plan Stakeholder Engagement process.)

This tool and technique requires an analysis of the items in the preceding list to make certain you're communicating information that's valuable to the stakeholders. Communicating valuable information doesn't mean you always paint a rosy picture. Communications to stakeholders might consist of either good or bad news—the point is that you don't want to bury stakeholders in too much information but you want to give them enough so that they're informed and can make appropriate decisions.

Project communication will always involve more than one person, even on the tiniest of projects. As such, communication network models have been devised to explain the relationships between people and the number or type of interactions needed between project participants. What you must remember for the exam is that network models consist of nodes with lines connecting the nodes that indicate the number of communication channels, also known as *lines of communication*. Figure 6.4 shows an example of a network communication model with six participants and 15 channels of communication.

The nodes are the participants, and the lines show the connections between them all. You'll need to know how to calculate the number of communication channels when you take the exam. You could draw them out as in this example and count up the lines, but there's an easier way. The formula for calculating the lines of communication is as follows: (number of participants × (number of participants less 1)) divided by 2.

FIGURE 6.4   Network communication model

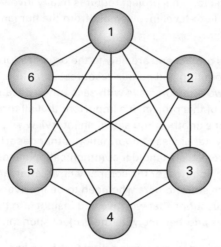

Nodes = Participants
Lines = Lines of Communication between Participants

Here's the calculation in mathematical terms:

$$n(n-1)/2$$

Figure 6.4 shows six participants, so let's plug that into the formula to determine the lines of communication:

$$6(6-1)/2 = 15$$

Exam Spotlight

I recommend you memorize the communications channel formula before taking the exam.

 **Real World Scenario**

Stakeholder Relationships

Bill is an information technology manager working on an enterprise resource planning project. He's one of the key stakeholders on this project. Bill reports to the CIO, who in turn reports to the executive vice president, who also happens to be the project sponsor.

Bill is close friends with the human resources director but doesn't get along so well with the accounting department director. This project requires heavy involvement from the accounting department and medium-level involvement from the human resources department.

You are the project manager for this project and are new to the organization. You know Bill's relationship with both the accounting and human resources directors. What you don't know is the relationship the two directors have with each other. Because all three stakeholders are key to the success of this project, it's important that all three communicate with you as well as with one another. You set up an interview with each of these stakeholders to determine several pieces of information: other departments that might need to be involved on the project, stakeholder communication needs and timing, external needs, timing of status updates for the company newsletter, and other department members aside from the stakeholders who need to be involved in the project. You also plant a few surreptitious questions that will give you insight into the relationships the stakeholders have with one another and with the project sponsor.

You discover that the human resources and accounting directors have known each other for several years and worked together at another organization prior to coming to working here. This tells you that if you can get one of them to buy in on project decisions, the other will likely follow suit. They both have the utmost respect for Bill and his technical capabilities, even though the accounting director doesn't care for his abrupt, direct communication style. You also learn that although they both have respect for the position of the executive vice president, they don't believe the person filling that role is competent to do the job. They question his decision-making ability—or lack thereof—and warn you that you need to write down his answers and direction so that he doesn't change his story halfway through the project. Although you won't formally document this valuable piece of information, you'll definitely put it into action right away.

Communication Technology

Communication technology examines the methods (or technology) used to communicate the information to, from, and among the stakeholders. Methods can take many forms, such as written, spoken, email, formal status reports, meetings, online repositories, online schedules, and so on. This tool and technique examines the technology elements that might affect project communications.

You should consider several factors before deciding what methods you'll choose to transfer information. The timing of the information exchange or need for updates is the first factor. The availability of the technology you're planning on using to communicate project information is important as well. Do you need to procure new technology or systems, or are systems already in place that will work? Staff experience with the technology is another factor. Are the project team members and stakeholders experienced at using this technology, or will you need to train them? Consider the duration of the project and the project environment. Will the technology you're choosing work throughout the life of the project, or will it have to be upgraded or updated at some point? How does the project team function? Are the members located together or spread out across several campuses or locations? Is the information you're distributing confidential or sensitive in nature? If so, consider what security measures should be put in place to protect the information and ensure that it is delivered only to the intended recipients.

The answers to these questions should be documented in the communications management plan.

Communication Models

Communication models depict how information is transmitted from the sender and how it's received by the receiver. According to the *PMBOK® Guide*, a basic sender/receiver communication model includes the following key components:

- Encode
- Transmit
- Decode

Encoding the message simply means putting the information or your thoughts or ideas into a language that the receiver will understand. The message is the result, or output, of the encoding.

The sender transmits the message using any number of methods, including written, oral, email, and so on. Barriers can exist that compromise this information, such as cultural differences, distance the message must travel, technology used to transmit and receive, and more. Any barrier, including those just mentioned, that keeps the message from being either transmitted or understood is called *noise*.

Decode is performed by the receiver, and it refers to translating the information that was sent.

An interactive communication model includes all the elements just described plus two more: acknowledge and feedback/response.

Acknowledge is when the receiver lets the sender know they have received the message. This is not an indication of agreement.

Last but not least, feedback/response is provided after the receiver has decoded the original message. Then, the receiver encodes their response and sends it back to the sender, starting the cycle all over again. The feedback/response element is important because it confirms that the communication was received and understood.

The sender is responsible for encoding the message, transmitting the message, and decoding the feedback message. They are also responsible for making certain the information is clear and complete, and they should ensure the receiver understands the information. The receiver is responsible for decoding the original message from the sender, ensuring they understand the message, and encoding and sending the feedback message.

Many factors will influence how messages are encoded, received, and understood. These factors include age, culture, nationality, and ethnicity, among others. It's the responsibility of both the sender and receiver to make certain the information is communicated in a way that both parties understand it and so that the information is clear and interpreted accurately.

Communication Methods

Communication methods refer to how the project information is shared among the stakeholders. According to the *PMBOK® Guide*, there are three classifications of communication methods. We'll briefly look at each of them:

Interactive Communication Interactive communication involves multidirectional communication where two or more parties must exchange thoughts or ideas. This method includes videoconferencing, phone or conference calls, meetings, and so on.

Push Communications Push communications is one way and refers to sending information to intended receivers. It includes methods such as letters, memos, reports, emails, voicemails, and so on. This method ensures that the communication was sent, but it is not concerned with whether it was actually received or understood by the intended receivers.

Pull Communications This is the opposite of push communications. The likely recipients of the information access the information themselves using methods such as websites, e-learning sites, knowledge repositories, shared network drives, and so on.

There are many methods you can use to communicate, including in person in a face-to-face setting (or on the telephone); in small groups; using a speaker to communicate to an audience of people; using mass communication methods such as newsletters, media, reports, and other publications; using network methods such as instant messaging, email, and videos; or using social networks.

 Remember that the communication management plan is used to manage the flow of information (according to the plan) in order to keep stakeholders engaged and informed.

Interpersonal and Team Skills

We've seen this tool and technique in other processes. In the Plan Communications Management process, you should consider the communication styles of your stakeholders and team members, the political environment of your organization, and the culture of your organization.

When assessing your stakeholders for the stakeholder register you created earlier, you could easily add information regarding their communication styles and their preferred method of communication. Some stakeholders like email, whereas some detest it and prefer to use the phone or meet in person. Make note of these preferences so that when you need to contact that stakeholder, particularly when you need their support on an issue, you use their preferred method of communication.

As we've discussed before, it's important that the project manager understands the stakeholder's position in the organization in terms of the power and influence they exercise (or in other words, political awareness). You may have stakeholders who are one level down (or more) in a functional hierarchy who actually have more power and influence than more senior managers or those higher up the org chart.

The culture of every organization is unique. Some examples of organizational culture that I've worked under include autocratic. where the big boss makes the decision and there is no discussion. You step up and salute and make things happen. I've also worked in organizations that have a culture of democracy where decisions are made by committee. All the senior leaders provide their input on the project or issue, each gets an equal say in the outcome, and decisions are made by consensus. This, in my opinion and experience, is often a difficult culture to work in. There is also a hybrid between these two extremes where a small handful of key executives, rather than one person at the top, have the power and authority to make decisions. Perhaps they seek input and consider other opinions before making the decision, but they have the authority to say "go" and set things in motion. Again, you must understand the culture of your organization so that you can navigate effectively and communicate appropriately.

Documenting the Communications Management Plan

The three outputs to the Plan Communications Management process are the communications management plan, project management plan updates, and project documents updates. The project management plan updates that may be required as a result of performing this process are contained in the stakeholder engagement plan. The project documents updates might include the project schedule and the stakeholder register. Let's take a closer look at the details of the communications management plan.

All projects require sound communication plans, but not all projects will have the same types of communication or the same methods for distributing the information. The *communications management plan* documents the types of information needs the stakeholders have, when the information should be distributed, how the information will be delivered, and how communications will be monitored and controlled throughout the project. This plan is developed by taking the organizational structure and stakeholder requirements into consideration. The answer to the pop quiz posed earlier in this chapter is that the communications management plan is a subsidiary plan of the project management plan.

 According to the *PMBOK® Guide*, the communications management plan defines how communications will be formulated, executed, and monitored for effectiveness throughout the project.

The type of information you will typically communicate includes project status, project scope statements and scope statement updates, project baseline information, risks, action items, performance measures, deliverables acceptance, and so on. What's important to know for this process is that the information needs of the stakeholders should be determined as early in the Planning process group as possible so that as you and your team develop project planning documents, you already know who should receive copies of them and how they should be delivered.

According to the *PMBOK® Guide*, the communications management plan typically describes the following elements:

- The communication requirements of each stakeholder or stakeholder group
- Purpose of communication
- Frequency of communications, including time frames for distribution
- Escalation process
- Name of the person responsible for communicating information along with the person responsible for authorizing the communication and the recipients
- Format of the communication and method of transmission
- Method for updating the communications management plan
- Flowcharts
- Glossary of common terms
- Constraints that might impact communication such as legislation, technology, or organizational policies

I've included only some of the most important elements of the communications management plan in the list of elements for it. I recommend you review the entire list in the *PMBOK® Guide*.

The information that will be shared with stakeholders and the distribution methods are based on the needs of the stakeholders, the project complexity, and the organizational policies. Some communications might be informal—a chat by the coffeemaker, for instance—whereas other communications are more formal and are kept with the project files for later reference. The communications management plan may also include guidelines for conducting status meetings, team meetings, and so on.

I've mentioned this a couple of times but it's worth repeating. Consider setting up an online project site and posting the appropriate project documentation there for the stakeholders to access any time they want. Also be certain to apply version controls to the documents so that there is no mistaking which document is the latest and greatest. Include a link to the project site in the communications management plan, and notify your stakeholders when updates or new communications are posted.

Exam Spotlight

For the exam, know that the communications management plan documents how the communication needs of the stakeholders will be met, including the types of information that will be communicated, who will communicate it, who receives the communication, the methods used to communicate, the timing and frequency, the method for updating this plan as the project progresses, the escalation process, and a glossary of common terms. The plan is used to manage the flow of information in order to keep stakeholders engaged and informed.

Communicating on an Agile Team

Agile teams have an inherent need to communicate on a continuous basis. The amount of change and uncertainty on an agile-based project requires quick and frequent communications so that all team members know and understand the new or emerging details. The agile team holds a series of meetings during the course of the project. We studied the iteration planning meeting earlier. As you'll recall, this is a meeting at the beginning of the iteration where items are chosen from the product backlog and placed in the iteration backlog to be worked on during the upcoming iteration. There are three other meetings in an agile methodology that we'll look at next:

Daily Stand-ups   Agile teams require frequent interactions and team checkpoints. Iteration-based agile teams use what's called a daily stand-up meeting for these checkpoints. *Daily stand-up* meetings are used to review the work of the iteration and make any changes or adjustments necessary to the iteration to ensure the work can be completed. It's also an opportunity to discuss any obstacles standing in the way of the team doing their work. Any changes that are needed might be made in the current iteration or documented for inclusion in the next one.

The daily stand-ups should be held at the same time and same place every day and should be time limited, usually no more than 15 minutes. These meetings are run by the team. It's a good idea to hold the stand-up meeting in front of the task board (or Scrum board) or burndown chart so that the tasks are easily referenced. Team members must come prepared to discuss the answers to three questions at each stand-up meeting:

1. What did I accomplish yesterday?
2. What will I work on today?
3. Do I have any roadblocks or issues preventing me from doing my work?

Stand-ups are an important element in the agile process. They keep the team informed and alert them to any obstacles standing in the way of completing tasks. The daily stand-ups in an iteration-based approach such as Scrum focus on the team member.

Daily stand-ups occur in a flow-based agile approach, such as Kanban, as well. These meetings focus on the workflow and capacity to produce work, not the team members. In this stand-up meeting, the team will review the task board (or Kanban board). Let's say the workflow columns on the task board are Plan, Design, Build, Test, and Deploy. Starting with the last column and working toward the left (from Deploy to Plan in our example), the team will answer the following questions:

- What is needed to advance this task?
- Are tasks being worked on that are not on the board?
- What is needed in order for us to finish as a team?
- Are any obstacles or bottlenecks blocking the workflow?

Daily stand-ups are not status meetings for the product owner or the stakeholders, nor are they problem-solving meetings. They are working meetings for the team members and the product owner to determine what work was accomplished and what will be worked on today. Problems may be identified during the meeting but should not be solved during the meeting. Address problem resolution in a separate meeting right after the stand-up.

Iteration Review   At the end of the iteration, agile teams hold regular stakeholder meetings known as *reviews*. Such a meeting includes team members, the Scrum master, the product owner, customers, stakeholders, and management. This is not a decision-making meeting. The purpose is to review the work that was accomplished or completed during the iteration and give the team a chance to show off their accomplishments. This may include a demonstration of functionality or a demonstration of the design work completed during the iteration. This meeting demonstrates the progressive completion and improvement of the deliverable during the iteration. Stakeholders will likely provide feedback during this meeting that the team can use to improve the deliverable in the next iteration.

If backlog items were not completed during the current iteration, you'll want to discuss this at the next planning meeting. The product owner will use information from this meeting to inform project stakeholders of the overall progress.

Retrospective   At the end of the iteration, another meeting is held called a *retrospective*. This meeting is held after the review meeting and includes the cross-functional team members, Scrum master, and product owner. A retrospective is held to determine the following:

- Overall progress of the work
- Changes that are needed to schedule or scope

- Ideas for improvement for the next iteration
- What worked well during this iteration
- Lessons learned to determine how the next iteration, and future iterations, can be improved

You've now seen the full circle of agile meetings. It all starts with the product owner developing user stories and placing them in the product backlog. The first meeting is the iteration planning meeting where items are chosen from the product backlog. Next, the daily stand-up meetings occur to review how the work of the iteration is proceeding. Then at the end of the iteration, a review meeting is held to demonstrate the features and receive feedback from stakeholders on the deliverable. Lastly, the retrospective is held to determine what went well and what could be improved for the next iteration, and to discuss any changes needed to scope.

Agile projects produce project documentation, just as predictive projects do, so be certain to post all project artifacts to the project site. According to the *Agile Practice Guide* (PMI®, 2017), posting project artifacts promotes transparency and promotes communication with the stakeholders and with management. Remember to use version controls on your artifacts so that when they are referenced, you can ensure everyone is reviewing the latest version.

A Closer Look at Adaptive Methodologies

As discussed in previous chapters, the project manager is responsible for tailoring the processes to match the project. This also includes determining the best life cycle and development methodology to use for the project. Let's take a deeper look at the various methodologies you've learned about so far and then, later in this section, figure out how to choose the appropriate methodology for your project.

We'll start with the adaptive methodology. Within an adaptive approach are two types of life cycle approaches:

Iterative Life Cycle Approach   The iterative life cycle approach allows for feedback on the work as it progresses, and before it's completed, so that improvements and modifications can be made. Iterative makes use of prototypes or mock-ups, which allows for feedback from the team and successive improvements to the prototype in each iteration. The team will act on the information received and incorporate modifications and new ideas into the next iterations. In this way, an iterative life cycle reduces uncertainty. This approach focuses on the iterative learning and discovery about the product, rather than speed of delivery. Use the iterative approach when the project is highly complex, there are frequent changes, and/or when the final product is not fully understood by all stakeholders.

Incremental Life Cycle Approach   The incremental life cycle approach delivers completed functionality or deliverables at the end of the iteration that the product owner can use immediately. This may include delivering a minimum viable product in cases where it's not possible to deliver a fully functioning deliverable. In this method, the team plans the deliverables before beginning the work, and they may plan future deliverables for future iterations as well. The incremental methodology increases the speed at which the team can produce deliverables. Rather than wait for every deliverable to be completed and then presented to the customer at the end of the project (as in a predictive approach), subsets of the overall product or result of the project are delivered incrementally throughout the life of the project. These small, frequent deliverables that consist of a finished piece of work characterize an incremental approach.

Agile Life Cycle Methodology   Agile focuses on value to the customer. When producing work with an agile approach, the work is seen and known by all team members and stakeholders throughout the project. In this way, agile is very transparent. This is a highly collaborative approach as well.

An agile life cycle is a combination of both iterative and incremental approaches. It incorporates the best of both iterative and incremental methodologies because feedback is received during the iteration, the team can make modifications, and they can deliver a functioning product or prototype or some portion of a deliverable at the end of the iteration. The idea of changing requirements throughout the project is incorporated in the agile methodology.

The *Agile Practice Guide* (PMI®, 2017) describes two types of agile approaches: the *iteration-based agile* approach and the *flow-based agile* approach. The iteration-based delivery method involves the team working collaboratively within time-bound periods to deliver completed features or functionality at the end of the time period. For example, your workflows may consist of plan, design, build, test, and deploy. Each iteration will repeat these workflows within its given time-bound period. This helps the team to discover missing or misunderstood requirements. All iterations, or time periods, are the same throughout the project. This is what the Scrum methodology is based on.

In the flow-based approach, items are pulled from the backlog based on capacity to start work. The work is managed on a task board, where each column represents a workflow such as plan, design, build, test, and deploy (similar to a Kanban board). A new feature, or a piece of functionality, or minimum viable product is chosen and worked on through the workflow stages until it is completed. The number of tasks in each of the workflows is limited by the capacity of the team to take on new work. For example, if the build workflow is at maximum capacity, no new work can be pulled into this workflow until existing work finishes and is pulled into test. The number of features that the team can work on is limited by the capacity of the team to take on new work within each workflow. This ensures the team can address issues quickly and prevents rework due to the limited number of tasks worked on at any one time. There is no time-bound period in which the work should be completed, and each feature may take more time or less time than others to complete, so the team and stakeholders must decide when to review the product, perform retrospectives, and plan for new user stories to begin.

Exam Spotlight

Iteration-based agile and flow-based agile both use daily stand-ups. The questions in an iteration-based agile approach are answered individually by each team member one after the other and are focused on working in a time-bound approach. The flow-based agile questions are answered by the team collaboratively as a team and are focused on team throughput.

Predictive Methodology   I have covered the predictive life cycle throughout this book. As noted previously, the *PMBOK® Guide* is heavily focused on the predictive life cycle methodology. That is, planning is fully vetted and completed before proceeding to performing the actual work of the project, and once work begins, it's monitored and controlled to adhere to the project plan. In a predictive methodology, requirements are well known and documented before the work of the project begins. The project teams are established in advance of and tend to remain stable throughout the project. Risk is lower in a predictive methodology because there is so much certainty about the project, the requirements, scope, schedule, and cost. These elements are well defined before the work begins, thereby reducing risk.

Hybrid Methodology   Hybrid is a combination of predictive, iterative, incremental, and agile methodologies. For example, perhaps the requirements and project scope are not well understood and there is some uncertainty about the final outcome that the project will produce. You could employ an adaptive approach for the Planning processes until the requirements are well understood and then switch to a predictive approach when performing the Executing processes to do the work of the project. Using a hybrid methodology is a common practice whereby you use adaptive methodologies for the uncertain parts of the project and predictive methods for the well-established processes.

Consider using a hybrid approach when there is uncertainty or complexity, or there are uncertain risks about the project requirements. Then switch to a predictive approach to perform the work of the project. I have used this approach when managing technology projects. The customer may have some idea about the functionality required but can't articulate what the final product should look like. They may express it this way: "I'll know it when I see it." Using an agile approach, we can produce a prototype and continually evolve the requirements as the project progresses. Feedback is provided during each iteration to help improve the overall product, and work continues in an agile fashion until the programming is completed. Then, the operations team steps in and may use a more predictive or traditional approach in rolling out the new system to the organization. The opposite scenario may apply as well, where the team uses a more predictive methodology to determine requirements, estimate activities and activity durations, and make work assignments, but incorporates some of the features of agile such as performing work in iteration and holding planning meetings, daily stand-ups, reviews, and retrospective meetings.

Hybrid is a great methodology to use to introduce the team and the organization to the principles of agile. You might start the team out on a simple project (one without a lot of risk to the organization), working in iterations and obtaining continual feedback from the customer. As the team learns and becomes increasingly comfortable with this methodology, you can graduate to using agile on more complex projects.

Table 6.2 shows the first four life cycle methodologies we discussed and their characteristics. Hybrid consists of any combination of the other methodologies.

TABLE 6.2   Characteristics of methodologies

	Predictive	Iterative	Incremental	Agile
Requirements gathering	Onetime effort to document all the requirements of the project	At the beginning of each iteration with modifications during the iteration	At the beginning of each iteration or at the beginning of the project	At the beginning of each iteration with modifications during the iteration
How Planning processes are performed	Planning is performed before the work begins. The plan drives the work.	Planning is performed at the beginning of each iteration and may consist of prototypes and mock-ups. Customer evaluates and provides feedback. The product is improved through continued enhancements to prototypes or mock-ups.	Planning involves decomposing work of the overall project into successive subsets of deliverables that can be delivered at the end of each iteration. Planning may occur for one or more deliverables before beginning the work.	Planning is performed at the beginning of each iteration.

	Predictive	**Iterative**	**Incremental**	**Agile**
How the work is performed	Work is performed in the Executing process group after all Planning processes are completed. Work is performed once, unless there are approved changes.	Work is performed in time-bound periods (such as an iteration) and is repeated, and then modified in future iterations until it meets specifications.	Work is performed once during an iteration. If corrections are needed, they are made in the next iteration. Each stage of work is repeated during each iteration (i.e., design, construct, verify, deliver).	Work is repeated until it meets specifications.
Deliverable frequency	Deliverables are produced in the Executing and Monitoring and Controlling processes of the project and presented to the customer at the end of the project.	One deliverable at the end of the iteration	Recurrent small deliverables throughout the project	Recurrent small deliverables throughout the project, typically delivered at the end of each iteration but could be completed midway through
Focus	Controlling scope, cost, and schedule	Learning optimization	Speed of delivery	Value to the customer
Benefits	Helps in managing cost and scope. Reduces complexity and uncertainty because much is known about the project.	Helps in ensuring the solution is correct and meets the need. Allows for feedback on partially completed work in order to modify or improve the work.	Helps in performing the work faster, speeding up the project. Usable deliverables are completed at the end of each workflow and are delivered to the customer.	Helps in delivering business value frequently and in receiving immediate feedback. Combines the benefits of iterative and incremental, allowing for modifications to the work during the iteration; allows for immediate feedback; and produces usable deliverables.

Exam Spotlight

Planning is required for all projects no matter which methodology you use. Planning is performed differently, at different times, and at different levels of complexity, depending on the methodology you use, but you must perform Planning processes on all projects.

Other Methodologies

There are other, lesser-known adaptive methodologies that you should know about for the exam. We'll take a look at each of them next.

Extreme Programming (XP)

I introduced XP in Chapter 1, "Building the Foundation," but I'd like to provide a little more detail here. XP was developed to improve software development projects and is based on frequent cycles—that is, delivering software when the customer needs it. XP started out with a set of 12 practices but evolved over time to incorporate other techniques. XP operates on a set of core values that will help improve software projects:

- Communication
- Simplicity
- Feedback
- Courage
- Respect

XP encourages teams to sit together in a colocated workspace. Programmers share code and make use of *refactoring* techniques whereby they can improve the quality of the code without changing its functionality. Refactoring reduces duplication and eliminates poor code. The idea is to start with a solid, simple design and test the code often, which will help the team increase the speed with which they can program.

XP delivers business value in each iteration and starts with creating story cards. Story cards are like user stories and contain requirements, features, and functionality. The story cards are designed in an incremental fashion, much like other agile methodologies. As the project progresses, more information is known and more requirements are developed. The user stories are worked on by pairs of programmers and as a group. There is never a single programmer working alone in this methodology. They perform what's known as *pair programming.* That is, there are always two developers working as a pair at the same computer. It's important that programmers use a consistent style and pattern when writing the code so that all team members can understand what's been done. Having two programmers work on the same code increases the quality of the code because there are two people writing and reviewing the code and one may see something the other does not.

As in other agile methodologies, XP relies heavily on feedback. One important form of feedback comes from *test-driven development*, or a test-first approach. As code is released by the pair programmers, it is tested rigorously. The tests for the new code are combined with existing unit tests, and all the tests are run each time there is a release. Each test is run against the entire program (including the new code) and all the tests, including the new test, must pass with no errors. Code may be released twice a day or more, so this provides a significant amount of feedback for the team.

XP also concerns continuous integration. Everyone on the team has a shared understanding of what the end product looks like. The XP process keeps all the work integrated continuously throughout the project. For example, once testing is completed for a particular piece of code, it is kept up and running correctly throughout the remainder of the iteration (and project) and becomes integrated into new code or functionality as the project progresses.

XP may use both release planning and iteration planning to accomplish work. At the end of each iteration, the team delivers a functioning piece of software to the customer that they can use and/or validate as having met the business value. There isn't a set time frame for iterations, but they can vary from daily to quarterly, depending on the project.

Here is a list of the benefits and techniques of the XP methodology:

- Delivers business value in each iteration.

- Based on frequent cycles, delivers software to the customer when they need it.

- Encourages collocation of team members.

- User stories are worked on by pair programmers at the same computer, which improves quality of the code.

- Refactoring reduces duplication and eliminates poor code.

- Test-driven development provides significant feedback to the team because tests for new code are combined with existing unit tests and all the tests are run each time there is a release.

- Tests must pass with no errors in every iteration.

- Keeps all of the work integrated continuously.

- May use both release planning and iteration planning.

 Stakeholders in the XP methodology are called customers.

Crystal Methods

This is a family of methodologies designed to scale to the project needs. The name Crystal comes from its geological counterpart (the stone), which has a geometrical shape that is multifaceted. The Crystal methodology is also multifaceted, and each face represents a core value. The core values of Crystal are:

- People

- Interaction

- Community
- Communication
- Skills
- Talents

The core values show that Crystal is people focused, not process focused. The processes can and should adapt and change based on the needs of the project. The people are the most important aspect of Crystal, and the processes should be adapted to meet the needs of the team.

Crystal examines three factors to determine the right methodology to use for the project: the criticality of the project, the priority of the project, and the number of people involved. Once you've evaluated these factors, you can tailor the processes and practices to meet the unique project needs. For example, large teams with multiple members and stakeholders means there are more lines of communication, which brings its own set of complexities to the project. The team will need rigorous communication practices in place when dealing with large teams.

Crystal also has seven properties that help ensure the delivery of a successful project:

- Easy access to expert end users.

- Personal safety, which allows team members to speak freely without fear of reprimand. This builds trust among the team members.

- Focus on the work, including uninterrupted work time set aside each day to focus on priorities.

- Agile technical development practices to speed delivery. This includes tests and integration.

- Frequent delivery of usable code.

- Reflective improvement that involves continual review and experimentation to improve the process.

- Osmotic communication.

Osmotic communication is an interesting concept. I consider it a polite form of eavesdropping. The term was created by Alistair Cockburn, one of the founders of agile and a signer on the *Agile Manifesto*. Osmotic communication occurs when there are conversations going on in the background but within earshot of the team. The team members overhear the conversation and may realize the topic has importance to the project. Or they may recall the conversation sometime later and revisit the topic of the conversation in order to realize benefits to the project. The key to osmotic communication being effective is to colocate agile teams.

There are four methodologies within Crystal: Crystal Clear, Crystal Yellow, Crystal Orange, and Crystal Red. You will choose the methodology based on the criticality of the project and the number of people involved. Crystal Clear is the most common Crystal methodology.

- Crystal Clear has the least amount of criticality and lowest number of team members, generally between one and four members.
- Crystal Yellow is critical and teams consist of between 6 and 20 members.
- Crystal Orange is critical and teams consist of between 20 and 40 members.
- Crystal Red is the most critical of all projects and teams consist of between 5 and 100 people.

Scrumban

This methodology is a hybrid between Scrum and Kanban. The idea is that the work is organized in sprints, as in traditional Scrum, but uses a Kanban board to display the work of the sprint and monitor work in progress. Daily stand-ups are held to review progress and determine whether any obstacles exist that stand in the way of the team. The Kanban board is a good visual backdrop when holding the stand-ups because the team can see the progress of the work and answer the three questions during the stand-up. The Scrum team roles can be used in this methodology (Scrum master, team members, and product owner).

Feature-Driven Development (FDD)

This methodology was developed to address large software development projects. It focuses on delivering usable, working software continually in a timely manner. According to the *Agile Practice Guide* (PMI®, 2017) it's based on five processes or features:

- Developing a model
- Creating a features list
- Planning based on features
- Designing based on features
- Building based on features

FDD is a simple, five-step approach that allows teams to develop code rapidly. The last three steps, plan, design, and build, are typically performed in short iterations of two weeks or less. If a feature takes more than two weeks to complete, it will be broken down in the planning step into smaller components that can be completed during the iteration.

FDD is performed using software development best practices. FDD assumes the team has a predefined set of standards for development and that by using these standards, they can move quickly. FDD works well for large projects but not so well for small development projects. This methodology relies on a more top-down approach to decision-making than other methodologies, due to the size of the project.

There are six roles in this methodology: project manager, chief architect, development manager, chief programmer, class owner, and domain expert. Team members may take on one or more of these roles at any one time. Stakeholders in the FDD methodology are called clients.

Dynamic Systems Development Method (DSDM)

This methodology was introduced in the mid-1990s as a software development methodology. It has been revised over the years so that the methodology can be used for any project, not just software development projects. This methodology is grounded in governance framework that combines the iterative and incremental approaches. At the onset of the project, DSDM establishes the cost of the project, the quality standards, and the time frame to completion and, as such, is constraint driven. Scope is prioritized to meet the cost, quality, and time frame constraints. According to the *Agile Practice Guide* (PMI®, 2017), there are eight guiding principles of DSDM:

- Focus on the business need

- Ensure on-time delivery

- Collaborate with team members and stakeholders

- Never compromise on quality

- Build incrementally

- Use iterative development techniques

- Continual communications that are clear and concise

- Demonstrate control

Agile Unified Process (AUP or AgileUP)

This methodology is also used in software development projects. AUP incorporates several other agile processes that help improve productivity: test-drive development, agile modeling, agile change management, and database refactoring. AUP is an iterative approach that includes feedback during the work cycle (before the end product is delivered). According to the *Agile Practice Guide* (PMI®, 2017), there are seven disciplines within AUP and the iterative approach is performed across these disciplines in each release. The disciplines are as follows:

- Model—Describes the problem the organization is trying to solve with this project (the "what") and provides a feasible solution.

- Implementation—Transforms the "what" into basic programming code that can be unit-tested.

- Test—Tests the code produced in implementation to verify that the program performs as designed. Defects are identified and repaired.

- Deployment—This is rolling out the new system to the end users.

- Configuration—Manages changes to code and project artifacts.

- Project management—Manages all aspects of the project work.

- Environment—Ensures that the team has all the tools needed to perform the work and that they understand any standards, guidelines, or compliance needs for the project.

The benefits of AUP are based on the following philosophies and disciplines:

- Team members who know what they're doing and can work on their own with occasional high-level guidance
- Simplicity in the process and documentation
- Agility by following agile processes and principles
- A focus only on activities that produce high value
- Tool independence, which allows team members to choose the tools they want to use to perform the work
- Tailoring of AUP techniques to fit the organization's and team's needs
- Situationally specific, much like tailoring, using AUP when appropriate and efficient

Combining Methodologies

We've learned so far that there are several approaches and methodologies to managing projects, including predictive, incremental, iterative, agile, and hybrid. Within the adaptive methodologies are a host of agile methods, including iteration-based approaches and flow-based approaches, all designed to fit a specific need.

We have looked at each of these methodologies independently. However, it is possible, and very popular, to combine the approaches. For example, combining an iteration-based approach with a flow-based approach (such as Scrum and Kanban), as we discussed in the "Scrumban" section earlier, brings the best of both approaches to the team. You can use time-bound iterations to produce work while using the Kanban board to visually see capacity constraints, bottlenecks, and where tasks are in the workflow.

It's also possible to combine Scrum, Kanban, and XP. Scrum utilizes the product backlog and all the roles on a traditional Scrum project, such as the product owner and Scrum master, while visually displaying the flow of the work on the Kanban board. Work can be managed using capacity limits in each workflow. The XP principles, such as using story cards, using test-driven development, working in a continuous integration fashion, and refactoring, all help to improve the team's effectiveness.

Each project the organization undertakes has its own set of unique characteristics. Team members from across the organization will participate on the project. However, it's not likely that all team members will work on all projects, so the team makeup will change depending on the project. The complexity, criticality, and scale of the project are also factors to consider when combining agile approaches. The team is free to tailor the methodologies and practices to fit the project. The key is that the team should be focused on delivering value to the customer on a regular basis. According to the *Agile Practice Guide* (PMI®, 2017), combining approaches will increase the effectiveness of the team. This creates a synergy that will exceed what any individual team member can contribute on their own.

The *Agile Practice Guide* (PMI®, 2017) outlines several factors to consider when tailoring processes to fit the project:

- Does the project require iterative or incremental delivery?
- What is the desired frequency of process improvement?

- How should the flow of work proceed? Does it start and stop or need to evolve continually?

- Does quality impact the project outcome?

- How many teams are required to complete the work of the project?

- What is the experience level of the team members in using agile?

Use any combination of adaptive methodologies on your next low-risk project to see what might work well for your teams.

 Real World Scenario

Project Case Study: New Kitchen Heaven Retail Store

After creating the first draft of the project schedule network diagram, you went back to each stakeholder to ask for cost estimates for each of the activities. Ricardo's estimates are shown here with the activities he gave you last time:

1. Procure the cabling vendor. This takes 30 to 45 days and will have ongoing costs of $3,000 per month. Procurement costs are covered in the monthly expense.

2. Run Ethernet cable throughout the building. The estimated time to complete is 16 hours at $100 per hour, which was figured using parametric estimating techniques.

3. Purchase the router, switch, server, and rack for the equipment room and four point-of-service terminals. The estimated cost is $17,000.

4. Install the router and test the connection. Testing depends on the cable vendor. The time estimate to install is eight hours. Ricardo's staff will perform this activity at an average estimated cost of $78 per hour.

5. Install the switch. Based on past experience, the time estimate to install is two hours. Ricardo's staff will perform this activity at an average estimated cost of $78 per hour.

6. Install the server and test. Based on past experience, the estimate to install is six hours at $84 per hour.

7. The web team will add the new store location and phone number to the lookup function on the Internet site. The time estimate is two hours at $96 per hour.

Jake and Jill have each written similar lists with time and cost estimates. Using this information, you create the activity cost estimates and are careful to document the basis of estimates. The following list includes some of the information you document in the basis of estimates:

- Ricardo's use of parametric estimates for his cost estimates

- Jake's use of both analogous and parametric estimating techniques

- Jill's use of reserve analysis to include contingencies for unplanned changes involving vendor deliveries

- Assumptions made about vendor deliveries and availability of the cable vendor and assumptions made regarding when lease payments begin

- The range of possible estimates is stated as plus or minus 10 percent.

You also document the cost management plan and the cost baseline along with the project funding requirements. Since this project will occur fairly quickly, only two funding requirement periods are needed.

The stakeholder engagement plan is written and describes the relationships between the stakeholders, their communication requirements, and the format, reasons, and timing of the distribution of information, as well as strategies used to engage them throughout the project.

The communications management plan is also complete, and you've asked the key stakeholder to review it before posting it to the intranet site for the project. You want to make certain you've identified stakeholder communication needs, the method of communication, and the frequency with which communications will occur.

Ruth is managing the website launch project and has provided estimates to you in terms of staffing costs. The team will be using a hybrid approach to perform the work of the project. The team likes the idea of time-bound iterations to perform the work. They want continual feedback on their deliverables and prefer to manage the workflow on a task board. The workflows represented on the board are user story, plan, design, build, test, and deploy. All team members will participate in the daily stand-ups, reviews, and retrospective meetings.

Project Case Study Checklist

The main topics discussed in the case study are as follows:

Plan Cost Management

Estimate Costs

Determine Budget
- Cost aggregation

- Reserve analysis

- Expert judgment

- Parametric estimates

- Cost baseline

- Project funding requirements

Plan Communications Management

Determine effective and efficient communications

Review stakeholder register and stakeholder management strategy

Communications requirements analysis

Communication technology

Communication models

Communication methods

Communications Management Plan

Stakeholder needs

Format and language for information

Time frame and frequency of communication

Person responsible for communication

Methods for communicating

Glossary of terms

Adaptive Methodologies

Iteration-based method

Flow-based method

Hybrid methodologies

Combining methodologies

Understanding How This Applies to Your Next Project

In a previous job working in state government, I had to perform the Plan Cost Management and Estimate Costs processes rather early in the project because of the long procurement cycle. Our cost management plan (the policies and procedures used to plan and execute project costs) was set in legislation and government rulings, so there wasn't a lot of flexibility around this process in my situation. The way our funding request process worked is that I had to have an estimated cost for the project before I could request funding. If funding was approved, the project was approved. If I was not awarded funding, the project died and we moved on to the next one.

I rely on expert judgment and parametric estimating techniques to determine activity and total project costs. I often engage vendors and my own project team to help determine the costs. When I have a large project that must go out for bid, I have a well-defined scope and some idea of schedule dates and overall costs, but I won't complete the schedule until after the contract is awarded. In an ideal world, I would prefer to create the schedule prior to determining cost estimates and budgets ... but we don't live in an ideal world and sometimes Planning processes have to be performed out of order.

The authorized project budget becomes one of the key measurements of project success. In Chapter 11, "Measuring and Controlling Project Performance" we'll talk about monitoring the budget to determine whether we're tracking with our estimates.

I generally create a stakeholder engagement plan and a communications management plan as one document (I call it a communications plan). The reason for this is that the stakeholder engagement plan contains a lot of information about communication needs, timing, distribution methods, distribution formats, and so on. I find it easier to keep this information together in one plan. This plan is a must-have for every project. I can't stress enough how often I've seen the root cause of project issues end up being communication problems. Never assume keeping the stakeholders informed or engaged is an easy job. Even if you know the stakeholders well, always create a communications plan. Document how you'll communicate status, baseline information, risks, and deliverables acceptance. That way, there's no question as to how information will be relayed, who's going to receive it, or when it will be delivered.

Agile methodologies have been used in the information technology field (the field I work in) for some time now. It's common practice to combine methodologies to fit the project needs and the organizational culture. Stakeholders who are used to the old way of doing things (that is, the waterfall approach) may be uncomfortable working in an agile environment. Using a hybrid approach in this case is ideal. You can work with stakeholders in a predictive fashion to document requirements, create a scope statement, and determine cost and resource needs. Once the planning is completed, you can switch to an agile methodology and include them in your daily stand-ups, reviews, and retrospectives. You are able to introduce them to agile processes while also giving them a sense of assurance by taking the time up front to document their needs and requirements. It's a win-win.

Summary

The cost management plan is the only output of the Plan Cost Management process. This documents the policies and procedures you'll use to plan and execute project costs as well as how you will estimate, manage, and control project costs. The Estimate Costs process determines how much the project resources will cost, and these costs are usually stated in monetary amounts. Some of the techniques used specifically for estimating costs are analogous estimating, parametric estimating, bottom-up estimating, three-point estimates, and reserve analysis. You can also use bottom-up estimating for total project cost estimates. This approach involves estimating the cost of each activity and then rolling them up to come up with a total work package cost. The primary outputs of this process are the cost estimates and the basis of estimates that details all the support information related to the estimates.

The tools and techniques of the Determine Budget process include cost aggregation, data analysis (reserve analysis), expert judgment, historical information review, funding limit reconciliation, and financing. These tools together help you produce the final, authorized project budget known as the cost baseline, which is an output of this process. You will use the cost baseline throughout the remainder of the project to measure project expenditures, variances, and project performance. The cost baseline is graphically displayed as an S curve.

The cost baseline is also known as a performance measurement baseline (PMB) when you're calculating earned value management formulas. PMBs are management controls that should change only infrequently. Examples of the performance measurement baselines you've looked at so far are the scope, schedule, and cost baselines. The completed project plan itself also becomes a baseline. If changes in scope or schedule do occur after Planning is complete, you should go through a formalized process to implement the changes.

The Plan Stakeholder Engagement process focuses on effectively engaging stakeholders, understanding their needs and interests, identifying the good and bad things they might bring to the project, and determining how the project will impact them. The primary output of this process is the stakeholder engagement plan.

The purpose of the communications management plan is to determine and document the communication needs of the stakeholders by defining the types of information needed, the format for communicating the information, how often it's distributed, and who prepares it. This plan is a subsidiary plan of the project management plan and is created in the Plan Communications Management process.

Communication on an agile team consists of a series of meetings and continual feedback throughout the project. The meetings are daily stand-ups, reviews, and retrospectives. The daily stand-ups for an iteration-based approach address what the team has accomplished and will accomplish, and what obstacles are in the way. Daily stand-ups for a flow-based approach address workflow and capacity and ask what's needed to advance the task, what's being worked on that isn't on the board, what is needed for the team to finish, and whether any bottlenecks in the process exist.

The iterative life cycle approach makes use of prototypes and focuses on the iterative learning and discovery of the product, rather than speed of delivery.

The incremental life cycle approach delivers completed functionality that the customer can use at the end of the iteration. This method increases the speed at which the team can produce deliverables. Small, frequent deliverables characterize an incremental approach.

The agile life cycle methodology is transparent and highly collaborative. There are two types of agile life cycles: iteration-based agile and flow-based agile. Iteration-based agile involves time-bound periods where the team works collaboratively to deliver features or functionality at the end of the time period. All iterations use the same time frame (two weeks, for example) throughout the project. Flow-based agile is capacity bound and pulls backlog items onto a task board based on the capacity for work. There are no time-bound periods of work, so the team and stakeholders must decide when to hold product reviews and retrospectives, and when to begin planning for new user stories as capacity is freed up.

Hybrid involves a combination of methodologies and is a great way to introduce agile practices to the organization.

Another adaptive methodology is Extreme Programming (XP), which is used in software development projects. It is based on frequent cycles and delivers software to the customer when they need it. XP delivers business value in each iteration. IT incorporates and operates with techniques such as refactoring, pair programming, and test-driven development. Stakeholders in the XP methodology are called customers.

Crystal methods are a family of methodology. Crystal is people focused, not process focused. Crystal processes should be adapted to meet the needs of the team. Osmotic communication is one of the seven properties that ensure a successful project using this methodology. Osmotic communication is a polite form of eavesdropping. Crystal has four methodologies that are based on the criticality of the project and the number of people involved: Crystal Clear, Crystal Yellow, Crystal Orange, and Crystal Red.

Scrumban is a combination of both Scrum and Kanban. Work is managed within time-bound periods, and the workflows are displayed on a task board.

Feature-Driven Development (FDD) is used on large software development projects. It relies on a top-down approach to decision making. Stakeholders in the FDD methodology are called clients.

The Dynamic Systems Development Method (DSDM) establishes the cost, quality standards, and time frame to completion at the onset of the project. This process is constraint driven. Scope is prioritized to meet the cost, quality, and time frame constraints.

The Agile Unified Process (AUP) is an iterative approach that includes feedback during the work cycle, before the end product is delivered. AUP disciplines include model, implementation, test, deployment, configuration, project management, and environment.

Combining methodologies is common and encouraged. Scrum and Kanban are often combined to realize the benefits of both methodologies. Scrum, Kanban, and XP are another example of combining methodologies. Combining methodologies increases the effectiveness of the team. The key is that the team is focused on delivering value to the customer on a regular basis.

Exam Essentials

Be able to state the purpose of the cost management plan. The cost management plan is the only output of the Plan Cost Management process. It establishes policies and procedures for planning and executing project costs and documents the processes for estimating, managing, and controlling project costs.

Be able to identify and describe the primary output of the Estimate Costs process. Cost estimates are the primary output of Estimate Costs. These estimates are quantitative amounts—usually stated in monetary units—that reflect the cost of the resources needed to complete the project activities.

Be able to identify additional general management techniques that can be used in the Project Cost Management Knowledge Area. Some of the general management techniques that can be used in this Knowledge Area are return on investment, discounted cash flow, and payback analysis.

Be able to describe the cost baseline. The cost baseline is the authorized, time-phased cost of the project when using budget-at-completion calculations. The cost baseline is displayed as an S curve.

Be able to describe project funding requirements. Project funding requirements are one output of the Determine Budget process. They detail the funding requirements needed on the project by time period (monthly, quarterly, annually).

Be able to describe the purpose of the Plan Stakeholder Engagement process. The Plan Stakeholder Engagement process concerns effectively engaging stakeholders, understanding their needs and interests, identifying how they may help or hurt the project, and determining how the project will impact them. The primary output of this process is the stakeholder engagement plan.

Be able to describe the primary purpose of the stakeholder engagement plan. The stakeholder engagement plan documents the engagement levels of the stakeholders and strategies for engaging stakeholders throughout the project.

Be able to describe the purpose of the communications management plan. The communications management plan determines the communication needs of the stakeholders. It documents what information will be distributed, how it will be distributed, to whom, and the timing of the distribution.

Be able to describe the difference between iterative and incremental life cycle approaches. Iterative approaches focus on discovery of the product, rather than speed of delivery. Incremental approaches deliver completed functionality at the end of the iteration that the customer can use. This increases the speed at which the team can deliver.

Be able to name the two types of agile life cycle methodologies. Iteration-based agile involves time-bound periods of work, and flow-based agile involves work based on capacity.

Be able to describe the hybrid methodology. Hybrid is a combination of methodologies, typically combining predictive approaches with agile approaches.

Be able to name several adaptive methodologies. Examples are Extreme Programming, Crystal, Scrumban, Feature-Driven Development, Dynamic Systems Development Method, and Agile Unified Process.

Review Questions

You can find the answers to the review questions in Appendix A. Be sure to download the Bonus Exams and Bonus Questions so that you'll have a broader exposure and more experience answering questions related to the topics in this chapter.

1. Which of the following are true regarding the stakeholder engagement plan? (Choose two.)

   A. The stakeholder engagement plan helps to define and manage the flow of information to the stakeholders.

   B. The stakeholder engagement plan is developed by analyzing the needs, interests, and potential impacts of the stakeholders. It documents strategies needed to promote stakeholder decision-making and execution.

   C. The stakeholder engagement plan captures the strategies needed to engage stakeholders throughout the project.

   D. The stakeholder engagement plan documents the types of information needs the stakeholders have, when the information should be distributed, and how the information will be delivered.

   E. The stakeholder engagement plan considers the organizational structure and stakeholder requirements.

2. When you're creating the cost management plan, which project elements should you consider and review? (Choose two.)

   A. The approved project charter and project scope

   B. The level of precision and accuracy

   C. The units of measure and control thresholds

   D. The project schedule and project resources

3. You are a project manager working for iTrim Central. Your organization has developed a new dieting technique that is sure to be the next craze. You're preparing your cost management plan. You know that all of the following are true regarding this plan except for which one?

   A. The WBS provides the framework for this plan.

   B. Units of measure should be described in the plan usually as hours, days, weeks, or lump sum.

   C. This plan is a subsidiary of the project management plan.

   D. Control thresholds should be described in the plan as to how estimates will adhere to rounding ($100 or $1,000, and so on).

4. You are a project manager working for iTrim Central. Your organization has developed a new dieting technique that is sure to be the next craze. One of the deliverables of your feasibility study was an analysis of the potential financial performance of this new product, and your executives are very pleased with the numbers. You will be working with several vendors to produce products, marketing campaigns, and software that will track customers' progress with the new techniques. For purposes of performing earned value measurements for project costs, you are going to place which of the following in the WBS?

    A. Chart of accounts

    B. Code of accounts

    C. Control account

    D. Reserve account

5. Which of the following options is the key component of determining cost estimates and should be completed as early in the project as possible?

    A. Scope definition

    B. Resource requirements

    C. Activity cost estimates

    D. Basis of estimates

6. You want to improve your cost estimates by taking into account estimation uncertainty and risk. Which of the following estimating techniques will you use?

    A. Analogous estimates

    B. Three-point estimating

    C. Parametric estimates

    D. Bottom-up estimates

7. When an agile team is communicating, several meetings occur including a daily stand-up. Which of the following options are true regarding the daily stand-up in an agile methodology? (Choose two.)

    A. Stand-ups are intended to examine the work of the project and note corrections that are needed for the next iteration.

    B. Iteration-based stand-ups focus on team members.

    C. Stand-ups are intended to examine what is going well, and not so well, with the iteration.

    D. Flow-based stand-ups focus on team capacity and workflow.

8. Your organization has historically performed projects using a predictive methodology. You'd like to give the agile process a try. You have a low-risk, low-priority project that is perfect for this experiment. However, the project sponsor of this project is nervous about using a new approach that's never been tried before. Which of the following actions should you take in this scenario?

    A. Make certain you are allocating time to mentoring stakeholders.

    B. Ensure that the sponsor is committed to the project and to the methodology.

    C. Mentor the sponsor and the stakeholders on the agile process by providing training to educate them on the agile processes.

    D. Consider using a hybrid approach to help the sponsor feel a little more at ease, because this will mix some of the processes with which they are familiar with agile processes.

    E. All of the above.

9. You are the project manager for a custom home–building construction company. You are working on the model home project for the upcoming Homes Tour. The model home includes smart home connections, talking appliances, and wiring for home theaters. You are working on the cost baseline for this project. Which of the following statements are true? (Choose three.)

   A. This process aggregates the estimated costs of project activities.

   B. The cost baseline will be used to measure variances and future project performance.

   C. This process assigns cost estimates for expected future period operating costs.

   D. The cost baseline is the time-phased budget at completion for the project.

10. Adaptive methodologies focus on eliminating waste, Lean thinking, and pull-based systems. When using this methodology, stakeholders are highly engaged with the project. Which of the following are true regarding positive stakeholder engagement in an adaptive methodology? (Choose three.)

    A. Their engagement and the methodology creates transparency on the project.

    B. Stakeholders have a neutral classification in an adaptive methodology.

    C. Stakeholders have a supportive classification in an adaptive methodology.

    D. Their engagement increases trust and reduces risk, increasing the changes for a successful project.

    E. Stakeholders have a resistant classification when using an adaptive methodology.

11. You are the project manager for a custom home–building construction company. You are working on the model home project for the upcoming Homes Tour. The model home includes smart home connections, talking appliances, and wiring for home theaters. You are working on the Determine Budget process. Which of the following statements are true? (Choose three.)

    A. You document the funding limit reconciliation to include a contingency for unplanned risks.

    B. You discover that updates to the risk register are needed as a result of performing this process.

    C. You document that funding requirements are derived from the cost baseline.

    D. The performance measurement baseline will be used to perform earned value management calculations.

12. Which of the following is displayed as an S curve?

    A. Funding requirements

    B. Cost baseline

    C. Cost estimates

    D. Expenditures to date

13. In a flow-based agile approach, which of the following questions would you ask at the daily stand-up meeting? (Choose two.)

    A. What did I accomplish yesterday?

    B. What is needed to advance this task?

    C. Are tasks being worked that are not on the board?

    D. What will I work on today?

    E. Are there any roadblocks preventing me from doing my work?

14. Which of the following statements are true regarding the iteration review meeting? (Choose three.)

    A. This meeting is a decision-making meeting.

    B. This meeting includes team members, the Scrum master, the product owner, customers, stakeholders, and management.

    C. This meeting occurs at the end of the iteration.

    D. This meeting includes the cross-functional team members, Scrum master, and product owner.

    E. This meeting may include a demonstration of functionality.

15. You have eight key stakeholders (plus yourself) to communicate with on your project. Which of the following is true?

    A. There are 36 channels of communication, and this should be a consideration when using the communications technology tool and technique.

    B. There are 28 channels of communication, and this should be a consideration when using the communications requirements analysis tool and technique.

    C. There are 28 channels of communication, and this should be a consideration when using the communications technology tool and technique.

    D. There are 36 channels of communication, and this should be a consideration when using the communications requirements analysis tool and technique.

16. When tailoring communications for the project, which of the following should you consider in order to ensure effective communications with and among the stakeholders? (Choose three.)

    A. Geographic locations of your team members

    B. Relationships between and among stakeholders

    C. Language that will be standard for the project

    D. Engagement strategies for stakeholder decision-making

    E. Technology to perform communications

17. You are preparing your communications management plan and know that all of the following are true except for which one?

    A. *Decode* means to translate thoughts or ideas so that they can be understood by others.

    B. *Transmit* concerns the method used to convey the message.

    C. *Acknowledgment* means the receiver has received and agrees with the message.

    D. Encoding and decoding are the responsibility of both the sender and receiver.

18. Match the following characteristics of development life cycles to their description.

Characteristics of methodologies

Methodology	Characteristic
A. Predictive	1. Deliverables are completed at the end of each iteration and can be turned over to the customer to use.
B. Iterative	2. Focuses on controlling scope, cost, and schedule.
C. Incremental	3. A combination of multiple life-cycle development approaches.
D. Agile	4. Work is repeated until it meets specifications.
E. Hybrid	5. Uses prototypes and mockups.

19. This agile methodology focuses on frequent cycles and delivering software to the customer when the customer needs it. The core values for this methodology are communication, simplicity, feedback, courage, and respect. Which methodology does this describe?

   A. Feature-Driven Development

   B. Crystal

   C. Dynamic Systems Development Method

   D. Extreme Programming

20. What is the definition of osmotic communication and what methodology is it associated with? (Choose two.)

   A. This is a polite form of eavesdropping.

   B. This is associated with the Feature-Driven Method.

   C. This is associated with the Crystal method.

   D. This is a type of communication transmission.

   E. This is associated with the Agile Unified Process.

Chapter 7

Identifying Project Risks

- 3.1.3 Determine potential threats to compliance

- 3.1.4 Use methods to support compliance

- 3.1.5 Analyze the consequences of noncompliance

- 3.1.6 Determine necessary approach and action to address compliance needs

Risk is evident in everything we do. When it comes to project management, understanding risk and knowing how to minimize its impacts (or take full advantage of its opportunities) on your project are essential for success. This chapter is dedicated to project risk. Five of the seven risk processes, all contained in the Project Risk Management Knowledge Area, fall in the Planning process group. I'll cover Plan Risk Management, Identify Risks, Perform Qualitative Risk Analysis, Perform Quantitative Risk Analysis, and Plan Risk Responses in this chapter.

Hold on to your hats! I'm going to cover a lot of material in this chapter, but it will go fast. I promise.

 The process names, inputs, tools and techniques, outputs, and descriptions of the project management process groups and related materials and figures in this chapter are based on content from *A Guide to the Project Management Body of Knowledge (PMBOK® Guide), Sixth Edition* (PMI®, 2017). The references to adaptive and hybrid methodologies, related materials, and figures in this chapter are based on content from the *Agile Practice Guide* (PMI®, 2017).

Understanding Risk

Every one of us takes risks on a daily basis. Just getting out of bed in the morning is a risk. You might stub your toe in the dark on the way to the light switch or trip over the dog and break your leg. These events don't usually happen, but the possibility exists. The same is true for your project. Risk exists on all projects, and the potential that a particular risk event will occur depends on the nature of the risk.

Risk, like most of the elements of the other Planning processes, changes as the project progresses and should be monitored throughout the project. As you get close to a risk event, that's the time to reassess your original assumptions about it and your plans to deal with it and make any adjustments as required.

Not all risks are bad. Risks can present future opportunities as well as future threats to a project. Risk events may occur due to one reason or several reasons, and they may have multiple consequences. Those consequences will likely impact one or more of the project objectives, and you'll need to know whether the consequences are positive or negative.

Risks may come about as an individual event (or individual risk), or the project itself as a whole could present a risk due to the uncertain nature of the project outcome. For example, your organization might be taking on more than it can accomplish, or the complexity of the project is such that your organization doesn't have resources with skills or experience to perform the project. This could cause negative impacts to the organization and should be considered when undertaking the project.

Failing to address individual or project risks could cause financial loss, damage the organization's reputation, extend the project timeline, or have any number of other unpleasant outcomes.

Risk is, after all, uncertainty. The more you know about risks and their impacts beforehand, the better equipped you will be to handle a risk event when it occurs. The processes that involve risk, probably more than any other project Planning process, concern balance. You want to find that point where you and the stakeholders are comfortable with the risk based on the benefits you can potentially gain. In a nutshell, you're balancing the action of taking a risk against avoiding its consequences or impacts, or against enjoying the benefits it may bring.

You'll need to tailor the risk processes according to the size, complexity, and importance of the project, as well as for the development life cycle you're using to manage the project. For example, if you're using an agile life cycle approach, you'll need to perform the risk planning processes at the beginning of every iteration. This exercise will be quicker and more condensed than performing the same processes in a waterfall life cycle approach. The changing nature of an agile project means you will have to make certain that knowledge is shared across the team. That way, team members understand the risks and their impacts so that they can manage them during each iteration. It also means that the team can easily update the requirements and adapt them as necessary to account for risks during each iteration.

In the waterfall approach, you will spend much more time identifying and analyzing risks during the Planning process group up front, attempting to identify everything possible, good or bad, that could occur on the project. This will be repeated throughout the life of the project but not to the extent you will go through in the Planning process group.

Creating the Risk Management Plan

Risks come about for many reasons. Some are internal to the project, and some are external. The project environment, the planning process, the project management process, inadequate resources, and so on can all contribute to risk. Some risks you'll know about in advance and plan for during this process; other risk events will occur unannounced during the project. The *Plan Risk Management* process determines how you'll prepare for and perform risk management activities on your project.

Exam Spotlight

The *PMBOK® Guide* contends that the Plan Risk Management process should begin as soon as the project begins and should be concluded early in the Planning processes. It may need to be updated or repeated if the scope of the project changes dramatically and/ or you are starting a new phase of the project. The risk management plan, the only output of this process, ensures that the appropriate amount of resources and the appropriate time are dedicated to risk management. "Appropriate" is determined based on the levels, the importance, and the types of risks. The most important function the risk management plan serves is that it's an agreed-on baseline for evaluating project risk.

Risks associated with a project are generally concerned with any combination of four project objectives—time, cost, scope, and quality. As you might have guessed, the project management plan is an input to this process, and it includes the scope statement, project schedule, project budget, and quality plan, all of which are considered when planning for and dealing with risks. You've seen all of the inputs to the process before. They include the project charter, project management plan, project documents (stakeholder register), enterprise environmental factors, and organizational process assets. Remember that organizational process assets include policies and guidelines that might already exist in the organization. Your organization's risk categories, risk statement formats, and risk templates should be considered when planning for risks. Also, when developing the risk management plan, consider the defined roles and responsibilities and the authority levels the stakeholders and project manager have for making decisions regarding risk planning.

There is one new element in the EEF input—risk attitude—that you'll want to know about for the exam. We'll look at that next.

Risk Attitude

The *risk attitude* of the organization and the stakeholders consists of two elements:

Risk Appetite   *Risk appetite* is the level of uncertainty the stakeholders are willing to accept in exchange for the potential positive impacts of the risk. For example, let's say your organization is a multinational manufacturing firm that is implementing a new inventory system. The end users are grumbling and have expressed their concerns about the new system. The old system does everything they want it to do, and they are not interested in this new technology. There is a potential for the represented employees to protest this new system, and such a reaction could impact production. Your stakeholders are willing to accept this risk even though they don't know if, or to what extent, production may be impacted because the benefits of the new system far outweigh the potential unknown impacts of employees protesting.

Risk Threshold   *Risk thresholds* are measures or levels of uncertainty or impact the organization is willing to operate within. For example, a monetary risk threshold might state that if the risk poses a threat that could cost more than 5 percent of the total project budget, the risk should not be accepted. If it's below 5 percent, it may be accepted.

Risk threshold is that balance I talked about earlier where stakeholders are comfortable taking a risk because the known benefits to be gained outweigh what could be lost—or just the opposite. They will avoid taking a risk because the cost or impact is too great given the amount of benefit that can be derived. Here's an example to describe risk threshold: Suppose you're a 275-pound brute who's surrounded by three bodyguards of equal proportion everywhere you go. Chances are, walking down a dark alley in the middle of the night doesn't faze you at all. That means your risk threshold for this activity is high. However, if you're a petite 90-pounder without the benefit of bodyguards or karate lessons, performing this same activity might give you cause for concern. Your risk threshold is low, meaning that you wouldn't likely do this activity. The higher your threshold for risk, the more you're willing to take on risk and its consequences.

Risk threshold is different from risk appetite because risk appetite concerns the amount of uncertainty you are willing to take on to gain a benefit, whereas risk threshold concerns the amount of risk and, hence, the potential consequences or benefits you might gain or lose if the risk event occurs.

Organizations and stakeholders, as well as individuals, all have different thresholds for risk. One organization might believe that the risk of a potential 7 percent cost overrun is high, whereas another might think it's low. However, either one of these organizations might decide to accept the risk if it believes that the risk is in balance with the potential rewards. It's important for the project manager to understand the threshold level that the organization and the stakeholders have for risk before evaluating and ranking risk.

Conducting Risk Meetings

The Plan Risk Management process has three tools and techniques: expert judgment, data analysis (stakeholder analysis), and meetings.

Data analysis in the form of stakeholder analysis in this case involves understanding stakeholder risk appetites and thresholds, developing a method for scoring risks, and determining the risk exposure of the project. We will cover these topics in more depth throughout the remainder of this chapter.

The purpose of meetings—which are held with project team members, stakeholders, functional managers, and others who might have involvement in the risk management process—is to contribute to the risk management plan. During these meetings, the fundamental plans for performing risk management activities will be discussed and determined and then documented in the risk management plan.

The key outcomes of performing these planning meetings are as follows:

- Risk cost elements are developed for inclusion in the project budget.
- Schedule activities associated with risk are developed for inclusion in the project schedule.
- Risk responsibilities are assigned.
- The risk contingency reserve process is established or reviewed.
- Templates for risk categories are defined or modified for this project.
- Definitions of terms (*probability*, *impact*, *risk types*, *risk levels*, and so on) are developed and documented.
- The probability and impact matrix is defined or modified for this project.

 Real World Scenario

Do We Need a Risk Management Plan?

Juliette is the project manager for a small project her department is undertaking. The project objective is to give customers the ability to take virtual tours of the properties her organization has listed for lease. Two programmers from the information technology department will be working on the updates to the website, programming the links for the tours, and so on. The project sponsor wants to fast-track this project. She'd like to skip most of the Planning processes, and she sees no need for a risk management plan. Juliette explains to the sponsor that the risk management plan for a project this size might be only a paragraph or two long. She emphasizes the importance of documenting how they'll identify risks, how they'll quantify them, and how they'll monitor the risks as the project progresses. Juliette has project management experience on projects of all sizes and knows firsthand that ignoring this step could bring some unexpected surprises to the sponsor later in the project. She describes a bad past experience where this step was ignored and then assures the sponsor that they can probably agree to the plan, identify and quantify the risks, and determine response plans in an hour and a half or less. The sponsor now understands the importance of documenting the risks and agrees to the meeting.

Ultimately, your goal for this process is documenting the risk management plan. This document is the basis for understanding the remaining risk processes. Because the risk management plan encompasses a wealth of information, I've given this topic its own section. Let's get to it.

Documenting the Risk Management Plan

The purpose of the Plan Risk Management process is to create a *risk management plan*, which describes how you will define, monitor, and control risks throughout the project. This process involves identifying risks, analyzing and prioritizing them, and developing risk response strategies so that you can manage both uncertainty and opportunity throughout the life of the project. The risk management plan is a subsidiary of the project management plan, and it's the only output of this process.

The risk management plan details how risk management processes (including Identify Risks, Perform Qualitative Risk Analysis, Perform Quantitative Risk Analysis, Plan Risk Responses, Implement Risk Responses, and Monitor Risks) will be implemented, monitored, and controlled throughout the life of the project. It details how you will manage risks but does not attempt to define responses to individual risks.

The risk management plan (the only output of this process) will also become part of the project management plan.

According to the *PMBOK® Guide*, the risk management plan should include the following elements:

- Risk strategy
- Methodology
- Roles and responsibilities
- Funding
- Timing
- Risk categories
- Definitions of risk probability and impact
- Probability and impact matrix
- Stakeholder risk appetite
- Reporting formats
- Tracking

Exam Spotlight

It's important to spend time developing the risk management plan because it's an input to every other risk-planning process and it enhances the probability of risk management success.

We'll take a look at most of these elements next. However, risk categories, probability and impact, and probability and impact matrix are pretty meaty topics, so I'll cover those in their own sections following this one.

Risk Strategy   Strategy refers to the approach you'll use in managing risks and risk events that occur on the project.

Methodology   Methodology is a description of how you'll perform risk management, including elements such as methods, tools, and where you might find risk data that you can use in the later processes.

Roles and Responsibilities   Roles and responsibilities describe the people who are responsible for managing the identified risks and their responses for each type of activity identified in the risk management plan. These risk teams might not be the same as the project team. Risk analysis should be unbiased, which might not be possible when project team members are involved.

Funding   The budget for risk management is included in the plan as well. With this category, you'll assign resources and estimate the costs of risk management and its methods, including contingency reserves. These costs are then included in the project's cost baseline.

Timing   Timing documents the timing of the risk management processes (including when and how often they'll be performed on the project) and includes the activities associated with risk management in the project schedule.

Stakeholder Risk Appetite   This is just as it implies. As you proceed through the risk management processes, you might find that risk appetites will change. Document those new risk threshold levels in the risk management plan.

Reporting Formats   Reporting formats describe the content of the risk register and the format of this document. (I'll talk more about the risk register later in this chapter.) Reporting formats also detail how risk management information will be maintained, updated, analyzed, and reported to project participants.

Tracking   This includes a description of how you'll document the history of the risk activities for the current project and how the risk processes will be audited. You can reference this information when you're performing risk-planning processes later in the current project or on future projects. This information is also helpful for lessons learned, which I'll cover in Chapter 10, "Sharing Information."

Risk Categories

Risk categories are a way to systematically identify risks and provide a foundation for understanding. When we're determining and identifying risks, the use of risk categories helps improve the process by giving everyone involved a common language or basis for describing risk.

Risk categories should be identified during this process and documented in the risk management plan. These categories will assist you in making sure the next process, Identify Risks, is performed effectively and produces a quality output. The following list includes some examples of the categories you might consider during this process (or modify based on previous project information):

- Technical, quality, and performance risks

- Project management risks

- Organizational risks

- External risks

You can go about describing categories of risk in a couple of ways. One way is simply listing them. You could, and should, review prior projects for risk categories and then tailor them for this project.

You could also construct a *risk breakdown structure (RBS)*, which lists the categories and subcategories. Figure 7.1 shows a sample of an RBS.

FIGURE 7.1   Risk breakdown structure

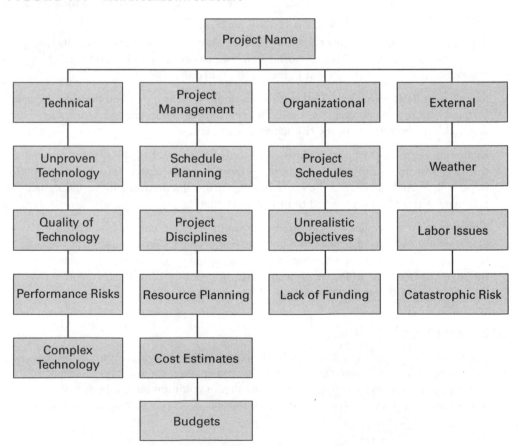

NOTE   The organizational process assets input might include an RBS that you can reference for this project. Don't forget your PMO because they might have templates or an RBS already developed for you to use.

Risk categories might reflect the type of industry or application area in which the project exists. For example, information technology projects will likely have many risks that fall into the technical category, whereas construction projects might be more subject to risks that fall into the external risks category. The categories do not have to be industry specific, however. Keep in mind that project management, for example, is a risk for every project in every industry. You can find a description of each of the categories next:

Technical, Quality, and Performance Risks   Technical, quality, and performance risks include risks associated with unproven technology, complex technology, or changes to technology anticipated during the course of the project. Performance risks might include unrealistic performance goals. Perhaps one of the project deliverables concerns a component manufactured to specific performance standards that have never been achieved. That's a performance risk.

Project Management Risks   The project management risk category includes improper schedule and resource planning, poor project planning, and improper or poor project management disciplines or methodologies.

Organizational Risks   The organizational risk category can include resource conflicts because of multiple projects occurring at the same time in the organization; scope, time, and cost objectives that are unrealistic given the organization's resources or structure; and lack of funding for the project or diverting funds from this project to other projects.

External Risks   The external risk category includes those aspects that are external to the project, such as new laws or regulations, labor issues, weather, changes in ownership, and foreign policy for projects performed in other countries. Catastrophic risks— known as *force majeure*—are usually outside the scope of Plan Risk Management and instead require disaster recovery techniques. Force majeure includes events such as earthquakes, meteorites, volcanoes, floods, civil unrest, terrorism, and so on.

In addition to risk categories, you'll want to consider other trends in risk management when planning for risks. According to the *PMBOK® Guide,* these trends include *non-event risks, project resilience,* and integrated risk management. Let's take a look at each.

Non-event Risks   This almost sounds like risks that won't occur, but that's not exactly right. Non-event risks are risks that are event based and generally outside the control of the organization (although this is not always true). For example, customers may decide to significantly modify the project requirements after seeing a prototype, or contractors may be acquired by another organization during the project. Non-event risks may also involve the skills and abilities of the team members, their knowledge, or their decision-making capabilities.

According to the *PMBOK® Guide,* there are two types of non-event risks: variability and ambiguity. Variability risk is uncertainty as it pertains to the event or activity itself. An example might include unexpected weather events or unexpected resource availability that impacts productivity.

Monte Carlo analysis is a great technique to use when dealing with variability risks. Monte Carlo analysis is a simulation technique that helps you predict the range of outcomes (or impacts) of a risk event and the probability of those outcomes should the risk event occur. This simulation is performed using a range of values. The first value (or impact in our case) is plugged into the model and the first result is produced. This simulation is repeated for hundreds or even thousands of values to determine a range of possible outcomes ranging from conservative to the most outlandish. Monte Carlo analysis is used in many industries.

Ambiguity is general uncertainty about the project due to the unknown or due to the lack of knowledge about how or what is required to fulfill the project objectives. Examples here include technology and governmental or industry regulations. Ambiguity risks are addressed using a gap analysis approach. For example, let's say the technical solution you're using has elements to it that are only partially known. You'll want to document what is known and then identify what is unknown or what's not understood about the technology. Work with experts in the industry, outside of the organization, and/or use benchmarking to help fill in those gaps in knowledge and come to a better understanding of the technology and the risks it may pose.

Resilience   Project resilience is the ability to cope with unknown risks as they occur while still fulfilling the goals of the project. Unknown risks, by their nature, are not known until they have occurred. Developing project resilience involves several skills, including these:

> Accounting for correct levels of budget for emerging risks and for schedule contingency
>
> Developing processes that are flexible enough to cope with risks yet still work toward completing the project goals
>
> Creating and following strong change management processes
>
> Allowing your team members the freedom to perform their activities as they see fit while adhering to the schedule and overall goals
>
> Making it a practice to watch for and identify emerging risks
>
> Working with stakeholders to adjust project scope where needed when emerging risks are present

Integrated Risk Management   If you work in an organization that uses a PMO to manage projects, you are probably already practicing integrated risk management. Risk exists on an individual level, the project level, the program level, the portfolio level, and an organizational level. Best practices encourage an integrated, enterprise-wide approach to managing risks across all these levels.

Defining Probability and Impact

When you're writing the risk management plan, you'll want to document the definitions for probability and impact as they relate to potential negative risk events and their impacts on the four project objectives. Probability describes the potential for the risk event occurring, whereas impact describes the effects or consequences the project will experience if the risk event occurs. This definition can be sophisticated or simple. For example, you might use numeric values to define probability and impact, or simply assign a high, medium, or low rating to each risk. What's important to note now is that you don't use these probability and impact definitions here. You use these definitions later in the Perform Qualitative Risk Analysis process. But you should define and document them here in the risk management plan.

Probability and Impact Matrix

A *probability and impact matrix* prioritizes the combination of probability and impact scores and helps you determine which risks need detailed risk response plans. For example, a risk event with a high probability of occurring and a high impact will likely need a response plan. This matrix is typically defined by the organization, but if you don't have one, you'll need to develop this now—during your planning meetings (a tool and technique of this process). You'll use this matrix in the Perform Qualitative Risk Analysis process that I'll cover in the section "Analyzing Risks Using Qualitative Techniques" later in this chapter. Again, you want to define (or modify) and document the probability and impact matrix in the risk management plan.

The key point about this process is that you'll define what the probability and impact tools look like now during Plan Risk Management so that the team has an agreed-on basis for evaluating the identified risks later during the Perform Qualitative Risk Analysis process.

To recap, the steps associated with these last few elements of the risk management plan are as follows:

1. Define the risk categories (these will assist the risk team in the Identify Risks process).

2. Determine how probability and impact will be defined (to be used in the Perform Qualitative Risk Analysis process).

3. Develop or modify the probability and impact matrix (to be used in the Perform Qualitative Risk Analysis process).

Doing all these steps, together with the other elements of the risk management plan, gives you and the risk management team a common understanding for evaluating risks throughout the remainder of the project.

Identifying Potential Risks

The *Identify Risks* process involves identifying all the risks that might impact the project, documenting them, and documenting their characteristics. Identify Risks is an iterative process that continually builds on itself. As you progress through the project, more risks might present themselves. Once you've identified or discovered a potential new risk, you should analyze it to determine whether a response plan is needed.

You can include several groups of people to help identify risks: project team members, risk team members, stakeholders, subject matter experts, users of the final product or service, and anyone else you think might help in the process. Perhaps in the first round of Identify Risks you could include just the project team and subject matter experts and then bring in the stakeholders or risk management team to further flesh out risks during the second round of identification. Risk events can occur at any time during the project, and all project participants should be encouraged to continually watch for and report potential risk events.

Remember that you are identifying both individual risks and overall project risks in this process.

Risks might or might not adversely affect the project. Some risks have positive consequences, whereas others have negative consequences. However, you should identify all risk events and their consequences. Here's a partial list to get you thinking about where risk might be found:

- Budgets/funding
- Schedules
- Scope or requirements changes
- Project plan
- Project management processes
- Technical issues
- Personnel issues
- Hardware
- Contracts and vendors
- Political concerns
- Business processes
- Legal processes and activities
- Environmental concerns
- Reputation

- Management policies, processes, and attitudes
- Compliance and regulatory requirements

This is by no means an exhaustive list. Remember that risk is uncertainty and realize that risk (uncertainty) is lurking almost anywhere on your project. It's your job to discover as many of the potential risks as possible using the tools and techniques of this process and to document these risks.

You will gather several inputs, most of which you've seen previously, to begin the Identify Risks process. Don't forget to refer to Appendix B to see the complete list of inputs, tools and techniques, and outputs for any process.

One of the inputs you haven't seen yet is the quality management plan. For purposes of the Identify Risks process, understand that the quality management process, identified in the quality management plan, has the potential to produce or prevent risks.

You should pay particular attention to the roles and responsibilities section of the risk management plan (another input of this process) and the budget and schedule for risk activities. Don't forget to examine the categories of risks as well. This is a great place to start when you get the team together and begin brainstorming your list of risks.

Another document that will help get everyone thinking about potential risks is the assumption log. You'll recall that assumptions are things believed to be true. During the risk-planning stages of your project and throughout the work of the project, it's imperative to revisit and revalidate your project assumptions. For example, maybe at the time you recorded an assumption about vendor deliveries, the vendor had a great track record and never missed a date. Months later on the project, that vendor merges with one of its competitors. Now you'll need to reexamine your assumptions about delivery times and determine whether the assumption is still valid or whether you have a risk on your hands.

The enterprise environmental factors input concerns aspects from outside the project that might help you determine or influence project outcomes. Be certain to check for industry information (commercial databases, checklists, benchmarking studies, and so on) or academic research that might exist for your application areas regarding risk information. Be certain to also check for regulatory or compliance issues that must be addressed on your project. This may include security, safety, health, or other regulatory issues.

As always, don't forget about historical information such as previous project experiences via the project files. You might find risk templates and lessons learned in these files that will help with the current project. Project team knowledge is another form of historical information.

Although the *PMBOK® Guide* doesn't mention it, I've found that other elements of your project are helpful when identifying risk, such as the work breakdown structure (WBS), the resource management plan, project team assignments, and resource availability. In practice, you should examine the outputs of most of the Planning processes when attempting to identify risks.

Data Gathering and Data Analysis Techniques for Identifying Risks

The Identify Risks process is undertaken using several tools and techniques. We'll cover three of them in this section: data gathering, data analysis, and prompt lists. I'll cover interpersonal and team skills in Chapter 9, "Developing the Project Team." One important note about expert judgment (a tool and technique of this process) is that you should include people who have experience working on similar projects as well as experience working in the business area for which the project is undertaken, or those with specific industry experience in the Identify Risks process. Keep in mind any bias your experts may have regarding the project or potential risk events and take it into account when performing the risk processes.

Data Gathering

Data gathering encompasses several techniques, including brainstorming, the Delphi technique (a form of brainstorming), checklists, and interviewing. The goal of these techniques is to end up with a comprehensive list of risks at the end of the meeting. Let's take a quick look at each of these techniques.

Brainstorming

Brainstorming is probably the most often used technique of the Identify Risks process. You've seen brainstorming before, but let's look at it from the perspective of this process. Brainstorming involves getting subject matter experts, team members, risk management team members, and anyone else who might benefit from the process in a room and asking them to start identifying possible risk events. The trick here is that one person's idea might spawn another idea, and so on, so that by the end of the session you've identified all the possible risks. The facilitator could start the group off by going through the categories of risks, and/or using an RBS to get everyone thinking in the right direction. Edward de Bono devised a method of brainstorming called Six Thinking Hats (based on his book by the same name) that you might want to investigate. You may recall that Edward de Bono is also noted for lateral thinking techniques, which we discussed in Chapter 4, "Developing the Project Scope."

Nominal Group Technique

The Nominal Group technique is not included as part of the information-gathering techniques in the *PMBOK® Guide*. However, it's possible you may see a question about it on the exam. The *Nominal Group technique* is a brainstorming technique, or it can be conducted as a mass interview technique.

This technique requires the participants to be together in the same room. Each participant has paper and pencil in front of them, and they are asked to write down what risks they think the project faces. Using sticky notes is a good way to do this. Each piece of paper should contain only one risk. The papers are given to the facilitator, who sticks them up on a wall or whiteboard. The panel is then asked to review all the risks posted, rank them and prioritize them (in writing), and submit the ranking to the facilitator. Once this is done, you should have a complete list of risks.

Delphi Technique

The *Delphi technique* is a lot like brainstorming, except that the people participating in the meeting don't usually know one another. In fact, the people participating in this technique don't all have to be located in the same place and usually participate anonymously. You can use email to facilitate the Delphi technique easily.

First, you assemble your experts, from both inside and outside the organization, and provide them with a questionnaire to identify potential risks. The questionnaire is often designed with forced choices that require the experts to select between various options. The questionnaire asks participants about risks associated with the project, the business process, and the product of the project, and it asks the readers to rank their answers in regard to the potential impacts of the risks. They in turn send their responses back to you (or to the facilitator of this process). All the responses are organized by content and sent back to the Delphi members for further input, additions, or comments. The participants then send their comments back one more time, and the facilitator compiles a final list of risks.

The Delphi technique is a great tool that allows consensus to be reached quickly. It also helps prevent one person from unduly influencing the others in the group and, therefore, prevents bias in the outcome because the participants are usually anonymous and don't necessarily know how others in the group responded.

Checklists

Checklists used during the Identify Risks process are usually developed based on historical information and previous project team experience. You may also obtain risk checklists from sources that are specific to your industry. If you typically work on projects that are similar in nature, begin to compile a list of risks. You can then convert this to a checklist that will allow you to identify risks on future projects quickly and easily. However, don't rely solely on checklists for Identify Risks because you might miss important risks. It isn't possible for a single checklist to be an exhaustive source for all projects. You can improve your checklists at the end of the project by adding the new risks that were identified.

Interviews

Interviews are question-and-answer sessions held with others, including other project managers, subject matter experts, stakeholders, customers, the management team, project team members, and users. These folks provide you with possible risks based on their past experiences with similar projects.

This technique involves interviewing those folks with previous experience on projects similar to yours or those with specialized knowledge or industry expertise. Ask them to tell you about any risks that they've experienced or that they think might happen on your project. Show them the WBS and your list of assumptions to help get them started thinking in the right direction.

Data Analysis

Data analysis also involves several techniques including root cause analysis (RCA), assumption and constraint analysis, SWOT analysis, and document analysis. We'll look at each as they pertain to this process.

Root Cause Analysis (RCA)

Did you ever hear someone say you're looking at the symptoms and not at the problem? That's the idea here. *Root cause analysis* involves digging deeper than the risk itself and looking at the cause of the risk. This helps define the risk more clearly, and it also helps you later when it's time to develop the response plan for the risk. A great technique you can use for RCA is called 5 Whys. This was originally developed by Toyota. The idea is that you keep asking "why" five times in succession to help understand the root cause of the problem. For example, let's say that your organization is considering reducing the amount of health insurance premiums it pays on behalf of employees. The first question to ask is "Why?" The answer is the current benefits package is expensive and the organization is paying 80 percent of the premium for its employees. The next question is "Why?" It's expensive because the rates have gone up consistently over the last few years. "Why?" Because there were more expensive claims filed over the past two years than previous years. "Why?" Because employees are not practicing healthy lifestyle habits. "Why?" Because the organization has experienced significant growth over the last few years and employees are working a significant number of hours and are not taking time off. Ah, we've arrived at the root cause of the problem. Employees are working so many hours that they are not taking care of themselves. This is leading to more health insurance claims than in past years—which is driving up the cost of health insurance. Once the root cause has been determined, you can brainstorm alternative solutions to address the root cause.

Assumption and Constraint Analysis

Assumption and constraint analysis is a matter of validating the assumptions and constraints you identified and documented during the course of the project-planning processes. Assumptions should be accurate, complete, and consistent. Examine all your assumptions for these qualities. Assumptions are also used as jumping-off points to further identify risks.

The important point to note about the project assumptions is that all assumptions are tested against two factors:

- The strength of the assumption or the validity of the assumption

- The consequences that might impact the project if the assumption turns out to be false

All assumptions that turn out to be false should be evaluated and scored just as risks.

Constraints are examined to determine if they are valid and accurate and to determine which ones could bring about risk. Examine the constraints that could bring about positive risks to see if the risks could be enhanced by softening the restrictions caused by the constraint or if the constraint should be eliminated altogether.

Strengths, Weaknesses, Opportunities, and Threats (SWOT)

Strengths, weaknesses, opportunities, and threats (also known as *SWOT analysis*) is a technique that examines the project from each of these viewpoints. It also requires examining the project from the viewpoint of the project itself and from project management processes, resources, the organization, and so on to identify risks, including those that are generated internally. Strengths and weaknesses are generally related to issues that are internal to the organization. Strengths are what your organization does well according to your customers or the marketplace. Weaknesses are areas the organization could improve on. Typically, negative risks are associated with the organization's weaknesses and positive risks are associated with its strengths. Opportunities and threats are usually external to the organization. SWOT analysis is sometimes known as internal-external analysis and can be used in combination with brainstorming techniques to help discover and document potential risks.

Document Analysis

Document analysis involves reviewing project plans, assumptions, procurement documents, and historical information about previous projects both from a total project perspective and at the individual deliverables and activities level. This review helps the project team identify risks associated with the project objectives. Pay attention to the quality of the plans (is the content complete, or does it seem to lack detail?) and the consistency between plans. An exceptionally documented schedule is great, but if the budget isn't as well documented, you might have some potential risks.

Prompt List

A *prompt list* is somewhat similar to a checklist in that you have a defined set of risk categories you can use to determine both individual risks and overall project risks. You can use the lowest level of the RBS to construct a prompt list for individual risks.

According to the *PMBOK® Guide,* you could consult strategic analysis tools such as the political, economic, social, technological, environmental, and legal tool known as PESTLE, the technical, economic, commercial, organizational, and political tool known as TECOP, or the volatility, uncertainty, complexity, and ambiguity tool known as VUCA to assist in determining overall project risk.

PESTLE analysis examines political, economic, social, technological, legal, and environmental factors to see whether they may have an overall project impact. For example, let's say your project involves new oil and gas drilling procedures. When examining risks from the perspective of PESTLE, environmental factors should jump right out as a potential area for overall project risk. The team should explore this area in depth to determine and identify the risks that could impact the project.

TECOP analysis looks at technical, environmental, commercial, operational, and political risks. As in the earlier example, you should examine each of these elements to determine whether overall project risks could impact the project.

VUCA analysis looks at volatility, uncertainty, complexity, and ambiguity.

Documenting the Risk Register

The purpose of the Identify Risks process is to produce the primary outputs, which are the *risk register* and the risk report.

Everything you've done in this process up to this point will get documented in the risk register. It contains the following elements:

- List of identified risks
- Potential risk owners
- List of potential responses

The risk register also often contains warning signs, or triggers, although they aren't listed as an official part of the register. We'll take a look at all of these elements next. Understand that all risks should be documented, tracked, reviewed, and managed throughout the project.

List of Identified Risks

Risks are all the potential events and their subsequent consequences that could occur as identified by you and others during this process. You might want to consider logging your risks in a risk database or tracking system to organize them and keep a close eye on their status. This can easily be done in spreadsheet format or whatever method you choose. List the risks, assign each risk a tracking number, and note the potential cause or event and the potential impact. This list gives you a means to track the risks, their occurrence, and the responses implemented.

Potential Risk Owners

Risk owners are the departments or individuals responsible for monitoring the risks, implementing risk response plans if the risk event occurs, and updating the project team on status.

List of Potential Responses

You might identify potential responses to risks at the same time you're identifying the risks. Sometimes just identifying a risk will tell you the appropriate response. Document those responses in the risk register. You'll refer to them again in the Plan Risk Responses process a little later in this chapter.

A sample risk register is shown in Table 7.1. As you progress through the risk planning processes, and through the project itself, more risks may be identified and more information will become known about the risks. You should update the risk register with new information as it becomes known.

TABLE 7.1   Risk register

ID	Risk	Trigger	Event	Cause	Impact	Owner	Response plan
1	Infrastructure team is not available when needed	Predecessor tasks not completed on time	Operating system upgrade delayed	Equipment was not delivered on time	Schedule delay	Brown	Compress the schedule by beginning tasks in the next milestone while working on operating system upgrade
2	Hardware components are not delivered on time	No shipping notice received the week before scheduled delivery	Delays all successor tasks	Vendor has supplier issues	Schedule delay	Taylor	Contact alternative vendor to determine availability of equipment for emergency delivery

Triggers

Triggers are warning signs or symptoms that a risk event is about to occur. For example, if you've ever suffered from a hay fever attack, you can't mistake the itchy, runny nose and scratchy throat that can come on suddenly and send you into a sneezing frenzy. Signals like this are known as *triggers*, and they work the same way when you're determining whether a risk event is about to occur. For example, if you're planning an outdoor gathering and rain clouds start rolling in from the west on the morning of the activity, you probably have a risk event waiting to happen. A key team member hinting about job hunting is a warning sign that the person might be thinking of leaving, which in turn can cause schedule delays, increased costs, and so on. This is another example of a trigger.

Triggers are *not* listed as one of the risk register elements until the Plan Risk Responses process is carried out, but in practice this is an appropriate time to list them. You will likely encounter questions on the exam about triggers, so don't say I didn't warn you. Also, throughout the remainder of the project, be on the alert for triggers that might signal that a risk event is about to occur.

Risk Report

The *risk report* is created and elaborated on as you work through the risk processes in the Planning process group. It documents all the sources of overall project risk, including those that have the highest potential for impacting the project. It also contains high-level information about the potential individual risks that exist on the project, including the number of risks, types of risk categories represented by the risks, and so on.

Exam Spotlight

This process focuses on individual risks, but you should keep in mind the overall project risk as well. Your organization might be taking on more than it can accomplish, or the complexity of the project is such that your organization doesn't have resources with skills or experience to perform the project. This could cause negative impacts to the organization and should be considered when undertaking the project.

Identifying Risks Using an Agile Approach

Projects with a high degree of uncertainty about the project requirements are typically high-risk projects. And uncertainty can increase the risk of rework. High uncertainty is an overall project risk, not an individual risk. Not knowing all the requirements and figuring things out as you go calls for an agile approach to managing the project. Obviously, you will need to identify this risk before deciding what approach you should use to manage the project. Projects with high uncertainty about the requirements and uncertainty about the knowledge or technology required to perform the project will create a significant amount of change and complexity as the project proceeds. Choosing an adaptive methodology that allows for small increments of work that produce frequent deliverables, along with frequent reviews of the work, will help manage uncertainty and decrease the chances of rework. The project team can determine what work to perform, receive continuous feedback, and better understand the requirements of the project so that they can fulfill the business value the customer expects.

According to the *Agile Practice Guide* (PMI®, 2017), there are three characteristics of uncertainty:

- Product specification. This entails the requirements of the project and determining whether the right product is being produced.

- Production capability. This involves understanding the technical feasibility of the product and the methods used to develop the product.

- Process suitability. This involves using the right people and the right processes to perform the work.

Exam Spotlight

The *Agile Practice Guide* (PMI®, 2017) notes that when both technical uncertainty (production capability) and requirements uncertainty (product specification) are high, the project moves from being complex to chaotic. You'll need to address, understand, and manage one or the other of these factors to ensure a successful project.

An agile approach produces accurate requirements in a shorter time frame than writing out all the requirements at the beginning of the project, only to discover when the work is completed that the requirements are not correct. Because the team is producing small increments of work that are tested and reviewed in each iteration, risk is more easily understood and managed. The benefits of producing a minimum viable product using an agile approach are that risk is reduced, the delivery of business value is maximized, and uncertainty is decreased.

According to the *Agile Practice Guide* (PMI®, 2017) the frequent reviews practiced in an agile approach help to increase knowledge among the cross-functional team members and helps them understand and manage risk. When using an agile approach, risks are identified, analyzed, and managed in each iteration. The team can reprioritize the work in each iteration throughout the life of the project based on changing requirements and continual risk assessment. This iterative way of identifying, analyzing, and managing risks improves the team's overall understanding of risk exposure. Risk exposure concerns the potential for loss or gain that might occur as a result of the risk event coming about. Exposure is quantified using the methods we'll discuss in the next two sections of this chapter.

Don't forget to analyze any compliance or regulatory issues that may impact your project. You'll need to know what regulations or standards you need to apply to the project before you begin the work, whether you use a predictive or an agile approach to manage the project.

Iterative-based agile, flow-based agile, and other agile methodologies like Six Sigma DMAIC, Extreme Programming, and others have risk assessment and risk management embedded within the approach due to their short, frequent cycles of delivering small, incremental units of work.

Analyzing Risks Using Qualitative Techniques

The *Perform Qualitative Risk Analysis* process involves determining what impact the identified risks will have on the project objectives and the probability they'll occur. It also ranks the risks in priority order according to their effect on the project objectives so that you can spend your time efficiently by focusing on the high-priority risks. This helps the team determine whether Perform Quantitative Risk Analysis should be performed or whether you can skip right to developing response plans. The Perform Qualitative Risk Analysis process also considers risk threshold levels, especially as they relate to the project constraints (scope, time, cost, and quality) and the time frames of the potential risk events.

The Perform Qualitative Risk Analysis process should be performed throughout the project. If you're using an agile approach, you'll perform this process at the beginning of each iteration. This process is the one you'll use most often when prioritizing project risks because it's fast, relatively easy to perform, and cost effective. The *PMBOK® Guide* notes that you should identify and manage the risk attitudes of those assisting with this process because risk perceptions will introduce bias. You'll need to watch for bias, evaluate it, and correct it if necessary. Usually, simply conducting the Perform Qualitative Risk Analysis process and evaluating the impact and probability of risks will help to keep bias at a minimal level.

The Perform Qualitative Risk Analysis process has four inputs: the project management plan (risk management plan), project documents (assumption log, risk register, stakeholder register), enterprise environmental factors, and organizational process assets.

You are familiar with all these inputs, but there a couple of key considerations you should know about for the exam. First, projects with high levels of uncertainty or that are more complex than what the team has undertaken before require more diligence during the Perform Qualitative Risk Analysis process. Second, as with the Identify Risks process, you should examine historical information and lessons learned from past projects as a guide for prioritizing the risks for this project. Risk databases from your industry or application area can be used here as well. The real key to this process lies in the tools and techniques you'll use to prioritize risks. Hold on tight because you're going in the deep end.

Performing Qualitative Risk Analysis

The tools and techniques in the Perform Qualitative Risk Analysis process are primarily concerned with discovering the probability of a risk event and determining the impact (or consequences) the risk will have if it does occur. Make certain that all the information you gather regarding risks and probability is as accurate as possible. It's also important that you gather unbiased information so that you don't unintentionally overlook risks with great potential or consequences.

The purpose of this process is to determine risk event probability and risk impact and to prioritize the risks and their responses. You'll use the tools and techniques of this process to establish risk scores, which is a way of categorizing the probability and risk impact. You'll record the risk scores in the risk register (part of the project documents updates output of this process).

The Perform Qualitative Risk Analysis process includes the following tools and techniques:

- Expert judgment
- Data gathering (interviews)
- Data analysis (risk data quality assessment, risk probability and impact assessment, assessment of other risk parameters)
- Interpersonal and team skills (facilitation)
- Risk categorization
- Data representation (probability and impact matrix, hierarchical charts)
- Meetings

We'll look at the new tools and techniques of this process next and touch one more time on expert judgment.

Expert Judgment

Because this process defines qualitative values, by its very nature you must rely on expert judgment to determine the probability, impact, and other information we've derived so far. The more knowledge and experience your experts have, the better your assessments will be. Interviews and facilitated workshops are two techniques you can use in conjunction with expert judgment to perform this process. As with the Identify Risks process, make certain to take into account any bias your experts have and to correct it when necessary.

Risk Data Quality Assessment (Data Analysis)

The risk data quality assessment involves determining the usefulness of the data gathered to evaluate risk. Most important, the data must be unbiased and accurate. You'll want to examine elements such as the following when using this tool and technique:

- The quality of the data used
- The availability of data regarding the risks
- How well the risk is understood
- The reliability and integrity of the data
- The accuracy of the data

Low-quality data will render the results from the Perform Qualitative Risk Analysis process almost useless. Spend the time to validate and verify the information you've collected about risks so that your prioritization and analysis are as accurate as they can be. If you find that the quality of the data is questionable, you guessed it—go back and get better data.

Risk Probability and Impact Assessment (Data Analysis)

This tool and technique assesses the probability that the risk events you've identified will occur, and it determines the effect their impacts have on the project objectives, including time, scope, quality, and cost. Analyzing risks in this way allows you to determine which risks require the most aggressive management. When determining probabilities and impacts, you'll refer to the risk management plan element called "definitions of risk probability and impact."

Probability

Probability is the likelihood that an event will occur. The classic example is flipping a coin. There is a 0.50 probability of getting heads and a 0.50 probability of getting tails on the flip. Note that the probability that an event will occur plus the probability that the event will not occur always equal 1.0. In this coin-flipping example, you have a 0.50 chance that you'll get heads on the flip. Therefore, you have a 0.50 chance you will not get heads on the flip. The two responses added together equal 1.0. Probability is expressed as a number from 0.0—which means there is no probability of the event occurring—to 1.0—which means there is 100 percent certainty the event will occur.

Determining risk probability can be difficult because it's most commonly accomplished using expert judgment. In non-project management terms, this means you're guessing (or asking other experts to guess) at the probability a risk event will occur. Granted, you're basing your guess on past experiences with similar projects or risk events, but no two risk events (or projects) are ever the same. It's best to fully develop the criteria for determining probability and get as many experts involved as you can. Carefully weigh their responses to come up with the best probability values possible.

Impact

Impact is the amount of pain (or the amount of gain) the risk event poses to the project. The risk *impact scale* can be a relative scale (also known as an ordinal scale) that assigns values such as high, medium, or low (or some combination of these) or a numeric scale known as a *cardinal scale*. Cardinal scale values are actual numeric values assigned to the risk impact. Cardinal scales are expressed as values from 0.0 to 1.0 and can be stated in equal (linear) or unequal (nonlinear) increments.

Table 7.2 shows a typical risk impact scale for cost, time, and quality objectives based on a high-high to low-low scale. You'll notice that each of the high-medium-low value combinations on this impact scale has been assigned a cardinal value. I'll use these in the section coming up on the probability and impact matrix.

TABLE 7.2   Risk impact scale

Objectives	Low-Low	Low	Medium	High	High-High
	0.05	0.20	0.40	0.60	0.80
Cost	No significant impact	Less than 6% increase	7–12% increase	13–18% increase	More than 18% increase
Time	No significant impact	Less than 6% increase	7–12% increase	13–18% increase	More than 18% increase
Quality	No significant impact	Few components impacted	Significant impact requiring customer approval to proceed	Unacceptable quality	Product not usable

When you're using a high-medium-low scale, it's important that your risk team understands what criteria were used to determine a high score versus a medium or low score and how they should be applied to the project objectives.

Assessing Probability and Impact

The idea behind both probability and impact values is to develop predefined measurements that describe what value to place on a risk event.

Exam Spotlight

For the exam, don't forget that you define probability and impact values during the Plan Risk Management process and document them in the risk management plan.

If the risk impact scale has not been previously defined, develop one for the project as early in the Planning processes as possible. You can use any of the techniques I talked about earlier in the section such as brainstorming or the Delphi technique, to come up with the values for probability and impact.

During the Perform Qualitative Risk Analysis process, you'll determine and assess probability and impact for every risk identified during the Identify Risks process. You could interview or hold meetings with project team members, subject matter experts, stakeholders, or others to help assess these factors. During this process, you should document not only the probability and impact but also the assumptions your team members used to arrive at these determinations.

Assessment of Other Risk Parameters (Data Analysis)

According to the *PMBOK® Guide,* there are several other factors you can consider in addition to probability and impact when determining how you'll prioritize the risks at the end of this process. These factors include the following:

Urgency   Determining how quickly a response needs to be implemented

Proximity   Determining how quickly the risk will impact one or more of the project objectives

Dormancy   The period of time between the risk occurrence and discovery of the risk

Manageability   How well the risk owner manages the risk event

Controllability   The ability of the risk owner to control the impact of the risk

Detectability   The ability to detect a risk trigger and understand a risk event is about to occur

Connectivity   The relationship between the individual risks

Strategic impact   The impact to the organization's strategic goals if the risk event occurs

Propinquity   The stakeholders' perception of the risk significance

Let's walk through an example of how you would use these risk parameters to prioritize risk. Perhaps you have a potential financial risk on the project that, if it occurs, could cost the organization hundreds of thousands in penalties and fees. Your response plan will adequately address the risk event and you will avoid the penalties and fees, provided the response plan is put into place within 12 hours of the risk event occurring. This describes a high urgency level for this risk, and you will likely prioritize this risk close to the top of your risk ranking to make certain it is being monitored and dealt with should it occur.

Risk Categorization

This tool and technique is used to determine the effects risk has on the project. You can examine not only the categories of risk determined during the Plan Risk Management process (and described in the RBS), but also those in the project phase and the WBS to determine the elements of the project that are affected by risk. By grouping risks into categories, you can see at a glance those that may have the greatest potential for impact on the project.

Probability and Impact Matrix (Data Representation)

Earlier in this chapter we talked about probability and impact values. The probability and impact matrix takes the probability and impact values one step further by assigning an overall risk score classification to each of the project's identified risks. The combination of probability and impact results in a classification usually expressed as high, medium, or low. Typically, high risks are considered a red condition, medium risks are considered a yellow condition, and low risks are considered a green condition. This type of ranking is known as an *ordinal scale* because the values are ordered by rank from high to low. (In practice, ordinal values might also include ranking by position. In other words, the risks are listed in order by rank as the first, the second, the third, and so on.)

Remember that probability and impact can be determined for the risk as a whole or for components of the project the risk may affect, such as the project schedule, cost, and quality.

Exam Spotlight

The *PMBOK® Guide* notes that risk rating rules and the probability and impact matrix values and steps are usually set by the organization and/or key stakeholders.

Now let's look at an example. You have identified a risk event that could impact project costs, and your experts believe costs could increase by as much as 9 percent. According to the risk impact rating matrix in Table 7.2, this risk carries a medium impact, with a value of 0.40. Hold on to that number because you're going to plug it into the probability impact matrix—along with the probability value—to determine an overall risk value next.

You'll remember from the discussion earlier that probability values should be assigned numbers from 0.0 to 1.0. In this example, the team has determined that there is a 0.2 probability of this risk event occurring. The risk impact scale shows a medium, or 0.40, impact should the event occur.

Now, to determine whether the combination of the probability and impact of this risk is high, medium, or low, you'll need to check the probability impact matrix. Table 7.3 shows a sample probability and impact matrix.

TABLE 7.3   Sample probability and impact matrix

Probability	Impact values* Low-Low .05	Low .20	Medium .40	High .60	High-High .80
0.8	**0.04**	**0.16**	*0.32*	*0.48*	*0.64*
0.6	**0.03**	**0.12**	0.24	*0.36*	*0.48*
0.4	0.02	**0.08**	0.16	0.24	*0.32*
0.2	0.01	0.04	0.08	**0.12**	**0.16**

*No formatting = low assignment or green condition; **bold** = medium assignment or yellow condition; ***bold italic*** = high assignment or red condition.

First, look at the Probability column. Your risk event has a probability of 0.2. Now, follow that row across until you find the column that shows the impact score of 0.40 (it's the Medium column). According to your probability and impact matrix values, this risk carries an overall score of 0.08 and falls in the low threshold, so this risk is assigned a low (or green condition) value.

The values assigned to the risks determine how Plan Risk Responses is carried out for the risks later during the risk-planning processes. Obviously, risks with high probability and high impact are going to need further analysis and formal responses. Remember that the values for this matrix (and the probability and impact scales discussed earlier) are determined prior to the start of this process and documented in the risk management plan. Also keep in mind that probability and impact do not have to be assigned the same values I've assigned here. You might use 0.8, 0.6, 0.4, and 0.2 for probability, for example, and assign 0.05, 0.1, 0.3, 0.5, and 0.7 for impact scales.

Real World Scenario

Screen Scrapers, Inc.

Screen Scrapers is a software-manufacturing company that produces a software product that looks at your mainframe screens, commonly called *green screens*, and converts them to browser-based screens. The browser-based screens look like any other Windows-compatible screen with buttons, scroll bars, and drop-down lists.

Screen Scrapers devised this product for companies that use mainframe programs to update and store data because entry-level workers beginning their careers today are not familiar with green screens. They're cumbersome and difficult to learn, and no consistency exists from screen to screen or from program to program. Pressing the F5 key in one program might mean go back one page, whereas pressing F5 in another program might mean clear the screen. New users are easily confused, make a lot of mistakes, and have to write tablets full of notes on how to navigate all the screens.

Your company has purchased the Screen Scraper product and has appointed you as the project manager over the installation. This project consists of a lot of issues to address, and you've made great headway. You're now at the Identify Risks and Perform Qualitative Risk Analysis stage. You decide to use the Delphi technique to assist you in identifying risk and assigning probability and impact rankings. Some experts are available in your company to serve on the Delphi panel, as well as some folks in industry organizations you belong to outside the company.

You assemble the group, set up a summary of the project, and send it out via email, requesting responses to your questions about risk. After the first pass, you compile the list of risks as follows (this list is an example and isn't exhaustive because your list will be project specific):

- Vendor viability (will the software company stay in business?)

- Vendor responsiveness with problems after implementation

- Software compatibility risk with existing systems

- Hardware compatibility risk

- Connection to the mainframe risk

- Training IT staff members to maintain the product

You send this list back to the Delphi members and ask them to assign a probability of 0.0 to 1.0 and an impact of high-high, high, medium, low, or low-low to each risk. The Delphi members assign probability and impact based on a probability scale and an impact scale designed by the risk management team. The values of the impact scale are as follows:

- High-high = 0.8

- High = 0.6

- Medium = 0.4
- Low = 0.2
- Low-low = 0.05

The results are compiled to determine the following probability and impact values:

- Vendor viability = 0.6 probability, high impact
- Vendor responsiveness = 0.4 probability, medium impact
- Software compatibility = 0.4 probability, medium impact
- Hardware compatibility = 0.6 probability, high-high impact
- Mainframe connection = 0.2 probability, high-high impact
- Training = 0.2 probability, low-low impact

The probability and impact matrix you used to assign the overall risk scores was derived from the probability and impact matrix shown in the following table.

Based on the probability and impact matrix thresholds, the project risks are assigned the following overall probabilities:

- Vendor viability = high
- Vendor responsiveness = medium
- Software compatibility = medium
- Hardware compatibility = high
- Mainframe connection = medium
- Training = low

PI Matrix for Screen Scrapers, Inc.

Probability	Impact Scores*.05	.20	.40	.60	.80
0.8	**0.04**	**0.16**	*0.32*	*0.48*	*0.64*
0.6	**0.03**	**0.12**	0.24	*0.36*	*0.48*
0.4	0.02	**0.08**	0.16	0.24	*0.32*
0.2	0.01	0.04	0.08	0.12	0.16

*No formatting = low assignment or green condition; **bold** = medium assignment or yellow condition; ***bold italic*** = high assignment or red condition.

Hierarchical Charts (Data Representation)

Hierarchical charts are a way to graphically represent risks that are categorized using more than two factors. Probability and impact, as you saw in the previous section, use two parameters to categorize risk. But what if you wanted to use some of the other risk assessment factors such as urgency, proximity, and impact? Figure 7.2 shows a sample bubble chart plotting urgency and proximity, where the bubbles represent impact. The values range from 1 to 10, where 1 is low and 10 is high. You can see that based on these factors, risks A and C have the highest impact on the project.

FIGURE 7.2   Risk bubble chart

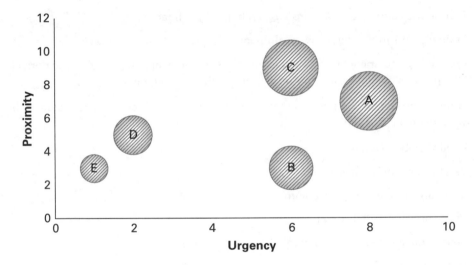

Ranking Risks in the Risk Register

The goal of the Perform Qualitative Risk Analysis process is to rank the risks and determine which ones need further analysis and, eventually, risk response plans. Project documents updates are the output of this process. This almost always involves updating the risk register and may include updates to the risk report, the assumption log, and the issue log. According to the *PMBOK® Guide*, you'll update the risk register with the following information:

- Risk priority level for the identified risks
- Risk scores
- Risk owner
- Updated probability and impact analysis

- Risk urgency information
- List of risks that need additional analysis and response
- Watch list of low-priority risks

Each element becomes a new entry in the risk register. For example, risk ranking assigns the risk score or priority you determined using the probability and impact matrix to the list of identified risks previously recorded in the risk register. I discussed the categories and the list of risks requiring near-term responses earlier. Note these in the risk register.

You should also note those risks that require further analysis (including using the Perform Quantitative Risk Analysis process). You'll create a list of risks that have low risk scores to review periodically, and you should note any trends in Perform Qualitative Risk Analysis that become evident as you perform this process.

The risk report should be updated to show the risks with the highest risk score along with a prioritized list of risks. The assumption log is another project document that should be updated as new information is discovered throughout this process. Assumptions and constraints can and do change throughout the course of the project, and the process of identifying and prioritizing risks may require updates to the assumption log and the issue log.

Quantifying Risk

The *Perform Quantitative Risk Analysis* process evaluates the impacts of risk prioritized during the Perform Qualitative Risk Analysis process. This process is not needed on every project, but when you do use it, it is typically performed for risks identified in Perform Qualitative Risk Analysis that could have a significant impact on the project. Perform Quantitative Risk Analysis quantifies the aggregate risk exposure for the project by assigning numeric probabilities to risks and their impacts on project objectives. This quantitative approach is accomplished using techniques such as Monte Carlo simulation and decision tree analysis. To paraphrase the *PMBOK® Guide*, the purpose of this process is to perform the following:

- Quantify the project's possible outcomes and probabilities.
- Determine the probability of achieving the project objectives.
- Identify risks that need the most attention by quantifying their contribution to overall project risk.
- Identify realistic and achievable schedule, cost, or scope targets.
- Make the best project management decisions possible when outcomes are uncertain.

Perform Quantitative Risk Analysis—like Perform Qualitative Risk Analysis—examines risk and its potential impact on the project objectives. You might choose to use both of these processes to assess all risks or only one of them, depending on the complexity of the project and the organizational policy regarding risk planning. The Perform Quantitative Risk Analysis process follows the Perform Qualitative Risk Analysis process. If you use this process, be sure to repeat it every time the Plan Risk Responses process is performed and as part of the Monitor Risks process so that you can determine whether overall project risk has decreased.

I've already covered most of the inputs to the Perform Quantitative Risk Analysis process. Risk databases and risk specialists' studies performed on similar projects are elements of the enterprise environmental factors you'll want to pay close attention to as an input to this process.

Performing Quantitative Risk Analysis

The Perform Quantitative Risk Analysis process includes five tools and techniques: expert judgment, data gathering (interviews), interpersonal and team skills (facilitation), representations of uncertainty, and data analysis (simulations, sensitivity analysis, decision tree analysis, influence diagrams). You already know expert judgment and learned about interpersonal and team skills (facilitation) in earlier chapters. We'll take a look at the remaining tools and techniques next.

Data-Gathering Techniques

The data-gathering technique used in this process is interviews. This technique is like the interviewing technique discussed earlier in the Identify Risks process. Project team members, stakeholders, and subject matter experts are prime candidates for risk interviews. Ask them about their experiences on past projects and about working with the types of technology or processes you'll use during this project.

> For the exam, remember that interviewing is a tool and technique of the Perform Quantitative Risk Analysis process. Although you can use this technique in the Identify Risks process, keep in mind that it's part of the data-gathering tool and techniques in both processes and not a named tool and technique itself.

When using this technique, you should first establish which methods of probability distribution (described next in the representations of uncertainty tool and technique) you'll use to analyze your information. The technique you choose will dictate the type of information you need to gather. For example, you might use a three-point scale that assesses the low, high, and most likely risk scenarios or take it a step further and use standard deviation calculations.

Make certain you document how the interviewees decided on the risk ranges, the criteria they used to place risks in certain categories, and the results of the interview. This information will help you later in developing risk responses.

Representations of Uncertainty

Representations of uncertainty deal with probability distributions that may contain a range of values, particularly when cost, resource estimates, and duration estimates for activities are ambiguous. It's beyond the scope of this book to delve into probability distributions and calculations, so I'll point out a few aspects of them that you should remember for the exam.

Continuous probability distributions (particularly beta and triangular distributions) are commonly used in Perform Quantitative Risk Analysis. According to the *PMBOK® Guide*, continuous probability distributions include normal, lognormal, triangular, beta, and uniform distributions. Distributions are graphically displayed and represent both the probability and time or cost elements.

Triangular distributions use estimates based on the three-point estimate (the pessimistic, most likely, and optimistic values). This means that during your interviews, you'll gather these pieces of information from your experts. Then you'll use them to quantify risk for each WBS element.

Normal and lognormal distributions use mean and standard deviations to quantify risk, which also require gathering the optimistic, most likely, and pessimistic estimates.

Discrete distributions represent possible scenarios in a decision tree (we'll discuss this in a bit), outcomes of a test, results of a prototype, and other uncertain events.

Data Analysis

For the exam, you should know the four data analysis and modeling techniques for this process: sensitivity analysis, decision tree analysis, simulations, and influence diagrams. Expected monetary value (EMV) analysis is not mentioned in the Data Analysis tool and technique list, but it is an important concept you may need to know for the exam. Let's take a brief look at each of them.

Sensitivity Analysis

Sensitivity analysis is a quantitative method of analyzing the potential impact of risk events on the project and determining which risk event (or events) has the greatest potential for impact by examining all the uncertain elements at their baseline values. One of the ways sensitivity analysis data is displayed is a *tornado diagram*. Figure 7.3 shows a sample tornado diagram.

FIGURE 7.3   Tornado diagram

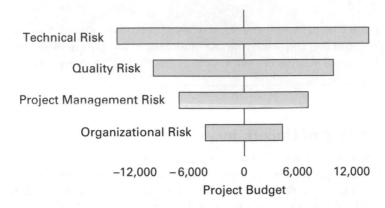

You can see by the arrangement of horizontal bars (each representing a sensitivity variable) how the diagram gets its name. The idea is that each sensitivity bar displays the low and high values possible for the element the bar represents. It's beyond the scope of this book to explain how these values are determined. The questions you might encounter on the exam are focused on the context of this type of analysis. The variables with the greatest effect on the project appear at the top of the graph and decrease in impact as you progress down through the graph. This gives you a quick overview of how much the project can be affected by uncertainty in the various elements. It also allows you to see at a glance which risks might have the biggest impacts on the project and will require carefully crafted, detailed response plans. You can use tornado diagrams to determine sensitivity in cost, time, and quality objectives or for risks you've identified during this process. Sensitivity analysis can also be used to determine stakeholder risk threshold levels.

Expected Monetary Value (EMV) Analysis

Expected monetary value is not a named tool and technique of this process, but you should have some understanding of this technique, since it can be used as a component of decision tree analysis and simulations.

Expected monetary value (EMV) analysis is a statistical technique that calculates the average anticipated future impact of the decision. EMV is calculated by multiplying the probability of the risk by its impact for two or more potential outcomes (for example, a good outcome and a poor outcome) and then adding the results of the potential outcomes together. EMV is used in conjunction with the decision tree analysis technique, which is covered next. I'll give you an example of the EMV formula in the next section. Positive results generally mean the risks you're assessing pose opportunities to the project, whereas negative results generally indicate a threat to the project.

Decision Tree Analysis

Unfortunately, this isn't a tree outside your office door that produces "yes" and "no" leaves that you can pick to help you make a decision. *Decision trees* are diagrams that show the sequence of interrelated decisions and the expected results of choosing one alternative over the other. Typically, more than one choice or option is available when you're faced with a decision or, in this case, potential outcomes from a risk event. The available choices are depicted in tree form, starting at the left with the risk decision branching out to the right with possible outcomes. Decision trees are usually used for risk events associated with time or cost.

Figure 7.4 shows a sample decision tree using expected monetary value (EMV) as one of its inputs.

FIGURE 7.4   Decision tree

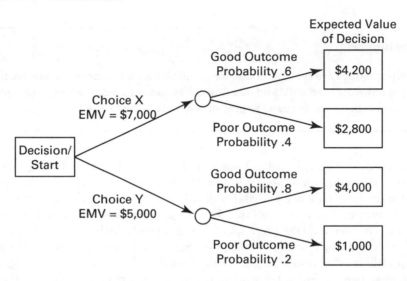

The expected monetary value of the decision is a result of the probability of the risk event multiplied by the impact for two or more potential outcomes and then summing their results. The squares in this figure represent decisions to be made, and the circles represent the points where risk events might occur.

The decision with an expected value of $7,000 is the correct decision to make because the resulting outcome has the greatest value.

Simulations

Simulation techniques are often used for schedule risk analysis and cost analysis. For example, simulations allow you to translate the potential risks at specific points in the project into their impacts so that you can determine how the project objectives are affected. Simulation techniques compute the project model using various inputs, such as cost estimates or activity durations, to determine a probability distribution for the variable chosen. (Cost risks typically use either a work breakdown structure or a cost breakdown structure as the input variable. Schedule risks always use the schedule network diagram and duration estimates as the input variables.) If you used simulation techniques to determine project cost and used the cost of the project elements as the input variable, a probability distribution for the total cost of the project would be produced after running the simulation numerous times. Simulation techniques examine the identified risks and their potential impacts on the project objectives from the perspective of the whole project.

Monte Carlo analysis is an example of a simulation technique. Monte Carlo analysis is replicated many times, typically using cost or schedule variables. Every time the analysis is performed, the values for the variable are changed using a probability distribution for each variable. Monte Carlo analysis can also be used during the Develop Schedule process.

Exam Spotlight

Simulation techniques are recommended for predicting schedule or cost risks because they're more powerful than EMV and less likely to be misused. For the exam, remember that simulation techniques are used to predict schedule or cost risks.

Simulation analysis is displayed as histograms or S-curve diagrams. Histograms show the number of times the risk or impact produced the same result during the simulation. S-curves display the cumulative probability distribution of the risk impact occurring over the life of the project. You may recall that S-curves also represent cost baselines because spending starts out small on the project, grows and expands during the project, and then tapers off at the end when the project wraps up. Simulation technique results will behave in a similar fashion when displayed graphically because risk is smallest at the beginning of the project and grows during the work of the project, where risk would have the most significant impact, and finally tapers off toward the end as the work completes.

Criticality analysis involves determining which risks have the most significant impact on the critical path activities of the project. You'll recall that the critical path is the longest full path through the project and is made up of activities with zero float. Risks that impact the critical path activities will affect the project schedule and could also impact project cost.

Criticality analysis is performed by calculating an index for each risk that impacts activities on the critical path. The more activities the risk impacts, the higher the index. This index is used as an input to the simulation (it's usually noted as a percentage calculated based on the number of times the risk impacts critical path tasks). The final result of criticality analysis highlights the risks that are likely to impact the overall schedule. The project team should focus their efforts on creating response plans to these risks, since they will have the most significant impact on the schedule.

Influence Diagrams

According to the *PMBOK® Guide*, *influence diagrams* typically show the causal influences among project variables, the timing or time ordering of events, and the relationships among other project variables and their outcomes. Simply put, they visually depict risks (or decisions), uncertainties, or impacts and how they influence each other. Monte Carlo analysis or other simulation techniques are used to analyze the influence diagram to determine which risks will have greatest impact on the overall project. Figure 7.5 shows an influence diagram for a product introduction decision. The weather is a variable that could impact delivery time, and delivery time is a variable that can impact when revenues will occur.

FIGURE 7.5   Influence diagram

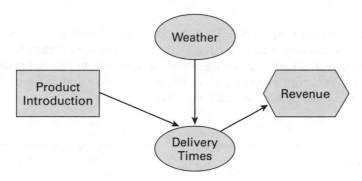

Updating the Risk Report

The output of the Perform Quantitative Risk Analysis process is—I'll bet you can guess—project documents updates. You'll record the following new elements in the risk report:

Assessment of Overall Project Risk Exposure   Using the tools and techniques of Perform Quantitative Risk Analysis allows you to create an overall probability of the project completing its goals along with a range of potential outcomes. You can also make a determination regarding the variability remaining for the project. This requires a thorough understanding of the current project objectives and knowledge of the risks. Overall project risk, just like individual risks, can have negative or positive consequences. Managing overall project risk helps ensure that the risk exposure for the organization is within an acceptable range. According to the *PMBOK Guide*, this is accomplished by minimizing those risks or elements that cause negative variations, maximizing those risks or elements that cause positive variations, and maximizing the probability of meeting all the project goals.

Probabilistic Analysis of the Project   Probabilistic analysis of the project is the forecasted results of the project schedule and costs as determined by the outcomes of risk analysis. These results are displayed as S-curves, tornado diagrams, and criticality analysis. You'll include a summary of the results here as well as determine the amount of contingency reserve (including a confidence level), identify the individual risks with the greatest impact on the critical path, and identify the overall project risks with greatest effect to the project overall and its outcomes.

Prioritized List of Risks   The risks included on this list are those identified during sensitivity analysis as having the greatest effect on the project outcomes. It should include risks that pose both the greatest threats and the greatest opportunities to the project.

Trends in Quantitative Risk Analysis Results   Trends in quantitative risk analysis will likely appear as you repeat the risk analysis processes. This information is useful as you progress, making those risks with the greatest threat to the project more evident, which gives you the opportunity to perform further analysis or go on to develop risk response plans.

Recommended Risk Responses   The risk report should include a high-level recommendation for those risks that pose the greatest threat or opportunity to the project. The recommendation may include high-level ideas on how to respond and/or identify those risks that require a response plan. The response plan itself is documented in the Plan Risk Response process, which is up next.

Exam Spotlight

Understand the differences between the Perform Qualitative Risk Analysis and Perform Quantitative Risk Analysis processes for the exam.

Developing a Risk Response Plan

The *Plan Risk Responses* process is the last process covered in this chapter. (I hear you cheering out there!) Plan Risk Responses is a process of deciding what actions to take to reduce threats and take advantage of the opportunities discovered during the risk analysis processes. This process also includes assigning risk owners; they may be departments or individual staff members who are responsible for carrying out the risk response plans you'll outline in this process.

> The more effective your risk response plans are, the better your chances for a successful project. Well-developed and well-written risk response plans will likely decrease overall project risk.

Generally, you'll want to develop risk response plans for risks with a combination of high probability of occurrence and significant impact to the project, those ranked high (or red) on the probability/impact matrix, or those ranked high as a result of

Perform Quantitative Risk Analysis. Developing risk response plans for risks of low severity or insignificant impact is not an efficient or good use of the project team's time. Spend your time planning responses that are appropriate given the impact the risk itself poses (or the opportunity the risk presents), and don't spend more time, money, or energy to produce a response than the risk event itself would produce if it occurred.

Several strategies are used in this process to reduce or control risk. It's important that you choose the right strategy for each risk so that the risk and its impacts are dealt with effectively. After deciding on which strategy to use, you'll develop an action plan to put this strategy into play should the risk event occur. You might also choose to designate a secondary or backup strategy.

Exam Spotlight

The rank of the risk will dictate the level at which Plan Risk Responses should be performed. For example, a risk with low severity wouldn't warrant the time it takes to develop a detailed risk response plan. Risk responses should be cost effective—if the cost of the response is more than the consequences of the risk, you might want to examine a different risk response. Risk responses should also be timely, agreed to by all the project stakeholders, and assigned to an individual (risk owner) who is responsible for monitoring and carrying out the risk response plan if needed.

Strategies for Creating Risk Responses

The Plan Risk Responses process has four new tools and techniques you'll need to know for the exam. They are strategies for threats, strategies for opportunities, contingent response strategies, and strategies for overall project risk. We'll look at each of these next.

Strategies for Threats

Five strategies exist to deal with negative risks or threats to the project objectives: escalate, avoid, transfer, mitigate, and accept. Accept is a strategy you can use for positive risks or opportunities as well.

Escalate

Risks that require escalation are generally outside the boundaries of the project or the risk response plan is beyond the authority of the project manager to implement and resolve. However, the project manager is responsible for notifying the appropriate person or business unit that there is a threat or opportunity they need to address.

Escalated risks are not managed at a project level. They are generally managed at a program or portfolio level. When a risk is escalated, it involves alerting the risk owner and handing off the management of the risk to the risk owner. Once a risk is escalated, it is no longer the responsibility of the project manager or project team and should be monitored and managed by the risk owner.

Avoid

To *avoid* a risk means you'll evade it altogether by eliminating the cause of the risk event or by changing the project management plan to protect the project objectives from the risk event. Let's say you're going to take a car trip from your home to a point 800 miles away. You know—because your friends who just took the same trip told you—that there is a long stretch of construction on one of the highways you're planning on using. To avoid the risk of delay, you plan the trip around the construction work and use another highway for that stretch of driving. In this way, you change your plans, avoid the risk of getting held up in construction traffic, and arrive at your destination on time.

With risk avoidance, you essentially eradicate the risk by eliminating its cause. Here's another example: Suppose your project was kicked off without adequate scope definition and requirements gathering. You run a high probability of experiencing scope creep—ever-changing requirements—as the project progresses, thereby impacting the project schedule. You can avoid this risk by adequately documenting the project scope and requirements during the Planning processes and taking steps to monitor and control changes to scope so it doesn't get out of hand.

Risks that occur early in the project might easily be avoided by improving communications, refining requirements, assigning additional resources to project activities, refining the project scope to avoid risk events, and so on.

This technique is generally used for risks with high impact and high probability of occurring.

Transfer

The idea behind a risk *transfer* is to transfer the risk and the consequences of that risk to a third party. The risk hasn't gone away, but the responsibility for the management of that risk now rests with another party. Most companies aren't willing to take on someone else's risk without a little cash thrown in for good measure. This strategy will impact the project budget and should be included in the cost estimate exercises if you know you're going to use it.

Transfer of risk can occur in many forms but is most effective when dealing with financial risks. Insurance is one form of risk transfer. You are probably familiar with how insurance works. Car insurance is a good example. You purchase car insurance so that if you come upon an obstacle in the road and there is no way to avoid hitting it, the cost to repair the damage to the car is paid by the insurance company ... okay, minus the deductible and all the calculations for the age of the car, the mileage, the color and make of the car, the weather conditions the day you were driving—but I digress.

Another method of risk transfer is contracting. Contracting transfers specific risks to the vendor, depending on the work required by the contract. The vendor accepts the responsibility for the cost of failure. Again, this doesn't come without a price. Contractors charge for their services, and depending on the type of contract you negotiate, the cost might be quite high. For example, in a fixed-price contract, which I'll talk more about in Chapter 7, the vendor (or seller) increases the cost of the contract to compensate for the level of risk they're accepting. A cost-reimbursable contract, however, leaves the majority of the risk with you, the buyer. This type of contract might reduce costs if project changes occur midway through the project.

Keep in mind that contracting isn't a cure-all. You might just be swapping one risk for another. For example, say you hire a driver to go with you on your road trip, and that person's job is to do all the driving. If the driver becomes ill or in some way can't fulfill their obligation, you aren't going to get to your destination on time. You've placed the risks associated with the trip on the contract driver; however, you've taken on a risk of delay because of nonperformance, which means you've just swapped one risk for another. You'll have to weigh your options in cases like this and determine which side of the risk coin your organization can more readily accept.

Other forms of transference include warranties, guarantees, and performance bonds.

Mitigate

When you *mitigate* a risk, you attempt to reduce the probability of a risk event occurring or reduce its impacts to an acceptable level. This strategy is a lot like defensive driving. You see an obstacle in the road ahead, survey your options, and take the necessary steps to avoid the obstacle and proceed safely on your journey. Seeing the obstacle ahead (identifying risk) allows you to reduce the threat by planning ways around it or planning ways to reduce its impact if the risk does occur (mitigation strategies).

According to the *PMBOK® Guide*, the purpose of mitigation is to reduce the probability that a risk will occur and/or reduce the impact of the risk to a level where you can accept the risk and its outcomes. It's easier to take actions early on that will reduce the probability of a risk event or its consequences than it is to fix the damage once it has occurred. Some examples of risk mitigation include performing more tests, using less complicated processes, creating prototypes, and choosing more reliable vendors.

Accept

The acceptance strategy is used when you aren't able to eliminate all the threats on the project and/or you are willing to accept the impacts of the risk. *Acceptance* of a risk event is a strategy that can be used for risks that pose either threats or opportunities to the project and is generally used for low-impact risk events. There are two alternatives to the acceptance strategy. *Passive acceptance* means you won't make any plans to try to avoid or mitigate the risk. You're willing to accept the consequences of the risk should it occur. This strategy is often used when it's more cost-effective to accept the impacts of the risk than to spend time or resources developing plans to deal with the consequences. Passive acceptance might also be used because the project team was unable to come up with an adequate response strategy and must accept the risk and its consequences. *Active acceptance* is the second strategy and might include developing contingency plans and reserves to deal with risks should they occur. (You'll look at contingency reserves in the Contingency Planning section.)

Let's revisit the road trip example. You could plan the trip using the original route and just accept the risk of running into construction. If you get to that point and you're delayed, you'll just accept it. This is passive acceptance. You could also go ahead and make plans to take an alternative route but not enact those plans until you actually reach the construction and know for certain that it is going to impede your progress. This is active acceptance and might involve developing a contingency plan.

Exam Spotlight

Understand all the strategies and their characteristics in each Plan Risk Response tool and technique for the exam.

 Real World Scenario

New Convention Center Wing

You work for a small hotel and resort in western Colorado. Your company has taken on a project to expand the convention center by adding another wing. This wing will add six more meeting rooms (four of which can be combined into a large room by folding back the movable walls to accommodate large groups). This is a popular resort, and one of the risks identified with this project was an increase in demand for conference reservations after the construction finishes. The marketing team members decide they'll mitigate this risk by contacting the organizations who've reserved space with them over the past three years and offer them incentives on their next reservation if they book their convention before the construction is completed. That way, their most important customers won't be turned away when the new reservations start pouring in.

Strategies for Opportunities

Five strategies exist to deal with opportunities or positive risks that might present themselves on the project: escalate, exploit, share, enhance, and accept. We already covered escalate and accept, so we'll look at the remaining strategies next.

Exploit

When you *exploit* a risk event, you're looking for opportunities for positive impacts. This is the strategy of choice when you've identified positive risks that you want to make certain will occur on the project. Examples of exploiting a risk include reducing the amount of time to complete the project by bringing on more qualified resources, providing even better quality than originally planned, or taking advantage of new technologies or technology upgrades that will reduce costs and allow you to complete the project sooner than planned.

Share

The *share* strategy is similar to transferring because you'll assign the risk to a third-party owner who is best able to bring about the opportunity the risk event presents. For example, perhaps what your organization does best is investing. However, it isn't so good at marketing. Forming a joint venture with a marketing firm to capitalize on a positive risk will make the most of the opportunities.

Enhance

The *enhance* strategy closely watches the probability or impact of the risk event to ensure that the organization realizes the benefits. The primary point of this strategy is to attempt to increase the probability and/or impact of positive risks. This entails watching for and emphasizing risk triggers and identifying the root causes of the risk to help enhance impacts or probability.

Contingency Planning

Contingent response strategy, better known as *contingency planning*, involves planning alternatives to deal with certain risks (such as those with accept strategies) should they occur. This is different from mitigation planning in that mitigation looks to reduce the probability of the risk and its impact, whereas contingency planning doesn't necessarily attempt to reduce the probability of a risk event or its impacts.

Contingency comes into play when the risk event occurs. This implies that you need to plan for your contingencies well in advance of the threat or opportunity occurring. After the risks have been identified and quantified, contingency plans should be developed and kept at the ready.

Contingency reserves are a common contingency response. We talked about this topic in the last chapter as well. Contingency reserve strategies for risk events are similar to those for cost and activity durations. This may include setting aside funding or resources or adding contingency time to the project schedule.

Fallback plans should be developed for risks with high impact or for risks with identified strategies that might not be the most effective at dealing with the risk. Fallback plans are not contingency plans. For example, in the information technology field, we often have high-impact risks associated with upgrading operating systems or implementing new applications. A contingency plan may include calling the vendor to assist in the middle of the night (that's always when things fall apart) when we can't figure out a solution. A fallback plan in this case may describe how we are going to bring the old system back online so that users can still do their work while we figure out what went wrong with the implementation.

In practice, you'll find that identifying, prioritizing, quantifying, and developing responses for potential threats might happen simultaneously. In any case, you don't want to be taken by surprise, and that's the point of the risk processes. If you know about potential risks early, you can often mitigate them or prepare appropriate response plans or contingency plans to deal with them.

Strategies for Overall Project Risk

Risks and risk strategies in practice are typically dealt with on most projects at an individual level. It is also important to consider overall project risks and consider developing response plans for high-impact, high-probability risks. You can use most of the same strategies we've discussed for both threats and opportunities when preparing responses to overall project risks. Avoid, exploit, transfer (or share), mitigate (or enhance), and accept are all appropriate strategies for overall project risk.

Remember that all of the risk processes we've discussed so far are performed in each iteration or workflow when using an adaptive approach to manage the project.

Documenting the Risk Responses Plan

As you've no doubt concluded, the purpose of the Plan Risk Responses process is to develop risk responses for those risks with the highest threat to or best opportunity for the project objectives. The Plan Risk Responses process has three outputs: change requests, project management plan updates, and project documents updates. Change requests may come about after performing this process that will impact the cost and schedule baselines. These changes, like all changes, need to be processed through the Perform Integrated Change Control process. Let's look at the other two outputs next.

Remember that you might need to revisit other Planning processes after performing Plan Risk Responses to modify project plans as a result of risk responses.

Project Management Plan Updates

As we've discussed throughout the book, the management plans created to document how the schedule, budget, quality, procurement, and human resources will be defined, managed, and controlled on the project become subsidiary plans to the project management plan. All of these management plans may require updates after you perform the risk processes. In addition, the WBS, the scope baseline, the schedule baseline, and cost baseline may require updates as well. Specifically, these project management plan updates are listed in the *PMBOK® Guide:*

- Schedule management plan
- Cost management plan

- Quality management plan
- Resource management plan
- Procurement management plan
- Scope baseline
- Schedule baseline
- Cost baseline

Project Documents Updates

The primary output of this process consists of updating the risk register by indicating whether a risk response exists for each risk and, most importantly, where the risk response plan is located. If the response plan can be documented in a few sentences, you could include the risk response plan in the risk register itself. You'll recall that the risk register lists the risks in order of priority (those with the highest potential for threat or opportunity first), so it makes sense that the response plans you have for the high-priority risks will be more detailed than those for lower-priority risks. Some risks might not require response plans at all, but you should put them on a watch list and monitor them throughout the project. Remember to implement version control for the risk register and each of the risk response plans so that you and your stakeholders always have the latest updates.

You may also need to update the assumption log, cost forecasts, lessons-learned register, project schedule, project team assignments, and risk report after completing this process.

Let's take a look at what the risk register should contain at this point. According to the *PMBOK® Guide*, after Identify Risks, Perform Qualitative Risk Analysis, Perform Quantitative Risk Analysis, and Plan Risk Responses are performed, the following elements should appear in the risk register:

- Prioritized list of identified risks, including their descriptions, what WBS element (or area of the project) they impact, categories (RBS), root causes, and how the risk impacts the project objectives
- Risk ranking
- Risk owners and their responsibility
- Agreed-on response strategies
- Actions needed to implement response plans
- Risk triggers
- Cost and schedule activities needed to implement risk responses
- Contingency plans
- Fallback plans
- List of residual and secondary risks
- Contingency reserves

The only elements in the preceding list I haven't talked about so far are residual and secondary risks. A *residual risk* is a leftover risk, so to speak. After you've implemented a risk response strategy—mitigation, for example—some minor risk might still remain. The contingency reserve is set up to handle situations like this.

Secondary risks are risks that come about as a result of implementing a risk response. The example given previously where you transferred risk by hiring a driver to take you to your destination but the person became ill along the way is an example of a secondary risk. The driver's illness delayed your arrival time, which is a risk directly caused by hiring the driver or implementing a risk response. When planning for risk, identify and plan responses for secondary risks.

You've gone to a good deal of work to document risk response plans. Remember that these are used to mitigate the risk event, or to reduce (or enhance) the consequences if the risk does occur. Once a risk event occurs, it is considered an issue and managed as an issue. It is no longer a risk. A risk is an unplanned or uncertain event that hasn't happened yet that exposes the organization to potential consequences. An issue is an actual event.

Exam Spotlight

According to the *PMBOK® Guide*, once a negative risk event occurs, it's considered an issue and is no longer a risk.

It's critical that everyone on the project be on the lookout for potential risks throughout the project life cycle. The project manager should include a continuous repetition of risk processes throughout all stages of the project life cycle. As you get closer to the end of the project, the likelihood of a risk event occurring lessens, but you should still be watching for them.

Risks exist on all projects, and risk planning is an important part of the project Planning processes. Just the act of identifying risks and planning responses can decrease their impact if they occur. Don't take the "What I don't know won't hurt me" approach to risk planning. This is definitely a case where not knowing something can be devastating. Risks that are easily identified and have planned responses aren't likely to kill projects or your career. Risks that you should have known about but ignored could end up costing the organization millions of dollars, along with causing schedule delays, loss of competitive advantage, or ultimately killing the project. There could be a personal cost as well, because cost and schedule overruns due to poor planning on your part are not easily explained. Time and again I've seen projects fail because the project team did not take the time to perform the risk processes. Don't let a risk devastate your career. Take the time to identify, analyze, and repeat the risk processes throughout the project.

Assessing Risks Using an Agile Approach

Agile methodologies were invented to deal with uncertainty and complexity, and to allow for changes to the requirements as the project progresses. Agile approaches can address many of the issues you may experience on a project, including uncertain requirements, changing requirements, complexity, risk, and more. According to the *Agile Practice Guide* (PMI®, 2017) the following is a list of some of the issues encountered on projects that could be overcome by using an agile methodology:

- Unclear team purpose and understanding
- Elusive requirements
- Poor user participation and past experiences on projects
- Unclear estimates and work assignments
- Poor quality
- Work and schedule delays
- Highly complex projects
- Long planning processes that lack flexibility for change
- Stakeholders with high expectations who are disappointed at the end of the project
- Lack of communication among team members
- Siloed teams that do not include members from all functional areas
- Stop and start workflows rather than continuous deliverables

Almost every item I listed could be a risk or lead to a risk event. For example, let's look at elusive requirements. If the stakeholders are unsure about what they want or they can't articulate the outcome in a way that can be easily understood and documented, it's likely that the project will not be successful because the requirements are so unclear. This is a risk that should be documented and assessed. Better yet, if you know about this risk at the beginning of the project (and you should!), you can choose a life cycle methodology, such as agile, that is better able to address this risk and reduce or eliminate its impacts. Table 7.4 shows the characteristics of assessing risk by life cycle methodology.

TABLE 7.4   Risk assessment in life cycle methodologies

Life cycle methodology	Description	Assessing risk	Objective
Predictive	Requirements are well known, and uncertainty is low due to the detail that goes into the Planning processes before the work begins. This reduces risk, uncertainty, and complexity. Changes are infrequent and must follow an established process.	Risk identification, assessment, and response plans are documented in detail early in the project. The risk processes must be performed consistently and repeatedly throughout the project. Risks that occur become issues.	Manage cost
Iterative	Requirements are defined when creating the product backlog and during each iteration. Feedback is provided on partially completed work so that future work can be improved or modified, thereby reducing risk. Changes are incorporated into each iteration.	Risk identification, assessment, and response plans (where appropriate and when needed) are performed during each iteration.	Deliverable is correct.
Incremental	Requirements are defined when creating the product backlog. Deliverables are completed and given to the customer at the end of the workflow process. Risk is reduced due to small, frequent deliverables. Changes are incorporated into each workflow stage.	Risk identification, assessment, and response plans (where appropriate and when needed) are performed during the workflow stages.	Speed of delivery
Agile	Requirements are defined when creating the product backlog and during each iteration. The goal is to deliver business value to the customer. Feedback is received on the work during the iteration and/or at the retrospective, which improves quality and reduces risk. Changes are incorporated into each iteration. The team delivers the work with the highest value to the customer first.	Risk identification, assessment, and response plans (where appropriate and when needed) are performed during each iteration.	Deliver highest business value deliverables to the customer in priority order using frequent deliveries and incorporating feedback.

Planning for Project Compliance

There is one last topic you should know about for the exam before we close out this chapter. It isn't a process in and of itself, but there are some important elements you should be familiar with for the exam. I've mentioned project compliance a few times throughout the book, but I'd like to take some time to examine this topic more fully here. I've included this conversation in this chapter because compliance (or the lack thereof) could be a risk to the project. *Compliance* involves adhering to laws, regulations, standards, or rules. Failure to adhere to compliance requirements can have devastating consequences to the organization in terms of fines and penalties, lawsuits, damage to reputation, and much more.

It could be argued that regulatory and compliance issues are requirements and should be included in the requirements documentation and added to the project scope statement. I'm okay with that approach and encourage you to add them as requirements when you know about them before the project starts. The one caveat to adding them as requirements is that the team may focus too narrowly on them as individual tasks and not consider how the regulations or standards may apply holistically to the project. For example, the Payment Card Industry (PCI) standards do not end when the project ends. Compliance to these standards needs to continue even after the project is turned over to operations. Standards and regulatory compliance issues are often considered part of the organizational process assets inputs to many processes. No matter how or where they are captured, the key is that you are aware that the project has regulatory compliance issues and that they are addressed.

Numerous categories of compliance requirements could impact your project, including but not limited to the following:

- Health and medical
- Safety
- Environment
- Security
- Workplace conditions
- Quality
- Social responsibility
- Financial
- Privacy
- Regulatory
- Technology
- Government
- Physics
- International Organization for Standardization (ISO)

- Weights and measures
- Sizing
- Securities

Work with the key stakeholders in your organization to determine compliance requirements or regulations that may apply to your project. You can use any number of tools and techniques we've discussed, such as brainstorming and interviewing, to discover and analyze compliance requirements. Document the compliance needs and then classify them using some of the examples I provided or by using classifications that apply to your industry or organization.

You will follow a process similar to the ones outlined in this chapter to document and classify compliance needs. The steps include the following:

- Identify and document compliance requirements.
- Rank and prioritize compliance requirements.
- Determine potential threats for noncompliance.
- Analyze the consequences of noncompliance.
- Quantify their impacts.
- Develop plans, actions, and approaches to address compliance.
- Audit compliance during the project and continue audits after the project is turned over to operations.

One of the key steps in this list is making certain you work with the stakeholders to determine what threats may present themselves to the organization if you do not comply with the standards, laws, regulations, or rules. Again, you could use any of the tools and techniques outlined in this chapter to help in defining those threats. Compliance issues may impact the project scope, schedule, or budget, so another important key is to develop action plans to deal with compliance.

 Real World Scenario

Project Case Study: New Kitchen Heaven Retail Store

Ricardo knocks on your office door and asks whether you have a few minutes to talk. "Of course," you reply, and he takes a seat on one of the comfy chairs at the conference table. You have a feeling this might take a while.

"I think you should know that I'm concerned about the availability of the cable vendor. I've already put in the call to get us on the list because, as I said last week, there's a 30- to 45-day lead time on these orders."

"We're only partway through the Planning processes. Do you need to order the cable install so soon? We don't even know the store location yet," you say.

"Even though they say lead time is 30 to 45 days, I've waited as long as five or six months to get cable installed in the past. I know we're really pushing for the January store opening, so I thought I'd get the ball rolling now. What I need from you is the location address, and I'll need that pretty quick."

"We're narrowing down the choices between a couple of properties, so I should have that for you within the next couple of weeks. Is that soon enough?"

"The sooner, the better," Ricardo replies.

"Great. I'm glad you stopped by, Ricardo. I wanted to talk with you about risk anyway, and you led us right into the discussion. Let me ask you, what probability would you assign to the cable installation happening six months from now?"

"I'd say the probability for six months is low. It's more likely that if there is a delay, it would be within a three- to four-month time frame."

"If they didn't get to it for six months, would it be a showstopper?" you ask. "In other words, is there some other way we could transfer Jill's data and process sales from the point of sale systems if it didn't get installed in time?"

"Sure, we could use other methods. Jill won't want to do that for very long, but work-arounds are available."

"Good. Now, what about the risk for contractor availability and hardware availability and delivery schedules for some of your other deliverables?" you continue.

You and Ricardo go on to discuss the risks associated with the IT tasks. Later, you ask Jill and Jake the same types of questions and compile a list of risks. In addition, you review the project information for the Atlanta store opening because it's similar in size and scope to this store. You add the risks from that store opening to your list as well. You divide some of the risks into the following categories: IT, Facilities, and Retail. A sample portion of your list appears as follows, with overall assignments made based on Perform Qualitative Risk Analysis and the probability and impact matrix:

- Category: IT
  - Cable vendor availability and installation. Risk score: Low
  - Contractor availability for hardware installation and burn-in. Risk score: Medium
  - POS and server hardware availability. Risk score: Medium
- Category: Facilities
  - Desirable location in the right price range. Risk score: High

- Contractor availability for build-out. Risk score: Low

- Availability of fixtures and shelving. Risk score: Low

- Category: Retail

  - Product availability. Risk score: Medium

  - Shipment dates for product. Risk score: Medium

After examining the risks, you decide that response plans should be developed for the last two items listed under the IT source, the first item under Facilities, and both of the risks listed under Retail.

Ricardo has already mitigated the cable installation risk by signing up several months ahead of the date when the installation is needed. The contractor availability can be handled with a contingency plan that specifies a backup contractor should the first choice not be available. For the POS terminals and hardware, you decide to use the transfer strategy. As part of the contract, you'll require these vendors to deliver on time, and if they cannot, they'll be required to provide and pay for rental equipment until they can get your gear delivered.

The Facilities risk and Retail risks will be handled with a combination of acceptance, contingency plans, and mitigation.

You've calculated the expected monetary value for several potential risk events. Two of them are detailed here.

Desirable location has an expected monetary value of $780,000. The probability of choosing an incorrect or less than desirable location is 60 percent. The potential loss in sales is the difference between $2.5 million in sales per year that a high-producing store generates versus $1.2 million in sales per year that an average store generates.

The expected monetary value of the product availability event is $50,000. The probability of the event occurring is 40 percent. The potential loss in sales is $125,000 for not opening the store in conjunction with the Home and Garden Show.

Project Case Study Checklist

- Plan Risk Management

- Identify Risks

  - Document analysis

  - Data-gathering techniques

- Perform Qualitative Risk Analysis

  - Risk probability and impact

  - Probability and impact rating

  - List of prioritized risks

- Perform Quantitative Risk Analysis
  - Interviewing
  - Expected monetary value
- Plan Risk Responses
  - Avoidance, transference, mitigation, and acceptance strategies
  - Risk response plans documented

Understanding How This Applies to Your Next Project

Risk management, and all the processes it involves, is not a process I recommend you skip on a project of any size. This is where the Boy Scouts of America motto, "Be Prepared," is wise advice. If you haven't examined what could be lurking around the corner on your project and come up with a plan to deal with it, then you can be assured you're in for some surprises. Then again, if you like living on the edge, never knowing what might occur next, you'll probably find yourself back on the job-hunting scene sooner than you planned (oh, wait, you didn't plan because you're living on the edge).

In all seriousness, as with most of the Planning processes I've discussed so far, risk management should be tailored to match the complexity and size of your project. If you're working on a small project with a handful of team members and a short timeline, it doesn't make sense to spend a lot of time on risk planning. However, it does warrant spending *some* time identifying project risk, determining impact and probability, and documenting a plan to deal with the risk.

My favorite Identify Risks technique is brainstorming. I like its cousin, the Nominal Group technique, too. Both techniques help you quickly get to the risks with the greatest probability and impact because, more than likely, these are the first risks that come to mind. Identify Risks can also help the project team find alternative ways of completing the work of the project. Further digging and the ideas generated from initial identification might reveal opportunities or alternatives you wouldn't have thought about during the regular Planning processes.

After you've identified the risks with the greatest impact to the project, document response plans that are appropriate for the risk. Small projects might have only one or two risks that need a response plan. The plans might consist of only a sentence or two, depending on the size of the project. I would question a project where no risks require a response plan. If it seems too good to be true, it probably is.

The avoid, transfer, and mitigate strategies are the most often used strategies to deal with risk, along with contingency planning. Of these, mitigation and contingency planning are probably the most common. Mitigation generally recognizes that the risk will likely occur and attempts to reduce the impact.

I have used brainstorming and the Nominal Group technique to strategize response plans for risks on small projects. When you're working on a small project, you can typically identify, quantify, and create response plans for risks at one meeting.

Identifying positive risk, in my experience, is fairly rare. Typically, when my teams perform Identify Risks, it's to determine what can go wrong and how bad the impact will be if it does. The most important concept from this chapter that you should apply to your next project is that you and your team should identify risks and create response plans to deal with the most significant ones.

Risk identification and assessment are fast and efficient when working in an agile environment because risks are identified, assessed, and managed during each iteration. Generally speaking, this means you'll rarely have risk events on an agile project of a nature that could kill the project because you're constantly on top of them, before they can get out of hand.

Summary

Congratulations! You've completed another fun-filled, action-packed chapter, and all of it on a single topic—risk. Risk is inherent in all projects, and risks pose both threats to and opportunities for the project. Understanding the risks facing the project better equips you to determine the appropriate strategies to deal with those risks and helps you develop the response plans for the risks (and the level of effort you should put into preparing those plans).

The Plan Risk Management process determines how you will plan for risks on your project. Its only output is the risk management plan, which details how you'll define, monitor, and control risks throughout the project. The risk management plan is a subsidiary of the project management plan.

The Identify Risks process seeks to identify and document the project risks using data-gathering and data analysis techniques such as brainstorming, the Delphi technique, interviewing, and root cause analysis. This list of risks gets recorded in the risk register, the primary output of this process.

Perform Qualitative Risk Analysis and Perform Quantitative Risk Analysis involve evaluating risks and assigning probability and impact values to them. Many tools and techniques are used during these processes, including expert judgment, interviews, risk data quality assessment, risk probability and impact assessment, assessment of other risk parameters, facilitation, risk categorization, probability and impact matrix, hierarchical charts, and meetings.

A probability and impact matrix uses the probability multiplied by the impact value to determine the risk score. The threshold of risk based on high, medium, and low tolerances is determined by comparing the risk score based on the probability level to the probability and impact matrix.

Monte Carlo simulation is a technique used to quantify schedule or cost risks in the Perform Quantitative Risk Analysis. Decision trees graphically display decisions and their various choices and outcomes, and they are typically used in combination with expected monetary value.

The Plan Risk Responses process is the last Planning process and culminates with an update to the risk register documenting the risk response plans. The risk response plans detail the strategies you'll use to respond to risk and assign individuals to manage each risk response. Risk response strategies for threats include escalation, avoidance, mitigation, and transference. Risk strategies for opportunities include escalate, exploit, share, and enhance. Acceptance is a strategy for both negative and positive risks.

Contingency planning involves planning alternatives to deal with risk events should they occur. Contingency reserves are set aside to deal with risks associated with cost and time according to the stakeholder threshold levels.

All the risk processes discussed in this chapter are performed in each iteration when using an agile methodology. The agile approach is best for projects with a high degree of uncertainty about the project requirements and high-risk projects. Uncertainty can increase the risk of rework and is an overall project risk, not an individual risk. Uncertainty is characterized by product specification, production capability, and process suitability.

Compliance involves adhering to laws, regulations, standards, or rules. Failure to adhere to compliance requirements can have devastating consequences to the organization in terms of fines and penalties, lawsuits, damage to reputation, and much more. Be certain to work with your stakeholders to identify, prioritize, quantify, assess, and develop action plans for compliance.

Exam Essentials

Be able to define the purpose of the risk management plan.   The risk management plan describes how you will define, monitor, and manage risks throughout the project. It details how risk management processes (including Identify Risks, Perform Qualitative Risk Analysis, Perform Quantitative Risk Analysis, Plan Risk Responses, Implement Risk Responses, and Monitor Risks) will be implemented, monitored, and controlled throughout the life of the project. It describes how you will manage risks but does not attempt to define responses to individual risks. The risk management plan is a subsidiary of the project management plan, and it's the only output of the Plan Risk Management process.

Be able to name the purpose of Identify Risks.   The purpose of the Identify Risks process is to identify all risks that might impact the project and then document them and identify their characteristics.

Be able to define the purpose of Perform Qualitative Risk Analysis.   Perform Qualitative Risk Analysis determines the impact the identified risks will have on the project and the probability they'll occur, and it puts the risks in priority order according to their effects on the project objectives.

Be able to define the purpose of Perform Quantitative Risk Analysis.   Perform Quantitative Risk Analysis evaluates the impacts of risk prioritized during the Perform Qualitative Risk Analysis process and quantifies risk exposure for the project by assigning numeric probabilities to each risk and their impacts on project objectives.

Be able to describe the purpose of the Plan Risk Responses process.   Plan Risk Responses is the process where risk response plans are developed using strategies such as escalate, avoid, transfer, mitigate, accept, exploit, share, enhance, develop contingent response strategies, and apply expert judgment. The risk response plan describes the actions to take should the identified risks occur. It should list all the identified risks, a description of the risks, how they'll impact the project objectives, and the people assigned to manage the risk responses.

Be able to define the risk register and some of its primary elements.   The risk register is an output of the Identify Risks process, and updates to the risk register occur as an output of every risk process that follows this one. By the end of the Plan Risk Responses process, the risk register contains these primary elements: identified list of risks, risk owners, risk triggers, risk strategies, contingency plans, and contingency reserves.

Be able to list the three characteristics of uncertainty.   They are product specification, production capability, and process suitability. Choosing an adaptive methodology that allows for small increments of work that produce frequent deliverables, along with frequent reviews of the work, will help manage uncertainty and decrease the chances of rework.

Be able to state the benefits of frequent reviews practiced in an agile approach as it pertains to risk.   Frequent reviews help to increase knowledge among the cross-functional team members, which helps them understand and manage risk.

Be able to name some of the common issues on projects that could be overcome by using an agile approach.   Such issues include unclear team purpose, elusive requirements, poor user participation, unclear estimates and work assignments, poor quality, work and schedule delays, highly complex projects, lack of flexibility, disappointed stakeholders, lack of communication, siloed teams, stop and start workflows.

Be able to describe the differences in assessing risk when using a predictive vs. an adaptive methodology.   The predictive approach requires risk identification, assessment, and preparation of response plans up front, before the work begins. These processes should be repeated throughout the project. Risk identification, assessment, and response planning are performed in each iteration, or workflow stage, of an agile project.

Be able to describe compliance and why it's important to the project.   Compliance involves adhering to laws, regulations, standards, or rules. Failure to adhere to compliance requirements can have devastating consequences to the organization in terms of fines and penalties, lawsuits, damage to reputation, and much more. Be certain to work with your stakeholders to identify, prioritize, quantify, assess, and develop action plans for compliance.

Review Questions

You can find the answers to the review questions in Appendix A. Be sure to download the Bonus Exams and Bonus Questions so that you'll have a broader exposure and more experience answering questions related to the topics in this chapter.

1. You are a project manager for Fountain of Youth Spring Water bottlers. Your project involves installing a new accounting system, and you're performing the risk-planning processes. You have identified a variability risk and an ambiguity risk. Which of the following is *not* true regarding these risks?

    A. Variability risk is uncertainty about the event or activity.

    B. Ambiguity risk is addressed using data analysis.

    C. These are types of non-event risks.

    D. Decisions, skills and abilities, and knowledge of team members is a type of non-event risk.

2. Which of the following address the purpose of the Perform Qualitative Risk Analysis process? (Choose three.)

    A. Quantifying the aggregate risk exposure for the project

    B. Assessing the probability and consequences of identified risks to the project objectives

    C. Using Monte Carlo simulation and decision tree analysis to determine risk exposure and responses

    D. Determining and assigning risk scores to each risk

    E. Prioritizing the list of risks according to their effects on the project objectives

3. Which of the following describe the purpose for the risk management plan? (Choose two.)

    A. The risk management plan describes the strategies for overall project risks.

    B. The risk management plan includes a description of the responses to risks and triggers.

    C. The risk management plan includes risk thresholds, risk scoring and risk interpretation methods, responsible parties, and budgets.

    D. The risk management plan is an input to all the remaining risk-planning processes and enhances the probability of risk management success.

4. You are using the representations of uncertainty technique of the Perform Quantitative Risk Analysis process. All of the following statements are true regarding this technique except for which one?

    A. Beta and triangular distributions are commonly used in this process.

    B. Normal and lognormal distributions use mean and standard deviation to quantify risks.

    C. Continuous probability distributions are graphically displayed and depict the impact of the risk event.

    D. Triangular distributions rely on optimistic, pessimistic, and most likely estimates to quantify risks.

5. You are performing the Perform Qualitative Risk Analysis process and want to display proximity and connectivity and their impacts on the project. Which of the following methods will you use?

    A. Tornado diagram

    B. Influence diagram

    C. Histogram

    D. Bubble chart

6. You are using criticality analysis to determine risk impact. You have calculated an index for each risk and realize the more activities the risk impacts, the higher the index. The index is used as an input to a simulation. Which risk does criticality analysis assess?

    A. Cost

    B. Resource

    C. Schedule

    D. Quality

7. Your project has a high degree of uncertainty about the project requirements. You decide to use an agile approach to manage the project in order to achieve which of the following?

    A. Decrease the risk of rework

    B. Produce accurate requirements in a faster timeframe than in a predictive methodology

    C. Increase knowledge among cross-functional team members, which helps them understand and manage risks

    D. Reduce risk and maximize business delivery

    E. All of the above

8. You've identified a risk event on your current project that could save $100,000 in project costs if it occurs. Which of the following is true based on this statement?

    A. This is a risk event that should be accepted because the rewards outweigh the threat to the project.

    B. This risk event is an opportunity to the project and should be exploited.

    C. This risk event should be mitigated to take advantage of the savings.

    D. This is a risk event that should be avoided to take full advantage of the potential savings.

9. You've identified a risk event on your current project that could save $500,000 in project costs if it occurs. Your organization is considering hiring a consulting firm to help establish proper project management techniques in order to ensure it realizes these savings. Which of the following is true based on this statement?

    A. This is a risk event that should be accepted because the rewards outweigh the threat to the project.

    B. This risk event is an opportunity to the project and should be exploited.

   C. This risk event should be mitigated to take advantage of the savings.

   D. This is a risk event that should be shared to take full advantage of the potential savings.

10. Your hardware vendor left you a voicemail saying that a snowstorm in the Midwest might prevent your equipment from arriving on time and they are watching the weather closely. She wanted to give you a heads-up and asked that you return the call. Which of the following statements is true?

   A. This is a trigger.

   B. This is a contingency plan.

   C. This is a residual risk.

   D. This is a secondary risk.

11. You are constructing a probability and impact matrix for your project. Which of the following statements are true? (Choose two.)

   A. The probability and impact matrix multiplies the risk's probability by the cost of the impact to determine an expected value of the risk event.

   B. The probability and impact matrix uses sensitivity analysis to analyze the potential impact of risk events and to determine which event has the greatest potential for impact.

   C. The probability and impact matrix prioritizes the combination of probability and impact scores using predetermined thresholds that are set by the organization, which helps determine which risks need response plans.

   D. The probability and impact matrix multiplies the risk's probability by the risk impact—which both fall between 0.0 and 1.0—and uses a predetermined threshold to determine the overall score.

12. Your stakeholders have asked for an analysis of the cost risk. Which of the following are true? (Choose three.)

   A. Monte Carlo analysis is the preferred method to use to determine the cost risk.

   B. Monte Carlo analysis is a simulation technique that computes project costs one time.

   C. A traditional work breakdown structure can be used as an input variable for the cost analysis.

   D. Monte Carlo usually expresses its results as probability distributions of possible costs.

13. Your hardware vendor left you a voicemail saying that a snowstorm in the Midwest will prevent your equipment from arriving on time. You identified a risk response strategy for this risk and have arranged for a local company to lease you the needed equipment until yours arrives. This is an example of which risk response strategy?

   A. Transfer

   B. Acceptance

   C. Mitigate

   D. Avoid

14. Risk attitude is an enterprise environmental factor that you should evaluate when performing the Plan Risk Management process. Risk attitude consists of which of the following elements? (Choose two.)

A. Risk urgency

B. Risk threshold

C. Risk manageability

D, Risk tolerance

E. Risk appetite

15. You work for a large manufacturing plant. You are working on a new project to release an overseas product line. This is the company's first experience in the overseas market, and it wants to make a big splash with the introduction of this product. The stakeholders are a bit nervous about the project and historically proceed cautiously and take a considerable amount of time to examine information before making a final decision. The project entails producing your product in a concentrated formula and packaging it in smaller containers than the U.S. product uses. A new machine is needed to mix the ingredients into a concentrated formula. After speaking with one of your stakeholders, you discover that this will be the first machine your organization has purchased from your new supplier. Which of the following statements is true given the information in this question?

A. The question describes risk threshold levels of the stakeholders, which should be considered when performing the Plan Risk Management process.

B. This question describes risk urgency, which should be considered when performing the Plan Risk Management process.

C. This question describes risk triggers that are derived using interviewing techniques and recorded in the risk register during the Perform Qualitative Risk Analysis process.

D. This question describes a risk that requires a response strategy from the positive risk category.

16. Your project team has identified several potential risks on your current project that could have a significant impact if they occurred. The team examined the impact of the risks by keeping all the uncertain elements at their baseline values. Which of the following will the team use to display this information?

A. S curve

B. Tornado diagram

C. Influence diagram

D. Histogram

17. You are using a defined set of risk categories to determine both individual risks and overall project risks. You will also use PESTLE, a strategic analysis tool to assist in determining overall project risk. You are also using the lowest level of the RBS to help in constructing which of the following in order to identify risks?

A. Checklist

B. Prompt list

C. RCA

D. Decision tree analysis

18. All of the following statements are true regarding the RBS except for which one?

  A. The RBS is contained in the risk management plan.

  B. It describes risk categories, which are a systematic way to identify risks and provide a foundation for understanding for everyone involved on the project.

  C. The lowest level of the RBS can be used as a prompt list, which is a tool and technique of the Identify Risks process.

  D. The RBS is similar to the WBS in that the lowest levels of both are easily assigned to a responsible party or owner.

19. Match the following life cycle methodologies with their objectives.

Life cycle methodologies

Methodology	Objective
A. Agile	1. Deliverable is correct
B. Incremental	2. Deliver highest business value deliverables in priority order
C. Predictive	3. Speed of delivery
D. Iterative	4. Manage cost

20. Which of the following are three characteristics of uncertainty according to the *Agile Practice Guide* (PMI®, 2017)? (Choose three.)

  A. Product ambiguity

  B. Process expertise

  C. Product specification

  D. Production capability

  E. Process suitability

Chapter

8

Planning and Procuring Resources

THE PMP® EXAM CONTENT FROM THE PEOPLE DOMAIN COVERED IN THIS CHAPTER INCLUDES THE FOLLOWING:

✓ **Task 1.4 Empower team members and stakeholders**

- 1.4.1 Organize around team strengths

✓ **Task 1.8 Negotiate project agreements**

- 1.8.1 Analyze the bounds of the negotiations for agreement

- 1.8.2 Assess priorities and determine ultimate objective(s)

- 1.8.4 Participate in agreement negotiations

- 1.8.5 Determine a negotiation strategy

✓ **Task 1.10 Build shared understanding**

- 1.10.2 Survey all necessary parties to reach consensus

THE PMP® EXAM CONTENT FROM THE PROCESS DOMAIN COVERED IN THIS CHAPTER INCLUDES THE FOLLOWING:

✓ **Task 2.1 Execute project with urgency required to deliver business value**

- 2.1.2 Examine the business value throughout the project

✓ **Task 2.5 Plan and manage budget and resources**

- 2.5.4 Plan and manage resources

✓ **Task. 2.7 Plan and manage quality of products/ deliverables**

- 2.7.1 Determine quality standard required for project deliverables

- 2.7.2 Recommend options for improvement based on quality gaps

✓ **Task 2.9 Integrate project planning activities**

  - 2.9.1 Assess consolidated project plans for dependencies, gaps, and continued business value

  - 2.9.4 Determine critical information requirements

✓ **Task 2.11 Plan and manage procurement**

  - 2.11.1 Define resource requirements and needs

  - 2.11.2 Communicate resource requirements

  - 2.11.4 Plan and manage procurement strategy

  - 2.11.5 Develop a delivery solution

✓ **Task 2.13 Determine appropriate project methodology/methods and practices**

  - 2.13.2 Recommend project execution strategy

  - 2.13.4 Recommend a project methodology/approach

  - 2.13.5 Use iterative, incremental practices throughout the project life cycle

THE PMP® EXAM CONTENT FROM THE BUSINESS ENVIRONMENT DOMAIN COVERED IN THIS CHAPTER INCLUDES THE FOLLOWING:

✓ **Task 3.2 Evaluate and deliver project benefits and value**

  - 3.2.1 Investigate that benefits are identified

  - 3.2.4 Evaluate delivery options to demonstrate value

✓ **Task 3.4 Support organizational change**

  - 3.4.3 Evaluate impact of the project to the organization and determine required actions

We're closing in on finishing up the Planning group processes. We're at a place where we need to address some processes that aren't necessarily related to one another but must be completed before you begin the Executing processes (unless you're working in an agile approach, and we'll talk about that in this chapter, too).

I'll start this chapter with the Plan Procurement Management process, which deals with managing how goods and services are obtained for the project. Procurement involves purchasing goods or services from external vendors. Sometimes the entire project is procured so that all aspects, from planning to implementation, including all of the physical resources needed to complete the project, are procured from an external vendor. Sometimes you may only procure physical resources or only consulting services, or other elements. Procurement will influence your project schedule and budget, so don't skip this process.

All projects require human resources to perform or oversee the activities needed to bring them to completion. The Plan Resource Management process is where you will develop the resource management plan that will help provide guidance for acquiring your project team members in the Executing processes.

Next we'll cover the Plan Quality Management process. This process focuses on determining the quality standards that are necessary for the project and for documenting how you'll go about meeting them.

We'll wrap up the chapter with a discussion on choosing an appropriate life cycle methodology for your project, be it predictive, adaptive, or hybrid. Let's get going.

The process names, inputs, tools and techniques, outputs, and descriptions of the project management process groups and related materials and figures in this chapter are based on content from *A Guide to the Project Management Body of Knowledge (PMBOK® Guide), Sixth Edition* (PMI®, 2017). The references to adaptive and hybrid methodologies, related materials, and figures in this chapter are based on content from the *Agile Practice Guide* (PMI®, 2017).

Procurement Planning

Procurement is sometimes a world unto its own. It has its own language, rules, regulations, and generally a small army of personnel to manage it all. Project managers need to work in close coordination with the procurement officers, and the roles and responsibilities of each should be clearly understood. In all but the smallest of organizations, a project manager does not usually have the authority to enter into a procurement agreement. In fact, after the procurement has been handed off to the procurement officer, the project manager isn't usually allowed to engage with the organizations (or people) who are bidding on the work until the process moves into the selection process. It's very important that you understand what authority you have as the project manager (if any) in a procurement process and who has the authority to sign and administer contract documents. I've seen more than one manager during the course of my career terminated for not following procurement rules.

Plan Procurement Management is a process of identifying what goods or services you're going to purchase from outside the organization and which needs the project team can meet. Part of what you'll accomplish in this process is determining whether you should purchase the goods or services and, if so, how much, when, and from which sellers. Keep in mind that I'm discussing the procurement from the buyer's perspective, because this is the approach used in the *PMBOK® Guide*.

The Plan Procurement Management process can influence the project schedule, and the project schedule can influence this process. For example, the availability of a contractor or special-order materials might have a significant impact on the schedule. Conversely, your organization's business cycle might have an impact on the Plan Procurement Management process if the organization is dependent on seasonal activity. The Estimate Activity Resources process can also be influenced by this process, as can make-or-buy decisions (I'll get to those shortly).

> You need to perform each process in the Project Procurement Management Knowledge Area (beginning with Plan Procurement Management and ending with Control Procurements) for each product or service that you're buying outside the organization. If you're procuring all your resources from within the organization, the only process you'll perform in this Knowledge Area is the Plan Procurement Management process.

Sometimes, you'll procure all the materials and resources for your project from a vendor. In cases like these, the vendor will have a project manager assigned to the project. Your organization might have an internal project manager assigned as well to act as the conduit between your company and the vendor and to provide information and monitor your organization's deliverables. When this happens, the vendor or contracting company is responsible for fulfilling all the project management processes as part of the contract. In the case of an outsourced project, the seller—also known as the *vendor, supplier, service provider,* or *contractor*—manages the project and the buyer becomes the stakeholder. If you're hiring a vendor, don't forget to consider permits or professional licenses that might be required for the type of work you need them to perform.

Several inputs are needed when planning for purchases. You'll look at them next.

Gathering Documents for the Procurement Management Plan

The Plan Procurement Management process requires a review of several planning documents you've already created, as well as identifying the enterprise environmental factors and organizational process assets that affect this process.

The important project management plan component for this process is the scope baseline. You will recall that the scope baseline includes the project scope statement that describes the need for the project and lists the deliverables and the acceptance criteria for the product or service of the project. Obviously, you'll want to consider these when thinking about procuring goods and services. You'll also want to consider the constraints (issues such as availability and timing of funds, availability of resources, delivery dates, and vendor availability) and assumptions (issues such as reliability of the vendor, availability of key resources, and adequate stakeholder involvement). The product scope description is included in the project scope statement as well and might alert you to special considerations (services, technical requirements, and skills) needed to create the project's product. As part of the scope baseline, the WBS and WBS dictionary identify the deliverables and describe the work required for each element of the WBS.

The risk register, which includes risk-related contract decisions, will guide you in determining the types of services or goods needed for risk management. For example, the transference strategy might require the purchase of insurance. You should review each of these elements when determining which goods and services will be performed within the project and which will be purchased.

Marketplace conditions are the key element of enterprise environmental factors you should consider for this process.

Many organizations have procurement departments that are responsible for procuring goods and services and writing and managing contracts. Some organizations also require that all contracts be reviewed by their legal department prior to signing. These are organizational process assets that you should consider when you need to procure goods and services.

Organizational Process Assets

The organization's guidelines and organizational policies (including any formal procurement policies), along with the organization's supplier system that contains prequalified sellers, are the elements of the organizational process assets you should pay attention to here.

It's important for the project manager to understand organizational policies because they might impact many of the Planning processes, including the procurement planning processes. For example, the organization might have purchasing approval processes that must be followed. Perhaps orders for goods or services that exceed certain dollar amounts need different levels of approval. As the project manager, you should be aware of policies like this so you're certain you can execute the project smoothly. It's frustrating to find out after the fact that you should have followed a certain process or policy and, now, because you didn't, you've got schedule delays or worse. You could consider using the "Sin now, ask forgiveness later" technique in extreme emergencies, but you didn't hear that from me. (By the way, that's not a technique authorized by the *PMBOK® Guide*.)

The project manager and the project team will be responsible for coordinating all the organizational interfaces for the project, including technical, human resources, purchasing, and finance. It will serve you well to understand the policies and politics involved in each of these areas in your organization.

I previously worked for an agency within state government and we were steeped in policy. (A government organization steeped in policy? Go figure!) It was so steeped in policy that we had to request the funds for large projects at least two years in advance. There were mounds and mounds of request forms, justification forms, approval forms, routing forms—you get the idea. My point is, if you miss one of the forms or don't fill out the information correctly, you can set your project back by a minimum of a year, if not two. Then once the money is awarded, there are more forms to fill out and policies to follow. Again, if you don't follow the policies correctly, you can jeopardize future project funds. Many organizations have a practice of not giving you all the project money up front in one lump sum. In other words, you must meet major milestones or complete a project phase before they'll fund your next phase. Know what your organizational policies are well ahead of time. Talk to the people who can walk you through the process and ask them to check your work to avoid surprises.

Keep in mind that if you are a seller, you may be managing the sale of your goods or services as a project. If that's the case, you will be following all the processes in all the Knowledge Areas and will want to ensure that you thoroughly understand the terms and conditions of the procurement documents (contracts, purchase orders, statement of work, and so on) associated with the project.

Teaming agreements are not an official input of any of the processes. However, teaming agreements by themselves are an input to the Planning process group. Teaming agreements are contractual agreements between multiple parties that are forming a partnership or joint venture to work on the project. Teaming agreements are often used when two or more vendors form a partnership to work together on a particular project. If teaming agreements are used on the project, typically the scope of work, requirements for competition, buyer and seller roles, and other important project concerns should be predefined.

Be aware that when teaming agreements are in force on a project, the planning processes are significantly impacted. For example, the teaming agreement predefines the scope of work, and that means elements such as the requirements and the deliverables may change the completion dates, thereby impacting the project schedule. Or they may affect the project budget, quality, human resources availability, procurement decisions, and so on.

Another component of organizational process assets you should be familiar with for the exam is contract types. We'll look at them next.

Contract Types

A *contract* is a compulsory agreement between two or more parties and is used to acquire products or services from outside the organization. Typically, money is exchanged for the goods or services. Contracts are enforceable by law and require an offer and an acceptance. Generally speaking, most organizations require a more extensive approval process for contracts than for other types of procurements. For example, contracts may require a signature from someone in the legal department, an executive in the organization, the chief financial officer, the procurement director, and the senior manager from the department that is having the work performed.

There are different types of contracts for different purposes. The *PMBOK® Guide* divides contracts into three categories:

- Fixed price
- Cost reimbursable
- Time and materials (T&M)

Within the fixed-price and cost-reimbursable categories are different types of contracts. We'll look at each in the following sections. Keep in mind that several factors will determine the type of contract you should use. The product requirements (or service criteria) might drive the contract type. The market conditions might drive availability and price—remember back in the dot.com era when trying to hire anyone with programming skills was next to impossible? Also, the amount of risk—for the seller, the buyer, and the project itself—will help determine contract type.

Exam Spotlight

Contract types help determine the risk the buyer and seller will bear during the life of the contract. The project manager should take this into consideration when purchasing goods and services outside the organization and make certain it is in keeping with the risk attitudes of the organization and stakeholders. There could always be an exam question or two regarding contract types, so spend some time getting familiar with them.

Fixed-Price Contracts

Fixed-price contracts can either set a specific, firm price for the goods or services rendered (known as a *firm fixed-price contract*, or *FFP contract*) or include incentives for meeting or exceeding certain contract deliverables.

Fixed-price contracts can be disastrous for both the buyer and the seller if the scope of the project is not well defined or the scope changes dramatically. It's important to have accurate, well-defined deliverables when you're using this type of contract. Conversely, fixed-price contracts are relatively safe for both buyer and seller when the original scope is well defined and remains unchanged. They typically reap only small profits for the seller and force the contractor to work productively and efficiently. This type of contract also minimizes cost and quality uncertainty. For the exam, you should know three types of fixed-price contracts:

Firm Fixed-Price (FFP)   In the FFP contract, the buyer and seller agree on a well-defined deliverable for a set price. The good news for the buyer is the price never goes up. However, if the deliverables are not well defined, the buyer can incur additional costs in the form of change orders. It's important that you clearly describe the work to avoid additional cost.

In this kind of contract, the biggest risk is borne by the seller. The seller—or contractor—must take great strides to assure they've covered their costs and will make a comfortable profit on the transaction. The seller assumes the risks of increasing costs, nonperformance, or other problems. However, to counter these unforeseen risks, the seller builds the cost of the risk into the contract price. This is the most common type of contract and the most often used.

Fixed-Price Incentive Fee (FPIF)   *Fixed-price incentive fee (FPIF) contracts* are another type of fixed-price contract. The difference here is that the contract includes an incentive—or bonus—for early completion or for some other agreed-on performance criterion that meets or exceeds contract specifications. The criteria for early completion, or other performance enhancements, are typically related to cost, schedule, or technical performance and must be spelled out in the contract so that both parties understand the terms and conditions. The fixed price, much like the FFP, is set and never goes up. The seller assumes the risk for completing the work no matter the cost.

Another aspect of fixed-price incentive fee contracts to consider is that some of the risk is borne by the buyer, unlike the firm fixed-price contract, where most of the risk is borne by the seller. The buyer takes some risk, albeit minimal, by offering the incentive to, for example, get the work done earlier. Suppose the buyer would like the product delivered 30 days prior to when the seller thinks they can deliver. In this case, the buyer assumes the risk for the early delivery via the incentive.

Fixed-Price with Economic Price Adjustment (FP-EPA)   There's one more type of fixed-price contract, known as a *fixed-price with economic price adjustment (FP-EPA) contract*. This contract allows for adjustments due to changes in economic conditions such as cost increases or decreases, and inflation. These contracts are typically used when the project spans many years. This type of contract protects both the buyer and seller from economic conditions that are outside of their control.

> ### Exam Spotlight
>
> The economic adjustment section of an FP-EPA contract should be tied to a known financial index.

Cost-Reimbursable Contracts

Cost-reimbursable contracts are as the name implies. The costs the seller incurs while producing the goods or services of the project are charged back to the buyer, and the buyer must reimburse the seller. It's important to note that only allowable costs associated with producing the goods or services are reimbursed. Allowable costs are outlined in the contract.

These contracts carry the highest risk to the buyer because the total costs are uncertain. If problems arise, the buyer has to shell out as much money as it takes to correct the problems, and this could quickly exceed the project budget. However, the advantage to the buyer with this type of contract is that scope changes are easy to make and can be made as often as you want—but it will cost you.

Cost-reimbursable contracts have a lot of uncertainty associated with them. The contractor has little incentive to work efficiently or be productive. This type of contract protects the contractor's profit because increasing costs are passed to the buyer rather than taken out of profits, as would be the case with a fixed-price contract. Be certain to audit your statements when using a contract like this so that charges from some other project the vendor is working on don't accidentally end up on your bill.

Cost-reimbursable contracts are used most often when the project scope contains a lot of uncertainty, such as for cutting-edge projects and research and development. They are also used for projects that have large investments early in the project life. Incentives for completing early, or not so early, or meeting or exceeding other performance criteria may be included in cost-reimbursable contracts much like the FPIF. We'll look at four types of cost-reimbursable contracts:

Cost Plus Fixed Fee (CPFF)   *Cost plus fixed fee (CPFF) contracts* charge back all allowable project costs to the buyer and include a fixed fee upon completion of the contract. This is how the seller makes money on the deal; the fixed fee portion is the seller's profit. The fee is always firm in this kind of contract, but the costs are variable. The seller doesn't necessarily have a lot of motivation to control costs with this type of contract, as you can imagine, and one of the strongest motivators for completing the project is driven by the fixed fee portion of the contract.

Cost Plus Incentive Fee (CPIF)   The next category of cost-reimbursable contract is *cost plus incentive fee (CPIF)*. This is the type of contract in which the buyer reimburses the seller for the seller's allowable costs and includes an incentive for meeting or exceeding the performance criteria laid out in the contract. An incentive fee actually encourages better cost performance by the seller, and a possibility of shared savings exists between the seller and buyer if performance criteria are exceeded. The qualification for exceeded performance must be written into the contract and agreed to by both parties, as should the definition of allowable costs; the seller can possibly lose the incentive fee if agreed-on targets are not reached.

There is moderate risk for the buyer under the cost plus incentive fee contract, and if well written, it can be more beneficial for both the seller and the buyer than a cost-reimbursable contract.

Cost Plus Percentage of Cost (CPPC)   In the *cost plus percentage of cost (CPPC)* contract, the seller is reimbursed for allowable costs plus a fee that's calculated as a percentage of the costs. The percentage is agreed on beforehand and documented in the contract. Because the fee is based on costs, it is variable. The lower the costs, the lower the fee, so the seller doesn't have a lot of motivation to keep costs low. This is not a commonly used contract type.

Cost Plus Award Fee (CPAF)   The *cost plus award fee (CPAF) contract* is the riskiest of the cost plus contracts for the seller. In this contract, the seller will recoup all the costs expended during the project but the award fee portion is subject to the sole discretion of the buyer. The performance criteria for earning the award are spelled out in the contract, but these criteria can be subjective and the awards are not usually contestable.

Time and Materials (T&M) Contracts

Time and materials (T&M) contracts are a cross between fixed-price and cost-reimbursable contracts. The full amount of the material costs is not known at the time the contract is awarded. This resembles a cost-reimbursable contract because the costs will continue to grow during the contract's life and are reimbursable to the contractor.

T&M contracts can resemble fixed-price contracts when unit rates are used, for example. Unit rates might be used to preset the rates of certain elements or portions of the project. For example, a contracting agency might charge you $250 per hour for a C++ programmer, or a leasing company might charge you $2,000 per month for the hardware you're leasing during the testing phase of your project. These rates are preset and agreed on by the buyer and seller ahead of time. T&M contracts are most often used when you need human resources with specific skills to work on the project and you are able to quickly and precisely define the scope of work. If you don't know the exact scope of work, you could get stuck with an extra-large bill at the end of the engagement as the scope continues to change and resources need more and more time to complete the work. It's a good idea to include a "not to exceed" clause in this type of contract to prevent a serious budget over-run. That way, you can bring in the resources and let them work as needed but when the contract approaches the ceiling amount stated in the contract, the contractors must roll off the project. The buyer bears the biggest risk in this type of contract.

Exam Spotlight

Understand the difference between fixed-price, cost-reimbursable, and time and materials contracts for the exam. Also know when each type of contract should be used, and know which party bears the most risk under each type of contract.

Source Selection Criteria

The Plan Procurement Management process consists of five tools and techniques: expert judgment, data analysis (make-or-buy analysis), data gathering (market research), source selection analysis, and meetings. I've already covered expert judgment and meetings, so we'll look at make-or-buy analysis followed by market research and then source selection analysis.

Make-or-Buy Analysis (Data Analysis)

The main decision you're trying to get to in *make-or-buy analysis* is whether it's more cost effective to buy the products and services or more cost effective for the organization to produce the goods and services needed for the project. Costs should include both direct costs (in other words, the actual cost to purchase the product or service) and indirect costs, such as the salary of the manager overseeing the purchase process or ongoing maintenance costs. Costs don't necessarily mean the cost to purchase. In make-or-buy analysis, you might weigh the cost of leasing items against the cost of buying them. For example, perhaps your project requires using a specialized piece of hardware that you know will be outdated by the end of the project. In a case like this, leasing might be a better option so that when the project is ready to be implemented, a newer version of the hardware can be tested and put into production during rollout.

Some of the analysis tools we discussed earlier in the book, such as discounted cash flow, net present value, internal rate of return, benefit–cost analysis, and return on investment, are useful in deriving a make-or-buy decision.

Other considerations in make-or-buy analysis might include elements such as capacity issues, skills, availability, and trade secrets. Strict control might be needed for a certain process, and therefore, the process cannot be outsourced. Perhaps your organization has the skills in-house to complete the project but your current project list is so backlogged that you can't get to the new project for months, so you need to bring in a vendor.

Make-or-buy analysis is considered a general management technique and concludes with the decision to do one or the other.

Market Research (Data Gathering)

Market research can consist of a variety of methods to assist you or the team in examining vendors and their capabilities and experience. Your first go-to should be your favorite search engine. This can reveal information about experience, market presence, customer reviews, and more. Conferences are another method of discovering new vendors or new services from vendors you already know. You should consider a few key items when engaging vendor services. First, get an understanding of their experience levels with your particular project or industry. You also want to know about the team members they are proposing for the project, including the depth and breadth of knowledge they have in the subject matter. You should also interview proposed team members, when appropriate, before they start work on the project. Another key factor is understanding the financial stability of the company. For example, if the vendor you are considering for your project is expected to support the project after go-live, you'll want to ensure that their finances are sound and that they are going to be in business long after your project is completed.

Source Selection Analysis

Source selection analysis, also known as evaluation criteria, are the factors you'll use to evaluate and assess vendor proposals. Selection criteria are usually determined when you create the procurement request. The criteria are typically included with the procurement documents so that your bidders know how you will evaluate the proposal and help choose the winning bid. According to the *PMBOK® Guide*, source selection analysis may include the following:

Low Cost   This one is obvious. The selection team will evaluate the proposals and choose the one with the lowest cost. This evaluation technique is generally used for commodity type items that are widely available in the marketplace.

Qualifications   Qualification criteria include elements such as the bidder's experience, references, and expertise in the area of the service or goods you are seeking. This evaluation technique is typically used for small procurements and when you don't need to, or do not have the time to, perform a full-fledged procurement process.

Quality-Based and Highest Technical Score   This evaluation method is based on the quality of the solution proposed by the bidders along with the highest technical score. This method is used when quality and technical criteria are more important than cost. However, cost may still be considered when using these evaluation criteria (and cost information should be requested from bidders) but would likely come into play only when two or more bidders have the same technical score and the quality of the proposed solution is relatively equal among the bidders.

Quality and Cost   This evaluation method should also be relatively obvious. The evaluation is based on the quality of the proposed solution along with cost.

Sole Source   A sole source evaluation method is used when you have justification for engaging only one vendor in the bidding process. For example, you may have a software system that requires an upgrade and your organization wishes to continue to use the existing system. A sole source could be used in this case since you need to ensure that the existing system is upgraded by the same vendor that already supports the system.

Fixed Budget   Fixed budget is just as it sounds. The total budget for the project or goods is published with the procurement documents. This alerts your bidders that they need to conform their bid, the scope of work, the quality of the solution, and resources to the available budget. This type of procurement requires a detailed statement of work and assurance that no changes to scope will be introduced during the project. Funding is limited in a fixed budget scenario and can't be exceeded.

Procurement Management Plan

The Plan Procurement Management process consists of 10 outputs. The first and primary output of this process is the procurement management plan. You've seen a few of the other outputs of this process before, so you're probably already ahead of me on this one. But hold the phone—I'll be covering some of the others you haven't seen: procurement strategy, bid documents, procurement statement of work, source selection criteria, make-or-buy decisions, and independent cost estimates.

Procurement Management Plan

The *procurement management plan* details how the procurement process will be managed. The procurement management plan is based primarily on the project scope, budget, and schedule, and it ensures that the resources you'll need throughout the project are available at the right time. According to the *PMBOK® Guide,* it includes the following information:

- The types of contract to use
- The authority of the project team and other stakeholders in the procurement process
- How the procurement process will be integrated with other project processes
- Where to find standard procurement documents (provided your organization uses standard documents)
- How the procurement process will be coordinated with other project processes, such as performance reporting and scheduling
- How the constraints and assumptions might be impacted by purchasing
- How independent estimates will be used during these processes
- Coordinating scheduled dates in the contract with the project schedule
- Timeline for procurement activities
- The currency to be used for payments
- The legal jurisdiction that presides over the contract
- Identification of prequalified sellers (if known)
- Risk management issues
- Procurement metrics for managing contracts and for evaluating sellers

The procurement management plan, like all the other management plans, is a component of the project management plan.

 Real World Scenario

Streamlining Purchases

Russ is a project manager for a real estate development company in Hometown, USA. Recently he transferred to the office headquarters to develop a process for streamlining purchases and purchase requests for the construction teams in the field. His first step was to develop a procurement management plan for the construction managers to use when ordering materials and equipment. Russ decided the procurement management plan could be used as a template for all new projects. That meant the project managers in the field didn't have to write their own procurement management plans when starting new construction projects. They could use the template, which had many of the fields prepopulated with corporate headquarters processes, and then they could fill in the information specific to their project. For example, the Types of Contracts section states that all equipment and materials purchases require fixed-price contracts. When human resources are needed for the project on a contract basis, a T&M contract should be used with the unit rates stated in the contract. A "not to exceed" amount should also be written into the contract so that there are no surprises as to the total amount of dollars the company will be charged for the resources.

Remember earlier in the chapter I said the procurement department has the primary role in procuring goods and services. The project team will be responsible for some steps (such as preparing the statement of work), and the procurement team will handle others. Your procurement management plan should include the action items needed to prepare the procurement, and I recommend including the responsible party and due dates. Table 8.1 includes the action items outlined in the *PMBOK® Guide* with sample responsibilities for the tasks. Be certain to check with your procurement department. You've been duly warned!

Procurement Strategy

Procurement strategy involves determining the contract type that will be used to procure the goods or services, the delivery methods to use, and the procurement phases you'll use on the project.

TABLE 8.1 Procurement timeline

Procurement steps	Responsible party	Due date
Create the SOW	Project team	May 6
Determine the cost estimate and budget	Project team	May 6
Advertise the procurement	Procurement officer	May 20
Identify qualified sellers	Project team and procurement officer	May 20
Prepare and post bid information	Procurement officer	May 30
Seller proposals submitted	Bidders and procurement officer	June 30
Evaluate technical aspects of proposals	Project team and procurement officer	July 15
Evaluate cost aspects of proposals	Project team and procurement officer	July 30
Evaluate combined technical and cost proposals and select a vendor	Project team and procurement officer	July 30
Contract negotiations and signatures	Procurement officer	Aug 30

The delivery methods for a services contract may include either using or restricting the use of subcontractors, coming to an understanding of what services the vendor will provide versus the organization procuring the services, and whether the seller can act as a representative of the buyer. It is common practice to engage professional services from a consultant who will act as a representative for your organization. Their job is to put the best interest of your organization first. They will help direct the work of the vendor that is performing the services requested in the contract and ensure that your organization is getting its money's worth and that the requirements in the statement of work are being fulfilled. The organization's representative and the vendor performing the work should not be employed by the same outside firm.

Delivery methods for construction or industrial projects may use industry-specific techniques such as design build, design bid operate, build own operate, and turnkey.

Procurement phases are defined so that the team knows the sequence for performing the procurement. This may also define procurement milestones, a plan for monitoring the phases, and criteria for moving from one phase to the next.

Bid Documents

Bid documents are used to solicit vendors and suppliers to bid on your procurement needs. You're probably familiar with some of the titles of procurement documents. They might be called request for proposal (RFP), request for information (RFI), invitation for bid (IFB), request for quotation (RFQ), and so on.

Bid documents should clearly state the description of the work requested, they should include the contract SOW, and they should explain how sellers should format and submit their responses. These documents are prepared by the buyer to ensure as accurate and complete a response as possible from all potential bidders. Any special provisions or contractual needs should be spelled out as well. For example, many organizations have data concerning their marketing policies, new product introductions planned for the next few years, trade secrets, and so on. The vendor will have access to this private information, and to guarantee that they maintain confidentiality, you should require that they sign a nondisclosure agreement (NDA).

There are a few terms that are sometimes used interchangeably in practice that have distinct definitions according to the *PMBOK® Guide*. For the exam, you should know that when your selection decision is going to be made primarily on price, the terms *bid* and *quotation* are used, as in invitation for bid or request for quotation. When considerations other than price (such as technology or specific approaches to the project) are the deciding factors, the term *proposal* is used, as in request for proposal.

Exam Spotlight

Understand the difference between *bid* and *quotation* and *proposal* for the exam. Bids or quotations are used when price is the only deciding factor among bidders. Proposals are used when there are considerations other than price.

Bid documents are posted or advertised according to your organizational policies. Most organizations post bid documents on their websites. Some archaic federal processes still require advertising in newspapers, so be sure you follow your organization's process.

Procurement Statement of Work

A *procurement statement of work (SOW)* contains the details of the procurement item in clear, concise terms. It includes the following elements:

- The project objectives
- A description of the work of the project and any post-project operational support needed

- Concise specifications of the products or services required
- The project schedule, time period of services, and work location

The procurement SOW might be prepared by either the buyer or the seller. Buyers might prepare the SOW and give it to the sellers, who in turn rewrite it so that they can price the work properly. If the buyer does not know how to prepare an SOW or the seller would be better at creating the SOW because of their expertise about the product or service, the seller might prepare it and then give it to the buyer to review. In either case, the procurement statement of work is developed from the project scope statement and the WBS and WBS dictionary.

Be aware that some organizations, particularly government agencies, do not allow a vendor to prepare the SOW on behalf of the buyer, even though the *PMBOK® Guide* states that this is an acceptable practice.

The seller uses the SOW to determine whether they are able to produce the goods or services as specified. It wouldn't hurt to include a copy of the WBS with the SOW if you have it. Any information the seller can use to properly price the goods or services helps both sides understand what's needed and how it will be provided.

Projects might require some or all of the work of the project to be provided by a vendor. The Plan Procurement Management process determines whether goods or services should be produced within the organization or procured from outside, and if goods or services are procured from outside, it describes what will be outsourced and what kind of contract to use and then documents the information in the SOW and procurement management plan. The SOW will undergo progressive elaboration as you proceed through the procurement processes. There will likely be several iterations of the SOW before you get to the actual contract award.

You used an SOW during the Develop Project Charter process. You can use that SOW as the procurement SOW during this process if you're contracting out the entire project. Otherwise, you can use just those portions of the SOW that describe the work you've contracted.

Another document similar to the SOW can be used when you are contracting for professional services. It's called a *term of reference (TOR)*. The TOR contains elements similar to those in the SOW. You'll want to include the tasks the vendor will complete, any standards that apply to the project, a description of what should be submitted for approval by the buyer, a schedule for reviews and approval, and so on.

Source Selection Criteria

Source selection criteria refer to the method your organization will use to choose a vendor from among the proposals you receive. The criteria might be subjective or objective. In some cases, price might be the only criterion, and that means the vendor that submits the lowest bid will win the contract. You should use purchase price (which should include costs associated with purchase price, such as delivery and setup charges) as the sole criterion only when you have multiple qualified sellers from which to choose.

Other projects might require more extensive criteria than price alone. In this case, you might use scoring models as well as rating models, or you might use purely subjective methods of selection. I described an example weighted-scoring method in Chapter 3, "Delivering Business Value." You can use this method to score vendor proposals.

Sometimes, the source selection criteria are made public in the procurement process so that vendors know exactly what you want in a vendor. This approach has pros and cons. If the organization typically makes known the source selection criteria, you'll find that almost all the vendors that bid on the project meet every criterion you've outlined (in writing, that is). When it comes time to perform the contract, however, you might encounter some surprises. The vendor might have done a great job of writing the bid based on your criteria, but in reality they don't know how to put the criteria into practice. On the other hand, having all the criteria publicly known beforehand gives ground to great discussion points and discovery later in the procurement processes.

The following list includes some of the criteria you can consider using for evaluating proposals and bids:

- Comprehension and understanding of the needs of the project as documented in the contract SOW
- Cost, up front as well as total cost of ownership over the life of the product or service
- Technical ability of vendor and its proposed team
- Technical approach
- Risk
- Experience on projects of similar size and scope, including references
- Project management approach
- Management approach
- Business type and size
- Financial stability and capacity
- Production capacity
- Warranty or guarantee
- Reputation, references, and past performance
- Intellectual and proprietary rights

You could include many of these in a weighted scoring model and rate each vendor on how well they responded to these issues.

When considering business type and size, also take into account whether the business is a small business, a disadvantaged or minority-owned business, or a women-owned business. Government organizations in particular may require awarding a certain percentage of procurements to businesses that fall within these categories.

The Customer Relationship Management System Response

Ryan Hunter is preparing the source selection criteria for an RFP for a customer relationship management (CRM) system. After meeting with key stakeholders and other project managers in the company who've had experience working on projects of this size and scope, he devised the first draft of the source selection criteria. A partial list is as follows:

- Successful bidder's response must detail how business processes (as documented in the RFP page 24) will be addressed with their solution.

- Successful bidder must document their project management approach, which must follow the *PMBOK® Guide* project practices. They must provide an example project management plan based on a previous project experience of similar size and scope to the one documented in the RFP.

- Successful bidder must document previous successful implementations, including integration with existing organization's PBX and network operating system, and must provide references.

- Successful bidder must provide financial statements for the previous three years.

Make-or-Buy Decisions

The *make-or-buy decision* is a document that outlines the decisions made during the process regarding which goods and/or services will be produced by the organization and which will be purchased. This can include any number of items, such as services, products, insurance policies, performance, and performance bonds.

Independent Cost Estimates

Your procurement department might conduct *independent estimates* (also known as *should cost estimates*) of the costs of the proposal and compare these to the vendor prices. If large differences between the independent estimate and the proposed vendor cost are revealed, one of two things is happening: either the statement of work (SOW) or the terms of the contract, or both, are not detailed enough to allow the vendor to come up with an accurate cost, or the vendor simply failed to respond to all the requirements laid out in the contract or SOW. Independent estimates can also be used to verify schedule estimates and other project estimates. The independent estimate can be prepared by internal resources, or you could engage outside experts or professional estimators to come up with the independent estimate.

Procurements in an Agile Environment

Procurements in an agile environment require close attention to timelines so that the purchases are lined up with the agile iterations. You can help mitigate timing and delivery issues by including the sellers as part of the agile team. This will occur naturally if you are contracting human resources to perform the work of the project. They will become members of the cross-functional agile team and should be offered the same respect and camaraderie as team members from within the organization. This collaboration between the vendor and the organization should be structured so that the vendor shares equally in the risk and rewards associated with the project. If the project falls behind schedule, for example, the vendor may incur penalties and/or be required to work overtime with the team to get the project back on track. If the project outperforms the schedule and exceeds expectations, the project team and the vendor may receive an incentive or bonus, or a special celebratory event.

Remember that large projects may benefit from a hybrid approach whereby deliverables that are unclear are produced using an adaptive approach, whereas others that are stable and clearly defined are produced using a predictive approach. Or perhaps all the development is performed in an agile fashion while training and handoff to operations follows a predictive approach. According to the *Agile Practice Guide* (PMI®, 2017), when you're using a hybrid approach to manage the project, you'll want to use a governing agreement such as a *master service agreement (MSA)*. An MSA is a contract type that outlines the specifications of the overall engagement along with some pre-agreed-on terms for future transactions or changes. For example, an agile methodology easily accommodates changes to requirements. When working under contract, those changes need to be incorporated into the contract. An MSA allows the two parties to quickly and easily negotiate the change and begin the work. This is because the MSA outlines the process for negotiating the changes and doesn't require rewriting the contract or drafting complicated change orders that could impact the contract.

Some organizations rely almost solely on vendors to perform project work. Be careful with this approach because when the vendor leaves, all the knowledge they've accumulated on the project goes with them. Using an agile approach and, in particular, the retrospective meetings at the end of each iteration will help with knowledge transfer and prevent the "brain drain" when vendors walk out the door.

When you're working with a vendor, you want to include them as part of the project team and work in a collaborative manner. You'll recall that one of the principles of the *Agile Manifesto* values customer collaboration over negotiation. Understandably you will have to perform negotiation even when using an agile approach, but the key is to foster a collaborative approach that shares both the risks and rewards across both organizations.

This is known as a *shared-risk-reward relationship*. The *Agile Practice Guide* (PMI®, 2017) notes that when contracted projects fail it is usually due to a breakdown in the customer–vendor relationship. In my experience, relationships do have a strong influence on the project outcome, but so does vendor performance. When vendor performance starts to wane, the relationship does as well. Nonetheless, for the exam, remember that project failure is due to a breakdown in relationship. There are several contracting techniques the *Agile Practice Guide* (PMI®, 2017) states can be used to help foster the shared-risk-reward relationship:

Multitiered Structure This entails breaking the contract up into individual documents, rather than having one large encompassing contract. Many of the common elements such as warranties, payment milestones, and arbitration proceedings can be contained in the master agreement. Other items such as the product description and rate structures for performing the work may be included in a schedule of services. Yet other items that are subject to change, such as the schedule, scope, and budget, could be finalized using a statement of work. This way, as changes to scope occur, only the statement of work needs to be modified. Using a multitiered approach gives the team greater flexibility for change.

Value-Driven Structure Many contracts are structured such that vendors receive payments at specific milestones or project phases. This is a controlled structure whereby the vendor must deliver or meet a milestone and then the amount associated with that milestone is paid. The amounts per milestone are agreed on in the contract. This approach encourages producing incremental pieces of work or artifacts that cannot be changed once they are delivered (unless you issue a change order and renegotiate terms and pricing). As you are aware, this is not conducive to an agile methodology. A better approach is to emphasize value-driven deliverables rather than simply meeting a milestone. Payment is released when the business value is realized and the stakeholders have a piece of work or artifact that they find useful. This will ensure value and quality of the deliverable rather than simply meeting a date or completing a milestone that produces little value.

Fixed-Price Increments This combines a fixed-price contract approach with the incremental deliverables produced when using an agile methodology. The work is broken down into user stories or releases, and prices are determined and fixed for those increments of work, rather than the entire project. This reduces financial risk to the seller because instead of obligating themselves to the whole project at once, they are agreeing to one deliverable, or increment of work, at a time.

Not-to-Exceed Time and Materials   I mentioned this one earlier in this chapter. Time and materials contracts outline a service rate for services and generally include a cost sheet for materials. For example, a program developer resource may cost $250 per hour. The not-to-exceed is stated in a lump sum. Let's say the estimate for the developer to complete the work is 1,000 hours; therefore, the not-to-exceed amount would be $250,000. A bit of contingency can help here and act as a buffer in case it takes a little bit longer than planned. You could bump up the not-to-exceed amount to $275,000 to provide that cushion. The advantage to not-to-exceed contracts are that the vendor has the freedom to perform the work the way they see fit without having to follow a rigid schedule and the buyer has the flexibility to add new requirements or make modifications to the project as they proceed. However, modifying or changing the work will require swapping out work originally planned with the newly planned work in order to stay within the not-to-exceed limits.

Graduated Time and Materials   Graduated time and materials is an approach similar to an incentive-based contract only with a catch. The vendor may be rewarded with a higher hourly rate when they finish the work before the contract date, but they may be penalized with a reduction in hourly rate when they deliver late. This approach allows for a shared financial risk between the buyer and the seller and helps ensure quality of the deliverable.

Early Cancellation   Changes occur frequently in an agile methodology, and the work may be completed faster than anticipated. The changes may include eliminating scope. If you've already contracted for that work, it could be difficult to get out of the contract. Having an early cancellation clause allows the buyer to cancel the contract when further work is no longer needed. This usually comes with a fee (agreed on in the contract). The buyer must pay the cancellation fee, but they aren't obligated to pay the remaining contract amount, which could be significant. The seller is also protected from loss because they receive the cancellation fee.

Dynamic Scope   This comes into play when the buyer has a fixed budget but they still want the option to modify scope at certain times in the project. The parties agree on specified points in the project where modifications can be made. The buyer has the flexibility to take advantage of and innovate changes to the project, and the seller is protected from overcommitment.

Team Augmentation   This option involves hiring a team of people with the skills and abilities needed to perform the work and allowing them to work collaboratively with the stakeholders. In this way, the buyer is purchasing team services rather than specific scope elements. The buyer also has the flexibility to direct the work of the team.

Favor Full-Service Suppliers   This option involves using more than one vendor on the project. This will diversify your risk because you have more than one supplier to rely on. If one supplier doesn't perform well, you have others that can pick up the slack. Be careful with this approach and spread the services or goods across the vendors. If you rely on one vendor for service "A" and another vendor for service "B," you haven't diversified your risk. Be certain to spread the services among the vendors and structure the contract to emphasize value and quality. This option works well on agile projects when you are engaging vendors that take responsibility for providing full value by having each vendor deliver independent feature sets.

Procurement departments may benefit from agile training in order to familiarize them with the processes. Agile projects require flexibility and emphasize value over rigid deadlines. The frequent changes on an agile project may require changes to the contract as well. Using some of the contracting techniques we just discussed can help minimize change, but it's helpful if your procurement team understands the principles of agile and they are willing to adjust procurement processes to fit the project needs.

Now we'll switch our focus to the human resource needs for the project and discuss the Plan Resource Management process next.

Developing the Resource Management Plan

All projects require resources, from the smallest project to the largest. The *Plan Resource Management* process documents project management activities relating to human resources, as well as how to acquire, estimate, and manage the physical resources, equipment, materials, and so on needed for the project.

Resources may be as varied and unique as your project. As the project manager, you'll want to tailor the resource processes to the size and complexity of the project. According to the *PMBOK® Guide*, you should also specifically consider the diversity of the team, their physical location during the project, and resources who may need to be acquired who have industry-specific knowledge and experience. You'll also tailor the acquisition of team resources to your organization's processes and will need to determine whether resources come from inside or outside the organization (or both), and whether you will need full-time or part-time resources. The management of the team should also be tailored to the team itself. This will depend on the diversity of the team members, training needs, organizational tools that may be needed, and so on.

The human resources portion of the resource management plan documents the roles and responsibilities of individuals or groups, and it documents the reporting relationships for each. Reporting relationships can be assigned to groups as well as to individuals, and the groups or individuals might be internal or external to the organization or a combination of both. Plan Communications Management goes hand in hand with Plan Resource Management because the organizational structure affects the way communications are carried out among project participants and the project interfaces.

When developing the resource management plan, the primary output of this process, you'll have to consider factors such as the availability of resources, skill levels, and training needs. Remember that availability of resources applies to physical resources as well as human resources. If you need a specialized piece of equipment to perform work on the project, for example, you'll want to ensure that it is available at the time indicated on the project schedule. If you are working with a contractor to obtain or lease the equipment, be sure to document the functions the equipment must perform and the specific time frame it's needed in the SOW. Each of these factors can have an impact on the project cost, schedule, and quality and may introduce risks not previously considered.

For the exam, remember that the Estimate Activity Resources process is part of the Project Resource Management Knowledge Area that we're discussing in this section. You will recall that activity resource estimates are needed to construct the project schedule.

Understanding Enterprise Environmental Factors

Plan Resource Management has five inputs: project charter, project management plan (quality management plan, scope baseline), project documents (project schedule, require-ments documentation, risk register, stakeholder register), enterprise environmental factors, and organizational process assets. You already know most of these inputs, so we'll look at the key elements of the EEFs and OPAs next.

Enterprise environmental factors play a key role in determining human resource roles and responsibilities. The type of organization you work in, the reporting relationships, and the technical skills needed to complete the project work are a few of the factors you should consider when developing the staffing management plan, which is a subset of the human resource management plan. Here is a list of some of the factors you should consider during this process:

Organizational Factors   Consider what departments or organization units will have a role in the project, the interactions between and among departments, organizational culture, and the level of formality among these working relationships.

Existing Resources and Availability   The existing base of resources that are employed by, or available to, the organization should be considered when developing the resource management plan as well as the availability of other resources, materials, and supplies.

Location and Logistics   Consider where the project team is physically located and whether they are all located together or at separate facilities (or cities or countries). Consider where facilities and materials and supplies are located and their logistics.

Marketplace Conditions   Marketplace conditions will dictate the availability of resources you're acquiring outside the organization and their costs or rate.

In addition to these factors, you should consider constraints that pertain to project teams, including the following:

Organizational Structures   Organizational structures can be constraints. For example, a strong matrix organization provides the project manager with much more authority and power than a weak matrix organization does. Functional organizations typically do not empower their project managers with the proper authority to carry out a project. If you work in a functional organization as I do, it's important to be aware that you'll likely face power struggles with other managers and, in some cases, a flat-out lack of cooperation. Don't tell them I said this, but functional managers tend to be territorial and aren't likely to give up control easily. Here's the best advice I have for you in this case:

- Establish open communications early in the project.
- Include all the functional managers with key roles in important decisions.
- Get the support of your project sponsor to empower you (as the project manager) with as much authority as possible. It's important that the sponsor makes it clear to the other managers that their cooperation on project activities is expected.

Collective Bargaining Agreements

Collective bargaining agreements are actually contractual obligations of the organization to the employees. Collective bargaining is typically associated with unions and organized employee associations. Other organized employee associations or groups might require specialized reporting relationships as well—especially if they involve contractual obligations. You will not likely be involved in the negotiations of collective bargaining agreements, but if you have an opportunity to voice opinions regarding employee duties or agreements that would be helpful to your project or future projects, by all means take it.

Economic Conditions

These conditions refer to the availability of funds for the project team to perform training, hire staff, and travel. If funds are severely limited and your project requires frequent trips to other locations, you have an economic constraint on your hands.

Exam Spotlight

For the exam, understand the key environmental factors (organizational factors, existing human resources and market conditions, personnel policies, technical factors, interpersonal factors, location and logistics, and political factors) and the three constraints (organizational structures, collective bargaining agreements, and economic conditions) that can impact the Plan Resource Management process.

Organizational Process Assets

You should consider several elements of the organizational process assets input during this process: human resource policies and procedures, templates and checklists, historical information, safety and security policies, and physical resource management policies and procedures. You may also consider escalation procedures for the team and the organization as a whole if your organization has these documented. Be certain you have an understanding of the personnel policies in the organization regarding hiring, firing, and tasking employees, along with the organization's holiday schedules, leave time policies, and so on.

The term *templates*, in this case, refers to documentation such as project descriptions, organizational charts, performance appraisals, and the organization's conflict management process. Checklists might include elements such as training requirements, project roles and responsibilities, skills and competency levels, and safety issues.

Exam Spotlight

Using templates and checklists is one way to ensure that you don't miss any key responsibilities when planning the project and will help reduce the amount of time spent on project planning.

Using Data Representation Techniques for Plan Resource Management

The Plan Resource Management process consists of four tools and techniques. Remember that your goal is to produce the resource management plan output of this process that includes a description of the team's roles and responsibilities, organizational charts, and a project team resource plan. You'll see that the tools and techniques of this process directly contribute to the components of the resource management plan. These tools are expert judgment, data representation (hierarchical-type charts, responsibility assignment matrix, text-oriented formats), organizational theory, and meetings. We've discussed expert judgment and meetings previously, so let's take a look at the remaining tools and techniques next.

Data Representation (Hierarchical-Type Charts, Responsibility Assignment Matrix, Text-Oriented Formats)

Data representation consists of hierarchical-type charts, responsibility assignment matrix (RAM), and text-oriented formats. We'll look at each of these next.

Hierarchical-Type Charts   Hierarchical charts, like a WBS, are designed in a top-down format. An organization chart is a type of hierarchical chart that lists your position, your boss, your boss's boss, your boss's boss's boss, and so on. Typically, the organization chart shows the organization or department head at the top, the management employees who report to the organization head are next, and so on, descending down the structure. An *organization breakdown structure (OBS)* is a form of organization chart that shows the departments, work units, or teams within an organization (rather than individuals) and their respective work packages.

A *resource breakdown structure (RBS)* is another type of hierarchical chart that breaks down the work of the project according to the types of resources needed. (RBS also stands for *risk breakdown structure*, as you learned in the previous chapter.) For example, you might have programmers, database analysts, and network analysts as resource types on the RBS. However, they won't all necessarily work on the project team. You might have programmers reporting to the project team, the finance department, and the customer service department, for example. An RBS can help track project costs because it ties to the organization's accounting system. Let's suppose you have programming resources in the RBS at the junior, advanced, and senior levels. Each of these levels of programmer has an average hourly salary recorded in the accounting system that makes it easy for you to track project costs. Ten senior programmers, 14 advanced programmers, and 25 junior-level programmers are easy to calculate and track.

Responsibility Assignment Matrix (RAM)   Matrix-based charts are used to show the type of resources and the responsibility they have on the project. Many times a project manager will use a *responsibility assignment matrix (RAM)* to graphically display this information. A RAM is usually depicted as a chart with resource names listed in each row (for example, programmers, testers, and trainers) and project phases or WBS elements listed as the columns. (It can also be constructed using team member names.) Indicators in the intersections show where the resources are needed. However, the level of detail is up to you. One RAM might be developed showing only project phases; another RAM might show level-two WBS elements for a complex project, with more RAMs subsequently produced for the additional WBS levels; or a RAM might be constructed with level-three elements only.

Exam Spotlight

The RAM relates the OBS to the WBS to ensure that every component of the work of the project is assigned to an individual.

Table 8.2 shows a type of RAM called an *RACI chart* for a software development team. In this example, the RACI chart shows the level of accountability each of the participants has on the project. The letters in the acronym RACI are the designations shown in the chart:

R = Responsible for performing the work

A = Accountable, the one who oversees producing the deliverable or work package, approves or signs off on the work, and makes decisions regarding the work

C = Consult, someone who has input to the work or decisions

I = Inform, someone who must be informed of the decisions or results

TABLE 8.2   Sample RACI* chart, a type of RAM

	Olga	Rae	Jantira	Nirmit
Design	R	A	C	C
Test	I	R	C/I	A
Implement	C	I	R/A	C
Review	R/A	C	C/I	C/I

*R = Responsible, A = Accountable, C = Consult, I = Inform

In this example, Olga is responsible for design, meaning she creates the software programming design document, but Rae is accountable and is the one who must make sure the work of the project is completed and approved. She will also make decisions regarding issues that arise involving this task. Jantira is both responsible and accountable for the implement task and should be consulted and informed regarding the test task. A RACI chart is a great tool because it shows at a glance not only where a resource is working but what that resource's responsibility level is on the project.

Keep in mind that this is only one type of RAM chart. You may choose to use other esignations in place of R-A-C-I.

Text-Oriented Formats   Text-oriented formats are used when you have a significant amount of detail to record. These are also called *position descriptions* or *role-responsibility-authority forms*. These forms detail (as the name implies) the role, responsibility, and authority of the resource, and they make great templates to use for future projects.

Organizational Theory

Organizational theory refers to all the theories that attempt to explain what makes people, teams, and work units perform the way they do. I'll talk more about motivation techniques (which are a type of organizational theory) in Chapter 9, "Developing the Project Team." Organizational theory improves the probability that planning will be effective and helps shorten the amount of time it takes to produce the Plan Resource Management outputs.

Although the topic is not specifically mentioned in the *PMBOK® Guide*, I recommend brushing up on your networking skills as well—that is, you know someone who knows someone and you can share information, learn new techniques, and interact with each other. Several types of networking activities exist, including proactive communication, lunch meetings (my personal favorite), informal conversations (ah, the information you learn by hanging out at the espresso machine), and trade conferences (another favorite because they get you out of the office). Networking might help when you have a specific resource need on the project but can't seem to locate someone with that set of skills.

Documenting the Resource Management Plan

The Plan Resource Management process has three outputs: the resource management plan, team charter, and project documents updates (assumption log, risk register). We'll cover the resource management plan and team charter in this section.

According to the *PMBOK® Guide*, the *resource management plan*, a subsidiary or component of the project management plan, documents how resources should be defined, staffed, managed and controlled, and released from the project when their activities are complete. It also helps in establishing an effective project team by defining the types of resources needed during the project, documenting when they're needed, and providing direction regarding how the resources should be managed.

This output has several components:

- Methods to identify and determine types and quantities of resources
- Acquiring resources
- Roles and responsibilities
- Project organizational charts
- Project team resource management
- Training
- Team development
- Resource control
- Recognition plan

The resource management plan will help you to create the project organizational structure. This structure will be documented using project organizational charts. I've already covered this in detail, so we'll look at a few of the other components of the resource management plan now. Several of these components are put into action during the Executing processes and have processes of their own. We'll discuss most of these elements in detail in Chapter 9.

Acquiring Resources

This describes how team members are acquired (from inside or outside the organization), where they're located, and the costs for specific skills and expertise. It also describes the process for acquiring the physical resource needs for the project.

Roles and Responsibilities

This component of the resource management plan is the list of roles and responsibilities for the project team. It can take the form of the RAM or RACI chart I talked about earlier, or the roles and responsibilities can be recorded in text format. According to the *PMBOK® Guide*, the following are the key elements you should include in the roles and responsibilities documentation:

Role   Describes the parts of the project for which the individuals or teams are accountable. This should also include a description of authority levels, responsibilities, and what work is not included as part of the role.

Authority   Describes the amount of authority the resource has to make decisions, dictate direction, and approve the work.

Responsibility   Describes the work the person is required to perform to complete the project activities.

Competency   Describes the skills and ability needed to perform the project activities.

Resource Management

This output refers to how you'll staff the team, manage the team, and release team members. Attention should be given to how you'll release project team members at the end of their assignments. You should have reassignment procedures in place to move folks on to other projects or back to assignments they had before the project. This reduces overall project costs because you pay them only for the time they work and then release them. You won't have a tendency to simply keep them busy between assignments or until the end of their scheduled end date if they complete their activities early. Having these procedures in place will also improve morale because everyone will be clear about how reassignment will occur. This should reduce anxiety about their opportunities for employment at the conclusion of the project or their assignments.

Resource management involves determining when resources are available to work on the project and when the recruitment process should begin. The resources can be described individually, by teams, or by function (programmers, testers, and so on). A resource histogram can be used to graphically display when resources are required. This is usually drawn in chart form, with project time along the horizontal axis and hours needed along the vertical axis. Figure 8.1 is an example of a histogram that shows the hours needed for an asphalt crew on a construction project.

FIGURE 8.1   Sample resource histogram

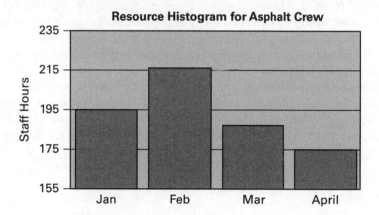

Training Needs and Team Development

This component describes any training plans needed for team members who don't have the required skills or abilities to perform project tasks. You should also consider and document team development techniques you'll use during the project to keep team members motivated and engaged.

Resource Control

This component deals with the management of physical resources on the project such as equipment, supplies, and materials, and makes certain that the resources are procured and available when needed throughout the project.

Recognition and Rewards

This component describes the systems you'll use to reward and reinforce desired behavior.

Exam Spotlight

Be sure you understand the roles and responsibilities along with the other elements of the resource management plan for the exam.

Team Charter

The *team charter* documents how the team will work together—how team members interact with one another—and it outlines team norms, ground rules, and more. It is best if it's developed by the team members themselves. They come up with the ground rules for team behavior, communication guidelines, codes of conduct, conflict resolution, and more. There are several benefits of having the team develop a set of team values and expectations and document them. It keeps team member disagreements or misunderstandings to a minimum, increases productivity, and ensures buy-in from the team members, and thus more likelihood of adhering to the agreements, because they developed the rules of engagement and agreements on their own.

Agile work teams benefit from a team charter so that they can establish standards for working together and standards for collaborating, and outline a clear set of working agreements. According to the *Agile Practice Guide* (PMI®, 2017) the team charter is known as the *team's social contract*. The key elements that should be included in the team charter are:

- Team values
- Working agreements
- Ground rules
- Group norms

The *Agile Practice Guide* (PMI®, 2017) notes that an agile team charter should also reference information about the project, including:

- Project vision, including why the organization is undertaking this project
- Project purpose, including the stakeholders who will benefit and how they will realize business value from the project
- Project success criteria and release criteria
- A description of how the team will work together, including workflows

> According to the *Agile Practice Guide* (PMI®, 2017) the goal of writing a team charter is to create an agile setting where team members can work collaboratively, to the best of their ability, to ensure a successful project.

No matter the life cycle methodology, the team should take the lead on creating this team charter. Drafting this charter does not need to be a formal process. One of the team members might take the lead by assuming a servant leadership role and coordinating the creation of the team charter. A servant leader is one who puts others' needs before their own and is one of the leadership styles you'll learn about in Chapter 9. These ideas for the social contract can get you started. The servant leader, working with the agile team, may identify other team norms they wish to document in the charter, so don't limit yourself to only these topics.

Exam Spotlight

Ground rules (a tool and technique of the Manage Stakeholder Engagement process) are defined in the team charter.

Resources on an Agile Project

As we've discussed throughout this book, agile teams are small, cross-functional teams that are self-organized and self-directed. Because they are small (ideally three to nine members), it's important that teams be organized according to their strengths and that there are varying strengths among the team members. However, when you have agile team members who have not worked on an agile project, don't expect that they will instantly know how to go about self-direction. Without training and guidance, this scenario could devolve into inaction and unfinished work. Inexperienced teams will require guidance and assistance with breaking down backlog items, estimating tasks, determining which tasks can be completed in an iteration, and how to interact with one another. One approach that will help with training and support for inexperienced teams is to establish agile centers of excellence (or centers of competencies), where best practices, training, access to experienced team members, and support are available to help build their knowledge in agile processes and to provide guidance.

Motivated team members who are experienced in agile techniques will benefit from a hands-off approach. Provide them support when needed but let them do what they do best.

Self-organized teams require members with strong communication, collaboration, and problem-solving skills. One of the critical values in the *Agile Manifesto* is individuals and interactions. Interactions cannot be accomplished without collaboration, which is the key to success on agile projects. Strong collaboration will help the team increase productivity, improve creative problem-solving, increase knowledge sharing, enable integration of work tasks, and enable flexible work assignments. Collaboration is also required to adapt to changes quickly, problem-solve, and determine what tasks to work on when. Because of the evolving requirements and frequent changes on an agile project, there is little time to use waterfall type techniques such as formal decision-making and vetting processes. Nor is there time for or benefit in understanding how to utilize team member skills before the work begins or by assigning all work tasks up front for the entire project. Agile teams will self-select work based on their skills and availability.

Agile methodologies are focused on delivering features to the customers frequently and quickly. This is known as *flow of value*. According to the *Agile Practice Guide* (PMI®, 2017) teams that focus on flow of value are more likely to collaborate, finish their work faster, and make the best use of their time because they are focused on delivering value.

Agile environments can sometimes involve engaging different team members for different iterations throughout the project. This will require close coordination with your human resources and/or procurement department so that resources are available at the right time. You'll need vendors who can meet your needs quickly and supply resources almost on a moment's notice.

In the next section, we'll shift our focus once again but to a completely new topic. We'll look at quality and its effect on the project planning processes.

Quality Planning

Quality is an important element of all projects. You can work hard, produce results, meet timelines, and stay on budget, but your stakeholders will immediately notice if quality is not up to expectations. Customer expectations are key in managing quality. The quality processes help you define, manage, and control all aspects of quality associated with the project. One of the aspects of quality involves continuous improvement. You'll recall we talked about the Plan-Do-Check-Act cycle in Chapter 1, "Building the Foundation." Plan-Do-Check-Act is one method you can use to improve quality. Other quality methodologies include total quality management, Six Sigma, and Lean Six Sigma.

Quality is affected by the triple constraints (scope, schedule, and cost), and quality concerns are found in all projects. Quality typically defines whether stakeholder expectations were met. Being on time and on budget is one thing; if you deliver the wrong product or an inferior product, on time and on budget suddenly don't mean much.

The *Plan Quality Management* process is concerned with targeting quality standards that are relevant to the project at hand and devising a plan to meet and satisfy those standards. It describes how quality will be managed and how quality will be verified throughout the course of the project. The quality management plan is an output of this process that describes how the quality policy will be implemented by the project management team during the course of the project. Everything discussed in this section, including the inputs and tools and techniques of this process, will be used to help develop this primary output.

Exam Spotlight

Plan Quality Management is a key process performed during the Planning processes and when developing the project management plan. It should be performed in conjunction with other Planning processes. According to the *PMBOK® Guide*, quality should be planned, designed, and built in—not just inspected in.

Preparing for Quality

The Plan Quality Management process has several inputs we've discussed previously. The new elements I'll cover are standards and regulations, which are part of the enterprise environmental factors input, and the quality policy, which is part of the organizational process assets input.

Standards and Regulations

The project manager should consider any standards, regulations, guidelines, or rules that exist concerning the work of the project when writing the quality plan. A *standard* is something that's approved by a recognized body and that employs rules, guidelines, or characteristics that should be followed. For example, the Americans with Disabilities Act (ADA) has established standards for web page designers that outline alternative viewing options of web pages for people with disabilities. PMI® guidelines regarding project management are another example of standards.

Standards aren't legally mandatory, but it's a good idea to follow them. Many industries have standards in place that are proven best practice techniques. Disregarding accepted standards can have significant consequences. For example, if you're creating a new software product that ignores standard protocols, your customers won't be able to use it. Standards can be set by the organization, independent bodies, organizations such as the International Organization for Standardization (ISO), and so on. ISO develops and publishes international standards on various subjects, including mathematics, information technology, railway engineering, construction materials and building, and much more. More than 160 countries have participating members (one member per country) in ISO's national standards institutes. ISO is not a government organization, but it works closely with government and the private sector to establish standards in many disciplines to meet the needs of business and society. You can find out more about ISO at www.iso.org. According to the *PMBOK® Guide*, the Project Quality Management Knowledge Area is designed to be in alignment with the ISO.

A *regulation* is mandatory. Regulations are almost always imposed by governments or institutions such as the American Medical Association. Organizations might have their own self-imposed regulations that you should be aware of as well. Regulations require strict adherence, particularly in the case of government-imposed regulations, or stiff penalties and fines could result—maybe even jail time if the offense is serious enough. Hmm, it might be tough to practice project management from behind bars—not a recommended career move.

If possible, it's a good idea to include information from the quality policy and any standards, regulations, or guidelines that affect the project in the quality management plan. If it's not possible to include this information in the quality management plan, then at least refer to the information and where it can be found. It's the project management team's responsibility to be certain all stakeholders are aware of and understand the policy issues and standards or regulations that might impact the project.

> Contracts might have certain provisions for quality requirements that you should account for in the quality management plan. If the quality management plan was written prior to the Plan Procurement Management process, you should update the quality management plan to reflect it.

Quality Policy

The quality policy is part of the organization's quality management system, which is part of the organizational process assets input. It's a guideline published by executive management that describes what quality policies should be adopted for projects the company undertakes. It's up to the project manager to understand this policy and incorporate any predetermined company guidelines into the quality plan. If a quality policy does not exist, it's up to the project management team to create one for the project.

Exam Spotlight

It is the responsibility of the project management team to ensure that all key project stakeholders are aware of and have received copies of the quality policy.

Developing the Quality Management Plan

The Plan Quality Management process has several tools and techniques to help develop the quality management plan. Some of them are familiar but have something unique you should know for this process, and some tools and techniques are new. We'll look at benchmarking, cost–benefit analysis, cost of quality, decision-making, data representation, and test and inspection planning.

Benchmarking

Benchmarking is a process of comparing previous similar activities to the current project activities to provide a standard to measure performance against. This comparison will also help you derive ideas for quality improvements on the current project. For example, if your current printer can produce 20 pages per minute and you're considering a new printer that produces 80 pages per minute, the benchmark is 80 pages per minute.

Cost–Benefit Analysis (Data Analysis)

You've seen the cost–benefit analysis technique before in the Initiating process group. In the case of quality management, you'll want to consider the trade-offs of the cost of quality.

It's cheaper and more efficient to prevent defects in the first place than to spend time and money fixing them later. The benefits of meeting quality requirements are as follows:

- Stakeholder satisfaction is increased.
- Costs are lower.
- Productivity is higher.
- There is less rework.

The primary cost of meeting quality requirements for a project is the expense incurred while performing project quality management activities.

Cost of Quality (Data Analysis)

The *cost of quality (COQ)* is the total cost to produce the product or service of the project according to the quality standards and/or the cost to make a product or service that does not meet the quality requirements. These costs include all the work necessary to meet the product requirements whether the work was planned or unplanned. It also includes the costs of work performed due to nonconforming quality requirements, assessment of whether the product or service meets requirements, and rework.

Three costs are associated with the cost of quality:

Prevention Costs   Prevention means keeping defects out of the hands of customers. *Prevention costs* are the costs associated with satisfying customer requirements by creating a product without defects. These costs are manifested early in the process and include aspects such as Plan Quality Management, training, equipment, documenting, and taking the time to do things right.

Appraisal Costs   *Appraisal costs* are the costs expended to examine the product or process and make certain the requirements are being met. Appraisal costs might include costs associated with aspects such as inspections and testing. Prevention and appraisal costs are often passed on to the acquiring organization because of the limited duration of the project.

Failure Costs   *Failure costs* are what it costs when things don't go according to plan. Failure costs are also known as cost of poor quality. Two types of failure costs exist:

> Internal Failure Costs   These result when customer requirements are not satisfied while the product is still in the control of the organization. Internal failure costs might include corrective action, rework, scrapping, and downtime.

> External Failure Costs   External failure costs, unfortunately, are when the customer determines that the requirements have not been met. Costs associated with external failure costs might include inspections at the customer site, warranty work, returns, liabilities, lost business, and additional customer service costs.

There are two categories of costs within COQ: the cost of conformance and the cost of nonconformance. Conformance costs are associated with activities undertaken to avoid failures, whereas nonconformance costs are those undertaken because a failure has occurred. All of the types of costs of quality we just covered fall into one of these categories. Table 8.3 is a quick reference.

TABLE 8.3   Cost of conformance and nonconformance

Conformance costs	Nonconformance costs
Prevention costs	Internal failure costs
Appraisal costs	External failure costs

The cost of quality can be affected by project decisions. Let's say you're producing a new product. Unfortunately, the product scope description or project scope statement was inadequate in describing the functionality of the product. The project team created the product exactly as specified in the project scope statement, the WBS, and other planning documents. Once the product hit the store shelves, the organization was bombarded with returns and warranty claims because of the poor quality. Therefore, your project decisions impacted the cost of quality. Recalls of products can also impact the cost of quality.

Cost of quality is a topic you'll likely encounter on the exam. The following sections will discuss some of the pioneers in this field. To make sure the product or service meets stakeholders' expectations, quality must be planned into the project and not added in after the fact.

Four people in particular are responsible for the rise of the modern quality management movement and the theories behind the cost of quality: Philip B. Crosby, Joseph M. Juran, W. Edwards Deming, and Walter Shewhart. Each of these men developed steps or points that led to commonly accepted quality processes that we use today and either developed or were the foundation for the development of quality theories such as Total Quality Management, Six Sigma, cost of quality, and continuous improvement. I'll also cover the Kaizen approach from a quality angle and discuss Capability Maturity Model Integration.

Philip B. Crosby

Philip B. Crosby devised the *zero defects* practice, which means, basically, do it right the first time. (Didn't your folks tell you that?) Crosby says that costs will increase when quality planning isn't performed up front, which means you'll have to engage in rework, thereby affecting productivity. Prevention is the key to Crosby's theory. If you prevent the defect from occurring in the first place, costs are lower, conformance to requirements is easily met, and the cost measurement for quality becomes the cost of nonconformance rather than the cost of rework.

Joseph M. Juran

Joseph M. Juran is noted for his *fitness for use* premise. Simply put, this means the stakeholders' and customers' expectations are met or exceeded. This says that conformance to specifications—meaning the product of the project that was produced is what the project set out to produce—is met or exceeded. Fitness for use specifically reflects the customers' or stakeholders' view of quality and answers the following questions:

- Did the product or service produced meet the quality expectation?
- Did it satisfy a real need?
- Is it reliable and safe?

Juran also proposed that there could be grades of quality. However, you should not confuse grade with quality. *Grade* is a category for products or services that are of the same type but have differing technical characteristics. *Quality* describes how well the product or service (or characteristics of the product or service) fulfills the requirements. Low quality is usually not an acceptable condition; however, low grade might be. For example, your new Dad's Dollars Credit Card software tracking system might be of high quality, meaning it has no bugs and the product performs as advertised, but of low grade, meaning it has few features. You'll almost always want to strive for high quality, regardless of the acceptable grade level.

Exam Spotlight

Understand the difference between quality and grade for the exam.

W. Edwards Deming

W. Edwards Deming suggested that as much as 85 percent of the cost of quality is management's problem and responsibility. Once the quality issue has hit the floor, or the worker level, the workers have little control. For example, if you're constructing a new highway and the management team that bid on the project proposed using inferior-grade asphalt, the workers laying the asphalt have little control over its quality. They're at the mercy of the management team responsible for purchasing the supplies.

Deming also proposed that workers cannot figure out quality on their own and, therefore, cannot perform at their best until they are shown what acceptable quality is. He believed that workers need to be shown what acceptable quality is and that they need to be made to understand that quality and continuous improvement are necessary elements of any organization—or project, in your case.

Many consider Deming to be a major contributor to the *Total Quality Management (TQM)* theory. TQM, like Deming, says that the process is the problem, not people. Every person and all activities the company undertakes are involved with quality. TQM stipulates that quality must be managed in and that quality improvement should be a continuous way of doing business, not a onetime performance of a specific task or process.

Exam Spotlight

There is some controversy surrounding who is the actual founder of TQM. Some say Deming, but others say Armand V. Feigenbaum. For the exam, I recommend knowing that Feigenbaum is the founder of TQM and Deming believes quality is a management issue. Six Sigma is a lean quality management approach that is similar to TQM and is typically used in manufacturing and service-related industries. We talked about Six Sigma in Chapter 1. From a quality perspective, Six Sigma is a measurement-based strategy that focuses on process improvement and variation reduction, which you can achieve by applying Six Sigma methodologies to the project. There are two Six Sigma methodologies. The first is known as DMADV (define, measure, analyze, design, and verify) and is used to develop new processes or products at the Six Sigma level. The second is called DMAIC (define, measure, analyze, improve, and control) and is used to improve existing processes or products. Another fact you should know about Six Sigma is that it aims to eliminate defects and stipulates that no more than 3.4 defects per million should be produced.

Walter Shewhart

Some sources say that Walter Shewhart is the grandfather of TQM, which was further popularized by Feigenbaum and Deming. Shewhart developed statistical tools to examine when a corrective action must be applied to a process. He invented control chart techniques (control charts are a tool and technique of the Control Quality process) and was also the inventor of the Plan-Do-Check-Act cycle that I talked about in Chapter 1.

Kaizen Approach

The Kaizen approach is a lean quality technique from Japan. You'll recall that *Kaizen* means *continuous improvement* in Japanese. With this technique, all project team members and managers should be constantly watching for quality improvement opportunities. The Kaizen approach states that you should improve the quality of the people first and then the quality of the products or service.

Continuous improvement involves everyone in the organization watching for ways to improve quality, whether incrementally or by incorporating new ideas into the process. This involves taking measurements, improving processes by making them repeatable and systemized, reducing variations in production or performance, reducing defects, and improving cycle times. TQM and Six Sigma are examples of continuous improvement.

A *Kaizen event* focuses on one process, or one area of the business, rather than the whole. It concerns breaking down the process into steps or parts, determining what elements are not necessary and eliminating them (or reducing their steps), and reformulating the process into a new, improved process. Kaizen events can be used in any business or industry and in any area you are considering improving. There are several steps you should follow (much like a project only on a smaller scale) when considering holding a Kaizen event. You need to determine the goal of the event, establish a team, communicate the goal to the team, select a team leader, and determine success criteria in the form of measurements if possible. Kaizen event teams are most successful when the team size is between six and ten people. The event itself should start with a review of the goal; then you'll document or draw the current state and analyze the parts that don't add value, create the future state, test it out, and implement.

Be careful when examining processes with this, or any Lean technique, that the team isn't recommending an antipattern as a solution. An *antipattern* is a common practice or procedure that appears to render a good result or provide a viable solution when, in fact, the pattern or process renders unfortunate consequences or unreliable results. The Kaizen event can help keep you from identifying an antipattern as a solution because you are breaking down each step in the process and analyzing the results.

Capability Maturity Model Integration

The *Capability Maturity Model Integration (CMMI)* is used to help organizations assess and improve performance. CMMI is used in many areas such as engineering, project management, and organizational development. CMMI models are based on five stages of development, ranging from almost no formal processes to the fifth stage, where a state of continuous, sustained improvements is reached. You may have heard about CMMI as it relates to project management. There are differing measures and stages of development depending on the industry, but most CMMI models have the following stages:

1. No formal processes are in place.

2. Basic processes exist but aren't standardized across the organization.

3. Best practices are in place and are standardized across the organization.

4. Best practices are in place and standardized across the organization, and they are measurable using quantifiable methods.

5. Continuous, sustained improvements are realized.

You can measure your organization's maturity regarding project management practices as a whole, or you can take it one step further and measure maturity in each of the 10 Knowledge Areas. Since we're discussing quality in this chapter, remember that CMMI can also be used to measure quality processes.

Exam Spotlight

For the exam, understand each of these theories on the cost of quality. Here's a key to help you remember:

- Crosby = Zero defects and prevention of rework results.

- Juran = Fitness for use, conformance. Quality by design.

- Deming = Quality is a managemont problem.

- Feigenbaum = Founder of TQM.

- Shewhart = Plan-Do-Check-Act cycle.

- TQM = Quality must be managed and must be a continuous process.

- Six Sigma = Six Sigma is a measurement-based strategy; no more than 3.4 defects per million opportunities.

- Kaizen = Continuous improvement; improve quality of people first.

- Continuous improvement = Watch continuously for ways to improve quality.

- CMMI = Assesses and improves performance by measuring the maturity levels of the organization.

Decision-Making

We've discussed multicriteria decision-making previously. This tool identifies alternatives that can be prioritized and applied to quality elements of the project.

Force field analysis is another decision-making technique that examines the drivers and resistors of a decision. You could use the old T-square approach and list all the drivers down the left column and all the resistors in the right. Determine which of the elements in the list are barriers to the project and which are enablers. Assign a priority or rank to each, and develop strategies for leveraging the strengths of the high-priority enablers while minimizing the highest-ranked barriers.

Data Representation (Flowcharts, Logical Data Model, Matrix Diagrams, Mind Mapping)

You will recall from Chapter 7, "Identifying Project Risks," that a flowchart graphically depicts the relationships between and among steps. They typically show activities, decision points, and the flow or order of steps in a process. Flowcharts may point out possible quality issues and are a great tool for the project team to use when reviewing quality results. They can also be helpful in determining the cost of quality by estimating the expected costs of conformance and nonconformance.

Matrix diagrams can show the relationships between various causes or other factors by examining the rows and columns in the matrix. Recall from earlier in this chapter that a RACI diagram is a matrix type chart. According to the *PMBOK® Guide,* several matrix diagrams are used in the quality processes, including an L, T, Y, X, C, and roof-shaped.

We've discussed mind mapping previously as well. This is a diagramming method that allows you to brainstorm ideas and then associate and group like ideas together. This is useful for determining quality requirements, identifying quality constraints, and determining relationships or dependencies that exist among and between quality elements and the project. These might be related to deliverables, quality policies, regulations, and so on.

Test and Inspection Planning

Test and inspection criteria are developed during the Plan Quality Management process and will be used throughout the remainder of the project to determine whether the quality standards are being met. Testing and inspection will vary depending on the type of project you're working on. For example, construction projects may require testing to determine weight limits or durability. Manufacturing projects may require inspection to determine whether parts are being produced according to the specifications. Software projects require testing throughout their development and a final user acceptance test to determine whether the system performs as designed.

Statistical sampling is a type of inspection technique that involves taking a sample number of parts from the whole population and inspecting them to determine whether they fall within acceptable variances.

Design of Experiments

Design of experiments (DOE) is not a named tool and technique of this process, but you should know this technique for the exam. DOE is a statistical technique that identifies the elements—or variables—that will have the greatest effect on overall project outcomes. It is used most often concerning the product of the project but can also be applied to project management processes to examine trade-offs. DOE designs and sets up experiments to identify the ideal solution for a problem using a limited number of sample cases. It analyzes several variables at once, allowing you to change all (or some) of the variables at the same time and determine which combination will produce the best result at a reasonable cost.

Exam Spotlight

For the exam, remember that the key to DOE is that it equips you with a statistical framework that allows you to change the variables that have the greatest effect on overall project outcomes all at once instead of changing one variable at a time.

Documenting the Quality Management Plan

Plan Quality Management uses many techniques to determine the areas of quality improvement that can be implemented, controlled, and measured throughout the rest of the project, as you've seen. These are recorded in the primary output of this process, which is the *quality management plan*. The other key output of this process is quality metrics. We'll look at both of these next.

Quality Management Plan

The quality management plan describes how the project management team will carry out the quality policy. It should document the resources needed to carry out the quality management plan, the responsibilities of the project team in implementing quality, and all the processes and procedures the project team and organization should use to satisfy quality requirements, including quality control, quality assurance techniques, and continuous improvement processes. This plan should also address the quality standards for both the project and the products of the project. As with other plans, you should consider project scope, risks, and requirements when writing this plan so that quality defects can hopefully be avoided, also allowing you to have better control of the cost of quality.

The project manager, in cooperation with the project staff, writes the quality management plan. You can assign quality actions to the activities listed on the WBS based on the quality plan requirements. Isn't that WBS a handy thing? Later in the Control Quality process, measurements will be taken to determine whether the quality to date is on track with the quality standards outlined in the quality management plan.

Exam Spotlight

The Project Quality Management Knowledge Area, which includes the Plan Quality Management, Manage Quality, and Control Quality processes, involves the quality management of the project as well as the quality aspects of the product or service the project was undertaken to produce. We'll discuss Quality Assurance and Quality Control in later chapters.

 Real World Scenario

Candy Works

Juliette Walters is a contract project manager for Candy Works. She is leading a project that will introduce a new line of hard candy drops in various exotic flavors: café latte, hot-buttered popcorn, and jalapeño spice, just to name a few.

Juliette is writing the quality management plan for this project. After interviewing stakeholders and key team members, she has found several quality factors of importance to the organization. Quality will be measured by the following criteria:

Candy Size   Each piece should measure 3 mm.

Appearance   No visible cracks or breaks should appear in the candy.

Flavor   Flavor must be distinguishable when the candy is taste tested.

Number Produced   The production target is 9,000 pieces per week. The current machine has been benchmarked at 9,200 candies per week.

Intensity of Color   There should be no opaqueness in the darker colors.

Wrappers   Properly fitting wrappers cover the candies, folding over twice in back and twisted on each side. There is a different wrapper for each flavor of candy, and they must match exactly.

The candy is cooked and then pulled into a long cylinder shape roughly 6′ long and 2″ in diameter. This cylinder is fed into the machine that molds and cuts the candy into drops. The cylinders vary a little in size because they're hand-stretched by expert candy makers, who then feed the candies into the drop maker machine. As a result, the end of one flavor batch—the café latte flavor—and the beginning of the next batch—the hot-buttered popcorn flavor—merge. This means the drops that fall into the collection bins are intermingled during the last run of the first flavor batch. In other words, the last bin of the café latte flavor run has some hot-buttered popcorn drops mixed in. There is no way to separate the drops once they've hit the bin. From there, the drops go on to the candy-wrapping machine, where brightly colored wrappers are matched to the candy flavor. According to the quality plan, hot-buttered popcorn drops cannot be wrapped as café latte drops. Juliette ponders what to do.

As she tosses and turns that night thinking about the problem, it occurs to her to present this problem to the company as an opportunity rather than as a problem. To keep production in the 9,000 candies per week category, the machines can't be stopped every time a new batch is introduced. So, Juliette comes up with the idea to wrap candies from the intermixed bins with wrappers that say "Mystery Flavor." This way, production keeps pace with the plan and the wrapper/flavor quality problem is mitigated.

Quality Metrics

A *quality metric*, also known as an *operational definition*, describes what is being measured and how it will be measured during the Control Quality process. For example, let's say you're managing the opening of a new restaurant in July of next year. Perhaps one of the deliverables is the procurement of flatware for 500 place settings. The operational definition in this case might include the date the flatware must be delivered and a counting or inventory process to ensure that you received the number of place settings you ordered. Measurements of this variable consist of actual values, not "yes" or "no" results. In our example, receiving the flatware is a "yes" or "no" attribute result (you have it or you don't), but the date it was delivered and the number of pieces delivered are actual values. Failure rates are another type of quality metric that is measurable, as are reliability, availability, test coverage, and defect density measurements.

> One of the results of the Plan Quality Management process is that your product or process might require adjustments to conform to the quality policy and standards. These modifications might result in cost changes and schedule changes. You might also discover that you'll need to perform risk analysis techniques for problems you uncover or when making adjustments as a result of this process.

Quality Planning for Agile Projects

Quality processes are essentially built into an agile project. Agile projects produce small quantities of work that are examined at the end of each iteration. The frequent reviews and retrospectives give the team an opportunity to understand what is working, examine the root cause of issues, and come up with modifications or new approaches for improving quality in the next iteration. This frequent review cycle allows the team to discover quality issues much more quickly than in a predictive or waterfall approach. Waterfall projects may take months or years to produce deliverables, thereby delaying the discovery of quality issues until late in the project. The cost to make changes at that stage of the project are usually significant. Agile projects, on the other hand, allow for frequent changes and quality issues are usually discovered early on in the project life cycle, thereby reducing the cost of change. For the exam, remember that iteration-based agile projects review all the finished work items at the end of the iteration. Flow-based agile projects review work when the team thinks there is enough functionality and features for the customer to examine. This is known as *fit for use*—that is, the product is usable and satisfies the customer's intended purpose. This review should occur at least once every two weeks according to the *Agile Practice Guide* (PMI®, 2017). This ensures that the team is getting timely feedback on quality and direction and can make corrections if needed. If the team cannot produce working functionality within a two-week period, they probably are not using agile techniques appropriately or may be caught up in antipatterns. Consider using an agile coach to provide the team with training and mentoring in agile practices.

You'll recall from Chapter 6, "Developing the Project Budget and Engaging Stakeholders," that we talked about Extreme Programming as an example of an agile methodology that leads to higher-quality software. XP uses refactoring techniques that help improve the quality of the code without changing its functionality. Refactoring reduces duplication and eliminates poor code. A new agile technique you haven't seen yet is called

Lean software development (LSD). It's another methodology that focuses on quality, speed of delivery, and business value for the customer. We talked about test-driven development, which is a technique in the XP methodology that provides significant feedback to the team. Test-driven development focuses on zero defects. In this approach, the team defines tests before they begin programming. Once programming commences, the tests are performed at the appropriate time in order to continuously validate the work. This may also include defining acceptance tests using the *acceptance test driven development (ATDD)* approach where the team works together to collaboratively create the acceptance tests before the programming activities begin. The tests are performed at the appropriate time in order to continuously validate the work.

Value stream and *value stream mapping* are Lean management techniques that help reduce waste. A value stream focuses on the flow of value to the customer in delivering products or services. Value stream mapping is a visual representation of the documented process that you'll analyze in order to eliminate waste and improve the flow of services or materials needed to produce the end product for the customer. The idea with a value-stream map is similar to the Kaizen event except that a value stream looks at the entire process from the beginning of the process all the way to delivery to the customer. You'll identify all the elements in the process that do not add value and reduce or eliminate waste, which, in turn, improves the efficiency of the value stream. The seven elements of waste that you're looking for with value stream mapping are as follows:

- Pace that's too fast
- Waiting
- Transport or conveyance
- Processing
- Excess stock
- Unnecessary motion
- Correcting mistakes

Project Planning Using Agile Methodologies

I've talked about the planning processes when using the agile methodologies throughout this book. From a high-level viewpoint, planning processes when using a predictive approach are performed before the work begins. In agile methodologies, planning occurs iteratively and continuously at the beginning of an iteration or workflow stage. As the *Agile Practice Guide* (PMI®, 2017) states, planning should always be performed no matter what methodology you are using. The key to planning is when and how much should be done based on the approach. Table 8.4 outlines the differences in the planning approaches for the various methodologies.

TABLE 8.4   Planning processes by life cycle approach

Life cycle approach	Planning characteristics
Predictive	Planning is performed prior to beginning the work. The plan is used to manage the work, verify the work, and ensure that the cost, schedule, and scope remain in alignment with the plan throughout the project.
Iterative	The iterative approach uses prototypes and proofs of concept for the customer to review and modify. Planning for the prototypes is performed early in the project and modified in each iteration as feedback is received from the customer.
Incremental	The incremental approach produces subsets of work that incrementally add up to the completed project. Planning may involve several successive deliveries or only one subset of work at a time.
Agile	Planning occurs at the beginning of an iteration and continues throughout the project. The team plans and completes work, the customer performs frequent reviews, and more is known about the final deliverable as the project progresses.

Exam Spotlight

Planning must be performed on all projects regardless of the development methodology you're using to manage the work. The methodology dictates the timing and depth of planning needed.

Bringing It All Together

Believe it or not, you have officially completed the Planning process group. You now have a completed project management plan, which is the approved, formal, documented plan that's used to guide you throughout the remainder of the project. After you have a completed plan, there are a couple of important steps to perform before obtaining sign-off. First, you'll want to assess each of the subcomponent plans for dependencies and gaps, and ensure they are aligned to provide continued business value. For example, the resource and procurement plans have dependencies on each other. It's difficult to procure resources that haven't yet been identified. Make certain there aren't gaps in the plan and that you've covered all the planning areas. Pay particular attention to the requirements, scope, and budget for any gaps that could derail the project. Also, ensure that all the subcomponent plans reflect the business value the project was brought about to produce. Next you'll need to identify and determine any critical information requirements of the stakeholders and the project team. These should be recorded in the communication management plan, but it's a good idea to make certain you keep the information needs handy so that you can keep the stakeholders updated on important information. Also, make certain to monitor their information requirements throughout the project.

The project management plan encompasses everything I've talked about up to now and is represented in a formal document, or collection of documents. This document consists of all the outputs of the Planning process group, including the component plans and baselines. It contains the project scope statement, deliverables, constraints, assumptions, risks, WBS, milestones, project schedule, resources, and more. It becomes the baseline you'll use to measure and track progress against. It's also used to help you control the components that tend to stray from the original plan so that you can get them back in line.

The project management plan serves several purposes, the most important of which is tracking and measuring project performance through the Executing and Monitoring and Controlling processes and making future project decisions. The project management plan is a communication and information tool for stakeholders, team members, and the management team. They will use the plan to review and gauge progress as well.

Exam Spotlight

Performance measurement baselines, when using predictive methodologies, are management controls that should change infrequently. Examples of the performance measurement baselines you've looked at so far are budget, scope, and schedule baselines. However, the project management plan itself also becomes a baseline. If changes in scope or schedule do occur after the Planning processes are complete, you should go through a formalized process to implement the changes.

Don't forget that sign-off on the project management plan is important to the project's success. It isn't necessary for the sponsor and key stakeholders to sign off on every individual project document, but you should obtain signatures on the project management plan. I would never embark on a project of any size without sign-off from at least the project sponsor as well as a few key stakeholders, the latter depending on the size, risk, and complexity of the project. If they've been an integral part of the Planning processes all along (and I know you know how important this is), obtaining sign-off on the project management plan should simply be a formality. Once you obtain sign-off, you have the approval to move into the Executing processes.

The project management plan consists of several components, and I've recapped them here for your reference. You can find the differences between project management plan components and project documents in a handy table (Table 4-1) in the *PMBOK® Guide* in Chapter 4:

- Change Management Plan
- Communications Management Plan
- Configuration Management Plan
- Cost Management Plan
- Cost Baseline
- Development Approach (adaptive, predictive, hybrid)
- Performance Measurement Baseline
- Procurement Management Plan
- Project Life Cycle Description
- Quality Management Plan
- Requirements Management Plan
- Resource Management Plan
- Risk Management Plan
- Schedule Baseline
- Schedule Management Plan
- Scope Baseline
- Scope Management Plan
- Stakeholder Engagement Plan

"But wait," I hear you saying, "we didn't discuss the configuration management plan." In a nutshell, configuration management is similar to change management, but configuration changes deal with the components of the product of the project, such as functional ability or physical attributes, rather than the project process itself. I will cover these topics in much more depth in Chapter 11, "Measuring and Controlling Project Performance."

Note that you will use all the management plans I discussed during the Planning processes—all those just listed—throughout the Executing process group to manage the project and keep the performance of the project on track with the project objectives. If you don't have a project management plan, you'll have no way of managing the project. You'll find that even with a project management plan, project scope has a way of changing. Stakeholders and others tend to sneak in a few of the "Oh, I didn't understand that before" statements and hope that they slide right by you. With that signed, approved project management plan in your files, you are allowed to gently remind them that they read and agreed to the project plan and you're sticking to it.

Choosing a Life Cycle Methodology

At this point in the project, you know a great deal of information about the project. You know the goal, the business value the project will bring about, the deliverables, the resources needed for the project, the activities, and the activity durations. It might be a good time to consider which project methodology (predictive, hybrid, or adaptive) is best for your project. This is the responsibility of the project manager with input from the project team. We've talked a good deal already about these methodologies along with their advantages and disadvantages.

According to the *PMBOK® Guide,* there are three main categories to consider when choosing a methodology to manage the work of the project:

Culture   Culture can be considered by examining *organizational bias.* This relates to the values and preferences that characterize the organization and its culture. According to the *Agile Practice Guide* (PMI®, 2017) the biases include exploring versus executing, speed versus stability, quality versus quantity, and flexibility versus predictability. I think you can agree that an organization that embraces values such as exploring, speed, quality, and flexibility would likely be successful at using adaptive or agile approaches in performing their projects.

Project Team   The skills and experience of the project team members are another factor when considering your life cycle approach. Teams who are trained in agile processes should be ready and willing to work on agile-based projects. If agile is new to the team or organization, consider agile training classes, using an agile coach to work with them individually and as a team, and blending approaches so that you can gradually introduce them to the concepts of agile. Perhaps start out with a predictive approach to define requirements and the goal of the project and then switch to an agile-based methodology with user stories and timeboxed periods of work. Other factors to consider in regard to the project team are the size and complexity of the project work, the skills and experience levels of the team members, and their ability to collaborate and work cohesively as a team.

Project Itself   The project itself can include many elements such as the complexity of the project, risk, deadlines, cost, business urgency to complete the project, and the experience of the organization or team working on projects of this nature. I will describe the complexity element shortly and provide several examples that work for this category as well.

The project team, along with the project manager, will determine the best methodology to use to manage the work of the project.

In addition to the factors we just discussed, you should take into account the needs of the project, its complexity, and its magnitude:

Needs   One element of needs is timing. Do you need to produce the product or result of your project quickly with interactive feedback from the stakeholders? Is the final result of the project not entirely clear at the beginning of the project and the stakeholders hope to refine scope as they go? If you answered yes to these questions, you'll likely want to use an adaptive methodology to manage the project. Projects with well-defined requirements and a clear vision, or those that will take a long time to complete, may benefit from a predictive or hybrid methodology. You might use the predictive approach to define the work of the project and complete the Planning processes, then switch to an agile approach when you begin the Executing process, or vice versa.

Other needs may include producing a mock-up of the final project so that feedback can be provided early in the project, or the need to change requirements or introduce new requirements during the project.

Complexity   Complexity of the project should be considered from a couple of angles. For example, complexity may be relative to the skills and ability of the project team. If you have a highly complex project with many unique deliverables and requirements that are critical to the success of the project, you'll want to consider whether your team is experienced at working on projects of this complexity. If they have little experience with projects like this, you might explore a predictive or hybrid methodology so that you can apply the rigor of planning all the project needs before beginning the work.

There are any number of elements you might consider when examining complexity, including project due dates, costs, resources, business urgency, risk, constraints, the power and influence of the stakeholders, and more. Complexity may also have to do with the complexity of the organization you're working for, the industry to which the organization belongs, and other elements such as compliance and regulatory issues.

You can plot the degree of certainty (or agreement) about the complexity elements such as requirements or team skills on a Stacey Complexity Model. The degrees are measured as low uncertainty or agreement, to high uncertainty or lack of agreement. This model displays levels of uncertainty from low to high on both the x- and the y-axis, or it may show levels of uncertainty on the x-axis and agreement levels on the y-axis. This two-dimensional model shows the progression of the certainty and agreement of complexity elements, and they are plotted into categories such as simple to chaotic. Looking at Figure 8.2, you'll see that requirements (an element of complexity) are plotted on a Stacey Complexity Model. Requirements that fall in the low uncertainty and full agreement area on the model are considered a simple project, and the predictive methodology works well for this approach. A higher degree of uncertainty and a higher degree of disagreement will land the project in the complicated or complex area, where an adaptive methodology is the best approach. Projects that fall in the chaotic area are extremely risky projects. You can use an adaptive methodology for these projects, but you'll need frequent inspections and frequent reviews to ensure you're on track.

FIGURE 8.2   Stacey Complexity Model

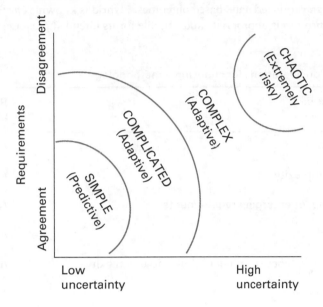

Magnitude   Magnitude has to do with the size of the project. You should consider the impact this may have on the organization. For example, large projects with multi-year timelines may benefit from a phased methodology approach. Then when each phase begins, you can assess which methodology may work best for the phase you're kicking off. You should also consider the skills and abilities of the project team, as we did with complexity.

Life Cycle Considerations

The methodology you choose to manage the project will depend on many factors. Some of those are listed in Table 8.5, along with a recommended approach. These aren't hard and fast rules, however. According to the *Agile Practice Guide* (PMI®, 2017) the key to any methodology is that the team is delivering business value to the organization using the method that will work best for the project and within the current business environment. As we discussed earlier, this may include culture, organizational structure, team capabilities, and the nature of the project. The primary focus is how to be successful, not which methodology to choose. Success for your project may mean delivering value incrementally or delivering value all at once at the end of the project because the nature of the work won't allow for incremental business value. Here are some of the considerations you should think about when choosing a life cycle or development methodology. I am using "agile" in this table, but do realize that you could substitute incremental or iterative or flow-based methods where you see agile. Keep in mind that it's also possible to blend agile approaches, such as Scrum and Kanban, as we discussed in previous chapters. A *blended agile* approach is considered an agile approach.

Hybrid is a blend of predictive and agile approaches and is not considered an agile approach; however, hybrid can produced agile-based outcomes. Hybrid is known as *fit for purpose*, meaning that the process is appropriate and suitable for its intended purpose.

TABLE 8.5   Considerations for choosing a methodology

Need or situation	Recommended methodology
Rapid deliverables	Agile
High degree of complexity	Agile
Uncertain, evolving, or emerging requirements	Agile
Uncertainty or high risk	Agile
Poor quality on past projects of similar scope and complexity	Agile
High rate of change	Agile
Difficulty describing goal or end product	Agile
Ability to deliver business value incrementally	Agile
Reduce waste and rework	Agile
Research and development are required	Agile
Feedback is needed as business value is produced in order to improve the product incrementally or iteratively	Agile
Customer-based deliveries on a frequent basis	Agile
Well-known requirements	Predictive
Low rates of change	Predictive
Agile principles don't fit the company culture	Predictive
Nature of the project work does not allow for incremental delivery	Predictive
High rates of change during requirements and scope definition; well-defined processes for performing work	Hybrid
Projects that have compliance or regulatory needs with well-defined deliverables	Hybrid

Need or situation	Recommended methodology
Project with medium risk and low degree of uncertainty	Hybrid
Organizations with multiple active projects ranging from small to large in magnitude	Predictive, agile, and/or hybrid
Organization culture is willing to try agile but has not used this approach before	Agile and/or hybrid
Culture exhibits a high degree of trust, adaptiveness to change, high risk tolerance, and buy-in from executives	Agile
Culture exhibits a low degree of trust, unwillingness to change, and low risk tolerance	Predictive
Culture exhibits a medium degree of trust, some adaptiveness to change, medium risk tolerance, and executives are open to agile	Hybrid

Exam Spotlight

According to the *Agile Practice Guide* (PMI®, 2017) you'll need to pick the methodology that works best for the organization according to the culture, the project itself, and the risks associated with the project. The methodology is not the end goal. The goal of project management, no matter which methodology you use, is to deliver business value in the best way possible within the current environment.

When you've determined an adaptive methodology is the best approach, you also need to decide whether the adaptive methodology will be iterative (as in Kanban) or incremental (as in Scrum) or a blended agile approach. You can use the elements we discussed earlier in this section, including the project needs, complexity, magnitude, culture, project team, and the project itself, to help determine which agile methodology to choose.

According to the *PMBOK® Guide,* the role of the project manager is the same whether you're using a predictive or an adaptive development methodology. However, when using an adaptive methodology to manage the project, project managers will need to understand and have experience using the tools and techniques in an adaptive approach to ensure project success.

Real World Scenario

Project Case Study: New Kitchen Heaven Retail Store

You're just finishing a phone conversation with Jill, and you see Dirk headed toward your office.

Dirk walks in, crosses his arms over his chest, and stands next to your desk with an "I'm here for answers" look.

"I thought I'd drop by and see whether you have signed a lease and gotten Jake started on that build-out yet," says Dirk.

"I just got off the phone with Jill," you reply. "The real estate agent found a great location, and we've set up a tour for tomorrow."

"What has been the holdup?" Dirk asks. "I thought we'd be ready to start the build-out about now."

"I've been working on the project plans."

"Project plans," Dirk interrupts. "We already have a plan. That schedule thing and risk stuff and the budget you drew up over the last couple of weeks spelled things out pretty clearly."

"I'm almost finished with the project plans. I'd like you to take a look at the human resource management plan after I review with you what we've done to date."

"I don't understand why you're wasting all this time planning. We all know what the objectives are."

"Dirk," you reply, "if we put the right amount of effort and time into planning, the actual work of the project should go pretty smoothly. Planning is probably one of the most important things we can do on this project. If we don't plan correctly, we might miss something very important that could delay the store opening. That date is pretty firm, I thought."

"Yes, the date is firm. But I don't see how we could miss anything. You and I have met several times, and I know you've met with Jill and Jake. They're the other key players on this."

"Let me finish," you reply. "I have met with all the key stakeholders, and after you review these last few documents I have for you, we're finished with the planning phase of this project. Ricardo drafted a procurement statement of work. He needs to hire some external resources—and I noted that in the human resource management plan by the way—to help run Ethernet cable, purchase some routers and switches, and set up the equipment. The purchase of the routers and switches will be accomplished using a fixed-price contract, and the human resources he needs will be procured using a time and materials contract. Jill also drew up a procurement statement of work for the gourmet and cookware lines purchase. I documented a lot of the resource management plan when we worked through the activity estimating and duration exercise and put some finishing touches on it yesterday. It includes an RACI chart for all our project key deliverables."

You sneak in a quick breath because you don't want Dirk to interrupt. "And the quality management plan is complete. After your review, I'll distribute it to the key stakeholders. The quality management plan describes how we'll implement our quality policy. You know the old saying, 'Do it right the first time.' I took the time to write down the specific quality metrics we're looking for, including the lease signing date (this one must start and finish on time) and the IT equipment specifications, and Jill has documented the gourmet products and the cookware line specifications. I also want to let you know that work on the website has already started. I briefed you a couple of weeks ago on the agile methodology we are using to manage that project. The user stories are documented and the team has started developing code. They will be using a two-week timebox to deliver incremental functionality that Jill will review at the end of each iteration."

Dirk looks impressed, but you can't tell for sure.

"After I look at the last of these planning documents, are we ready to actually start working?"

"Yes, as soon as the lease is signed. I anticipate that happening by the end of this week. Tomorrow's tour is the fifth location Jill will look at. She's very happy with the third property she saw but wants to look at this last property before she makes her final decision."

"Great. Let's take a look at that human plan or whatever it is."

After reviewing the last of the planning documents, Dirk signs off on the project plan.

Project Case Study Checklist

Project management plan completed including these elements:

- Procurement

- Fixed-price contract

- Time and materials contract

- Procurement documents prepared

- Procurement statement of work prepared

- Plan Resource Management

- Roles and responsibilities documented

- RACI chart developed

- Organizational chart developed

- Quality Management Plan and Quality Metrics

- Lease signing start and end dates

- IT equipment specifications

- Gourmet products—availability rates and defects

- Cookware products—availability rates

- Organizational chart developed

- Obtain approval and sign-off on project plan

Understanding How This Applies to Your Next Project

In my organization, the Plan Procurement Management process comes right after finalizing project scope because it takes a great deal of time and effort to procure goods and services. That means we have to start procuring resources as early in the project as possible in order to meet the project deadlines.

In all the organizations I've worked in, someone has always been responsible for procurement—whether it was a single person or an entire department. Typically, the procurement department defines many elements of the procurement management plan. Sure, the project team determines how many vendors need to be involved and how they'll be managed along with the schedule dates, but many other elements are predetermined, such as the type of contract to use, the authority of the project team regarding the contract, how multiple vendors will be managed, and the identification of prequalified sellers.

The procurement department also determines what type of procurement document you should use depending on the types of resources you're acquiring and the amount of money you're spending. Typically, they'll have a template for you to use with all the legalese sections prepopulated, and you'll work on the sections that describe the work or resources you need for the project, milestones or schedule dates, and evaluation criteria.

Don't make the mistake of thinking your procurement department will take care of all the paperwork for you. At a minimum, you will likely be responsible for writing the statement of work, writing the RFP, writing the contract requirements (as they pertain to the work of the project), creating the vendor selection criteria, and determining the schedule dates for contract work.

Plan Resource Management is a process you might not need to complete, depending on the size and complexity of the project. I typically work with the same team members over and over again, so I know their skills, capabilities, and availability. However, if you're hiring contract resources for the project or you typically work with new team members on each project, I recommend creating a resource management plan.

The quality management plan is another important element of your project management plan. You should take into consideration the final result or product of the project and the complexity of the project to determine whether you need a multi-page document with detailed specifications or if the plan can be more informal and broad in nature. Again, depending on the project complexity, the measurements or criteria you'll use to determine the quality objective could be a few simple sentences or bullet items or a more formal, detailed document. The quality baseline should be documented during this process as well.

Choosing a development methodology requires examining several factors. You need to know and understand the culture of your organization and your team members because this will influence whether the methodology will be successful. I use a hybrid methodology most often. My organization has a formal project intake process that requires some basic requirements-gathering, analysis, and vetting in front of our steering committee. After the project is approved to move forward, we undergo a formal requirements-gathering session, document the requirements, and obtain sign-off. After requirements are completed, we switch to an agile approach and requirements are converted into user stories, development is completed in iterations, and reviews and retrospectives are held. No matter which methodology we choose, all projects have a project manager. Our project managers are trained in agile methodologies so that they can move between the predictive and adaptive methodologies without hesitation.

Summary

This chapter's focus was on planning for project resources. Several aspects are involved in these planning activities, including procuring goods and services, planning human resources, and defining the activities in which human resources will be involved.

This chapter started with the Plan Procurement Management process. This process identifies the goods or services you're going to purchase from outside the organization and determines which goods or services the project team can make or perform. This involves tools and techniques such as make-or-buy decisions, expert judgment, source selection analysis, and contract types. The procurement management plan is one of the outputs of this process and describes how procurement services will be managed throughout the project. The procurement SOW (another output of this process) describes the work that will be contracted.

In our discussion of contract types, we covered fixed-price, cost plus, and time and materials contracts and the benefits and risks of using them. We also discussed contract types that work well in an agile environment, including multitiered structure, value-driven structure, fixed-price increments, not-to-exceed time and materials, graduated time and materials, early cancellation, dynamic scope, team augmentation, and favor full-service suppliers. Procurements in an agile environment require close attention to timelines so that the purchases line up with the agile iterations.

The Plan Resource Management process identifies and assigns roles and responsibilities and reporting relationships. Many times the roles and responsibilities assignments are depicted in a responsibility assignment matrix (RAM) or an RACI chart. The resource management plan describes how resources should be defined, staffed, managed and controlled, and released from the project when their activities are complete. Agile teams are self-directed and self-managed, and they focus on delivering features to the customers frequently and quickly. This is known as flow of value.

Plan Quality Management targets the quality standards that are relevant to your project. The quality management plan outlines how the project team will enact the quality policy. Quality in an agile project is built in due to the frequent deliveries, reviews, and retrospectives. Addressing quality issues is accomplished faster and more cheaply using an adaptive approach. You need to take into account the cost of quality when considering stakeholder needs. Four men led to the rise of the cost of quality theories. Crosby is known for his zero defects theory, Juran for the fitness for use theory, Deming for attributing 85 percent of cost of quality to the management team, and Shewhart for the Plan-Do-Check-Act cycle. The Kaizen approach says that the project team should continuously be on the lookout for ways to improve the process and that people should be improved first and then the quality of the products or services. TQM and Six Sigma are examples of continuous improvement techniques.

Cost of quality involves three types of costs: prevention, appraisal, and failure costs; the latter is also known as the cost of poor quality. Failure costs include the costs of both internal and external failures.

Cost–benefit analysis considers trade-offs in the Plan Quality Management process. Benchmarking compares previous similar activities to the current project activities to provide a standard to measure performance against. Design of experiments is an analytical technique that determines what variables have the greatest effect on the project outcomes. This technique equips you with a statistical framework, allowing you to change all the important variables at once instead of changing one variable at a time.

According to the *PMBOK® Guide*, there are three main categories to consider when choosing a methodology to manage the work of the project: the culture of the organization, the project team, and the project itself. In addition, you should also consider needs, complexity, and magnitude. The Stacey Complexity Model plots complexity elements to determine a methodology for the project.

Exam Essentials

Be able to describe the purpose of the Plan Procurement Management process.   The purpose of the Plan Procurement Management process is to identify which project needs should be obtained from outside the organization. Make-or-buy analysis is used as a tool and technique to help determine this.

Be familiar with the contract types and their usage.   Contract types are a tool and technique of the Plan Procurement Management process and include fixed-price and cost-reimbursable contracts. Use fixed-price contracts for well-defined projects with a high value to the company, and use cost-reimbursable contracts for projects with uncertainty and large investments early in the project life. The three types of fixed-price contracts are FFP, FPIF, and FP-EPA. The four types of cost-reimbursable contracts are CPFF, CPIF, CPF (or CPPC), and CPAF. Time and materials contracts are a cross between fixed-price and cost-reimbursable contracts.

Be familiar with the contract types that help foster a shared-risk-reward relationship.   Contract types that work for an agile project help foster a shared-risk-reward relationship between the vendor and the buyer. They include multitiered structure, which entails breaking the contract up into individual documents. Value-driven structure emphasizes having value-driven deliverables rather than meeting a milestone. Fixed-price increments breaks the work down into user stories or releases, and pricing is based on this. Not-to-exceed time and materials work on a time and material basis but have an overall contract limit that can't be exceeded. Graduated time and materials rewards the vendor with a higher hourly rate when they exceed contract dates, or reduce the hourly rate when delivery is late. Early cancellation allows the buyer to cancel when further work is no longer needed and typically includes a cancellation fee that is paid to the vendor. Dynamic scope specifies times or points in the project where changes can be made. Team augmentation involves hiring teams of people with the skills needed to work on the project. Favor full-service suppliers involves hiring more than one vendor to work on the project or supply resources to the project.

Be able to name two of the key elements of procurements on an agile project.   Agile projects require close attention to timelines so that purchases are lined up with the agile iterations. Make the vendor resources part of the agile team to help mitigate timing and delivery issues.

Be able to name the purpose of the Plan Resource Management process.   Plan Resource Management involves documenting both human and physical resources needed for the project. Regarding human resources, it documents roles and responsibilities and reporting relationships for the project, and creates the resource plan, which describes how team members are acquired and the criteria for their release. Physical resource needs and time frames are documented here as well.

Be able to describe an agile team and describe an approach that may help them with training and support.   Agile teams are small, cross-functional teams that range from three to nine members. Teams are organized according to strengths, and they are self-organized and self-managed. Agile teams require strong collaboration to help increase productivity, improve creative problem solving, increase knowledge sharing, enable integration of work tasks, and enable flexible work assignments. Establishing agile centers of excellence will help team members with training and support in agile practices.

Be able to list the benefits of meeting quality requirements.   The benefits of meeting quality requirements include increased stakeholder satisfaction, lower costs, higher productivity, and less rework and are discovered during the Plan Quality Management process.

Be able to define the cost of quality.   The COQ is the total cost to produce the product or service of the project according to the quality standards. These costs include all the work necessary to meet the product requirements for quality. The three costs associated with cost of quality are prevention, appraisal, and failure costs (also known as cost of poor quality).

Be able to name four people associated with COQ and some of the techniques they helped establish.   The four are Crosby, Juran, Deming, and Shewhart. Some of the techniques they helped to establish are TQM, Six Sigma, Lean Six Sigma, cost of quality, and continuous improvement. The Kaizen approach concerns continuous improvement and says people should be improved first.

Be able to describe why quality is built into an agile project.   Retrospectives occur at the end of each completed unit of work. The customer examines the work frequently, and inconsistencies and quality issues are identified early on when changes can be made at a lower cost, rather than at the end of the project.

Be able to name six elemental categories to consider when choosing a life cycle methodology for the project.   The six categories are culture, project team, the project itself, needs, complexity, and magnitude.

Be able to describe complexity.   Complexity may involve the skills and ability of the project team, the experience of the project team, unique deliverables or requirements, risk, due dates, costs, business urgency, constraints, and the power and influence of the stakeholders.

Review Questions

You can find the answers to the review questions in Appendix A. Be sure to download the Bonus Exams and Bonus Questions so that you'll have a broader exposure and more experience answering questions related to the topics in this chapter.

1. You are the project manager for an upcoming outdoor concert event. You're working on the procurement management plan for the computer software program that will control the lighting and screen projections during the concert. You're comparing the cost of purchasing a software product to the cost of your company programmers writing a custom software program. You are engaged in which of the following?

   A. Procurement planning

   B. Using expert judgment

   C. Creating the procurement management plan

   D. Make-or-buy analysis

2. You are the project manager for an outdoor concert event scheduled for one year from today. You're working on the procurement documents for the computer software program that will control the lighting and screen projections during the concert. You've decided to contract with a professional services company that specializes in writing custom software programs. You want to minimize the risk to the organization, so you'll opt for which contract type?

   A. FPIF

   B. CPFF

   C. FFP

   D. CPIF

3. You are the project manager for the Heart of Texas casual clothing company. Your company is introducing a new line of clothing called Black Sheep Ranch Wear. You will outsource the production of this clothing line to a vendor. The vendor has requested a procurement SOW. Which of the following options are true regarding the SOW?

   A. The procurement SOW contains a description of the new clothing line.

   B. As the purchaser, you are required to write the procurement SOW.

   C. The procurement SOW contains the objectives of the project.

   D. The vendor requires a procurement SOW to determine whether it can produce the clothing line given the detailed specifications of this product.

   E. Options A, C, D

   F. All of the above

4. You are the project manager for the Heart of Texas casual clothing company. Your company is introducing a new line of clothing called Black Sheep Ranch Wear. You will outsource the production of this clothing line to a vendor. You want to use a hybrid approach to manage this project. Which of the following contract types will you use?

    A. Dynamic scope

    B. Value-driven

    C. Multitiered

    D. Master service agreement

    E. Graduated time and materials

5. Which of the following statements are true regarding Plan Resource Management?

    A. You must closely coordinate the Plan Communications Management process with this process because the organization structure affects the way communications are carried out among project participants.

    B. The resource management plan created in this process describes how and when resources will be acquired and released.

    C. A RAM (or RACI chart) is produced in this process, and it allows you to see all the people assigned to an activity and their various responsibilities.

    D. Agile teams create and write a team charter that outlines the team values, ground rules, working agreements, and group norms.

    E. B, C, D

    F. A, B, C, D

6. Sally is a project manager working on a project that will require a specially engineered machine. Only three manufacturers can make the machine to the specifications Sally needs. The price of this machine is particularly critical to this project. The budget is limited, and there's no chance of securing additional funds if the bids for the machine come in higher than budgeted. She's developing the source selection criteria for the bidders' responses and knows all of the following are true except for which one? (Choose three.)

    A. Sally will use understanding of need and warranties as two of the criteria for evaluation.

    B. Sally will review the project management plan, requirements documents, and risk register as some of the inputs to this process.

    C. Sally will base the source selection criteria on price alone because the budget is a constraint.

    D. Sally will document an SOW, the desired form of response, and any required contractual provisions in the RFP.

    E. Sally will use a sole source evaluation method.

7. Which of the following are constraints that you should consider when documenting the resource management plan?

    A. Organizational structure

    B. Collective bargaining agreements

    C. Economic conditions

    D. Location and logistics

    E. Options B, C, D

    F. Options A, B, C, D

8. You have been hired as a contract project manager for Grapevine Vineyards. Grapevine wants you to design an Internet wine club for its customers. Customers must register before being allowed to order wine over the Internet so that legal age can be established. You know this project will require new hardware and an update to some existing infrastructure. You will have to hire an expert to help with the infrastructure assessment and upgrades. You also know that the module to verify registration must be written and tested using data from Grapevine's existing database. This new module cannot be tested until the data from the existing system is loaded. You are going to hire a vendor to perform the programming and testing tasks for this module to help speed up the project schedule. The vendor will be reimbursed for all their costs, and you want to use a contract type that will allow you to give the vendor a little something extra if you are satisfied with the work they do. You know all of the following apply in this situation except for which one?

   A. Contract type is determined by the risk shared between the buyer and seller.

   B. You'll use an FPIF contract for the programming vendor.

   C. Fixed-price contracts can include incentives for meeting or exceeding performance criteria.

   D. Each procurement item needs an SOW.

9. You are the project manager for BB Tops, a nationwide toy store chain. Your new project involves creating a prototype display at several stores across the country. You are using an RACI chart to display individuals and activities. What does RACI stand for? (Choose two.)

   A. Responsible, accountable

   B. Responsible, assigned

   C. Control, identify

   D. Consult, inform

10. Which process has the greatest ability to directly influence the project schedule?

    A. Plan Resource Management

    B. Plan Procurement Management

    C. Plan Communications Management

    D. Plan Quality Management

11. You are the project manager for BB Tops, a nationwide toy store chain. Your new project involves creating a prototype display at several stores across the country. You are using an agile methodology to manage your project and will be hiring contract services to help with this project. It's critical to the organization that this project succeed. You know which of the following are true?

    A. You could use a fixed price increment contract to foster a shared-risk-reward relationship.

    B. The key is to foster a collaborative approach that shares the risks and rewards with the vendor and the organization. This is known as a shared-risk-reward relationship.

    C. Project failures are due to a breakdown in the customer-vendor relationship, so you know it's important to build a strong relationship with them and use a collaborative approach.

    D. You could use a value-driven structure so that the vendors receive payments at specific milestones.

    E. A, B, C, D

    F. B, C

    G. A, B, C

12. Which of the following are true regarding planning for quality?

    A. Quality is built into an agile project because of frequent reviews and retrospectives.

    B. The Plan Quality Management is one of the key processes performed during the Planning process group and during the development of the project management plan.

    C. Changes to the product as a result of meeting quality standards might require cost or schedule adjustments.

    D. Agile projects release features that are fit for use at a minimum once every four weeks.

    E. A, B, C

    F. A, B, C, D

13. Four people are responsible for establishing cost of quality theories. Crosby is one and Juran is another. Which of the following describes their theories? (Choose two.)

    A. Grades of quality

    B. Total quality management

    C. Zero defects

    D. Fitness for use

14. The theory that 85 percent of the cost of quality is a management problem is attributed to _____.

    A. Deming

    B. Shewhart

    C. Juran

    D. Crosby

15. Which of the following are benefits of meeting quality requirements?

    A. An increase in stakeholder satisfaction

    B. Less rework

    C. Lower risk

    D. Higher productivity

E. Lower costs

F. A. B, D, E

G. A, B, C, D, E

16. Which of the following describes the cost of quality associated with scrapping, rework, and downtime?

   A. Internal failure costs

   B. External failure costs

   C. Prevention costs

   D. Appraisal costs

17. The quality management plan documents how the project team will implement the quality policy. It should address at least which of the following? (Choose three.)

   A. The resources needed to carry out the quality management plan

   B. The quality metrics and tolerances and what will be measured

   C. The responsibilities the project team has in implementing quality

   D. The processes to use to satisfy quality requirements

18. You work for a furniture manufacturer. Your project is going to design and produce a new office chair. The chair will have the ability to function as a regular chair and also the ability to move its occupant into an upright, kneeling position. The design team is trying to determine the combination of comfort and ease of transformation to the new position that will give the chair the best characteristics while keeping the costs reasonable. Several different combinations have been tested. This is an example of which of the following techniques?

   A. Benchmarking

   B. Quality metrics

   C. COQ

   D. DOE

19. Which of the following best characterizes Six Sigma as it relates to quality?

   A. Stipulates that quality must be managed in

   B. Focuses on process improvement and variation reduction by using a measurement-based strategy

   C. Asserts that quality must be a continuous way of doing business

   D. Focuses on improving the quality of the people first, then improving the quality of the process or project

20. Your organization is embarking on a long-term project that will require additional human resources on a contract basis to complete the work of the project. Since the project will span several years, you know one vendor probably can't supply all the resources you'll need over the course of the contract. However, you want to work with only one vendor throughout the project to minimize the number of procurement documents you'll have to produce. So, you'll specify in your procurement documents that contractors will have to form partnerships to work on this project. You know all of the following are true regarding this question except for which one?

A. You will use an RFP, which is part of the procurement documents output of the Plan Procurement Management process, as your procurement document for this project.

B. You'll use an FP-EPA contract because this project spans several years.

C. You should consider teaming agreements, which are an input to the Plan Procurement Management process.

D. Some of the quality metrics you'll use for this project include on-time performance, failure rates, budget control, and test coverage. Quality metrics are an output of the Plan Quality Management process.

Chapter

9

Developing the Project Team

THE PMP® EXAM CONTENT FROM THE PEOPLE DOMAIN COVERED IN THIS CHAPTER INCLUDES THE FOLLOWING:

✓ **Task 1.2 Lead a team**

- 1.2.1 Set a clear vision and mission
- 1.2.2 Support diversity and inclusion
- 1.2.3 Value servant leadership
- 1.2.4 Determine an appropriate leadership style
- 1.2.5 Inspire, motivate, and influence team members/ stakeholders
- 1.2.6 Analyze team members and stakeholders' influence
- 1.2.7 Distinguish various options to lead various team members and stakeholders

✓ **Task 1.3 Support team performance**

- 1.3.1 Appraise team member performance against key performance indicators
- 1.3.2 Support and recognize team member growth and development
- 1.3.3 Determine appropriate feedback approach
- 1.3.4 Verify performance improvements

✓ **Task 1.4 Empower team members and stakeholders**

- 1.4.1 Organize around team strengths
- 1.4.2 Support team task accountability
- 1.4.3 Evaluate demonstration of task accountability
- 1.4.4 Determine and bestow level(s) of decision-making authority

✓ **Task 1.5 Ensure team members/stakeholders are adequately trained**

- 1.5.1 Determine required competencies and elements of training

- 1.5.2 Determine training options based on training needs

- 1.5.3 Allocate resources for training

- 1.5.4 Measure training outcomes

✓ **Task 1.6 Build a team**

- 1.6.1 Appraise stakeholder skills

- 1.6.2 Deduce project resource requirements

- 1.6.3 Continuously assess and refresh team skills to meet project needs

- 1.6.4 Maintain team and knowledge transfer

✓ **Task 1.7 Address and remove impediments, obstacles, and blockers for the team**

- 1.7.1 Determine critical impediments, obstacles, and blockers for the team

- 1.7.2 Prioritize critical impediments, obstacles, and blockers for the team

- 1.7.3 Use network to implement solutions to remove impediments, obstacles, and blockers for the team

- 1.7.4 Re-assess continually to ensure impediments, obstacles, and blockers for the team are being addressed

✓ **Task 1.11 Engage and support virtual teams**

- 1.11.1 Examine virtual team member needs

- 1.11.2 Investigate alternatives (e.g., communication tools, colocation) for virtual team member engagement

- 1.11.3 Implement options for virtual team member engagement

- 1.11.4 Continually evaluate effectiveness of virtual team member engagement

✓ **Task 1.14 Promote team performance through the application of emotional intelligence**

- 1.14.1 Assess behavior through the use of personality indicators

- 1.14.2 Analyze personality indicators and adjust to the emotional needs of key project stakeholders

THE PMP® EXAM CONTENT FROM THE PROCESS DOMAIN COVERED IN THIS CHAPTER INCLUDES THE FOLLOWING:

✓ **Task 2.1 Execute project with urgency required to deliver business value**

- 2.1.1 Assess opportunities to deliver value incrementally

- 2.1.3 Support the team to subdivide project tasks as necessary to find the minimum viable product

✓ **Task 2.13 Determine appropriate project methodology/methods and practices**

- 2.13.2 Recommend project execution strategy

- 2.13.4 Recommend a project methodology/ approach

- 2.13.5 Use iterative, incremental practices throughout the project life cycle

This chapter begins the project Executing process group. I'll cover four of the processes in this chapter: Direct and Manage Project Work, Acquire Resources, Develop Team, and Manage Team. I'll cover the remaining Executing processes in the next chapter.

Direct and Manage Project Work is the action process. This is where you'll put the plans into action and begin working on the project activities. Execution also involves keeping the project in line with the original project management plan and bringing wayward activities back into alignment.

The Acquire, Develop, and Manage Team processes are all interrelated, as you can imagine, and work together to help obtain the best project team available. People are people whether they are working on a predictive project or an adaptive project. There are some specific topics for agile teams that we'll cover, but most of the information in this section relates to all teams.

Several things happen during the Executing processes. The majority of the project budget will be spent during this process group, and often the majority of the project time is expended here as well. The greatest conflicts you'll see during the project Executing processes are schedule conflicts. In addition, the product description will be finalized here and contain more detail than it did in the Planning processes. Are you ready to dive into Executing? Let's go.

The process names, inputs, tools and techniques, outputs, and descriptions of the project management process groups and related materials are based on content from *A Guide to the Project Management Body of Knowledge (PMBOK® Guide), Sixth Edition* (PMI®, 2017). The references to adaptive and hybrid methodologies, related materials, and figures in this chapter are based on content from the *Agile Practice Guide* (PMI®, 2017).

Directing and Managing Project Work

The *PMBOK® Guide* emphasizes process and provides a means for you to organize work and logically think through a progression of steps to manage the project. You'll need to have some process knowledge for the exam. However, the biggest job any project manager has is interacting with and managing people. Stakeholders are people, executives are people, team members are people, vendors are people. Personalities, cultures, and attitudes will influence your project to a much greater degree than any process ever will. You can have a perfect schedule and clearly defined scope, but if you don't have team members, stakeholders, and vendors who work collaboratively, believe in the goals of the project, avoid strife, and communicate well, you won't likely have a successful project. The exam will focus on the interactions of process and people. You'll need to have some process knowledge for the exam, and we've spent a good deal of time in this book discussing process. Keep in mind that processes can't be performed without people, and you'll need to understand the dynamics of teams, personalities, interactions, and the application of emotional intelligence in assessing teams for the exam. I'll cover all of these topics in this chapter.

The purpose of the *Direct and Manage Project Work* process is to carry out the project management plan and perform the work of the project. This is where your project comes to life and the work of the project happens. The work is authorized to begin and activities are performed. Resources are committed and people carry out their assigned activities to create the product, result, or service of the project. Funds are spent to accomplish project objectives. Performing project activities, training, selecting sellers, collecting project data, utilizing resources, and so on are all integrated with or are part of this process.

Direct and Manage Project Work is where the rubber meets the road. If you've done a good job planning the project, things should go relatively smoothly for you during this process. The deliverables and requirements are agreed on, the resources have been identified and are ready to go, and the stakeholders know exactly where you're headed because you had them review, agree to, and approve the project plan.

Some project managers think this is the time for them to kick back and put their feet up. After all, the project management plan is done, everyone knows what to do and what's expected of them, and the work of the project should almost carry itself out because your plan is a work of genius, right? Wrong! You must stay involved. Your job now is a matter of overseeing the actual work, producing deliverables, communicating, issuing change requests, implementing approved changes, managing the schedule, managing risks, managing stakeholders, staying on top of issues and problems, and keeping the work lined up with the project management plan. Direct and Manage Project Work will be performed from now throughout the remainder of the project.

Exam Spotlight

The project management plan serves as the project baseline. During the Executing processes, you should continually compare and monitor project performance against the baseline so that corrective actions can be taken and implemented at the right time to prevent disaster. This information will also be fed into the Monitoring and Controlling processes for further analysis.

One of the most difficult aspects of this process is coordinating and integrating all the elements of the project. Although you do have the project management plan as your guide, you still have a lot of balls in the air. You'll find yourself coordinating and monitoring many project elements, occasionally all at the same time, during the course of the Direct and Manage Project Work process. You might be negotiating for team members at the same time you're negotiating with vendors at the same time you're working with another manager to get a project component completed so that your deliverables stay on schedule. You should monitor risks and risk triggers closely. The Plan Procurement Management process might need intervention or cause you delays. The organizational, technical, and interpersonal interfaces might require intense coordination and oversight. Of course, you should always be concerned about the pulse of your stakeholders. Are they actively involved in the project? Are they throwing up roadblocks now that the work has started?

According to the *PMBOK® Guide*, this process also requires implementing corrective actions to bring the work of the project back into alignment with the project management plan, preventive actions to reduce the probability of negative consequences, and defect repairs to correct product defects discovered during the quality processes.

As you can see, your work as project manager is not done yet. Many elements of the project require your attention, so let's get to work.

Later in this chapter I'll also talk about Develop Team because this is an integral part of the Direct and Manage Project Work process. You'll want to lead and develop the project team and monitor their performance, the status of their work, and their interactions with you and other team members as you execute the project management plan in order to achieve the project deliverables.

Direct and Manage Project Work Inputs

The Direct and Manage Project Work process has five inputs: the project management plan, project documents, approved change requests, enterprise environmental factors, and organizational process assets.

The project management plan documents the collection of outputs of the Planning processes and describes and defines how the project should be executed, monitored, controlled, and closed. The project management plan documents the goals of the project and the actual work you and the team will execute in order to meet those goals. Once the project management plan is complete, you will know all the work and actions needed to meet the deliverables of the project and will have a plan for executing those actions. Most components of the project management come into play in this process.

The project documents listed as inputs to this process will assist in determining when resources are needed, milestone due dates, what changes may impact the work already completed or soon to be performed, and the project communications you'll be reporting on during this process. Let's take a brief look at the important elements of the other inputs next.

Approved Change Requests

Approved change requests come about as a result of the approved change requests output of the Perform Integrated Change Control process. (We'll cover this process in Chapter 11, "Measuring and Controlling Project Performance.") Change requests are approved or denied during this process, and once the decision has been made, approved changes come back through the Direct and Manage Project Work process for the project team to implement.

Approved changes might either expand or reduce project scope and may also cause revisions to project budgets, schedules, procedures, the project management plan, and so on. Change requests can be internal or external to the project or organization. For example, you may need to make changes because of a new law that affects your project. Keep in mind that changes on an agile project come about and are managed during the iteration. The change control process is integrated within the agile process so that the approved change requests happen at the same time the work, and the change, are occurring.

Organizational Culture

When performing this process, you will need to consider several enterprise environmental factors, including the organizational structure, the facilities available to the project team, risk tolerance levels of your stakeholders, and management structures. Organizational culture will influence the methodology you'll use to manage the project. Those organizations that are open to new ideas, willing to try new things, and support honesty and transparency will likely have good success with adaptive methodologies.

Peter Drucker is noted for the saying, "Culture eats strategy for breakfast." Strategy is typically viewed over a three- to five-year time horizon. It outlines who the organization is now and where the organization wants to be over the coming years. Culture, on the other hand, is almost tangible. It's the feeling, or the energy you experience working in the company. It's also the attitudes and actions of management toward staff, and staff's reactions to those actions. Does management trust their staff to do the work, or do they manage every detail? Do they take credit for work and ideas staff members produce, or do they give credit where credit is due? Culture has significant influence on the organization. For example, if your organization's culture is that "anything is possible, we can make it work, we're in this together," and your management team trusts staff members to run with the ball after it's been handed off, they probably can make anything work. Several years ago, I worked for a young start-up company that I used to joke only employed "A students." In other words, they had a knack for hiring only the best and the brightest with strong work ethics and a strong desire for success. The employees were passionate about the company vision. They viewed their contributions as meaningful, they worked their hearts out, the management team recognized and rewarded their hard work, and the company itself had success after success because of this culture. There was a natural symbiosis between the attitudes and actions of employees (including management) and the outcomes the organization experienced. I've also had the unfortunate experience of working in the opposite culture. Management second-guessed your every move, no one was to be trusted, staff members were not allowed to speak up on any topic, taking risks was forbidden, and anyone who dared question a management member didn't stay employed long. This meant employees had little passion for their work or their organization, and they were apathetic about their jobs. Needless to say, the organization itself didn't accomplish much and had issue after issue with its products and services. However, if organizations with this type of culture would practice being more trusting and transparent, establish and use KPI type measurements, and allow teams to experiment on noncritical projects and learn from failure, they could transform the culture and the methodologies they use to manage projects. The key takeaway about organizational culture is that it can limit the options and the ability to choose the methodology that suits the nature of the project. Agile, predictive, and hybrid can all work in either of the cultures I described. Obviously, an open, trusting culture is more likely to produce a successful project, no matter what development methodology you use.

Exam Spotlight

The organization's culture will always influence the life cycle methodology you'll use to manage the projects. If you are using an agile approach, culture will also influence which agile methodology you'll use.

Culture may present some interesting dilemmas. Perhaps your organization is fast paced and quick to market but they want to ensure quality in their products. Or maybe your organization serves the public, and ensuring the most good for the most people in a predictable fashion defines your culture. Choosing the right agile methodology to match these differing needs should be considered by assessing the organizational culture. The *Agile Practice Guide* (PMI®, 2017) emphasizes that an assessment should be performed by the project manager to determine culture. You should interview the stakeholders, team members, and executive management to determine organizational tendencies. Another tool you could use is shown in Figure 6.2 in the *Agile Practice Guide* (PMI®, 2017). It weighs several factors on a sliding scale, such as quantity versus quality, flexibility versus predictability, and more. Organizations that welcome experimenting with new processes, are interested in speed of delivery, and are flexible and transparent are ideal for agile projects.

After the interview and/or analysis process, the project manager plots where the organization falls between each of these factors on a sliding scale. If the organization tends more toward execution than exploration, the scale is weighted toward execution, and so on. The key is to spend the time understanding the culture, regardless of the tool you use. Understanding the culture will help you to choose the correct agile methodology to use for the project, to examine and determine the trade-offs, and to choose the right techniques for your organization.

Historical Information and Measurement Databases

As with many of the other processes we've covered, historical information from past projects, organizational guidelines, and work processes are some of the organizational process assets you should consider when performing this process. In addition, measurement databases and issue and defect databases can be used in this process to compare past projects to the current project and to capture information about the current project for future reference. You might be required to upload measurement data on the project management processes and those for the product of the project to your organization's performance measurement database for reporting purposes. Change control procedures and risk control procedures should be reviewed as you enter the Executing processes.

Project Management Information System

The tools and techniques of the Direct and Manage Project Work process are expert judgment, project management information system, and meetings. Your experts should have experience or knowledge of cost and budget management, procurement processes, regulations that impact your industry, and legal processes. The *project management information system (PMIS)* is an automated system used to document the project management plan and subsidiary plans, to facilitate the feedback process, and to revise the documents. It incorporates the configuration management system and the change control system, both of which I'll cover in Chapter 11. The PMIS can be used to control changes to any of the plans. When you're thinking about the PMIS as a tool and technique, think of it just that way—as a tool to facilitate the automation, collection, and distribution of data and to help monitor processes such as scheduling, resource leveling, budgeting, and web interfaces. A PMIS is typically an off-the-shelf software system that will manage all your project artifacts in one place and can automatically generate reports, track project performance, maintain and archive historical information for future reference, perform resource planning, and much more.

Some of the key tools and techniques you'll use in the Executing process group are consulting and meeting with stakeholders, professionals, and others; employing the project management methodology you developed in the Planning processes; and using the project management information system to update and track progress.

Deliverables and Work Performance Data

The outputs of the Direct and Manage Project Work process that you should know for the exam are deliverables, work performance data, issue log, and change requests. Project management plan updates, project document updates, and organizational process assets updates are other outputs of this process. Almost every process you've performed up to this point has defined and outlined what the work of the project entails and what the final results should look like. Now you're ready to begin performing the work and producing results. Let's look at the primary outputs that will help you document progress.

Deliverables

During Direct and Manage Project Work, you'll gather and record information regarding the outcomes of the work, including activity completion dates, milestone completions, the status of the deliverables, the quality of the deliverables, costs, schedule progress and updates, and so on. Deliverables aren't always tangible, but they are always unique and verifiable. For example, perhaps your team members require training on a piece of specialized equipment. Completion of the training is recorded as a work result. Capabilities required to perform a service that's described in the project management plan are also considered a deliverable. Capabilities may include any number of elements such as measurements, throughput, temperature, capacity, speed, timing, and much more.

Predictive methodologies typically produce deliverables toward the end of the project. We talked about this in previous chapters. Planning and Executing could be separated by months or years in a predictive project methodology. Deliverables in an adaptive approach are produced quickly, often, and in small units. Deliverables in an adaptive approach may also consist of minimally viable products where just enough features and functionality are produced with the least amount of effort possible so that the product owner and stakeholders can provide feedback to the team on what needs to be improved or changed and what works well.

Executing and Monitoring and Controlling are two process groups that work hand in hand. As you gather the information from work results, you'll measure the outputs and take corrective actions where necessary (performed during the Monitoring and Controlling processes). This means you'll loop back through the Executing processes to put the corrections into place. The *PMBOK® Guide* breaks these processes up for ease of explanation, but in practice, you'll work through several of the Executing and Monitoring and Controlling processes together. Keep in mind that adaptive methodologies incorporate Executing and Monitoring and Controlling processes within each iteration or release.

Work Performance Data

Work performance data concerns observing, gathering, documenting, and recording the status of project activities. The types of information you might gather during this process include some of the following:

- Key performance indicators
- Schedule status and progress
- Status of deliverable completion
- Progress and status of schedule activities, including start and end dates
- Percent of overall work complete
- Adherence to quality standards
- Number of change requests
- Status of costs (those authorized and costs incurred to date)

- Schedule activity completion estimates for those activities started
- Schedule activities percent complete
- Actual durations of activities that are completed
- Lessons learned
- Resource consumption and utilization
- Technical performance measures

Work performance data is a way of examining business value, and it becomes an input to a few of the Monitoring and Controlling processes where you'll perform further analysis on the data. It's important that you document this information so that when you get to the Monitoring and Controlling processes, you don't have to backtrack. Work performance data is useful for both predictive and adaptive projects. This data becomes historical information you can use to help improve future projects.

Issue Log

The issue log will come into play in the Executing processes in a big way. As your team performs the work of the project, issues will arise. I promise this isn't an "if" they come up, it's a "when" issues come up. You'll want to capture issues in an issue log so that you can track and manage them, report on them during status meetings, assign owners, and understand their status.

An issue log can be easily constructed using your PMIS or a good old handy spreadsheet. Table 9.1 shows a sample issue log with some of the information you might consider capturing.

TABLE 9.1   Sample issue log

Issue #	Issue type	Description	Priority	Issue reporter	Issue owner	Target date	Status	Solution
1	Schedule	Tester is out on unexpected leave	High	Eric	Tiffany	May 15	Open—researching temporary personnel to assist	

A *RAID log* is another place you might capture risks. RAID stands for risks, action items, and decisions. The idea is to capture all of these in one place, rather than having separate logs for risks, issues, and action items. The log may look similar to Table 9.1 with some changes to the headings and with an additional column to indicate whether it's a risk, issue, or action item. It's also handy to group items in each category together so that all the risks are in one section of the log, followed by action items and decisions.

Change Requests

As a result of working through activities and producing your product, service, or result, you will inevitably come upon things that need to be changed. Changes can also come about from stakeholder requests, external sources, technological advances, and so on. These change requests might encompass schedule, scope, requirement, or resource changes. The list could go on. Your job as project manager, if you choose to accept it, is to collect the *change requests* and make determinations about their impact on the project.

Exam Spotlight

According to the *PMBOK® Guide*, a change request is a formal proposal to bring about a change that will require revising a document, a project baseline, a deliverable, or some combination of all three.

Implementation of change requests may incorporate four types of action: corrective actions, preventive actions, defect repairs, or updates. Each of these topics is described next.

Corrective Actions   In my organization, a corrective action means an employee is in big trouble. Fortunately, this isn't what's meant here. *Corrective actions* are taken to get the anticipated future project outcomes to align with the project management plan. Maybe you've discovered that one of your programmers is adding an unplanned feature to the software project because they're friends with the user. The corrective action may entail redirecting them to the activities assigned originally (and have them stop working on the unplanned features) to avoid schedule delays. Perhaps your vendor isn't able to deliver the laboratory equipment needed for the next project phase. You'll want to exercise your contract options (let's hope there's a clause in the contract that says the vendor must provide rental equipment until they can deliver your order), put your contingency plan into place, and get the lab the equipment that's needed to keep the project on schedule.

Preventive Actions   A *preventive action* involves anything that will reduce the potential negative impacts of risk events should they occur. Contingency plans and risk responses are examples of preventive actions. I described these and other risk responses while talking about the Plan Risk Response process in Chapter 7, "Identifying Project Risks." You should be aware of contingency plans and risk responses so that you're ready to implement them at the first sign of trouble.

Defect Repairs   A defect occurs when a project component does not meet the requirements or specifications. Defects might be discovered when conducting quality audits in the Manage Quality process or when performing inspections during the Control

Quality process. For the exam, you should understand the difference between a validated defect repair and a defect repair. A validated defect repair is the result of a reinspection of the original defect repair. In other words, you found a problem with the product during the Quality processes, you corrected the problem (defect repair), and now you're reinspecting that repair (validated defect repair) to make certain the fix is accurate and correct and resolved the problem.

Updates   Changes may require updates to the project documents so that the new information is captured and recorded. For example, some of the documents that might require an update could include the scope statement, budget documents, the schedule, risk management plan or risk response plans, and so on.

I'll discuss change requests more in the coming chapters as well. Change requests are an output of several processes, including the Direct and Manage Project Work process and the Monitor Communications process in the Monitoring and Controlling process group. Remember that the Executing process group outputs and the Monitoring and Controlling process group outputs feed each other as inputs. In this particular case, approved change requests are an input, and change requests are an output of the same process.

Exam Spotlight

Direct and Manage Project Work is where the work of the project is performed and the project management plan is put into action and carried out. In this process, the project manager is like an orchestra conductor signaling the musicians when to begin their activities, monitoring what should be winding down, and keeping that smile going to remind everyone that they should be enjoying themselves. I recommend that you know the outputs of the Direct and Manage Project Work process for the exam.

Real World Scenario

We All Scream for Ice Cream

Heather is a pharmaceutical salesperson who is fed up with the rat race. She ran the numbers, decided to quit her day job, and bought an ice cream shop in a quaint tourist town. Having been involved in a few research and development projects, she understands the value of project management planning and using that plan as her guide to perform the work of the project.

Heather documented the deliverables needed to prepare for opening day in her scope statement. Some of those deliverables are as follows: remodel, develop staffing plan, procure equipment, and procure materials. Confident in her planning, Heather hired a contractor and began remodeling the shop. Then real life happened. The contractors discovered a water problem in the storage room. They installed a sump pump, which took care of the water, but discovered an even bigger problem when they moved the storage shelves. Mold was growing up the drywall. The drywall had to be removed, as did the insulation behind it, and the mold remaining on permanent fixtures had to be eliminated. Then new insulation and drywall had to be installed. The drywall had to be primed and painted. Because that portion of the storage room was getting a fresh coat of paint, Heather decided the contractors might as well paint the entire room.

All of these actions required another pass through the Planning processes. The schedule didn't require much modification because other work could be started while the water problem was being addressed, but the budget needed to be modified as a result of the additional work. To avoid more surprises, Heather requested that the contractor perform a thorough inspection of the property and determine whether there were any other hidden issues. Armed with the inspection report, Heather could knowledgeably plan corrective action for other items that needed to be addressed.

Directing Project Work on Agile Projects

We've discussed how the agile process works previously in this book. As a refresher, iteration-based agile uses timeboxes, or equal periods of time typically two to four weeks in length, to produce completed features. These features are reviewed at the end of the iteration and issues are identified and corrected as the work progresses. This allows for incremental improvements and experimentation to see what processes work best for the team and what features fulfill the business value the customer is expecting. Flow-based agile is based on the team's capacity to accept new work. Work is managed on a task board with different categories of work managed in columns on the board. Work is completed in stages as the team has capacity to accept the work, and some features may take more time than others to complete. These small units of work in progress make it easier for the team to identify issues early and minimize rework.

Agile project teams are self-directed and so is the work of the project. The product owner or customer works with the team to direct the work of the project. The steps described next outline a high-level view of the agile process. These steps can apply to both iteration-based and flow-based methodologies.

1. The initial product backlog is defined.

2. Planning meetings occur to determine which user stories to work on in the upcoming iteration. User stories are pulled into the backlog or onto the work columns on the task board.

3. Daily stand-ups are conducted by a team member. The project manager attends but should refrain from conducting these meetings.

4. Backlog refinement is conducted midway through the iteration for upcoming user stories to help the team understand the context of the user stories and how they relate to each other. (I'll discuss this in more detail in the next section.)

5. Review meetings are held for the customer to examine the features and provide feedback.

6. Retrospectives are held at the end of an iteration or unit of work, midway through the project, and/or at the end of the project to examine the process, determine what worked well (or not), and make improvements.

The real benefit of agile is that teams can identify and diagnose issues early and often. The continuous feedback from the customer, along with regular retrospectives, helps them to correct the issues and improve the process.

Exam Spotlight

According to the *Agile Practice Guide* (PMI®, 2017), the retrospective is the most important practice you can conduct. Retrospectives allow the team to learn what went well and what didn't go well, improve on the things that didn't go well, and modify and adapt the process to work best for the project and the team. This is not a blame-storming session. It's time to reflect on the work and focus on making improvements.

Remember that agile teams work through all of the project process groups within an iteration or a column on the task board. Work is planned at the beginning, work is performed (executed) when a card is pulled from the task board or the iteration begins, the features or products are examined, learning occurs through retrospectives, corrections and adaptations are made based on retrospectives and feedback, the work for this iteration ends, and then the next cycle begins again. This encompasses the full Initiating, Planning, Executing, Monitoring and Controlling, and Closing process group life cycle.

Backlog Refinement

Backlog refinement typically occurs in the middle of an iteration, or when the next card is pulled from the task board to start work. It's a meeting conducted between the product owner and the team to help them learn about upcoming stories for the next iteration or task, understand the context of the user stories, and identify how the user stories relate to one another and the project as a whole. The idea is for the product owner to explain their ideas to the team so that challenges or problems with the user stories can be discussed. This gives the team an opportunity to determine alternatives, refine the user stories, and break them down into small enough units of work that the team can continue to produce features and results during an iteration or as cards are worked on the task board. Teams should not spend more than one hour a week refining the user stories for the next period or category of work. The *Agile Practice Guide* (PMI®, 2017) points out that teams who need more than one hour of backlog refinement per week may not be skilled enough to evaluate and refine the user stories, or the product owner may be overpreparing for the meeting. Conversely, teams who are new to agile or aren't familiar with the business area or domain of the project may benefit from more frequent backlog refinement meetings until they assimilate more skills and knowledge.

A technique known as *impact mapping* can help the team see how the user stories fit into the overall project. The product owner typically conducts this exercise, although a servant leader may do so also. Impact mapping is a technique that allows the team to see the big picture and is similar to mind mapping. The impact map visualizes the who, what, why, and how of the project and you can define these categories for your project in any number of ways. A typical impact map starts with the project objective and defines the actors who have specific outcomes in mind; next, the impacts or expectations the actors are hoping for are mapped; and finally the deliverables that will produce the expectations are outlined. This is a collaborative technique that involves the agile team members and the product owner. Impact maps are a way to visualize assumptions, help reduce waste by preventing scope creep, and help ensure business value is realized.

Figure 9.1 depicts an impact map for a mileage tracker app for our organization. We have three actors: the end users, the accounting department, and the fleet manager. The fleet manager expects that this app will help them schedule maintenance for the vehicles more efficiently and will help decrease fuel costs (these are the expected impacts). To help in scheduling maintenance, they'd like to see the odometer readings at the end of every trip and they want an automatic notification sent when the vehicle is within 300 miles of its next scheduled maintenance.

FIGURE 9.1   Impact map

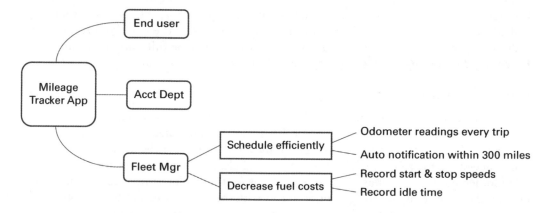

Tailoring the Approach

Agile approaches, just like project processes, should be tailored to meet the needs of the project, the organization, and the team. Agile, or even hybrid methodologies, can be a big change for an organization. Tailoring can help you introduce agile and determine the approach that will work best. Tailoring is a complex subject that could take up an entire book on its own. I'll highlight a few key points to keep in mind when tailoring agile methodologies in this section.

The first concept is that tailoring is a collaborative effort and should be performed by the team. Perhaps your team is new to agile and you are considering a hybrid approach. You'll want the people who will be performing the work to help in tailoring this approach. That way, you can ensure buy-in of the process and encourage participation and open feedback so that the team can continue to expand their skills while also tailoring the process to meet their needs.

I recommend starting with an established process such as Scrum. Follow the steps I outlined in the earlier section such as daily stand-ups, reviews, retrospectives, and so on. Allow the team enough time to learn the process and only tailor the approach after they've mastered it. Don't eliminate or tailor any of the steps until you and the team have a solid understanding of the process and what consequences could occur by not performing them. Under no circumstances should you ever eliminate retrospectives. This is the most important aspect of any agile process.

You may have projects with very large project teams performing the work. As you already know, agile teams typically consist of three to nine members, so large teams will not work if you want to use or tailor an agile approach. You could consider a couple of ways to deal with this scenario. You could break down the large project into smaller, individual projects with agile-sized teams. These projects could then be managed by using program management techniques. You could consider working on individual features and releasing them frequently. This would also require paring down the size of the team. Consider identifying the most critical team members needed to produce the features and assign them to small teams working on individual features. Tailoring large projects for an agile approach will likely require reducing the size of the teams. However, this will result in cost savings and reduce turnover. Small teams form camaraderie more easily than large teams and, because of these bonds, will likely commit to staying with the project and the team.

You may work for an industry that is heavily regulated. This could include any number of areas, from healthcare to transportation and more. Tailoring in this environment may involve a hybrid approach using multiple agile methodologies to improve communications while ensuring the compliance requirements are met. This often includes lots of documentation to prove compliance and that adequate controls are in place. In this case, the documentation is part of the deliverable completed by the end of the iteration.

Sometimes tailoring may be as simple as using language that your business units and executives understand. Break down the terminology in a way that they can understand and buy into the benefits of an agile approach. If they haven't been trained, they won't know what a "retrospective" entails. Instead, talk about how the team can quickly assimilate lessons learned and continually improve the process. Be mindful of the organizational culture here so that you are using terms that are professional and appropriate for your organization.

You can adapt the approaches we've discussed throughout this book (iteration-based, flow-based, hybrid, Lean, Extreme Programming, test-driven development, feature-driven development, and so on) by taking the best practices from a few of them and tailoring a process that works for you. Consider a small trial project to try out your newly tailored approach and refine it until the team is comfortable taking it to more complex projects.

Executing Practices for Delivering Project Work

Executing project work on an agile project has some of the same considerations as executing project work on a predictive project. On an agile project, executing the work is performed in smaller time frames with smaller deliverables. Predictive projects typically work on deliverables for an extended period of time and deliver them all at once toward the end of the project. The benefit of agile is the continuous work, feedback, and improvement or modification loop that's inherent in an agile project. The *Agile Practice Guide* (PMI®, 2017) outlines a few technical practices that may help agile teams deliver results faster, thereby maximizing the speed at which the team can work. As it notes, many of these practices come from Extreme Programming (XP). The practices include the following:

Continuous Integration   Extreme Programming is based on continuous integration. The idea is that as new features are delivered, the entire program is tested (with all the features delivered to date) to ensure everything is working. This testing happens every time new features are released, thereby incorporating the work into the whole program or product.

Testing at All Levels   Several types of testing can occur on agile-based software projects. *Smoke testing* is a high-level test designed to identify simple failures that could jeopardize the software program or prevent it from being released to production. These tests typically look at the most critical functions of the program and expose issues and problems early in the coding process.

Unit testing is performed on individual modules or units of source code. Earlier I gave an example of the Mileage Tracker app. A unit test in this case would involve only one of the modules, for example, ensuring that the start and stop speeds of the vehicle are recorded. None of the other functions are tested during this test (such as record idle time or record odometer readings). Each of these functions will be tested with independent unit tests.

Integration testing involves combining software modules and testing them as a group. This may also involve testing programs or modules that need to interact with one another. An integration test typically occurs after unit testing is completed. The integration test will group modules together that have been unit-tested previously and perform a series of tests to determine whether the modules are delivering the right results. In our Mileage Tracker app example, recording the odometer reading is needed so that the maintenance notification can be determined and sent. Grouping these two units of code together to test these functions is considered an integration test.

End-to-end testing is a system-level test that is just as it sounds. This test involves using the software from start to end (and everything in between) to ensure the application is working correctly. The end-to-end test for the Mileage Tracker app would involve logging in, performing the functions for each of the actors (end user, accounting department, and fleet manager), and following all the impacts through to ensure they deliver the expected features or functionality. This test typically occurs after integration testing and after all the code is written.

Regression testing is performed after changes are made to the code, when software configurations are modified, or after maintenance activities are performed on the hardware the code resides on. Regression testing ensures that the software works the same way it did prior to the change or maintenance activity. Regression tests help to maintain product quality and ensure agile teams can maintain cadence in producing deliverables.

Acceptance Test-Driven Development (ATDD)   Acceptance tests determine whether the customer is satisfied with the end product and are typically performed by the end user of the system. When using an agile approach, the team members (including the product owner) work together to determine the acceptance criteria for the system or work product. The agile team then creates tests to satisfy these criteria. After developing the tests, the agile team goes to work writing just enough code to perform the test and meet the criteria. Once the test is performed, the team receives feedback and makes modifications to the code and/or continues with the next unit of code.

Test-Driven Development (TDD)   Test-driven development is an XP agile methodology that provides significant feedback to the team. Each code release is tested thoroughly, iteratively, and repeatedly throughout the project to continuously validate the work. Often, the act of creating the tests will help the team to understand what the end product (or work component) is intended to produce or perform.

Spikes   *Spikes* are short periods of time for the team to research, experiment, and learn. Spikes should be timeboxed and not go on indefinitely. The work of the project must go on, after all. The *Agile Practice Guide* (PMI®, 2017) notes that spikes may be useful for developing acceptance criteria, estimating upcoming work, and learning about the anticipated actions and workflows a user may take through the system. For example, the fleet manager in our Mileage Tracker app expects to produce reports that will show mileage and odometer readings. A spike could be used by the team to walk through this workflow and understand how it should perform.

The Executing process group is about performing the work of the project, and in order to do that, you need resources. We'll look at two processes, Acquire Resources and Develop Team, in the next few sections.

Acquiring the Project Team and Project Resources

The *Acquire Resources* process involves acquiring and assigning human resources to the project and obtaining equipment, supplies, facilities, and other items you might need to perform the work of the project. This process is performed whenever you need to acquire new or additional resources, whether they come from inside or outside the organization.

Project staff might be hired specifically for the project from inside the company or brought in as contract help. In any case, it's your job as the project manager to ensure that resources are available and skilled in the project activities to which they're assigned. However, in practice, you might find that you don't always have control over the selection of team members. Someone else, the big boss for example, might handpick the folks they want working on the project. It's up to you to assess their skills and decide where they best fit on the project, and to determine their training needs and ensure that they receive the training.

> One thing is certain: you need to be sure that resources (both human and equipment or materials) are available and show up to the project according to schedule. If you don't have resources at the right time, you could impact the schedule or the budget, introduce risk, or put the project at risk of not finishing. This could be a career-limiting move, so take this process seriously. Work closely with your procurement department to ensure that resources with the right qualifications and/or specifications are available at the right time.

The Resource Management Plan

The resource management plan is one of the primary inputs of this process. This plan was created in the Plan Resource Management process. It describes the team's roles and responsibilities and includes project organization charts and other information on resource requirements for the project. This plan will be your guide in both acquiring and managing your project resources. Keep in mind that you may not have direct authority over the project resources due to matrix managed organizations, collective bargaining agreements, the use of contractors and subcontractors, and so on. You should document this in the resource management plan.

Resource calendars are used in this process because they describe the time frames in which the resources will be needed on the project and when the recruitment process should begin.

Enterprise environmental factors in this process account for project activities that might require special skills or knowledge in order to be completed. These may also consider personal interests, cost rates, prior experience, and availability of potential team members before making assignments. For example, consider the previous experience of the staff member you're thinking of assigning to a specific activity. Have they performed this function before? Do they have the experience necessary for the level of complexity this project activity requires? Are they competent and proficient at these skills?

Personal interests and personal characteristics play a big role as well. If the person you're thinking of just isn't interested in the project, that person isn't likely to perform at their best. Assign someone else in a case like this if you are able. Unfortunately, some people just don't play well with others. When you're assigning staff, if at all possible, don't put the only two people in the whole company who can't get along together on the same project. If a staff member you need has a skill no one else has, or they can perform a function like no one else can, you might not have a choice. In this case, you'll have to employ other techniques to keep the team cohesive and working well together despite the not-so-friendly relationship between the two staff members.

Here's one final consideration: check on the availability of key team members. If the team member you must have for the activity scheduled in February is on their honeymoon, you probably aren't going to win the toss.

Exam Spotlight

Remember that the availability, experience levels, interests, cost, and abilities of your resources are considered part of the enterprise environmental factors input. You should understand these inputs and their importance to the Acquire Resources process for the exam.

Recruitment practices are one example of the organizational process assets to watch for in this process. When hiring and assigning staff, make certain you're following the organization's recruitment procedures and processes. You should also note that organizational policies that dictate recruitment practices are constraints.

Techniques for Acquiring Resources

Acquiring resources involves using decision-making skills, negotiating for team members, understanding virtual teams, understanding dispersed teams, and working with teams that are new to agile projects. We'll look each of these next.

Multicriteria Decision Analysis   Selecting team members, just like selecting vendors, requires some type of criteria and analysis to determine whether the potential candidate is a fit. You could use any number of factors, including experience, education, skills, availability, knowledge, attitude, geographic location, and more, to rank and score candidates and choose among them. Although agile teams are self-directed, you still need to help in the selection of the team members and the overall makeup of the team based on the criteria described here. For example, too many team members with the same skill sets and no team members with other critical skills won't work. You will help select candidates and form the team by ensuring that the selection criteria are followed and that you have the resources needed to complete the work.

Negotiation   As the project manager, you will use the negotiation technique a lot, so brush up on those skills every chance you get. You'll have to negotiate with functional managers and other organizational department managers (and sometimes with the vendor if you want to get their best people) for resources for your project and for the timing of those resources.

Availability is one part of the negotiating equation. You'll have to work with the functional manager or other project managers to ensure that the staff members you're requesting are available when the schedule says they're needed. Agile projects need team members available for every iteration that requires their skills to produce the work. Some iterations may require a full 40-hour week commitment, whereas other iterations may only need afternoons for the duration of the project. Ideally for agile projects, it's best to have a fully committed, cross-functional team that's dedicated to the project.

The second part of the equation is the competency level of the staff members functional managers are assigning to your project. I remember hearing someone once say that availability is not a skill set. Be wary of functional managers who are willing to offer up certain individuals "any time" whereas others are "never available." Be certain your negotiations include discussions about the skills and personal characteristics of the team members you want on your project. Agile team members who are colocated, particularly those in a bullpen area or in a conference room, will need to play nicely with one another. You'll want to ensure that team members have easygoing personalities and that you don't have people who cannot get along with each other.

Preassignment   *Preassignment* can happen when the project is put out for bid and specific team members are promised as part of the proposal or when internal project team members are promised and assigned as a condition of the project. When staff members are promised as part of the project proposal—particularly on internal projects—they should be identified in the project charter.

Virtual Teams   *Virtual teams* typically consist of team members who work in disparate locations. This may be across campus, across town, or across the globe. Although virtual teams don't usually work in the same location, their members all share the goals of the project and each member has a role to fulfill. Virtual teams are not always geographically dispersed; there may be team members who work different hours or shifts than the other team members, those with mobility limitations who need to work from home or at a specific location, and so on. The idea is that virtual teams can meet the objectives of the project and perform their roles without meeting face to face. One way to do this is by using a virtual workspace. In today's wonderful world of technology, team members can create virtual workspaces by using the Internet, videoconferencing, web conferencing, teleconferencing, and more to meet, communicate, and collaborate on a regular basis.

If possible, it's a great idea to occasionally bring the team together at the same physical location. Teams on predictive projects with long timelines might benefit from a quarterly meeting whereas agile project teams may benefit from physically meeting together at every retrospective. This helps build camaraderie and improves communication. Virtual teams need to pay particular attention to the importance of communication. Make certain all team members are aware of the protocols for communicating in a virtual team environment, that they understand expectations, and that they are clear on how the decision-making processes work.

The *Agile Practice Guide* (PMI®, 2017) notes two techniques used for agile virtual teams that help manage communication. The first is a *fishbowl window*. This works by establishing videoconferencing links in every location where team members reside. At the beginning of the day, each team member logs in to the videoconference and stays logged in for the entire workday. Team members can see one another and naturally engage in conversation, ask questions, and improve collaboration despite the geographic differences. The next technique is *remote pairing*. Remote pairing also involves videoconferencing and team members logging in at the beginning of the day. Remote pairing goes one step further and includes screen sharing, video links, and voice communication.

Work with the team to investigate alternative solutions that will work for them, and within the company's culture, to ensure effective and open communication. Experiment with different techniques and tools and discuss ways that will improve their engagement. Check in with virtual team members on a regular basis to ask how they are doing and what ideas they may have to help improve participation. This is imperative on a virtual team because they don't have the ability to drop by your office or see you in the lunchroom and start up a conversation.

Exam Spotlight

Be certain to continually evaluate the effectiveness of team member engagement on a virtual team.

It's vital in this type of team structure that you, as the project manager, give credit to the appropriate team members for their performance and actions on the project. You might be the only one who fully understands the contributions individual team members have made. When teams are colocated, members have the opportunity to see for themselves the extraordinary efforts others are making on the project. Virtual team members don't necessarily know what their teammates have contributed to the project (or the level of effort they've exerted), so it's up to you to let everyone know about outstanding performance.

Dispersed Teams   Dispersed teams are similar to virtual teams because team members may work in different locations, offices, or even different floors in the same building. They may also employ virtual workspaces to help with communication. If possible, set up face-to-face meetings once the project is kicked off to help encourage camaraderie and build trust among team members. People who engage face-to-face at least part of the time are more likely to be open and honest in their communication because higher levels of trust are established when meeting someone in person. Virtual teams also benefit from meeting one another in person at the beginning of the project and periodically throughout if possible. Video meetings can be effective for both virtual and dispersed teams, but it's often difficult to read body language and facial expressions on a computer monitor. This can also increase the chances that introverts will not speak up. As the project leader, you should ensure that each person gets an opportunity to speak and voice their decision by calling each team member by name and asking for their input. One last tip that will help with both virtual and dispersed teams is to consider using iteration-based agile approaches. This is structured around small deliverables or units of work in timeboxed periods and allows team members from different time zones to more easily participate in the short, structured stand-up meetings and retrospectives. Because of the limited work and short time frames, communication is more focused. You could also encourage the team to meet independently with just a couple of team members on a frequent basis to keep the communication lines open.

Teams New to Agile   It's important that project teams (in both predictive and agile methodologies) consist of team members with different skills and backgrounds. Agile teams are highly collaborative and communication is critical to their success. Agile project managers should focus on establishing teams that are cross-functional. If all your team members have the same skills, the same experience, and the same personalities, it will be difficult to improve because everyone sees the situation the same way. All agile teams should organize around the strengths of the individual team members so that there are experts who represent every cross-functional area involved on the project. New agile teams should be fully dedicated to the project and not distracted with other functional duties or other projects. This will help them grow their skills in agile processes.

Teams that are new to agile will require education and training in agile practices. It may help to start out using a hybrid approach whereby planning is performed using predictive techniques while Executing processes are performed using agile techniques, including holding stand-up and retrospective meetings. Also make certain that business and executive managers are trained and educated in agile techniques. Use terms and language the business understands to promote the benefits of agile. If the executive team is not keen on the idea of agile, show them how areas related to their strategic goals and plans would directly benefit from agile practices.

Agile teams, particularly those new to this methodology, will need time to hone their estimating techniques. They may experience frustration at first trying to work on user stories that are too large or work on too many user stories in an iteration. The product owner can help by paring down the size of the user stories and reducing the number of them in an iteration. Once the team develops a cadence of completing a user story or two in each iteration, they will gain confidence in their success and their estimating skills will continue to improve.

No matter the team structure, maturity level of the team, or development methodology you're using, project teams that consist of cross-functional members with different skills and functions make the best teams. The *Agile Practice Guide* (PMI®, 2017) notes that agile teams are built using cross-functional team members, and even though we've talked about several techniques for choosing team members in this section, it's better for agile teams to self-select without management involvement whenever possible. In my experience, this doesn't usually work as well as the project manager having at least some influence over the team makeup. Small organizations or companies that only do development work may benefit from allowing teams to self-select. Large, functionally structured organizations usually require, at a minimum, an informal process for choosing and assigning team members. The *Agile Practice Guide* (PMI®, 2017) also notes that the company's compensation system should be structured to reward teams using agile approaches. When team members' paychecks are reliant on frequent, workable, quality products, it helps motivate them to succeed.

Project Team Assignments

There are three primary outputs of the Acquire Resources process. We'll look at each of them next.

Physical Resource Assignments Physical resources may encompass a vast array of equipment, materials, facilities, leases, office supplies, and so on. Some of these resources may be required at specific times during the project, and others may be needed for the length of the project. Let's say you're working on a project involving medical research. At times on the project, your team needs access to medical research equipment that is shared by the entire organization. You'll want to make certain the research equipment is assigned to the project at the right time so that the team can complete their work and the project schedule remains on track.

Project Team Assignments Your ability to influence the selection of resources (using the negotiating technique) will impact the project team assignments output. After determining elements such as the roles and responsibilities, reviewing recruitment practices, and negotiating for staff, you will assign project team members to project activities. Once assignments are assured, prepare and publish a project team directory listing the names of all project team members and stakeholders. Don't forget to also include team member names in project organization charts, RAM charts, and other planning documents if their names weren't known at the time you created those documents.

Resource Calendars Resource calendars show the team members' availability and the times they are scheduled to work on the project. A composite resource calendar includes availability information for potential resources as well as their capabilities and skills. (Resource calendars are an input to the Estimate Activity Resources process.) This comes in handy when you're creating the final schedule and assigning resources to activities.

Agile teams will benefit from resource calendars as well. This will help the product owner and the team schedule user stories for upcoming iterations. If a team member has vacation plans, for example, the number of user stories in the next iteration will need to be decreased to account for the absence of this team member. Resource calendars will help you and the product owner know when iterations may need to be adjusted to account for team availability.

Now that you have the team, what do you do with them? You'll look at topics such as motivation, rewards, and recognition in the next process, Develop Team.

 Real World Scenario

The Only Candidate

"Hey, did you hear?" your friend Story asks. "Roger has been assigned to the project team."

"Over my dead body," you reply, pushing away from your computer screen. You head straight for the project manager's office and don't wait for a response from Story.

Ann sets the phone into the cradle just as you walk through the door. Fortunately for you, Ann's door is always open, and she welcomes drop-ins.

"Seems like something is on your mind," Ann says. "What can I help with?"

"Story just told me that Roger has been assigned to the project team. I can't work with Roger. He's arrogant and doesn't respect anyone's work but his own. He belittles me in front of others, and I don't deserve that. I write good code, and I don't need Roger looking over my shoulder. I want to be on this team, but not if Roger is part of it."

Ann thinks for a minute and replies, "I want you to work on this project; it's a great opportunity for you. But there isn't anyone else who can work on the analysis phase of this project except Roger. He's the only one left who has a solid understanding of the legacy code. Unfortunately, those old programs were never documented well, and they've evolved over the years into programs on top of programs. Without Roger's knowledge of the existing system, we'd blow the budget and time estimates already established for this project. Since I need both of you on this project, here's what I propose: I will clearly outline the roles and responsibilities for all the key team members at the kickoff meeting. I'll also make it clear that negative team interactions won't be allowed. If you have a problem with Roger that you can't resolve on your own, you should get me involved right away."

Developing the Project Team

Projects exist to create a unique product, result, or service within a limited time frame. Projects are performed by people, and most projects require more than one person to perform all of the activities. If you've got more than one person working on your project, you've got a team. If you've got a team, you've got a wide assortment of personalities, skills, needs, and issues in the mix. Couple this with part-time team members, teams based in functional organizations whose loyalty lies with the functional manager, teams based in matrix organizations that report to you for project-related activities and another manager for their functional duties, or teams with members who are scattered around the globe, and you could have some real challenges on your hands. Good luck! Okay, I won't leave you hanging like that.

The *Develop Team* process is about creating an open, encouraging environment for your team and developing it into an effective, functioning, coordinated group. Projects are performed by individuals, and the better they work together, the smoother and more efficient the execution of the project will be. I'm sure you have had the experience of working with a team who pitched in and shared workloads when the work became unbalanced. I'm also sure you have worked with teams who didn't do this—teams whose members took on a "me first" attitude and couldn't care less about the plight of their fellow team members. I'd much rather work with a team like the first example.

The proper development of the team is critical to a successful project, and as the project manager, you should know that developing the team is one of your most important duties. Because teams are made up of individuals, individual development becomes a critical factor to project success. Individual team members need the proper development and training to perform the activities of the project or to enhance their existing knowledge and skills. The development needed will depend on the project. Perhaps you have a team member who's ready to make the jump into a lead role but she doesn't have any experience at lead work. Give her some exposure by assigning her a limited amount of activities in a lead capacity, provide her with some training if needed, and be available to coach and mentor where needed. The best option is to work with the management team to provide this person with the development she needs prior to the start of the project (if you're lucky enough to know early on who your resources might be and what their existing skills are).

Developing a team requires a lot of time and dedication on the part of you and the project manager, and guess which skill you'll need to employ effectively, efficiently, and clearly in order to make this work? You guessed it—communication. According to the

PMBOK® Guide, there are several reasons for spending time developing the team. They include the following:

- Team members skills and knowledge improve, which has an effect on the project of lowering costs, improving the schedule, and meeting or exceeding quality standards.

- Trust among team members is improved, which in turn improves morale, decreases conflicts among team members, and increases teamwork.

- The team becomes close-knit, practices cooperation, and improves productivity.

- Teams take ownership of the work, arc more efficient and effective, and are empowered to make decisions.

Generational Diversity

The workforce in today's environment is diverse and includes more generations working together than ever before in history. For example, in my organization we have people from the Silent generation, baby boomers, Gen X, millennials, and Gen Z. That includes people born starting in the 1940s through today. We have people 80+ years young and those barely entering their twenties working side by side. As a project manager, it's important to understand the uniqueness and sensitivities associated with each generation. That's not to say that every person within a specific generation identifies with the broad generalities associated with that generation, but it helps you form a picture of the team strengths, their work ethics, soft skills, and approach to life and work. Let's take a brief look at some of the unique characteristics of each generation.

The Silent Generation   These folks were born between 1925 and 1945. They lived through World War II, tend to conform to social standards and norms, and are often frugal. The war and the great depression caused a shortage of food, goods, and services. This brought about rationing of food, gas, and other everyday items we take for granted today. This influenced many of them to become strong savers. This generation of workers typically worked for (or still works for) the same company for life.

Baby Boomers   Baby boomers were born between 1946 and 1964. They were born after the war and for many years were the largest generation alive. Baby boomers tend to share characteristics of the generations around them. Baby boomers experienced affluence and an abundance of food and goods growing up and are sometimes said to be spoiled. This generation brought about social change in many areas, including music, personal expression, lifestyles, the civil rights movement, and more. They tend to reject the traditional values of the Silent generation and are more physically active and wealthier than the Silent generation. Some research groups predict that this age group (those 55 years of age and older) will be the fastest-growing segment of the workforce between now and 2025. Baby boomers are choosing to postpone retirement or come back to work once they have retired.

Gen X   This generation was born between 1965 and 1976, although some researchers show them being born as late as the mid-1980s. This generation was known as the "latchkey" generation because they grew up in households where both parents worked. They often came home to an empty household. As they reached their teens, they were also nicknamed the "MTV Generation." This generation tends to be entrepreneurial, no doubt as a result of having to fend for themselves as youngsters. They were the first to introduce the concept of work-life balance. Having watched baby boomer parents work themselves to death, only to then experience layoffs and loss of pensions, Gen Xers determined to enjoy both work and personal life and not allow work to consume them.

Millennials or Gen Y   Millennials were born between 1977 and 1995, although some researchers put this generation closer to the early 1980s through the early 2000s. Millennials grew up experiencing the advent of technology in everyday life. The Internet was dawning, games such as Xbox and PlayStation came about, and they found themselves becoming dependent on technology as it emerged. They are comfortable using technology and engaging in social media. Millennials are sometimes characterized as narcissistic, but this likely refers to their other characteristics such as confidence and assertiveness. They value tolerance and collaborative teams. Millennials are the largest generation in the workplace today.

Gen Z   This generation was born between 1996 (or the early 2000s according to some) and 2012. Some of them are just now entering the workplace. This generation grew up immersed in technology and they don't know any other way of life. Try describing a telephone that hangs on the wall to a Gen Zer and you'll get a blank stare. Partly due to their tech savvy and reliance, they are open to new ideas and new ways of doing things. They are heavily influenced by social media. They sometimes have short attention spans (also due to technology) and often prefer texting to talking. They are hardworking and often more stressed than other generations due to social media dependence, school shootings, political infighting, and other social and cultural issues prevalent today that were rare in past generations.

As you can see, each of these generations had a unique set of characteristics growing up from World War II to the arrival of social media and each experience informs their behaviors, beliefs, and actions. Generational sensitivity should be considered when choosing and developing teams. This also helps in developing cross-functional agile teams. With a wide diversity of age, experience, and outlooks comes a significant amount of creativity and thinking outside the box.

Diversity and inclusion are also important in both acquiring and developing teams. Diversity involves hiring team members who come from different generations, cultures, ethnicities, religious beliefs, sexual orientation, and more. Diverse teams are more apt to explore new ideas and develop approaches to issues and challenges because of the variance in backgrounds, experiences, and beliefs on the team. Diversity is only one side of the coin, however; inclusion is the other. Inclusion means that the team is collaborative, supportive, and respectful of one another, and that all team members have a voice and are expected to participate on the project. It's giving your team members a safe place to be who they really are and ensuring that you are working toward a set of equitable outcomes. This helps ensure the participation of all team members while preventing a domineering team member from overtaking the group. The project manager or team leader needs to manage this by calling on team members one by one and encouraging each one to express their thoughts. Vernā Myers is a diversity and inclusion consultant and TED Talk speaker who is credited with saying, "Diversity is being invited to the party; inclusion is being asked to dance."

Tools and Techniques to Develop the Team

There are several tools and techniques you can use to develop the project team. We will discuss several: colocation, interpersonal skills, team-building skills, training, individual and team assessments, and recognition and rewards.

Colocation

Colocation, also known as *tight matrix*, brings team members together in one physical location for the entire project or for important periods during the project life cycle. You should consider including funding in the project budget to bring the team together at the same location, at least occasionally. Another way to achieve colocation is to set aside a common meeting room, sometimes called a *war room*, for team members to meet and exchange information. Colocation is ideal for most teams, especially agile teams. According to the *Agile Practice Guide* (PMI®, 2017), there are several benefits of colocating agile teams, including:

- Improved communication
- Increased knowledge sharing
- Commitment among team members to one another and to the project
- Effective team dynamics
- Continuous learning environment at a reduced cost

In reality, it's not always possible to colocate team members. Please refer to working with virtual and dispersed teams and utilizing virtual workspaces, discussed earlier in this chapter, when colocation is not an option for your team.

Interpersonal Skills

Interpersonal skills are often referred to as soft skills. *Soft skills* include such things as leadership, influence, negotiation, communications, motivation, empathy, conflict management, team building, and creativity. For example, it's important for you as the project manager to understand your project team members' attitudes and opinions about their work. Bad attitudes, as the saying goes, are contagious. It doesn't mean the person who has the attitude is bad, but if you're paying attention to your team members and taking the appropriate amount of time to listen to their legitimate concerns and issues, and taking action on them, you can go a long way toward stemming bad attitudes.

Soft skills can be learned, but in my experience they are more often inherent in project managers' personalities. However, just because certain soft skills may not be in your nature, it doesn't mean you can't observe this behavior in others and incorporate those skills into your management techniques.

Team-Building Activities

Many times, project teams consist of folks who don't know one another. They aren't necessarily aware of the project objectives and might not even want to be a part of the team. Or the project manager might not have previously worked with the people assigned to the project team. Does this sound like a recipe for disaster? It's not. Thousands of projects are started with team members and project managers who don't know one another, and those projects come to a successful completion. How is that done? It's a result of the project manager's team-building and communication skills.

The project manager's job is to bring the team together, get its members all headed in the right direction, and provide motivation, reward, and recognition to keep the team in tip-top shape. This is done using a variety of team-building techniques and exercises. *Team building* is simply getting a diverse group of people to work together in the most efficient and effective manner possible. Achieving this might involve events organized by the management team or individual actions designed to improve team performance. You can find entire volumes on this subject, and it's beyond the scope of this book to go into all the team-building possibilities. The exam tends to focus more on the theories behind team building and the characteristics of effective teams, so that's what you'll spend your time exploring.

Bruce Tuckman and Mary Ann Jensen developed a model that describes how teams develop and mature. According to Tuckman and Jensen, all newly formed teams go through five stages of development:

1. Forming
2. Storming
3. Norming
4. Performing
5. Adjourning

Tuckman and Jensen originally devised this theory using the first four stages of development. Based on later research by the Tuckman-Jensen team, a fifth stage of development was added called adjourning. You've probably seen this model elsewhere, but because these stages might show up on the exam, you'll want to memorize them. Let's take a brief look at each of them:

Forming   This one is easy. Forming is the beginning stage of team formation, when all the members are brought together, introduced, and told the objectives of the project. This is where team members learn why they're working together. During this stage, team members tend to be formal and reserved and take on an "all-business" approach.

Your role as the project manager in this stage of development is communication. If the team is small, I recommend meeting with each of the members one on one and as a group. In my experience, team members who clearly understand why they are assigned to the project, what's expected of them regarding individual and team deliverables, and how to inform the project manager of their needs and issues will generally outperform their peers who do not have or understand this information.

Storming   Storming is where the action begins. Team members become confrontational with one another as they're vying for position and control during this stage. They're working through who is going to be the top dog and jockeying for status.

Your role as the project manager during this stage is to remind the team of the project goals and keep everyone centered on those goals. Conflicts aren't bad in this case; they're necessary to get the team into the next stage. During this stage, it is best if you can limit your intervention and let team members resolve their own issues as often as possible. Team members need to get a feel for where they stand, where the extent of their responsibility lies, and how they'll accomplish their tasks working with the other personalities of the team, and that usually involves some tussles. Questioning and conflict help clarify the goals of the project for everyone on the team, not just the person in conflict, so encourage your team members to ask questions and discuss conflicts openly. However, you won't progress to the next stage until the team has resolved the conflicts.

Norming   Now things begin to calm down. Team members know one another fairly well by now. They're comfortable with their positions in the team, and they begin to deal with project problems instead of people problems. In the norming stage, they confront the project concerns and problems instead of each other. Decisions are made jointly at this stage, and team members exhibit mutual respect and familiarity with one another.

As the project manager, you should continue to hold team meetings, especially during this stage, because team members can fall back into the storming stage if left to their own devices. During this stage, you should intervene more often when conflicts arise to keep the team moving forward. Monitor each team member's participation, and encourage the team to continue to remain focused on the project's goals and alert you of any problems as soon as they arise.

Teams in the norming stage are efficient, functioning teams. If your team has progressed to this stage, they'll likely be productive and work effectively toward meeting the project goals. They still aren't performing at their absolute best, though—that happens in the next stage.

Exam Spotlight

Different teams progress through the stages of development at different rates. When new team members are brought on to the team, the development stages start all over again. It doesn't matter where the team is in the first four phases of the development process—a new member will start the cycle all over again.

Performing Ahh, perfection. Well, almost, anyway. This stage is where great teams end up and where team members are productive and effective. The level of trust among team members is high, and great things are achieved. This is the mature development stage.

Your role as project manager during this stage should be more focused on the project management processes than on the team itself. Teams in this stage are usually self-directed and will hum along smoothly, provided you continue to update them on project progress and keep the lines of communication open.

Adjourning As the name implies, this phase refers to the breakup of the team after the work is completed.

As the project manager, you need to realize that many team members may experience a sense of loss at the end of the project, particularly long-term projects. Guide the team through a closure process. Team celebrations at the conclusion of the project are one way to accomplish this. Acknowledge their contributions and let them know you are grateful for their efforts and for any sacrifices they've made during the course of the project.

According to Tuckman and Jensen, leaders adapt their leadership styles as the teams develop maturity and progress through the development stages. For example, early in the forming stage, leaders take on a direct style of leadership. As the team progresses, their leaders will employ a coaching, participating, and then delegating style of leadership to match the level of development the team has achieved.

Let's take a closer look at focusing your team members throughout these stages of development, along with some of the characteristics of effective teams.

Team Focus

It's paramount that the team members know and understand the goals and objectives of the project. They should all understand the direction you're headed and work toward that end. After all, that's the reason they were brought together in the first place. Keep in mind that people see and hear things from their own perspective. A room full of people attending a speech will each come away with something a little different because what was said speaks to their particular situation in life at the time. In other words, their own perceptions filter what they hear. It's your job as project manager to make sure the team members understand the project goals and their own assignments correctly. I suggest you use solid communication skills to get your point across, and don't forget to employ your emotional intelligence skills as well. Ask your team members to tell you in their own words what they believe the project goals are. This is a great way to know whether you've got everyone on board and a great opportunity for you to clarify any misunderstandings regarding the project goals.

Effective Team Characteristics

Effective teams are typically very energetic teams. They often are characterized as high-performance teams and are motivated by results and the successful completion of tasks. Their enthusiasm is contagious, and it feeds on itself. They generate a lot of creativity and become good problem solvers. Teams like this are every project manager's dream. Investing yourself in team building as well as relationship building—especially when you don't think you have the time to do so—will bring many benefits. Here's a sample of the benefits:

- Better conflict resolution
- Commitment to the project
- Commitment to the project team members and project manager
- High job satisfaction
- Enhanced communication
- A sense of belonging and purpose
- Enhanced feelings of trust
- Lower project costs
- Improved productivity
- Improved quality
- A successful project

Dysfunctional teams will typically produce the opposite results of the benefits just listed. Dysfunctional teams don't just happen by themselves any more than great teams do. Sure, sometimes you're lucky enough to get the right combination of folks together right off the bat. But usually, team building takes work and dedication on the part of the project manager. Even in the situations where you do get that dynamite combination of people, they will benefit from team-building exercises and feedback.

Unfortunately, sour attitudes are just as contagious as enthusiasm. Watch for these symptoms among your team members, and take action to correct the situation before the entire team is affected:

- Lack of motivation or "don't care" attitudes
- Project work that isn't satisfactory

- Status meetings that turn into whining sessions
- Poor communication
- Lack of respect for and lack of trust in the project manager

 No amount of team building will make up for poor project planning or ineffective project management techniques. Neglecting these things and fooling yourself into thinking that your project team is good enough to make up for the poor planning or poor techniques could spell doom for your project. Besides that, it's not fair to your project team to put them in that position.

Training

Training is a matter of assessing your team members' skills and abilities, assessing the project needs, and providing the training necessary for the team members to carry out their assigned activities. Training might include a certain skill, like how to drive a tractor trailer, or learning a new process such as agile. Training can be formal or informal. Formal training may include classroom training, online training, or training performed on the job. Informal training might occur by observing others, by asking others how to perform a task, or by job shadowing. The following steps will help you in defining training for your team members and stakeholders.

1. Determine the required competencies the team needs and tailor the training to those needs. You can perform self-assessments, interviews, or simply ask team members about their training needs.

2. Determine the training options that will work best within your culture and project budget. If budget allows, you could send team members to training out of town or out of state. Alternatively, you could bring an instructor to your organization to train the team. Multiple options are available for self-study, including online training courses.

3. Make certain you allocate resources for training. This may include adding funding to the project budget for travel and training. It may also include funding to pay for an instructor on site (including their travel accommodations if needed). If you're performing training on site, make certain you have a dedicated room with computers, equipment, and other tools needed for teaching and learning.

4. Measure the training outcomes to ensure the concepts are being applied and the competencies you set out to gain are realized.

Training can sometimes be a reward as well. In the software industry, programmers seek out positions that offer training on the latest and greatest technologies, and they consider it a benefit or bonus to attend training on the company dollar and time. If you know early in the Planning processes that training is necessary, include the details of this in the staffing management plan. During the course of the project, you might observe team members who need training, or they might ask for training. Update the staffing management plan with this information.

Individual and Team Assessments

Assessment tools help highlight the strengths and weaknesses of the team by assessing various aspects of the team such as communication techniques, interpersonal skills and preferences, organizational skills, decision-making skills, and more. There are many tools available that can help in assessing work preference traits such as Emergenetics, Myers-Briggs MBTI, DiSC, and many more. Surveys, interviews, focus groups, ability exams, and other tools can also be used to assess your team's abilities, strengths, and weaknesses. The key to these assessments is understanding your team members' personality and behaviors and adjusting to their emotional needs. It's not a bad idea to perform these assessments for key stakeholders as well so that you can understand their emotional preferences. And while you're at it, be sure to use those emotional intelligence skills on your stakeholders to understand how they work and help determine when the project might need adjusting.

Recognition and Rewards

I have quite a bit of ground to cover with recognition and rewards. As I said earlier, you could see several exam questions regarding team building, so dig out all your favorite memorization techniques and put them to use.

Team building starts with project planning and doesn't stop until the project is completed. It involves employing techniques to improve your team's performance and keeping team members motivated. Motivation helps people work more efficiently and produce better results. If clear expectations, clear procedures, and the right motivational tools are used, project teams will excel.

Motivation can be extrinsic or intrinsic. Intrinsic motivators are specific to the individual. Some people are just naturally driven to achieve—it's part of their nature. (I suspect this is a motivator for you since you're reading this book.) Cultural and religious influences are forms of intrinsic motivators as well. Extrinsic motivators are material rewards and might include bonuses, the use of a company car, stock options, gift certificates, training opportunities, extra time off, and so on.

Recognition and rewards are important to team motivation and are an example of an extrinsic motivator. They are formal ways of recognizing and promoting desirable behavior and are most effective when carried out by the management team and the project manager. You should develop and document the criteria for rewards, especially monetary awards. Although rewards and recognition help build a team, they can also kill morale if you don't have an established method or criteria for handing them out. Track who is receiving awards throughout the project. For example, if you have consistent overachievers on the team, you could kill morale by consistently rewarding the same one or two people repeatedly. It could also be perceived that you're playing favorites. If team members believe the rewards are win-lose (also known as *zero-sum*) and that only certain team members will be rewarded, you might end up hurting morale more than helping. If you find yourself in this position, consider team awards. This is a win-win because all team members are recognized for their contributions. Agile projects utilize self-directed work teams, and team awards are most appropriate when using this methodology. Recognition and rewards should be proportional to the achievement. In other words, appropriately link the reward to the performance. For example, a project manager who has responsibility for the project budget and the procurement process and keeps the costs substantially under budget without sacrificing the results of the project should be rewarded for this achievement. However, if these responsibilities are assigned to a functional manager in the organization, it wouldn't be appropriate to reward a project manager who was not the one responsible for keeping the costs in line.

Baker's Gift Baskets

You're a contract project manager for Baker's Gift Baskets. This company assembles gift baskets of all styles and shapes with every edible treat imaginable. The company has recently experienced explosive growth, and you've been brought on board to manage its new project. The owners of the company want to offer "pick-your-own" baskets that allow customers to pick the individual items they want included in the baskets. In addition, they're introducing a new line of containers to choose from, including items such as miniature golf bags, flowerpots, serving bowls, and the like. This means changes to the catalog and the website to accommodate the new offerings.

The deadline for this project is the driving constraint. The website changes won't cause any problems with the deadline. However, the catalog must go to press quickly to meet holiday mailing deadlines, which in turn are driving the project deadline.

Your team members put their heads together and came up with an ingenious plan to meet the catalog deadline. It required lots of overtime and some weekend work on their part to pull it off, but they met the date.

You decide this is a perfect opportunity to recognize and reward the team for their outstanding efforts. You've arranged a slot on the agenda at the next all-company meeting to bring your team up front and praise them for their cooperation and efforts to get the catalog to the printer on time. You'll also present each of them with a gift certificate for a dinner with their friends or family at an exclusive restaurant in the city.

Team members should be rewarded for going above and beyond the call of duty. Perhaps they put in a significant amount of overtime to meet a project goal or spent nights round-the-clock babysitting ill-performing equipment. These types of behaviors should be rewarded and formally recognized by the project manager and the management team. On the other hand, if the ill-performing equipment was a direct result of mistakes made or if it happened because of poor planning, rewards would not be appropriate, obviously.

Consider individual preferences and cultural differences when using rewards and recognitions. Some people don't like to be recognized in front of a group; others thrive on it.

Some people appreciate an honest thank-you with minimal fanfare, and others just won't accept individual rewards because their culture doesn't allow it. Keep this in mind when devising your reward system.

There are many theories on motivation. As a project manager, it's important to understand them so that you can tailor your recognition and rewards programs to take into account the reasons people do what they do. You might encounter questions on these theories on the exam, so we'll discuss their primary points in the following sections.

Motivational Theories

Motivational theories came about during the modern age. Prior to today's information and service jobs and yesterday's factory work, the majority of people worked the land and barely kept enough food on the table to feed their families. No one was concerned about motivation at work. They worked because they wouldn't have anything to eat if they didn't. Today we have a new set of problems in the workplace. Workers in the service- and knowledge-based industries aren't concerned with starvation—that need has been replaced with other needs, such as job satisfaction, a sense of belonging and commitment to the project, good working conditions, and so on. Motivational theories present ideas on why people act the way they do and how you can influence them to act in certain ways to get the results you want. Again, there are libraries full of books on this topic, so I'll only cover four of those theories here.

MASLOW'S HIERARCHY OF NEEDS

You have probably seen this classic example of motivational theory. Abraham Maslow theorized that humans have five basic needs arranged in hierarchical order. The first needs are physical needs, such as the need for food, clothing, and shelter. The idea is that these needs must be met before the person can move to the next level of needs in the hierarchy, which includes safety and security needs. Here, the concern is for the person's physical welfare and the security of their belongings. Once that need is met, they progress to the next level, and so on.

Maslow's hierarchy of needs theory suggests that once a lower-level need has been met, it no longer serves as a motivator and the next higher level becomes the driving motivator in a person's life. Maslow conjectures that humans are always in one state of need or another. Here is a recap of each of the needs, starting with the highest level and ending with the lowest:

Self-Actualization   Performing at your peak potential

Self-Esteem Needs   Accomplishment, respect for self, capability

Social Needs   A sense of belonging, love, acceptance, friendship

Safety and Security Needs   Your physical welfare and the security of your belongings

Basic Physical Needs   Food, clothing, shelter

The highest level of motivation in this theory is the state of self-actualization. Several years ago, the United States Army had a slogan that I think encapsulates self-actualization very well: "Be all that you can be." When all the physical, safety, social, and self-esteem needs have been met, you reach a state of independence where you're able to express yourself and perform at your peak. You'll do good work just for the sake of doing good work. Recognition and self-esteem are the motivators at lower levels; now the need for being the best you can be is reached.

> In Maslow's later work, he discussed three additional aspects of motivation: cognitive, aesthetic, and transcendence. The five key needs are the ones you'll most likely need to know for the exam, but it wouldn't hurt to be familiar with the names of the three additional motivational levels. Also note that people are not "stuck" at a certain level forever. You will traverse the pyramid, up and down, throughout your life and career.

HYGIENE THEORY

Frederick Herzberg came up with the *Hygiene Theory*, also known as the *Motivation-Hygiene Theory*. He postulates that two factors contribute to motivation: hygiene factors and motivators. Hygiene factors deal with work environment issues. The thing to remember about hygiene factors is that they prevent dissatisfaction. Examples of hygiene factors are pay, benefits, the conditions of the work environment, and relationships with peers and managers. Pay is considered a hygiene factor because Herzberg believed that over the long term, pay is not a motivator. Being paid for the work prevents dissatisfaction but doesn't necessarily bring satisfaction in and of itself. He believed this to be true as long as the pay system is equitable. If two workers performing the same functions have large disparities in pay, then pay can become a motivator.

Motivators deal with the substance of the work itself and the satisfaction one derives from performing the functions of the job. Motivators lead to satisfaction. The ability to advance, the opportunity to learn new skills, and the challenges involved in the work are all motivators, according to Herzberg.

Exam Spotlight

For the exam, remember that Herzberg was the inventor of the Hygiene Theory and that this theory claims that hygiene factors (pay, benefits, and working conditions) prevent dissatisfaction whereas motivators (challenging work, opportunities to learn, and advancement) lead to satisfaction.

Expectancy Theory

The *Expectancy Theory*, first proposed by Victor Vroom, says that the expectation of a positive outcome drives motivation. People will behave in certain ways if they think there will be good rewards for doing so. Also note that this theory says that the strength of the expectancy drives the behavior. This means that the expectation or likelihood of the reward is linked to the behavior. For example, if you tell your two-year-old to put the toys back in the toy box and you'll give her a cookie to do so, chances are she'll put the toys away. This is a reasonable reward for a reasonable action. However, if you promise your project team members vacations in Hawaii if they get the project done early and they know there is no way you can deliver that reward, they have little motivation to work toward it. Also make certain you are using rewards that motivate your team members. If you make a trip to Hawaii the reward, and you can make good on that promise in this example but your team members are deathly afraid of flying, the reward won't have a motivating effect.

This theory also says that people become what you expect of them. If you openly praise your project team members and treat them like valuable contributors, you'll likely have a high-performing team on your hands. Conversely, when you publicly criticize people or let them know that you have low expectations regarding their performance, they'll likely live up (or down as the case might be) to that expectation as well.

ACHIEVEMENT THEORY

The *Achievement Theory*, attributed to David McClelland, says that people are motivated by the need for three things: achievement, power, and affiliation. The achievement motivation is obviously the need to achieve or succeed. The power motivation involves a desire for influencing the behavior of others, and the need for affiliation is relationship oriented. Workers want to have friendships with their coworkers and a sense of camaraderie with their fellow team members. The strength of your team members' desire for each of these will drive their performance on various activities.

Exam Spotlight

Make certain you understand the theories of motivation and their premises for the exam. Here's a summary to help you memorize them:

Maslow's Hierarchy of Needs   Abraham Maslow. Needs must be satisfied in a hierarchical order.

Hygiene Theory   Frederick Herzberg. Work environment (pay, benefits, and working conditions) prevents dissatisfaction.

Expectancy Theory   Victor Vroom. Expectation of positive outcomes drives motivation.

Achievement Theory   David McClelland. People are motivated by achievement, power, and affiliation.

I'll cover two more theories in the leadership section, which is next. They deal specifically with how leaders interact with their project team members.

Leadership vs. Management

Chapter 1, "Building the Foundation," introduced the differences between leaders and managers. I'll add a bit more information here regarding leadership theories and the types of power leaders possess, but first I'll recap leadership and management.

Leadership is about imparting vision and rallying people around that vision. Leaders motivate and inspire and are concerned with strategic vision. Leaders have a knack for getting others to do what needs to be done. Two of the techniques they use to do this are power and politics. *Power* is the ability to get people to do what they wouldn't do ordinarily. It's also the ability to influence behavior. *Politics* imparts pressure to conform regardless of whether people agree with the decision. Leaders understand the difference between power and politics and when to employ each technique. I'll talk more about power shortly.

Good leaders have committed team members who believe in the vision of the leader. Leaders set direction and time frames and have the ability to attract good talent to work for them. Leaders inspire a vision and get things done through others by earning loyalty, respect, and cooperation from team members. They set the course and lead the way. Leaders use delegation techniques to bring about their vision. While they set the vision, it's usually others who perform the work. A good leader delegates the vision (and the deliverables associated with it) to managers and then trusts them to get the work done. Good leaders are directive in their approach but allow for plenty of feedback and input. Good leaders commonly have strong interpersonal skills and are well respected.

Managers are generally task oriented and concerned with issues such as plans, controls, budgets, policies, and procedures. They're generalists with a broad base of planning and organizational skills, and their primary goal is satisfying stakeholder needs. They should possess motivational skills and have the authority to recognize and reward behavior. Managers also use delegation techniques and will assign tasks to staff members to help realize the goals of the organization.

Project managers need to use the traits of both leaders and managers at different times during a project. On large projects, a project manager will act more like a leader, inspiring the subproject managers to get on board with the objectives. On small projects, project managers will act more like managers because they're responsible for all the planning and coordinating functions and delegating tasks.

Leadership Theories

I'll discuss six theories regarding leadership and management. They are Douglas McGregor's Theory X and Theory Y, Dr. William Ouchi's Theory Z, the Contingency Theory, the Tannenbaum and Schmidt Continuum Management Theory, and the Situational Leadership Theory. Then I'll discuss the types of power leaders use and leadership styles.

THEORY X, THEORY Y, AND THEORY Z

Douglas McGregor defined two models of worker behavior, Theory X and Theory Y, that attempt to explain how different managers deal with their team members. *Theory X* managers believe most people do not like work and will try to steer clear of it; they believe people have little to no ambition, need constant supervision, and won't actually perform the duties of their jobs unless threatened. As a result, Theory X managers are like dictators and impose very rigid controls over their people. They believe people are motivated only by punishment, money, or position. Unfortunately for the team members, Theory X managers unknowingly also fall victim to the Expectancy Theory. If they expect people to be lazy and unproductive and treat them as such, their team members probably will be lazy and unproductive.

Theory Y managers believe people are interested in performing their best, given the right motivation and proper expectations. These managers provide support to their teams, are concerned about their team members, and are good listeners. Theory Y managers believe people are creative and committed to the project goals, that they like responsibility and seek it out, and that they are able to perform the functions of their positions with limited supervision.

Theory Z was developed by Dr. William Ouchi. This theory is concerned with increasing employee loyalty to their organizations. It came about in Japan in the 1980s when jobs were often offered for life. This theory results in increased productivity, it puts an emphasis on the well-being of the employees both at work and outside of work, it encourages steady employment, and it leads to high employee satisfaction and morale.

CONTINGENCY THEORY

The *Contingency Theory* builds on a combination of Theory Y behaviors and the Hygiene Theory. The Contingency Theory, in a nutshell, says that people are motivated to achieve levels of competency and will continue to be motivated by this need even after competency is reached.

TANNENBAUM AND SCHMIDT CONTINUUM MANAGEMENT THEORY

Robert Tannenbaum and Warren Schmidt developed a leadership theory, called the *Tannenbaum and Schmidt Continuum Management Theory,* that describes the level of authority a manager exerts on the team versus the freedom a team has to make decisions (under the guidance of the manager). They outline seven levels of delegated freedom ranging from the manager making all decisions and announcing them to the team to the manager allowing the team to identify the problem, determine alternatives, and make the final recommendation regarding the action needed to solve the problem. The level of freedom you use depends on the maturity and experience of the team and the manager. As the team progresses, their decision-making matures, and more and more freedom can be delegated. Managers are always engaged at all levels of this model, but their authority level will decrease as they delegate decision-making responsibility to the team.

SITUATIONAL LEADERSHIP THEORY

Paul Hersey and Ken Blanchard developed the *Situational Leadership Theory* during the mid-1970s. This theory's main premise is that the leadership style you use depends on the situation. A leader needs to use the appropriate leadership style for the circumstances so that team members are motivated properly. For example, perhaps you have a new employee fresh out of school and she is learning a new task. Obviously, this employee will need a lot more guidance and direction than an employee who has been with the organization for some time and knows how to perform the task at hand. Hersey and Blanchard theorized that leaders fell into one of four styles that reflected the level of experience with the tasks (known as task behavior) and the style (known as relationship behavior) the leader should use with the team. These styles are known as telling/directing, selling/coaching, participating/supporting, and delegating. Let's take a brief look at each.

Telling/Directing   Team members are new to the task and need specific instruction on how to perform the task. Leaders make the decisions and inform the team members what to do and how to do it.

Selling/Coaching   Team members are able to perform the task but still need some instruction and oversight on the task. The leaders use a coaching style to answer questions and give gentle guidance rather than a directive approach.

Participating/Supporting   Team members are experienced at the task and the leader lets them make most of the decisions. The leader helps improve the confidence level of the team members by encouraging them and providing effective feedback.

Delegating   Team members are experienced at the task and willing and able to take on the work. They may choose the tasks and fulfill them with little oversight.

Hersey and Blanchard broke apart and formed their own individual companies and each went on to develop other situational leadership models. This field has expanded over the years to include many theories.

The Power of Leaders

As stated earlier, power is the ability to influence others to do what you want them to do. Power can be used in a positive manner or a negative one. But that old saying of your grandmother's about attracting more flies with honey than with vinegar still holds true today.

Leaders, managers, and project managers use power to convince others to do tasks in a specific way. The kind of power they use to accomplish this depends on their personality, their personal values, and the company culture.

A project manager might use several forms of power. I've already talked about reward power, which is the ability to grant bonuses or incentive awards for a job well done. Here are a few more:

Punishment Power   Punishment, also known as *coercive* or *penalty power*, is just the opposite of reward power. The employee is threatened with consequences if expectations are not met.

Expert Power   Expert power occurs when the person being influenced believes the manager, or the person doing the influencing, is knowledgeable about the subject or has special abilities that make them an expert. The person goes along just because they think the influencer knows what they're doing and it's the best thing for the situation.

Legitimate Power   Legitimate, or formal, power comes about as a result of the influencer's position. Because that person is the project manager, executive vice president, or CEO, they have the power to call the shots and make decisions.

Referent Power   Referent power is inferred to the influencer by their subordinates. Project team members who have a great deal of respect and high regard for their project managers willingly go along with decisions made by the project manager because of referent power.

Punishment power should be used as a last resort and only after all other forms have been exhausted. Sometimes you'll have to use this method, but I hope much less often than the other three forms of power. Sometimes you'll have team members who won't live up to expectations and their performance suffers as a result. This is a case where punishment power is enacted to get the employees to correct their behavior.

Leadership Styles

Extensive research has been done in the area of leadership styles. I will highlight a few of the well-known styles for exam purposes, but I encourage you to read more about leadership on your own. You can train almost anyone to follow the principles and practices of sound project management, but you won't have much of a team to lead if you haven't mastered the art of great leadership.

Autocratic   Autocratic leaders are essentially dictators. All decisions are made by the leader with little to no input from the team.

Laissez-faire   This leadership style is the opposite of the autocratic style. The leader allows the team to drive decisions and recommend actions and has little involvement in the process.

Democratic   Democratic, or participative, leaders gather all the facts and ask for input from the team before making a decision. In this style, all team members participate in the decision-making process.

Situational   As we discussed earlier, the Blanchard theory of situational leadership has four styles. Directing is used when a team member needs to know the step-by-step procedures for the problem. Coaching is used with team members who have limited experience with the task at hand. They can perform some minor functions of the task but need direction with the majority of the task. Supporting is used with team members who have completed the same types of tasks in the past and are able to complete the majority of the task at hand on their own. They may need to ask a question or two to obtain guidance along the way. Delegating is used when team members have performed the same tasks in the past and are capable of making decisions regarding unexpected issues that may occur. Delegating involves little to no input from the leader.

Transactional and Transformational   Transactional and transformational leadership styles were first developed by Bernard Bass, who was a professor emeritus in the School of Management at Binghamton. He describes transactional leaders as autocratic, activity focused, and autonomous, and they use contingent reward systems and manage by exception.

Transformational leaders tend to focus on relationships rather than activities; they are collaborative and influential and inspire and motivate their teams to perform. Bass describes transformational leaders as empowering and concerned with social justice, equity, and fairness.

Servant Leadership   *Servant leadership* is a style used on agile projects to empower team members. You'll recall that agile teams are self-directed and don't have a leader; however, all agile teams need servant leadership. Servant leadership's main goal is to serve the team and enable the maximum performance possible by the team. This is not a power grab but rather an opportunity to put the needs of others first. The servant leader serves the team members, rather than the other way around. A servant leader may also act as the team facilitator and utilize skills such as coaching, leadership, facilitation, and removing obstacles out of the way of the team. According to the *Agile Practice Guide* (PMI®, 2017), one of the roles of a servant leader is to lead the team in learning and maturing agile practices. They also note that servant leaders do this in three steps in this order: purpose, people, and process. Purpose involves working with the team to define the purpose for the work. People involves encouraging the team's success. Process involves delivering finished work that provides business value. The focus is not on a perfect process but rather delivering value that satisfies the business need.

Servant leadership is not exclusive to agile; it can be practiced on any project. However, it does integrate well with the agile mindset and approach. The *Agile Practice Guide* (PMI®, 2017) notes that servant leaders have the following characteristics: promote emotional intelligence and self-awareness, are good listeners, put the needs of others first, help team members to improve their skills, coach instead of dictate or control, encourage safety, encourage respectful behaviors, build trust among the team, and promote the skills and intelligence of others. Servant leaders are those who step up to assist the team for the betterment of the project and ensure that obstacles that impede work are removed or resolved. Servant leaders do this without additional pay and without the benefit of a title, and they should be recognized and valued by the organization.

Exam Spotlight

Know the difference between leaders and managers, the motivational and leadership theories, and the types of power for the PMP® exam. Here's a summary to help you memorize them:

Leaders   Leaders motivate, inspire, and create buy-ins for the organization's strategic vision. Leaders use power and politics and delegation to accomplish the vision.

Managers   Managers are task oriented and concerned with satisfying stakeholder needs.

Theory X – McGregor   Most people don't like work.

Theory Y – McGregor   People are motivated to perform their best given proper expectations and motivation.

Theory Z – Ouchi   The implementation of this theory increases employee loyalty and leads to high satisfaction and morale.

Contingency Theory   People are motivated to achieve levels of competency and will continue to be motivated after competency is reached.

Tannenbaum and Schmidt Continuum Management Theory   You use seven levels of delegated freedom when working with the team.

Situational Leadership Theory   Hersey and Blanchard developed this theory, which states that different situations call for different leadership styles. Blanchard describes the styles in the Situational Leadership II Model as directing, coaching, supportive, and delegating.

Reward Power   You reward desirable behavior with incentives or bonuses.

Punishment Power   You threaten team members with consequences if expectations are not met (also known as penalty or coercive power).

Expert Power   The person doing the influencing has significant knowledge or skills regarding the subject.

Legitimate Power   This is the power of the position held by the influencer (the president or vice president, for example), also known as *formal* power.

Referent Power   This is power that's inferred to the influencer.

Servant Leadership   This is a person who puts the needs of others and the project above their own. They work to help the team learn and mature agile practices through purpose, people, and process (in that order).

Developing Agile Teams

We've already discussed several techniques throughout this chapter for acquiring and developing agile teams. There are a few more concepts you should know for the exam that we'll cover in this section. The first is the Shu Ha Ri model.

Shu Ha Ri

The *Shu Ha Ri model* comes from Aikido, a Japanese martial art form. It concerns progressive learning and improving of skills and can be applied to almost any area. The idea with this model is that each state, Shu, Ha, and Ri, helps the team progress from a strict adherence to the agile process to a state of deviation from the rules to a state of continuous improvement. Let's take a brief look at each of the stages as it relates to a team learning an agile methodology.

- The Shu stage means to obey or to protect. In this state, the agile team facilitator establishes the agile process, teaches the processes and the rules to the team members, and ensures that they do not deviate from it. This allows the team to build a solid foundation for learning and understanding the agile process.

- The Ha stage means to break free or digress. After the team has a solid understanding of the agile process, they can begin to break free a bit from the strict enforcement of the rules, begin to experiment with alternative ways to perform some functions, and apply what they've learned in more creative ways.

- The Ri stage means to separate or leave. At this stage, the agile team has learned everything they can from the team facilitator. Team members may break away and perhaps form their own teams or work collaboratively to create a new custom process that will benefit the team, the organization, and future projects. At this stage, they can tailor the process to fit the needs of the project, the organization, or the skills of the team.

Shu Ha Ri is practiced in a somewhat progressive fashion, but there could be some overlap. For example, you start with Shu and learning, move to Ha and improving, maybe overlapping some learning in the Ha stage, and then progress to Ri, where Shu and Ha (learning and improving) converge to enable you to create new processes.

Pairing, Swarming, and Mobbing

This title sounds like something you might conduct in a boxing ring but pairing, swarming, and mobbing are techniques that agile teams can use to help collaborate on the project.

Pairing   You may recall that we discussed *pairing* before in terms of two developers working side by side on the same programming code, which is called pair programming. This improves the quality of the code and may increase the speed of delivery. Pairing, paired work, and paired programming all refer to the same concept.

Swarming   Think of *swarming* as "two heads are better than one." Multiple team members work collaboratively and focus on a single issue or obstacle that is preventing progress. Together, they can brainstorm or use other techniques to come up with a solution to resolve the obstacle.

Mobbing   *Mobbing* is like swarming, where multiple team members put their heads together to focus on an issue, but mobbing involves bigger concepts than a single issue or obstacle. Mobbing might entail multiple team members focusing on a work item or user story.

Empowering Team Members and Stakeholders for Success

Empowering employees is not a new concept, but it is essential to success in an agile project. The product owner is empowered to speak on behalf of the business department. They define the objectives of the project, define user stories, and determine the priority of the user stories. It's the product owner's responsibility to ensure that they are communicating with the stakeholders and with the project team. The agile team members determine estimates for the iteration, work with the product owner to perform backlog refinement, and are accountable for delivering workable functionality at the end of the iteration. The project manager or team facilitator should support team accountability for completing user stories. Retrospectives are one way that the team can demonstrate accountability. They report to one another and to the product owner on what was accomplished, or not, in the iteration. Decision-making authority should be clearly defined and documented in the team charter and as part of the ground rules. In an agile team, most decisions should be pushed to the lowest level possible; however, there are times when decisions will need to be escalated. Outlining the criteria for this in the team charter and team ground rules will help alleviate conflict when challenges arise that require decision escalation.

Agile Team Success

Agile teams should consist of cross-functional team members with varying skills, experiences, and backgrounds. The *Agile Practice Guide* (PMI®, 2017) notes there are two types of people who make up a cross-functional team: those who are "I-Shaped" people and those who are "T-Shaped" people. *I-Shaped people* have a depth of knowledge in a single domain and rarely have knowledge outside their domain. I-Shaped people are also known as specialists. They will speak up and contribute when it comes to their expertise but likely stay silent when other topics are being discussed. A specialist is focused on one area, and they are not flexible enough to jump in and help out in other domains. This constraint can sometimes cause bottlenecks on an agile project because this methodology requires the entire team to focus on delivering finished work.

T-Shaped people have a breadth of knowledge or experience in several domains. T-Shaped people are also known as generalists. They have the ability and aptitude for helping and contributing in many areas of the project. Many agile teams are made up of T-Shaped people because the work is fast paced and requires a high degree of collaboration. T-Shaped people can jump in and assist other team members. This flexibility helps reduce hand-offs and alleviates the constraint of one and only one person who can perform the function. I-Shaped people who work on an agile team often evolve into T-Shaped people as they become knowledgeable and gain experience in other domains.

Cross-functional teams are one element of successful agile projects. The contributions of both generalists and specialists help the team to integrate all of the work, deliver value often, and provide and receive feedback. Here are several other characteristics, according to the *Agile Practice Guide* (PMI®, 2017):

- Dedicated team members who are focused on the project

- Small teams (three–nine members), which enables better productivity

- Cross-functional team members that deliver finished work frequently
- Better communication, knowledge sharing, and collaboration
- Variety of skills and talents, including I-Shaped and T-Shaped people
- Stable working environment where team members can rely on one another and agree on the working approach

Another key to any successful project team, including agile teams, is you as the project manager. One of your primary responsibilities, especially on an agile team, is to recognize, address, and remove obstacles, impediments, and blockers from the team. If obstacles exist, you'll want to prioritize them and address them in order of importance or according to the impact they have on the team or project. Make certain to continually reassess the situation and ensure that the impediments are resolved and don't reappear over time. Obstacles and impediments can include an unlimited number of examples, from confusion about a user story, to the temperature in the room the team is working in, to a lack of skills on the team. Blockers might also be obstacles or impediments, but a blocker is something that stops the work. You will need to determine and prioritize the impediments, obstacles, and blockers for the team and implement solutions to resolve them or remove them where possible.

Keep in mind that impediments and blockers may be expressed as symptoms of the problem and that you may need to put some root cause analysis skills to work to determine the real cause. Here's a bit of an oversimplification. Your team members are yawning and grousing about how tired they feel. If you use some RCA tools such as the 5 Whys or other questioning techniques, you may discover the root cause is the temperature in the room. Or maybe your team is spending too much time at work and not getting enough downtime. Make certain to engage your team members in assessing and implementing solutions. They are, after all, the ones being affected by the impediments, so they may have some great ideas on how to resolve them.

You'll need to continually assess and reassess impediments, obstacles, and blockers and ensure that they are being addressed. You will recall that one of the three questions asked at a daily stand-up meeting is one directly related to this issue: is there anything standing in your way of completing the work? If the answer is "yes," get to work discovering the root cause and remove the obstacle from their path.

Team Performance Assessments

You're now ready to close out the Develop Team process. The primary output of this process is team performance assessments, and I'll cover this topic next. Several updates may come about as a result of this process as well. For example, after you've acquired resources, their availability may impact the schedule, so the project schedule may need to be updated. Resource calendars will need to be updated when new team members are added to the project. Enterprise environmental factors may also need updates to include skills assessments and development plan records for the employees you've acquired. Change requests may come about as team members are added or released from the project. I'll discuss change requests at the end of this section.

Assessing Performance

Team performance assessments involve determining a team's effectiveness. As a result of positive team-building experiences, you'll see individuals improving their skills, team behaviors and relationships improving, conflict resolutions going smoothly, reduced turnover, and team members recommending ways to improve the work of the project. Make certain to recognize individual team members for their growth and development as well as the team's growth. I talked about effective team characteristics earlier in this chapter. Assessing these characteristics will help you determine where (or whether) the project team needs improvements.

One of the ways to assess performance is by developing key performance indicators (KPIs) or other metrics at the beginning of the project (or assessment period) that can be used to measure performance. KPIs are quantifiable and verifiable. Some example metrics may include meeting project milestone dates, delivering according to the quality plan, achieving customer satisfaction, completing tasks on time, staying within or under budget, and many more. Be certain to choose KPIs that are meaningful and measurable. At the end of the project, you'll use the KPIs to verify performance improvements.

Another way to assess team member performance is by observing it. I hope you've also learned how important communication is to the success of the project. This includes communicating with your team members. I know project managers who are reticent to engage their teams in conversation unless it's official project business. I've even known project managers who've instructed their administrative assistants to give specific directions to other team members. It's difficult to understand a team member's attitude or viewpoint toward the project if you're communicating through someone else. Establish an open door policy with your team members and live up to it. The benefits are so great that it's worth a few minutes a day of chitchat to establish that feeling of trust and camaraderie. If your team perceives you as open, honest, and willing to listen, you'll be the first person they come to when issues arise.

 Real World Scenario

What Not to Do

Tina is a newly minted project manager. She has worked on many projects as the assistant project manager, but this is the first time she has led the charge. Tina is so shy she finds it difficult to give team members any kind of direction or to assign tasks, so she has her administrative assistant do it for her. Tina tells her administrative assistant what needs to be done and who needs to do it and leaves it to the assistant to inform the appropriate team members.

As the project progresses, schedule milestone dates are missed, and Tina discovers tasks that haven't started that were scheduled to begin two weeks ago. Coming in from lunch one day she sees several project team members huddled around her administrative assistant's desk. From what she overhears, they are discussing a risk event that occurred on the project.

Fortunately for Tina, one of the project managers she has worked for in the past understands what is happening. Because the administrative assistant is the one who has established relationships with the team and is in effect giving the orders, the team is treating her as the project manager instead of Tina. Tina's friend has a one-on-one coaching session with Tina about her management style and the importance of conversation and observation. Together they are able to get the project back on track.

Individual team member performance assessments are typically annual or semiannual affairs where managers let their employees know what they think of their performance over the past year and rate them accordingly. They also offer the perfect time to review an employee's job description and clarify roles and responsibilities. Appraisals are usually manager-to-employee exchanges but can incorporate a *360-degree review*, which takes in feedback from just about everyone the team member interacts with, including stakeholders, customers, project manager, peers, subordinates, and the delivery person if they have a significant amount of project interaction. I'm not a fan of 360-degree reviews because they make most non-manager types uncomfortable. "I don't want to rate my peer," is a typical response. I also find 360-degree reviews are biased. At best you'll get a response like this: "Oh, Joe is great, just great. No problems—a good guy." Or you'll get exactly the opposite if the person you're speaking with doesn't like the team member you're reviewing. Performance appraisal should be a bit more constructive than this. Nonetheless, understand the 360-degree concept for the exam.

No matter what type of appraisal is conducted, project managers should contribute to the performance appraisals of all project team members. You should be aware of potential loyalty issues when you're working in a matrix organizational structure. The team member in this structure reports to both you (as the project manager) and a functional manager. If the project manager does not have an equal say, or at least some say, about the employee's performance, it will cause the team member to be loyal to the functional manager and show little loyalty to the project or project manager. Managing these dual-reporting relationships is often a critical success factor for the project, and it is the project manager's responsibility to ensure that these relationships are managed effectively.

Performance appraisal time is also a good time to explore training needs, clarify roles and responsibilities, set goals for the future, and so on.

Assessing Agile Team Performance

Assessing performance for an agile team involves the entire team. Because of the collaboration needed on this type of team to deliver finished work and the efficiencies needed to keep the flow of work moving, and because the goal of an agile team is to optimize the output of the entire team, you'll want to evaluate the team's performance as a whole, rather than individual contributors' performance. Some assessment of team performance naturally occurs on an agile project because of the retrospective meeting. It is a look back at what went well and what might need to be improved for the next or future iterations. It is an immediate assessment that occurs throughout the life of the project. You could consider another assessment at the end of the project with all the team members present to talk about the entire life cycle of the project and the agile process itself. Examine KPIs such as estimating techniques, customer satisfaction, business value delivery, success at adopting agile processes, throughput per iteration, experience levels of team members, training needs, and obstacles and their resolution. KPIs will help you verify performance improvements. Work with the team to determine the appropriate feedback approach for delivering the team assessment so that they feel safe participating in the feedback. Don't forget to recognize the team for their growth and development, especially if they are new to agile. They need encouragement and support during the project and recognition of a job well done when the project is completed as expected.

If you work in a functional organization, it's likely you'll be required to perform individual team member assessments (or contribute to their performance reviews). There also may be times when you have an employee who struggles to connect with the team or engage in the process. In this circumstance, you'll need to conduct an individual assessment and perhaps offer training or coaching, or work with the functional manager to return them back to their original job.

Change Requests

Change requests may come about when team members are added or removed from the project. During the Develop Team process, you may find that the skills or knowledge needed for certain tasks are not present on the existing team. This may require hiring outside resources to supplement the team, and this will generate a change request. Or perhaps some team members are not good team players and you'll need to assign those folks to other tasks with less direct team interaction or remove them from the team. Team members may leave of their own accord, and this could generate a change request.

Project managers wear a lot of hats. This is one of the things that make this job so interesting. You need organization and planning skills to plan the project. You need motivation and sometimes disciplinary skills to execute the project plans. You need to exercise leadership and power where appropriate—and all the while, you have a host of relationships to manage, involving team members, stakeholders, managers, and customers. It's a great job and brings terrific satisfaction.

Managing Project Teams

The *Manage Team* process is concerned with tracking and reporting on the performance of individual team members. During this process, performance appraisals are prepared and conducted, issues are identified and resolved, and feedback is given to the team members. Some team behavior is also observed during this process, but the main focus here is on individuals and their performance.

Exam Spotlight

Take note that the *PMBOK® Guide* states that one of the outcomes or results of the Manage Team process is an update to the resource management plan, which is part of the project management plan updates output of this process. Other outcomes to note are that issues are resolved and information is provided for performance appraisals and lessons learned.

With the exception of work performance reports, you've seen all the inputs to this process before. Work performance reports document the status of the project compared to the forecasts, including but not limited to cost control, scope validation, schedule control, and quality control. Keep in mind that work performance reports are an output of the Monitor and Control Project Work process and an input to this one (an Executing process). We'll talk more about these reports in Chapter 11 when we look at the Monitoring and Controlling processes.

Emotional Intelligence and Other Tools for Managing Teams

The tools and techniques for this process may seem at first glance to be the same as others you've seen. Don't skip studying them because you'll likely see exam questions on these topics. In this process, we'll cover decision-making and emotional intelligence.

Decision-Making

Decision-making, influencing, and leadership are the three interpersonal skills the *PMBOK® Guide* points out are used most often in this process. We've already covered leadership and influencing earlier in this chapter. Effective decision-making involves making decisions in a timely manner and making decisions that reflect and support the goals of the project and support the team. Effective decisions should bring about a good result for the project, the stakeholders, and the team members. They also help you take advantage of opportunities and minimize negative risks. As project managers and good leaders, we have a responsibility to put the good of the project and the organization over our own needs, so use sound judgment when making decisions.

In the course of my career, I have seen many project managers drag their feet when it comes to making a decision or downright refuse to make a decision. No decision is a decision—in effect, you're choosing to do nothing. This can have disastrous consequences for your project. Sometimes, it's better to make a misguided decision than no decision at all. Be sure to examine all the information known at the time, consult your experts, and finally, make the decision.

Emotional Intelligence

Emotional intelligence is an interpersonal skill that involves knowing and understanding yours, and others', emotions. It's an awareness of how your actions and emotions affect those around you and adjusting and acknowledging when others are experiencing tension, not participating at their usual pace, or showing other signs of stress. It also refers to the ability to control your emotions and interpret and respond to others. It considers the level of ability you have to work cooperatively with others and to understand what motivates you and your team members. Emotional intelligence is a critical skill project managers must possess because it will help you in promoting team performance. Knowing how to read people, understanding how your words and actions come across, and being adept at seeing that a team member is struggling with an issue or challenge by simply reading their facial expressions or body language can help you stay ahead of issues and keep your team members calm and focused. Ignoring signs that a team member is troubled or not realizing that you may be inflicting verbal pain or making someone uncomfortable can be disastrous for the team and the project.

Team members can put emotional intelligence to use for one another as well and help identify when others have concerns and when you should lend a listening ear, anticipate others' actions, and recognize and acknowledge others' concerns and issues.

 Real World Scenario

Underdeveloped Emotional Intelligence

I have a friend who just got a new boss. The new boss insists on training her the way he wants the work to be performed, even though she's been in this job for a long time and knows it inside and out. It's clear he is concerned about the exact actions she takes to perform the work, not the end results. This is a mistake in this particular job because the end results are what counts. He ignores my friend's body language, he is not adept at understanding how this is making her feel, and he doesn't want to hear what she has to say. This is demoralizing and will likely result in my friend looking for work elsewhere. If he had some level of emotional intelligence, he'd perceive her emotions, realize what he was doing, and change his approach. Unfortunately, managers like this are a dime a dozen. They assume that their title or role gives them the authority to bully others, put them down, and ignore their emotional needs. Don't be that guy.

Lessons Learned Managing Teams

The outputs of the Manage Team process are the result of the conversations, performance appraisals, and conflict resolution I've talked about previously. Change requests might come about in this process as a result of a change in staffing, corrective actions might come about because of disciplinary actions or training needs, and preventive actions might be needed to reduce the impact of potential human resource issues. Any of these actions might cause changes to the resource management plan, which means you should update the project management plan as well.

Lessons learned encompasses everything you've learned about the human resources aspect during this project, including documentation that can be used as templates on future projects (such as org charts, position descriptions, and so on), techniques used to resolve conflict, the types of conflict that came up during the project, when and how virtual teams were used on the project and the procedures associated with them, the resource management plan, special skills needed during the project that weren't known about during the Planning processes, and the issue log.

 **Real World Scenario**

Project Case Study: New Kitchen Heaven Retail Store

You are in Dirk's office giving him some good news.

"The lease is signed and the work of the project has started. Ricardo has several of his staff members assigned to perform tasks related to the information technology deliverables, as do Jill and Jake for their areas.

"I held a kickoff meeting with all the key project team members. We started out with some team-building exercises, and I explained the five stages of team development. It's normal to have some conflict as we're starting out, and I let them all know my door is always open and if they have issues they can't resolve, they can come to me directly.

I explained the goals of the project, laid some ground rules for team interaction, and talked with them about the conflict resolution techniques we'll use as we get further into the project."

"I'm just glad to hear we're finally doing something," Dirk replies.

"Even though Gomez Construction doesn't start until next week, they sent their crew leader to the kickoff meeting. I was impressed with that."

Dirk asks, "Why isn't Gomez starting work now?"

"They aren't scheduled to start until August 20 and we need to get our procurement documents signed. We have a week to finish up the signatures before they get here, so we're in fine shape. But Ricardo's group has already prepared their procurement documents to purchase the switches and other equipment they need to start work. Bryan, Ricardo's team lead, finished his other project sooner than anticipated and he started setup activities today."

The key stakeholder from the marketing department peeks her head in Dirk's door. "I saw you both in here and thought I'd ask you when someone from the project team is going to work with me on the website announcement. I haven't heard anything, and I don't want to cut this so close that we put something subpar on the website. The 50th anniversary deserves a little splash."

"Okay," you reply. "I'll set up a meeting with you to get more information, and then I'll work with Ricardo to determine who the best fit is. I've noted we need to assign someone to this activity in the issue log. The person he thought he was going to assign to this task is out on family leave, and we don't know when he's expected back."

"Thanks," the stakeholder replies. "I also heard you lost a valuable team member last week. I was really sorry to see Madelyn go. What happened? And will her loss impact the project?"

"I don't want to go into all the details, but she violated our Internet acceptable use policy. She was placed on disciplinary action on this very issue once before. This may impact the project schedule because her activities were on the critical path. I've already interviewed two internal candidates who've expressed interest in working on the project. I believe either one would work out nicely. They have the skills we need and are very interested. However, some ramp-up time is needed. I've added this to the risk list because we could have an impact to the schedule if we don't get Madelyn replaced by the end of this week. I've got a change request ready also, in case there is a schedule impact. I won't know more until next week."

Dirk glances at his desk clock.

You stand and on your way to the door tell him, "Next week I'll hold a formal status meeting with all the stakeholders and will begin distributing written weekly status reports."

Project Case Study Checklist

The processes and concepts discussed in the case study are as follows:

- Direct and Manage Project Work
  - Deliverables
  - Work results
  - Work performance data
- Develop Team
  - Interpersonal skills
  - Team building
  - Team performance assessments
- Acquire Resources
  - Negotiation
  - Acquisition
  - Resource calendars
- Manage Team
  - Change requests

Understanding How This Applies to Your Next Project

The topics in this chapter are some of my favorites because this is where project management shines—dynamic teams working under the direction of a capable, responsible leader who can effectively balance the needs of the team with the needs of the project (and ultimately the organization) and pull it all off successfully. There aren't many things better in an organization than a high-performing team working together to accomplish a well-understood goal. It doesn't matter whether the team members are all in the same company, department, or country. When they're working toward a common goal and functioning at the performing level of team performance, there's almost nothing they can't accomplish.

So, how does this apply? As the project manager, it's your responsibility, and dare I say duty, to acquire the best team members possible for your project. In my experience, this doesn't always mean all my team members are highly qualified. To me, team fit and team dynamics are as important as the team members' skills. I know some will disagree with me on this next point, but I believe it's easier to train someone on a new skill (given they have the aptitude) than it is to take on a team member with an abrasive personality who is eminently qualified but can't get along with anyone else on the team. Sometimes you don't have much choice when it comes to picking team members, as referenced in the sidebar "The Only Candidate" earlier in this chapter. When you find yourself in this situation, I recommend you lay down clear ground rules for communication, problem escalation, work assignments, and so on.

Make it a habit to read at least a couple of leadership books a quarter. You may already be familiar with the topic and think there is nothing new to learn. However, staying current on the topic will reinforce concepts that you already know and will remind you of other points that you forgot about and haven't yet developed but know you should. Occasionally, you will pick up a gold nugget of information that is new and immediately applicable to your situation.

Leadership skills are invaluable, but communication skills are just as important. In my opinion, it's difficult to be an effective leader without also being an effective communicator. My guess is that if you take a close look at the leaders you respect and admire, you'll discover they are also good communicators—and communication is mostly listening, not talking. I make it a habit to practice active listening. It's amazing what people will tell you when you smile politely and ask an open-ended question or two.

Let me stress again that you cannot successfully manage a project team without communicating with them on a regular basis. The last thing you want is for a stakeholder to follow you into the elevator to inform you about a major problem with the project that you weren't aware of. That will happen if you haven't established a relationship with your team. If they don't believe you're trustworthy or they don't know you well enough to know whether you'll stand by them, you'll be one of the last people to find out what's happening. I know managers and project managers who subscribe to the "don't get too close to your team" theory, and this approach often creates barriers between the manager and the team members. I subscribe to the "all things in moderation" theory. You do want to establish relationships and prove your loyalty to the team, but you also have to know where to draw the line. When it comes time to hold a team member accountable, it can be difficult to do if you have become very close on a personal basis. However, I advocate erring on the side of developing a relationship with the team. My teams have to trust me to the point that they know they can come to me—at any time, with all types of news, good or bad—and I'll help them resolve the problem.

Summary

This chapter described four processes from the Executing process group: Direct and Manage Project Work, Acquire Resources, Develop Team, and Manage Team.

In Direct and Manage Project Work, the project management plan comes to life; activities are authorized to begin; and the product, result, or service of the project is produced. Culture will always influence the life cycle methodology you'll use to manage the project, so project managers should assess the culture and its potential effects on the project. Work performance data is gathered and documented in this process and used to report status of the project. Agile project teams are self-directed and the retrospectives are the most important meeting you conduct on an agile project. Agile projects identify and diagnose issues early and often and are able to continually improve the process.

Acquire Resources involves negotiation with other functional managers, project managers, and organizational personnel to obtain human resources to complete the work of the project. The project manager might not have control over who will be a part of the team. Availability, ability, experience, interests, and costs are all enterprise environmental factors that should be considered when you are able to choose team members. Acquiring team members and resources on agile projects may also involve negotiating with functional managers. Ideally, it's best to have fully committed, cross-functional team members who are dedicated to the agile project. Virtual teams have special considerations and needs on any project. Be certain to continually evaluate the effectiveness of team member engagement when using virtual teams.

Develop Team involves creating an open, inviting atmosphere where project team members will become efficient and cooperative, increasing productivity during the course of the project. It's the project manager's job to bring the team together into a functioning, productive group. Generational diversity, diversity, and inclusion are important elements of the Develop Team process.

According to the Tuckman-Jensen model, team development has five stages: forming, storming, norming, performing, and adjourning. All groups proceed through these stages, and the introduction of a new team member will always start the process over again.

Colocation, also known as tight matrix, is physically placing team members together in the same location. This might also include a common meeting room or gathering area, known as a war room, where team members can meet and collaborate on the project.

Several motivational theories exist, including reward and recognition, Maslow's hierarchy of needs, the Hygiene Theory, the Expectancy Theory, and the Achievement Theory. These theories conjecture that motivation is driven by several desires: physical, social, and psychological needs; anticipation of expected outcomes; and needs for achievement, power, or affiliation. The Hygiene Theory proposes that hygiene factors prevent dissatisfaction.

Leaders inspire vision and rally people around common goals. Theory X leaders think most people are motivated only through punishment, money, or position. Theory Y leaders think most people want to perform the best job they can. The Contingency Theory says that people naturally want to achieve levels of competency and will continue to be motivated by the desire for competency even after competency is reached. The Tannenbaum-Schmidt Continuum Management Theory involves seven levels of delegated freedom regarding decision-making and problem-solving. The Blanchard version of the Situational Leadership Theory describes four leadership styles to use depending on the situation.

Leaders exhibit five types of power: reward, punishment, expert, legitimate, and referent power. There are several leadership styles, including autocratic, laissez-faire, democratic, situational, transactional and transformational, and servant leadership. Servant leadership should be highly valued. These leaders are individual team members who step up to the task and help the team to achieve the project goals while getting obstacles out of their way. A servant leader's first priorities are the needs of the project and the needs of others.

Shu Ha Ri is a method of developing team skills through progressive learning and practicing and helps agile teams progress from learning agile to tailoring and creating their own agile processes.

Manage Team involves tracking and reporting on project team member performance. Performance appraisals are conducted during this process, and feedback is provided to the team members. Emotional intelligence is an interpersonal skill used throughout the project that involves knowing and understanding yours, and others', emotions. It is a critical skill all project managers must possess.

Exam Essentials

Be able to identify the distinguishing characteristics of Direct and Manage Project Work.   Direct and Manage Project Work is where the work of the project is performed, and the majority of the project budget is spent during this process.

Name the steps to direct project work on an agile project.   The steps are define initial backlog, develop user stories, hold planning meeting, perform daily stand-up meetings, hold review meetings, and conduct retrospective meetings. Retrospectives are the most important practice on any agile project.

Be able to describe virtual teams.   Virtual teams consist of members who work in disparate locations. Virtual workspaces can be used to help engage the teams. The fishbowl window and remote pairing techniques can be used for virtual agile teams. Project managers must continually evaluate the effectiveness of team member engagement on virtual teams.

Be able to describe generational diversity.   Generational diversity involves team members from one of five different generations who may be employed in the workplace today. This includes Silent generation, baby boomers, Gen X, millennials (the largest generation in the workforce today), and Gen Z.

Be able to describe diversity and inclusion.   Diversity involves hiring team members from different generations, cultures, ethnicities, religious beliefs, sexual orientation, and more. Inclusion involves ensuring that the team is collaborative, supportive, and respectful of one another and that all team members participate on the project and have a voice.

Be able to name the five stages of group formation.   The five stages of group formation are forming, storming, norming, performing, and adjourning.

Be able to define Maslow's highest level of motivation.   Self-actualization occurs when a person performs at their peak and all lower-level needs have been met.

Know the five types of power.   The five types of power are reward, punishment, expert, legitimate, and referent.

Be able to describe a servant leader.   Servant leaders are most common on agile projects. A servant leader leads the team in learning and maturing agile practices by using three steps: purpose, people, and process. A servant leader's main goal is to serve the team and enable the maximum performance possible by the team.

Know the Shu Ha Ri model.   Shu means to obey or protect. Ha means to break free or digress. Ri means to separate or leave. Shu Ha Ri is practiced in somewhat of a progressive fashion.

Know the difference between "T-Shaped people" and "I-Shaped people."   T-Shaped people have a breadth of knowledge and/or experience in several domains. They are also called generalists. I-Shaped people have a breadth and depth of knowledge in a single domain and are called dedicated experts or specialists.

Be able to describe emotional intelligence.   Emotional intelligence is an interpersonal skill that involves knowing and understanding yours and others' emotions. It's a skill all project managers must possess to help promote team performance.

Review Questions

You can find the answers to the review questions in Appendix A. Be sure to download the Bonus Exams and Bonus Questions so that you'll have a broader exposure and more experience answering questions related to the topics in this chapter.

1. You are a project manager for a growing dairy farm business. It offers organic dairy products regionally and is expanding its operations to the West Coast. It is in the process of purchasing and leasing dairy farm businesses to get operations underway. You are in charge of the network operations part of this project. An important deadline that depends on the successful completion of the testing phase is approaching. You've detected some problems with your hardware in the testing phase and discover that the hardware is not compatible with other network equipment. You are empowered to take corrective action and exchange the hardware for more compatible equipment. Which of the following statements is true?

   A. This is not a corrective action because corrective action involves human resources, not project resources.

   B. Corrective action is taken here to make sure the future project outcomes are aligned with the project management plan.

   C. Corrective action is not necessary in this case because the future project outcomes aren't affected.

   D. Corrective action serves as the change request to authorize exchanging the equipment.

2. You are a project manager for a growing dairy farm business. It offers organic dairy products regionally and is expanding its operations to the West Coast. It's in the process of purchasing and leasing dairy farm businesses to get operations underway. The subproject manager in charge of network operations has reported some hardware problems to you. You're also having problems coordinating and integrating other elements of the project. Which of the following statements is true?

   A. You are in the Direct and Manage Project Work process.

   B. Your project team doesn't appear to have the right skills and knowledge needed to perform this project.

   C. You are in the norming stage of team development.

   D. Your project team could benefit from some team-building exercises.

3. You are considering using an agile approach to conduct your project. The organization's culture is open to new ideas and supports honesty and transparency. They use KPI measurements to evaluate success. Which of the following statements are true?

   A. Culture has a significant influence on the organization.

   B. Peter Drucker stated, "Culture eats strategy for breakfast."

   C. Agile works well in the type of culture described in the question.

   D. Culture directly influences the type of agile methodology you'll choose to manage the project.

    E. Organizations that use KPI measurements and allow teams to experiment on noncritical projects and learn from failure can transform the culture and the methodologies used to manage projects.

    F. All of the above.

4. Your team members and stakeholders are new to agile techniques. Which of the following are important steps you should take to ensure that they receive the training they need to be successful with this new methodology?

    A. Determine the required competencies they need and tailor the training to these needs.

    B. Determine training options, including onsite, offsite, video training, or others.

    C. Allocate resources for training, including funding for travel to the training site if needed, funding to bring an instructor on site, a dedicated room with computers and other tools needed for learning, and more.

    D. Measure training outcomes to ensure that the concepts are learned and the competencies are realized.

    E. A, B, C

    F. A, B, C, D

5. You are in the Executing process and observing and gathering information on project elements such as schedule status, deliverables completion, lessons learned, KPIs, technical performance measures, and resource utilization. Which of the following statements is true regarding this question?

    A. This question describes project management plan updates.

    B. This question describes topics you might discuss in a review meeting.

    C. This question describes some of the updates that may bring about preventive action.

    D. This question describes work performance data.

6. Your team is working on an agile project developing an app that alerts airline passengers to gate changes for their flights once they've checked in. They are having a difficult time understanding how the user stories work together toward the big picture. Which of the following should you do? (Choose two.)

    A. Create a burndown chart.

    B. Perform one-hour backlog refinement meetings midway through the upcoming iterations.

    C. Tailor the approach to help team members gain a better understanding of the user stories.

    D. Create an impact map.

7. During a recent team meeting, you reached a resolution to an issue that's been troubling the team for several weeks. It turned out that there was a problem with one of the manufactured parts required for the project. Once this was corrected, the remaining production run went off without a hitch. You took responsibility for searching out the facts of this problem and implemented a change request to resolve the issue. Which of the following is true regarding this question?

> **A.** This is a defect repair, which is performed during the Direct and Manage Project Work process.
>
> **B.** This is a corrective action, which is performed during the Direct and Manage Project Work process.
>
> **C.** This is a preventive action, which is performed during the Direct and Manage Project Work process.
>
> **D.** This is an unapproved change request and must first go through the Perform Integrated Change Control process.

8. You are the project manager for a cable service provider. Your team members are amiable with one another and are careful to make project decisions jointly. Which of the following statements is true?

   A. They are in the forming stage of team development.

   B. They are in the norming stage of team development.

   C. They are in the storming stage of team development.

   D. They are in the adjourning stage of team development.

9. You are the project manager for a cable service provider. Your project team is researching a new service offering. They have been working together for quite some time and are in the performing stage of team development. A new member has been introduced to the team. Which of the following statements is true?

   A. The team will start over again with the storming stage.

   B. The team will continue in the performing stage.

   C. The team will start over again with the forming stage.

   D. The team will start over again at the storming stage but quickly progress to the performing stage.

10. You are the project manager for a cable service provider. Your project team is researching a new service offering. They have been working together for quite some time and have a good understanding of the task at hand. They will not likely need any direction from you to complete this task. According to Blanchard, what type of leadership style does this describe?

    A. Supporting

    B. Autocratic

    C. Laissez-faire

    D. Delegating

11. You've promised your team two days of paid time off plus a week's training in the latest technology of their choice if they complete their project ahead of schedule. This is an example of which of the following?

    A. Achievement Theory

    B. Expectancy Theory

    C. Maslow's Theory

    D. Contingency Theory

12. Your team is dispersed and they are physically split between two buildings on opposite sides of town. As a result, the team isn't very cohesive because the members don't know each other very well. The team is still in the storming stage because of the separation issues. Which of the following statements are true regarding this question? (Choose two.)

    A. You should consider colocating the team.

    B. You should consider using conflict resolution techniques.

    C. You should consider training the team members.

    D. You should consider using virtual workspaces.

13. You work for a nonprofit organization and one of their charity efforts involves donating sewing machines to women in impoverished countries. The sewing machines come with a training program that instructs them on how to create simple patterns, use the sewing machine to create products, and, most importantly, how to manage a business. Your project involves developing the app-based training program that instructors will use when working with these women. Your team used a waterfall approach to gather requirements, develop the training module content, and create a draft-level project plan. They will be using the Scrum methodology to develop the app. The team members have not used Scrum before. Which of the following options are true regarding this question?

    A. Ensure that the team members have a cross-functional set of skills, experiences, and personalities.

    B. You will help influence team member selection.

    C. Ensure that the team members have proper training in the Scrum methodology. Ideally, you should measure and assess their training success.

    D. They may feel some frustration at first as they embark on the Scrum methodology and in estimating user story work, so the product owner should work with the team to break down user stories that are too large or to reduce the number per iteration, and over time, the team will gain confidence in their estimating techniques.

    E. This question describes a hybrid methodology, even though the team will be using Scrum to develop the app.

    F. A, B, C, D

    G. A, B, C, D, E

14. You are a fabulous project manager, and your team thinks highly of you. You are well respected by the stakeholders, management team, and project team. When you make decisions, others follow your lead as a result of which of the following?

    A. Referent power

    B. Expert power

    C. Legitimate power

    D. Punishment power

15. Theory Y managers believe which of the following?

    A. People are motivated only by money, power, or position.

    B. People will perform their best if they're given proper motivation and expectations.

    C. People are motivated to achieve a high level of competency.

    D. People are motivated by the expectation of good outcomes.

16. You work for a nonprofit organization and one of their charity efforts involves donating sewing machines to women in impoverished countries. The sewing machines come with a training program that instructs them on how to create simple patterns, use the sewing machine to create products, and most importantly, how to manage a business. Your project involves developing the app-based training program that instructors will use when working with these women. Your team is using the Extreme Programming methodology to perform this project so that you can deliver results faster and maximize the speed at which the team can work. Which of the following practices are associated with this question?

    A. Continuous integration

    B. Testing at all levels

    C. Acceptance test-driven development

    D. Spikes

    E. A, B, C

    F. A, B, C, D

17. You are preparing project performance appraisals and have decided you'd like each team member to receive feedback regarding their performance from several sources, including peers, superiors, and subordinates. Which of the following is true regarding this question?

    A. This is called *360-degree review* and is used for individual team member assessments.

    B. Project managers should contribute to the performance appraisal of all project team members.

    C. Performance appraisal time is a good time to explore training needs and clarify roles and responsibilities.

    D. This is called *360-degree review* and is often used for agile team assessments.

    E. A, B, C

    F. A, B, C, D

18. Your project team will consist of members from inside the organization and consultants. When the vendor presented their project proposal, they specified specific team members who would be assigned to your project. You also worked with the functional managers in your organization to ensure that resources from within the organization would be available once the project kicked off. Which of the following statements are true regarding this question? (Choose two.)

    A. This is known as preassignment, and it happens when you are working on the Acquire Resources process.

    B. Team members promised as part of the project proposal should be noted in the project charter.

   C. This is part of the project team assignments activity, and you should also include team member names in the project organization charts, RAM charts, and other planning documents.

   D. Preassignment pertains to human resources promised as part of the project proposal, not physical resources.

19. Diversity and inclusion are important elements to consider when acquiring team members. Which of the following statements are true regarding diversity and inclusion?

   A. Diverse teams are more apt to explore new ideas and develop ideas to combat challenges and issues.

   B. Generational sensitivities involves understanding the differences in the generations in the workforce today and helps to build a team with diverse strengths, work ethics, and skills.

   C. Millennials, or Gen Y, are the largest generation of workers in the workforce today.

   D. Inclusion involves hiring team members from different religious backgrounds, cultures, ethnicities, and experiences.

   E. A, B, C

   F. A, B, C, D

20. Match the following types of tests that are used on agile-based software projects to their description.

Testing at all levels

Name of test	Description
A. Smoke test	1. Testing the software from the start to the end to ensure the application is working correctly.
B. Unit test	2. A high-level test designed to identify simple failures that could jeopardize the software program.
C. Integration test	3. This test is performed after changes are made to the code or when maintenance activities are performed on the hardware the code resides on to ensure that the software works the same way it did before the change.
D. End-to-end test	4. This test combines software modules and tests them as a group.
E. Regression test	5. This test is performed on individual modules or individual components of source code.

Chapter 10

Sharing Information

THE PMP® EXAM CONTENT FROM THE PEOPLE DOMAIN COVERED IN THIS CHAPTER INCLUDES THE FOLLOWING:

✓ **Task 1.1 Manage conflict**
- 1.1.1 Interpret the source and stage of the conflict
- 1.1.2 Analyze the context for the conflict
- 1.1.3 Evaluate/recommend/reconcile the appropriate conflict resolution solution

✓ **Task 1.2 Lead a team**
- 1.2.1 Set a clear mission and vision

✓ **Task 1.7 Address and remove impediments, obstacles, and blockers for the team**
- 1.7.1 Determine critical impediments, obstacles, and blockers for the team
- 1.7.2 Prioritize critical impediments, obstacles, and blockers for the team
- 1.7.2 Use network to implement solutions to remove impediments, obstacles, and blockers for the team
- 1.7.4 Reassess continually to ensure impediments, obstacles, and blockers for the team are being addressed

✓ **Task 1.8 Negotiate project agreements**
- 1.81. Analyze the bounds of the negotiations for agreement
- 1.8.2 Assess priorities and determine ultimate objective(s)
- 1.8.3 Verify objective(s) of the project agreement is met
- 1.8.4 Participate in agreement negotiations
- 1.8.5 Determine a negotiation strategy

✓ **Task 1.9 Collaborate with stakeholders**
- 1.9.1 Evaluate engagement needs for stakeholders
- 1.9.2 Optimize alignment between stakeholder needs, expectations, and project objectives

- 1.9.3 Build trust and influence stakeholders to accomplish project objectives

✓ **Task 1.12 Define team ground rules**

- 1.12.1 Communicate organizational principles with team and external stakeholders

- 1.12.2 Establish an environment that fosters adherence to the ground rules

- 1.12.3 Manage and rectify ground rule violations

THE PMP® EXAM CONTENT FROM THE PROCESS DOMAIN COVERED IN THIS CHAPTER INCLUDES THE FOLLOWING:

✓ **Task 2.2 Manage communications**

- 2.2.1 Analyze communication needs of all stakeholders

- 2.2.2 Determine communication methods, channels, frequency, and level of detail for all stakeholders

- 2.2.3 Communicate project information and updates effectively

- 2.2.4 Confirm communication is understood and feedback is received

✓ **Task 2.3 Assess and manage risks**

- 2.3.1 Determine risk management options

- 2.3.2 Iteratively assess and prioritize risks

✓ **Task 2.4 Engage stakeholders**

- 2.4.3 Engage stakeholders by category

- 2.4.4 Develop, execute, and validate a strategy for stakeholder engagement

✓ **Task 2.7 Plan and manage quality of products/ deliverables**

- 2.7.1 Determine quality standard required for project deliverables

- 2.7.2 Recommend options for improvement based on quality gaps
- 2.7.3 Continually survey project deliverable quality

✓ **Task 2.11 Plan and manage procurement**

- 2.11.1 Define resource requirements and needs
- 2.11.2 Communicate resource requirements
- 2.11.3 Manage suppliers/contracts
- 2.11.4 Plan and manage procurement strategy
- 2.11.5 Develop a delivery solution

✓ **Task 2.12 Manage project artifacts**

- 2.12.1 Determine the requirements for managing the project artifacts
- 2.12.2 Validate that the project information is kept up to date and accessible to all stakeholders
- 2.12.3 Continually assess the effectiveness of the management of the project artifacts

✓ **Task 2.13 Determine appropriate project methodology/methods and practices**

- 2.13.1 Assess project needs, complexity, and magnitude
- 2.13.2 Recommend project execution strategy
- 2.13.4 Recommend a project methodology/ approach
- 2.13.5 Use iterative, incremental practices throughout the project life cycle

This chapter wraps up the Executing process group. We'll look at six processes in this chapter: Implement Risk Responses, Conduct Procurements, Manage Quality, Manage Communications, Manage Project Knowledge, and Manage Stakeholder Engagement. We'll also discuss large-scale agile practices and the types of agile frameworks you can use to manage them.

Most of these processes aren't related to each other, but all are necessary to conduct the work of the project. They work in coordination with all the other Executing processes and are extensions of the Planning processes that preceded them (such as Plan Procurement Management, Plan Quality Management, and so on). Conduct Procurements and Manage Quality, in particular, work hand in hand with their Monitoring and Controlling processes (Control Procurements and Control Quality) to implement, measure, and report services and results.

Implement Risk Responses, oddly enough, is about putting those risk response plans you created in the Planning processes into action. The Conduct Procurements process is where sellers respond to the bids prepared in the Plan Procurement Management process. Contract awards are made here also. In Manage Quality, we'll look at several techniques to audit the project's quality requirements against the quality results. Manage Communications will cover more communication topics, including status reports. Manage Project Knowledge involves sharing organizational as well as project knowledge along with creating new knowledge that can be shared in the future. Last but not least, we'll talk about Manage Stakeholder Engagement, which is critical during the Executing processes. Grab your favorite beverage and let's get started.

The process names, inputs, tools and techniques, outputs, and descriptions of the project management process groups and related materials and figures in this chapter are based on content from *A Guide to the Project Management Body of Knowledge (PMBOK® Guide), Sixth Edition* (PMI®, 2017). The references to adaptive and hybrid methodologies, related materials, and figures in this chapter are based on content from the *Agile Practice Guide* (PMI®, 2017).

Implementing Risk Responses

The *Implement Risk Responses* process concerns putting the agreed-on risk response plans into action when needed. Risk response plans help to address overall project risk exposure and will help protect the project from threats (or reduce their impact) and/or take advantage of opportunities. Remember the risk owners we talked about in Chapter 7, "Identifying Project Risks"? It is their responsibility to monitor for risks and then implement the risk response plan if needed. The response plan isn't of much value if no one is paying attention to the risk symptoms or triggers and taking action to manage the risks before they occur.

As the *PMBOK® Guide* makes a point of stating, the project team spends a good deal of time and effort in working all the risk processes in the Planning process group and if no one actually takes action to manage the risks, you won't be able to reduce the impact of threats or take advantage of opportunities.

The key to this process is implementing the agreed-on risk response plans. You'll need to use influencing skills in this process when risk owners are not part of the project team. Given that the project is probably not their top priority, the project manager may have to nudge them and remind them to monitor the risks they own and put their action plans into place.

Conducting Procurements

Many times project managers must purchase goods or services to complete some or all of the work of the project. Sometimes the entire project is completed on contract. The *Conduct Procurements* process is concerned with obtaining responses to bids and proposals from potential vendors, selecting a vendor, and awarding the contract.

The procurement management plan is an input to this process and it's based on the project scope, budget, and schedule. This plan ensures that the resources you'll need throughout the project are available at the right time so that you can fulfill the project requirements. Procurement documents are used in the process, and you'll recall that they include requests for proposals (RFPs), requests for information (RFIs), requests for quotations (RFQs), and so on. The responses to those requests are known as seller proposals, and they are inputs to this process. Most organizations have a preferred format for seller responses so that the selection committee has a little easier time comparing proposals.

One element of the organizational process assets input you might consider during this process is a *qualified sellers list*. Qualified sellers lists are lists of prospective sellers who have been preapproved or prequalified to provide contract services (or provide supplies and materials) for the organization. For example, your organization might require vendors to register and maintain information regarding their experience, offerings, and current prices on a qualified sellers list. Usually, vendors must go through the procurement department to be placed on the list. They will also need to go through a vetting process with the procurement department to ensure their qualifications and standards meet those of the organization. Project managers are then required to choose their vendors from the qualified sellers list published by the procurement department. Qualified sellers lists are especially helpful on agile projects because of the speed of delivery required on this type of project. The vendors are preapproved, and you can issue a work order or contract with them quickly and they can be ready to assist on the next iteration. However, not all organizations have qualified sellers lists. If a list isn't available, you'll have to work with the project team to come up with your own requirements for selecting vendors.

The Conduct Procurements process is used only if you're obtaining goods or services from outside your own organization. If you have all the resources you need to perform the work of the project within the organization, you won't use this process.

In the following sections, you'll examine some new tools and techniques that will help vendors give you a better idea of their responses, and then you'll move on to the outputs of this process, two of which are the selected sellers and agreements.

Evaluating Proposals

Remember that the purpose of the Conduct Procurements process is to obtain responses to your RFP (or similar procurement document), select the right vendor for the job, and award the contract. The tools and techniques of this process are designed to assist the vendors in getting their proposals to you. We'll look at advertising, bidder conferences, proposal evaluations, and negotiation next.

Advertising

Advertising is letting potential vendors know that an RFP is available. Advertising can be used as a way of expanding the pool of potential vendors, or it may be a requirement, such as in the case of government projects. The company's Internet site, professional journals, and newspapers are examples of where advertising might appear. In reality, most organizations, including government organizations, have replaced advertising in the newspaper with advertising on their own websites.

Bidder Conferences

Bidder conferences (also known as vendor conferences, prebid conferences, and contractor conferences, according to the *PMBOK® Guide*) are meetings with prospective vendors or sellers that occur prior to the completion of their response proposal. You or someone from your procurement department arranges the bidder conference. The purpose is to allow all prospective vendors to meet with the buyers to ask questions and clarify issues they have regarding the project and the RFP. The facilitator of the meeting should ensure that the meeting is orderly and that all questions and answers are restated in front of the entire group; that way, all potential vendors have an equal chance at the bid. The meeting is held once, and all vendors attend at the same time. The bidder conference is held before the vendors prepare their responses so that they are sure their RFP responses will address the project requirements.

Proposal Evaluation

You might use several techniques to evaluate proposals. Simple procurements may not require much more than a quick check against the statement of work or a price comparison. Complex procurements may need several rounds of evaluation to begin narrowing the list of sellers. This is where you can use one or more of the following evaluation techniques to get to the final winner.

You can use source selection criteria as one method of rating and scoring proposals. We first discussed source selection criteria in Chapter 8, "Planning and Procuring Resources." In this process, source selection criteria are part of the procurement documentation input. You may recall that these criteria include several elements such as an understanding of the work, costs, technical capability, risk, warranty, and so on. We'll get into more detail on some of these criteria here. Keep in mind this is an input to the Conduct Procurements process, not a tool and technique, even though you'll be using the information as you would a tool and technique.

The types of goods and services you're trying to procure will dictate how detailed your evaluation criteria are. (Of course, if your organization has policies in place for evaluating proposals, then you'll use the format or criteria already established.) The selection of some goods and services might be price driven only. In other words, the bidder with the lowest price will win the bid. This is typical when the items you're buying are widely available.

When you're purchasing goods, you might request a sample from each vendor in order to compare quality (or some other criteria) against your need. For example, perhaps you need a special kind of paper stock for a project you're working on at a bank. This stock must have a watermark; it must have security threads embedded through the paper; and when the paper is used for printing, the ink must not be erasable. You can request samples of stock with these criteria from the vendors and then test them to see whether they'll work for your project.

It's always appropriate to ask the vendor for references, especially when you're hiring contract services. It's difficult to assess the quality of services because it's not a tangible product. References can tell you whether the vendor delivered on time, whether the vendor had the technical capability to perform the work, and whether the vendor's management approach was appropriate when problems surfaced. Create a list of questions to ask the references before you call.

You can request financial records to assure you—the buyer—that the vendor has the fiscal ability to perform the services they're proposing and that the vendor can purchase whatever equipment is needed to perform the services. If you examine the records of the company and find that it's two steps away from bankruptcy, that company might not be a likely candidate for your project. Remember those business acumen skills? Here's another example where they come into play.

One of the most important criteria is the evaluation of the response itself to determine whether the vendor has a clear understanding of what you're asking them to do or provide. If they missed the mark (remember, they had an opportunity at the bidder conferences to ask clarifying questions) and didn't understand what you were asking them to provide, you'll probably want to rank them very low.

Now you can compare each proposal against the criteria and rate or score each proposal for its ability to meet or fulfill these criteria. This can serve as your first step in eliminating vendors that don't match your criteria. Let's say you received 18 responses to an RFP. After evaluating each one, you discover that six of them don't match all the evaluation criteria. You eliminate those six vendors in this round. The next step is to apply measurement or scoring techniques, or analytical techniques to further evaluate the remaining 12 potential vendors.

Weighting Systems

Weighting systems assign numerical weights to evaluation criteria and then multiply them by the weights of each criterion factor to come up with a total score for each vendor. This technique provides a way to quantify the data and assist in keeping personal biases to a minimum. Weighting systems are useful when you have multiple vendors to choose from because they allow you to rank the proposals to determine the sequence of negotiations.

You'll find an example of a weighted scoring system in Chapter 3, "Delivering Business Value." These systems are commonly used to evaluate vendor proposals.

Screening Systems

Screening systems use predefined performance criteria or a set of defined minimum requirements to screen out unsuitable vendors. Perhaps your project requires board-certified engineers. One of the screening criteria would be that vendors propose project team members who have this qualification. If they don't, they're eliminated from the selection process.

Screening systems can be used together with some of the other tools and techniques of this process to rank vendor proposals.

Seller Rating Systems

Seller rating systems use information about the sellers—such as past performance, delivery, contract compliance, and quality ratings—to determine seller performance. Your organization might have seller rating systems in place, and you should check with your procurement department to see whether they exist for the bidders on your project. Part of the Control Procurements process concerns gathering and recording this type of information. Don't use seller rating systems as your sole criterion for evaluating vendors.

Analytical Techniques

Analytical techniques are used to help research potential vendors' capabilities, get an idea of cost, analyze past performance, forecast future performance, and learn from others who have undergone similar projects. An Internet search is an analytical technique that has several useful features. You can use Internet searches to find vendors, perform research on their past performance, and compare prices. You can also use the Internet to purchase items that are readily available and are generally offered for a fixed price. Internet searches are probably not useful when you're conducting high-risk or complex procurements, but you could do some research on company performance and reputation to help in choosing a vendor.

You can also use independent cost estimates (they are an output of the Plan Procurement Management process) to analyze whether the proposed cost is in alignment with the project team's expectations on cost. Remember that if there are large differences between the independent estimate and the proposed vendor cost, one of two things is happening: either the statement of work (SOW) or the terms of the contract or both are not detailed enough to allow the vendor to come up with an accurate cost, or the vendor simply failed to respond to all the requirements laid out in the contract or SOW.

Proposal evaluation techniques are a combination of all the techniques I've discussed in this section. All techniques use some form of expert judgment and evaluation criteria—whether it's objective or subjective criteria. The evaluation criteria are usually weighted, much like a weighted scoring system, and those participating as reviewers provide their ratings (usually to the project manager) to compile into a weighted proposal to determine an overall score. Scoring differences are also resolved using this technique. Some or all of the evaluation techniques described here can be used in combination with the remaining tools and techniques of the Conduct Procurements process to evaluate seller responses.

Other data analysis techniques in this process can be used to help research potential vendors' capabilities, get an idea of cost, analyze past performance, forecast future performance, and learn from others who have undergone similar projects. Internet searches are an analytical technique that has several useful features. You can use Internet searches to find vendors, perform research on their past performance, and compare prices. You can also use the Internet to purchase items that are readily available and are generally offered for a fixed price. Internet searches are probably not useful when you're conducting high-risk or complex procurements, but you could do some research on company performance and reputation to help in choosing a vendor.

Negotiating Strategy

Negotiation in the Conduct Procurements process refers to procurement negotiation. In *procurement negotiation*, both parties come to an agreement regarding the contract terms. Negotiation skills are put into practice here as the details of the contract are ironed out between the parties. At a minimum, contract language should include price, responsibilities, regulations or laws that apply, and the overall approach to the project.

The complexity of the contract will determine how extensive the contract negotiations will be. Simple contracts may have predetermined nonnegotiable elements that require only seller acceptance. Complex contracts may encompass any number of elements, including financing options, overall schedule, proprietary rights, service-level agreements, technical aspects, and more. In either case, once agreement is reached and the negotiations are finished, the contract is signed by both buyer and seller and then it is executed.

Project managers don't typically lead the contract negotiations, but they should be participants in the process. First and foremost, make certain you are well informed about every aspect of the project, the deal, and the vendor. Next is to have an end goal in mind and know where your line is. For example, you may have a timeline or budget in mind that can't be exceeded. If you know going into the negotiations that your budget cannot exceed $2,500,000, you will stick to this figure and may need to negotiate other elements to stay within this budget. Perhaps quality is a constraint, so you'll negotiate contract terms that guarantee the quality you're looking for. You might want to consider including a condition called liquidated damages when negotiating contracts with hard-driving constraints such as time, scope, or quality. Liquidated damages imposes a penalty on the vendor if they don't meet the quality (or other) criteria outlined in the contract and/or may withhold pre-agreed-on percentages of upcoming milestones.

Another negotiating technique is called the *best alternative to a negotiated agreement (BATNA)*, and it's used when the negotiations fail. Again, you need to do your homework and understand what aspects of the deal you're willing to give on and which you're not. If you are not able to reach full agreement on all the aspects of the deal and the negotiations fail, you'll need an alternative plan. This is where the BATNA comes into play. A BATNA often consists of walking away from the deal. It could also include engaging another party to help with the negotiations, suspending the negotiations for an agreed-on time, or going to court to settle the dispute. Make certain you know your BATNA before the negotiations begin.

You might consider starting the negotiation with a draft agreement. This might include your organization's standard contract template with the typical terms and conditions they use on the type of contract you're negotiating. This way, there is something on the table to begin with and you aren't starting from a blank slate. Keep in mind that the first party to introduce an offer establishes what's known as *anchoring bias*. This is because the first offer, no matter how valid or crazy it is, has a significant impact on the negotiations that follow. Protect yourself from anchoring bias by being the first to make an offer.

Contingent agreements are a way to continue negotiations while agreeing to disagree on certain terms. Contingent agreements are a way to test if the vendor can perform the way they promise. Contingent contracts may offer incentives for meeting the original require-ments and penalties or liquidated damages for missing the requirements. If your project is time constrained, for example, your vendor may easily agree to meet the date you've spec-ified in the SOW. If you begin negotiating penalties for missing the date and the vendor starts to balk, you'll know they are being unrealistic about meeting the date.

Making multiple offers at the same time is a technique that can help you learn about your partner more quickly. You can ask them which one they are more in favor of, which will help you better understand their negotiating position. It also shows that you are flex-ible and willing to accommodate reasonable changes. It's a good tactic to use to learn the other party's preferences.

Remember that being prepared and well informed is the most important aspect of contract negotiation.

Simple communication techniques such as active listening, staying focused, and minding your body language will serve you well in negotiations. Remain focused and confident, and don't let your body language say something different than your words. Silence is another communication technique that can work well for you. Silence is golden, as the saying goes, and silence in negotiations can work in your favor, especially when you are experiencing a lack of agreement. For most people, silence is a bit uncomfortable, so the other party may speak up just to break the silence. And you never know what they might offer in the wake of that awkward silence.

You might see the term *fait accompli* show up on the exam. Fait accompli tactics are used during contract negotiation when one party tries to convince the other party discuss-ing a particular contract item that it is no longer an issue. It can be a distraction technique, when the party practicing fait accompli tactics is purposely trying to keep from negotiating an issue and claims the issue cannot be changed. For example, during negotiations the vendor tells you that the key resource it's assigning to your project must start immediately or you'll lose that resource and the vendor will assign that resource work elsewhere. How-ever, you don't know—because the vendor didn't tell you—that the vendor can reserve this resource for your project and hold it until the start date. In this instance, the vendor used fait accompli tactics to push you into starting the project, or hiring this resource, sooner than you would have otherwise.

Exam Spotlight

Procurement negotiation can be performed as a process itself with inputs and outputs.

Real World Scenario

Vendor Selection for Fitness Counts HR System

Amanda Jacobson is the project manager for Fitness Counts, a nationwide chain of gyms containing all the latest and greatest fitness equipment, aerobics classes, swimming pools, and such. Fitness Counts is converting its human resources data management system. The RFP addressed several requirements, including the following:

- The new system must run on a platform that's compatible with the company's current operating system.

- Hardware must be compatible with company standards.

- Data conversion of existing HR data must be included in the price of the bid.

- Fitness Counts wants to have the ability to add custom modules using internal programmers.

- Training for the Fitness Counts programmers must be included in the bid.

The project team is in the Conduct Procurements process and has received bids based on the RFP published earlier this month. Fitness Counts is using a combination of selection criteria and a weighted scoring model to choose a vendor.

One of the evaluation criteria states that the vendor must have prior experience with a project like this. Four vendors met that criterion and proceeded to the weighted scoring selection process.

Amanda is one of the members of the selection committee. She and four other members on the committee rated the four vendors who met the initial selection criteria. They read all of the proposals and rated the criteria using factors they had predetermined for each; for example, vendors who proposed a SQL database as part of the "Platform" criteria (along with the other predetermined factors) should receive a total weighted score of 5. The following table shows their results.

	Platform	Hardware	Data conversion	Custom modules/training	Totals
Weighting factor *	5	4	5	4	
Vendor A					
Raw score	3	3	3	4	
Weighted score	15	12	15	16	58

	Platform	Hardware	Data conversion	Custom modules/training	Totals
Vendor B					
Raw score	2	3	4	3	
Weighted score	10	12	20	12	54
Vendor C					
Raw score	4	4	4	3	
Weighted score	20	16	20	12	68
Vendor D					
Raw score	3	3	2	4	
Weighted score	15	12	10	16	53

*1–5, with 5 being highest

Vendor C is the clear winner of this bid. Based on the weighted scoring model, its responses to the RFP came out ahead of the other bidders. Amanda calls Vendor C with the good news and also calls the other vendors to thank them for participating in the bid. Vendor C is awarded the contract, and Amanda moves on to the Contract Administration process.

Creating Procurement Agreements

The Conduct Procurements concerns selecting sellers or vendors for the project and creating procurement agreements. Change requests may also come about as a result of this process. Change requests require agreement between the parties regarding the context of the changes. Changes that cannot be agreed on are called *contested changes*. These take the form of disputes, claims, or appeals. They might be settled among the parties directly, through a court of law, or through arbitration.

According to the *PMBOK® Guide*, senior management signatures may be required on complex, high-risk, or high-dollar contracts. Be sure to check your organization's procurement policies regarding the authority level and amounts for which you are authorized to sign. In my organization, not only do our contracts require senior management signatures, they also require signatures from two other external departments. We have to account for the multiple reviews, signatures, and question-and-answer sessions required to obtain the contract signatures. This has a definite impact on the project schedule and must be accounted for in time estimates for these tasks.

Elements of a Procurement Agreement

Procurement agreements take different forms depending on what you are procuring, who you are procuring it from, and the nature of the goods or services themselves. You might use a purchase order, a contract, a work order against an existing contract, or other forms. You might recall that a contract is a legally binding agreement between two or more parties, typically used to acquire goods or services. Contracts have several names, including agreements, memorandums of understanding (MOUs), subcontracts, and purchase orders.

The type of contract you'll award will depend on the product or services you're procuring and your organizational policies. I talked about the types of contracts—fixed price, cost reimbursable, and so on—in Chapter 8. If you need a refresher, you can refer back to that chapter. If your project has multiple sellers, you'll award contracts for each of them.

The contract or procurement agreement should clearly address the elements of the SOW, time period of performance, pricing and payment plan, acceptance criteria, warranty periods, dispute resolution procedures, status or performance reporting procedures, limitations of liabilities, change request process, penalties and incentives, and so on.

Because contracts are legally binding and obligate your organization to fulfill the terms, they'll likely be subject to some intensive review, often by several different people. Be certain you understand your organization's policies on contract review and approval before proceeding.

Contracts, like projects, have a life cycle of their own. You might encounter questions on the exam regarding the stages of the contract life cycles, so we'll look at this topic next.

Contract Life Cycles

The contract life cycle consists of four stages:

- Requirement
- Requisition
- Solicitation
- Award

These stages are closely related to the following Project Procurement Management Knowledge Area processes from the *PMBOK® Guide*:

- Plan Procurement Management
- Conduct Procurements

A description of each of the contract life cycles follows:

Requirement   The requirement stage is the equivalent of the Plan Procurement Management process I discussed in Chapter 8. You establish the project and contract needs in this cycle, and you define the requirements of the project. The SOW defines the work of the project, the objectives, and a high-level overview of the deliverables. You develop a work breakdown structure (WBS), a make-or-buy analysis takes place, and you determine cost estimates.

The buyer provides the SOW to describe the requirements of the project when it's performed under contract. The product description can serve as the SOW.

Requisition   In the requisition stage, the project objectives are refined and confirmed. Solicitation materials such as the request for proposals (RFP), request for information (RFI), and request for quotations (RFQ) are prepared during this phase. Generally, the project manager is responsible for preparing the RFP, RFI, and RFQ. A review of the potential qualified vendors takes place, including checking references and reviewing other projects the vendors have worked on that are similar to your proposed project. Requisition occurs during the Plan Procurement Management process.

Solicitation   The solicitation stage is where vendors are asked to compete for the contract and respond to the RFP. You can use the tools and techniques of the Conduct Procurements process during this contract stage. The resulting output is the seller proposals. Solicitation occurs during the Conduct Procurements process.

Award   Vendors are chosen and contracts are awarded and signed during the award stage. The Conduct Procurements process is the equivalent of the award phase.

The project manager—or the selection committee, depending on the organizational policy—receives the bids and proposals during the award phase and applies evaluation criteria to each in order to score or rank the responses. After ranking each of the proposals, an award is made to the winning vendor, and the contract is written.

Once you have a contract, someone has to administer it. In large organizations, this responsibility will fall to the procurement department. The project manager should still have a solid understanding of administering contracts because that person will work with the procurement department to determine the satisfactory fulfillment of the contract.

Project Management Plan Updates

Several components of the project management plan may require updates after awarding a contract. For example, vendors may propose changes to the quality processes, they may introduce changes to the requirements or timeline, they may introduce new risks, and so on, all of which require updates to the associated plans. Once the vendor is on board, they also need to be added to the communications management plan so that they are receiving project information and are informed about when status reports and other project documentation is required of them.

Conducting Procurements on Agile Projects

When you're working on a predictive, or waterfall project, extensive planning is conducted up front and comprehensive procurement processes are completed before the work of the project begins. Iteration-based agile projects plan for deliveries or releases in advance and may involve procurement activities for that iteration or release. The tasks performed during the iteration will help shape the work and procurement needs of future iterations or releases. Since agile involves frequent deliveries and you may not know much about future iterations at the beginning of the project, you might want to consider using a master service agreement (MSA) contract when engaging a vendor to work on the project. This contract may outline service rates and warranties in one section while allowing elements such as scope, budget, and timeline to be added as needed. For example, scope details for each iteration or release might be included in the MSA as an appendix, allowing scope to be fluid and flexible. If your organization is new to agile, your procurement department may need to develop contracts that are agile based so that you can quickly move through procurement processes as iterations or releases are completed.

 Remember that customer collaboration on an agile project should be valued more highly than contract negotiation, according to the *Agile Manifesto*.

Now we're going to switch course and discuss the quality assurance aspects of the project. We discussed the Plan Quality Management process in Chapter 8, which prepared you for the Manage Quality process that's conducted during the Executing process group. We'll look at it next.

Laying Out Quality Assurance Procedures

The Plan Quality Management process laid out the quality standards for the project and determined how those standards are to be satisfied. The *Manage Quality* process involves performing systematic quality activities and uses quality audits and failure analysis to determine which processes should be used to achieve the project requirements, to meet the quality objectives of the project, and to ensure that activities and processes are performed efficiently and effectively. Manage Quality is also known as *quality assurance* and involves all the activities that go with quality assurance.

The project team members, the project manager, and the stakeholders are all responsible for the quality of the project. Continuous process improvement can be achieved through this process, bringing about improved process performance and eliminating unnecessary actions.

A quality assurance department or organization may be assigned to the project to oversee these processes. In that case, quality assurance might be provided to (rather than by) the project team. The project manager will have the greatest impact on the quality of the project during this process.

You've heard about the inputs to Manage Quality before, so we'll move right into the tools and techniques of this process. But wait! One reminder for you before we move on: the quality management plan is used to ensure that the project work is performed according to the quality standards laid out in the plan.

Managing Quality with Data and Audits

The Manage Quality process is concerned with continuous improvement and quality improvements. Quality reports and testing, and evaluation documents are the primary outputs of this process. To meet the goal of quality improvements and produce these outputs, you'll need to use some new tools and techniques you haven't seen before. We will discuss process analysis, diagrams and flowcharts, quality audits, Design for X, and problem solving next.

Process Analysis

Process analysis involves examining elements such as constraints, problems, and activities that don't add value to the project in order to identify process improvements. While you're examining problems and constraints, you should also look for what's causing the underlying issue. Essentially, you're performing root cause analysis at the same time as identifying potential improvements. The result of this exercise will allow you to understand the root cause of the problem, recommend improvements, and develop preventive actions for problems that are similar to the one you're examining.

Diagrams and Flowcharts

The official name of this tool and technique is "data representation." There are several new tools and techniques embedded within this description. We'll cover them one by one.

Affinity Diagrams   Affinity diagrams, also known as KJ methods, are used to group and organize thoughts and facts and can be used in conjunction with brainstorming. After you've gathered all ideas possible with brainstorming, you group similar ideas together on an affinity diagram. You may be familiar with mind-mapping techniques, where the primary concept is captured in the middle of the diagram and related ideas are branched off the middle into their own segments. This is an example of an affinity diagram. This technique is useful when decomposing the scope of the project. Put a brief scope statement in the middle circle or bubble, and then identify all the deliverables that make up the scope on independent branches (and bubbles). You can use this technique to decompose each deliverable as well.

Cause-and-Effect Diagrams   Cause-and-effect diagrams (also known as fishbone diagrams) show the relationship between the effects of problems and their causes and help identify the root cause of the problem. This diagram depicts every potential cause and subcause of a problem and the effect they have on the problem. The causes of the problem make up the fish bones on this diagram, and the effect is shown as the head of the fish. These diagrams can be depicted horizontally (with the "head or effect" to the right) or vertically (with the "head or effect" on top). Each of the branches that extend from the spine lists a cause. All of the issues that contribute to that cause are shown as offshoots on the branch. Engage your subject matter experts in a brainstorming session to outline the causes. This visual representation of cause-and-effect is helpful when you don't have quantitative data available to help analyze the problem. This diagram is also called a why-why diagram, or an Ishikawa diagram after its developer, Kaoru Ishikawa. Figure 10.1 shows an example cause-and-effect diagram.

FIGURE 10.1 Cause-and-effect diagram

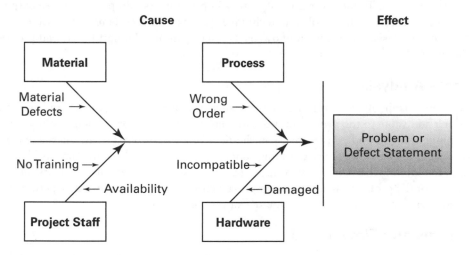

Flowcharts   The system or process flowchart shows the logical steps needed to accomplish an objective, how the elements of a system relate to each other, and what actions cause what responses. This flowchart is probably the one with which you're most familiar. It's usually constructed with rectangles and parallelograms that step through a logical sequence and allow for "Yes" and "No" branches (or some similar type of decision). Figure 10.2 shows a flowchart to help determine whether risk response plans should be developed for the risk.

Histograms   We looked at a resource histogram in Chapter 8. They are typically bar-type charts that show distributions of numerical data. Typically, histograms display continuous data such as weight, time, and amounts. In the quality processes, they can represent defects, the number of times processes are not in compliance, and other distributions regarding product or process defects.

FIGURE 10.2   Flowchart diagram

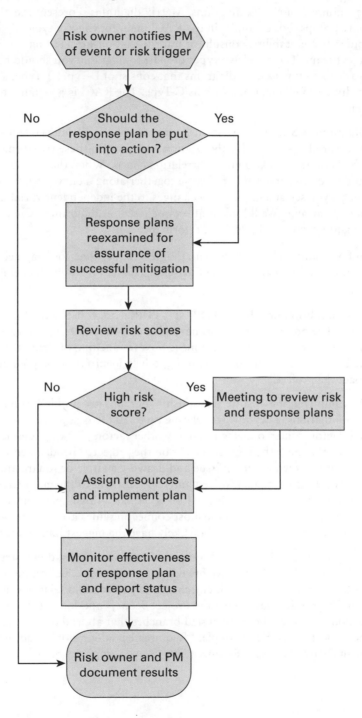

Matrix Diagrams   In Chapter 8 we also talked about RAMs and RACI charts, which are examples of matrix diagrams. Matrix diagrams represent the correlation of two or more groups. For example, in the RACI diagram you are typically relating the responsibility, accountability, consult, or inform relationships among several members of the project team. There are two types of matrix diagrams you should be aware of for the exam: two-dimensional diagrams that consist of L-Type, T-Type, and X-Type, and three-dimensional diagrams such as C-Type. The RACI is an example of an L-Type matrix.

Scatter Diagrams   Scatter diagrams use two variables—an independent variable and a dependent variable—to display the relationship between these two elements as points on a graph. They are also known as correlation charts. When the two variables correlate, the points on the graph will fall close together along a curve or line. The scatter diagram helps you see at a glance how changes to the independent variable will change the dependent variable. We'll look at an example scatter diagram in Chapter 11, "Measuring and Controlling Project Performance."

There are a few additional data representation techniques that you may need to know for the exam but that aren't named in the *PMBOK® Guide*. I recommend being familiar with them.

Process Decision Program Charts (PDPCs)   PDPCs are constructed like organizational charts. The project name or primary idea is in the top box, and related ideas are branched off from the main box, with more boxes branching off from them. This technique may be used for contingency planning to help identify all the possible results that could impact the project.

Interrelationship Digraphs   Interrelationship digraphs are problem-solving tools used for complex situations. They involve relating up to 50 factors, processes, or areas of focus. Each factor is interconnected to many other factors, making it easier to identify those elements that have the biggest impact on the project. The idea behind this tool is to start by writing down the main issue and drawing a circle or square around it. From there, identify all the elements involved and connect them to the main issue and to each other where a relationship exists. It can look a bit like a misshapen spider web when you are finished. The factors with the most connecting lines are the ones you'll want to pay particular attention to because they likely have the biggest impact on the project.

Tree Diagrams   Examples of tree diagrams include work breakdown structures, risk breakdown structures, and objective breakdown structures. The common theme with these diagrams is that they depict hierarchies and parent-child relationships. If you flip back to Chapter 4, "Developing the Project Scope," and look at the WBS for Billy Bob's Bassoon, you'll see there are nested branches that all lead to a single point (ultimately the project itself). For example, "Requirements definition" requires both "Game Requirements" and "Software Requirements" to be completed before this deliverable is complete. And both of these deliverables have work packages that must be completed. Uncomplicated structures such as this allow you to easily determine expected values for any level on the structure. (We talked about expected values in Chapter 5, "Creating the Project Schedule.") Tree diagrams are also known as systematic diagrams.

Prioritization Matrices   We've talked quite a bit about prioritization matrices, including probability and impact matrix and weighted scoring systems. Prioritization matrices are also applicable to the Manage Quality process.

Activity Network Diagrams   We talked about network diagrams in Chapter 5. They include both activity on nodes and activity on arrow diagrams and are used to help construct the project schedule.

Quality Audits

Audits, or more specifically, *quality audits*, are independent reviews performed by trained auditors or third-party reviewers. The purpose of a quality audit is the same as the purpose of the Manage Quality process—to identify ineffective and inefficient activities or processes used on the project. These audits might examine and uncover inefficient processes and procedures as well as processes that are not in compliance with organizational practices. You can perform quality audits on a regular schedule or at random, depending on the organizational policies. Quality audits performed correctly will provide the following benefits:

- The product of the project is fit for use and meets safety standards.

- Applicable laws and standards are followed.

- Corrective action is recommended and implemented where necessary.

- The quality plan for the project is followed.

- Quality improvements are identified.

- The implementation of approved change requests, corrective actions, preventive actions, and defect repairs are confirmed.

- Gaps or shortcomings in the process are identified.

- Good and best practices are implemented and shared within the organization or industry.

Quality improvements come about as a result of the quality audit. During the course of the audit, you might discover ways of improving the efficiency or effectiveness of the project, thereby increasing the value of the project and more than likely exceeding stakeholder expectations.

Quality improvements are implemented by submitting change requests and/or taking corrective action. Quality improvements interface with the Monitoring and Controlling processes because of the need to submit change requests. You'll look at change requests and change request procedures in Chapter 11.

Exam Spotlight

Remember in Chapter 7, "Identifying Project Risks," that we talked about the risk management plan. You should review the risk management plan when implementing approved corrective actions in order to take advantage of opportunities and/or minimize risk impacts.

Experienced specialists generally perform quality audits. The specialist's job is to produce an independent evaluation of the quality process. Some organizations are large enough to have their own quality assurance departments or quality assurance teams; others might have to hire contract personnel to perform this function. Internal quality assurance teams report results to the project team and management team of the organization. External quality assurance teams report results to the customer.

Design for X

Design for X (DfX) is also known as design for excellence. It is used to examine elements of the product design and product development to determine if quality standards are being met. The "X" refers to characteristics in the design of the product in order to ensure that particular element of the design is optimized. Element "X" may also improve the final product overall. "X" may represent almost any element of product development, including but not limited to usability, cost, service, safety, reliability, and most importantly to this process, quality.

Problem Solving

Problem solving looks at solving issues and finding answers to problems and challenges. According to the *PMBOK® Guide*, problem solving identifies and defines the problem, investigates and analyzes the data, determines possible solutions, implements the best solution, and checks the solution to ensure the problem is fixed. You'll also circle back around to verify that the solution was effective.

Quality Reports and Test and Evaluation Documents

The primary outputs of this process are quality reports and test and evaluation documents. Quality reports provide the stakeholders with information about whether the quality standards for the project are on track. The quality reports include findings from the Control Quality process, recommendations regarding corrective actions, and/or process or product improvements that may improve the quality results.

Test and evaluation documents are used to determine whether the quality objectives have been met. These documents may already exist within the organization in the form of templates or checklists. During this process, any recommended corrective actions, whether they are a result of a quality audit or process analysis, should be acted on immediately and processed through the change control process, which is governed by the change management plan. Let's say you're manufacturing parts for one of the deliverables of your project. Obviously, the moment you discover that the parts are not correct, you'd correct the process by calibrating the machine perhaps, or by using different raw materials, to make certain the parts are produced accurately.

Quality Improvements

Although it isn't stated as an output, one of the overarching goals of the Manage Quality process is to provide a foundation for continuous process improvements. As the name implies, continuous improvements are iterative. This process of continuous improvements sets the stage, so to speak, for improving the quality of all the project processes. This can mean project management processes, but it also means the processes and activities involved in accomplishing the work of the project.

The advantage of continuous process improvement is that it reduces the time the project team spends on ineffective or inefficient processes. If the activities don't help you meet the goals of the project or, worse yet, they hinder your progress, it's time to look at ways to improve them.

Conducting Quality Assessments on an Agile Project

The cross-functional team structure on agile projects is critical to quality. Cross-functional teams deliver completed work faster, and in much shorter periods of time, than teams using a predictive methodology. Due to the iterative nature and speed of delivery on an agile project, the team must have a high level of attention to quality or they won't produce working functionality at the end of each iteration. Quality criteria are part of the acceptance criteria that define what a completed work unit, whether iteration based or flow based, looks like. Quality on an agile team is higher due to the shortened time frames, limiting the amount of work in process, automated testing, and because dependencies from outside the project are eliminated. Predictive projects often have dependencies that are external to the project, or external to the project team. For example, your project may require cybersecurity testing. In a predictive methodology, the cybersecurity resources are not likely project team members. You'll need to coordinate with an external team to perform the testing and ensure it fits within the schedule. An agile team would have a member of the cybersecurity team as one of the cross-functional team members, and their tasks would be performed within each iteration.

Retrospective meetings, the most important practice on any agile project, helps to ensure quality criteria are met and maintained. If issues do arise, the team quickly goes about troubleshooting and performing root cause analysis to correct the error or adjust the process. They may also engage in alternative analysis in terms of trying out new ideas or approaches that will improve quality.

There are several agile methodologies we've looked at previously in this book that all contribute to increased quality. One of them is the dynamic systems development method (DSDM). One of the principles of DSDM is that quality should never be compromised. The Kanban methodology limits the amount of work in process, thereby increasing quality. Scrum delivers small units of completed functionality frequently, thereby increasing quality. Extreme Programming uses a technique called refactoring that allows the team to improve the quality of the code without impacting its functionality. Refactoring reduces duplication and eliminates poor code. Lean software development (LSD) is another software development methodology whose principles and practices incorporate quality results, as well as speed of delivery and alignment with customer expectations. Pervasive testing techniques such as acceptance test-driven development and test-driven development contribute to producing high-quality code due to the continuous testing and validating of the work. *Automated code quality analysis* is a method of testing code using scripts that review code line by line to identify bugs or vulnerabilities in the programming code. This analysis also works as a continuous improvement effort so that code is continually examined to ensure performance, readability, and consistency of the code, and to uncover bugs or vulnerabilities.

Once again we're going to shift gears and talk about a new process. We'll examine the Manage Project Knowledge process and then look at best practices regarding sharing and distributing information.

Managing Project Knowledge

Have you heard this saying, generally attributed to Francis Bacon: "Knowledge is power"? That sums up this process in a nutshell, so let's move on.

Wait, not so fast. There are some specific things you should know for the exam about this process, so let's spend a little time uncovering them.

Manage Project Knowledge involves sharing organizational as well as project knowledge and creating new knowledge that can be shared in the future. The purpose of sharing the knowledge is to increase the chance of success for your project and to improve project results (now that's power). This is much more than simply collecting project artifacts and documenting lessons learned. It involves the people who have experienced that knowledge actively sharing it with those who are about to embark on a similar project.

The *PMBOK® Guide* says there are two types of knowledge, explicit and tacit. Explicit information is easily conveyed with words, pictures, or numbers and is generally fact based. If I told you the moon is 238,900 miles from the earth, I have shared explicit information with you. Tacit information is based on experiences, beliefs, or insights. If I told you that I believe the moon is made of green cheese, I've shared tacit information with you—albeit it's incorrect information, but that's a story for another book.

The idea is to use both explicit and tacit information together to improve project results and to create new information that can be shared on future projects. Information is shared person to person and *information management tools* are how you'll capture and share that information.

The tools and techniques you'll use in the Manage Project Knowledge process that you haven't seen before are knowledge management and information management. We'll look at both next.

Knowledge Management

Knowledge management deals with sharing tacit information among the project team members (and others) so that new knowledge can be created and also to collaborate and take advantage of the knowledge and experiences of everyone involved. This can involve many forms of communication, including face-to-face or virtual sessions, interviews, networking, focus groups, workshops, storytelling, training sessions, and more. *Job shadowing* is a way to share information firsthand. This involves working with people side by side, watching them perform their tasks, and then trying out the tasks for yourself.

Information Management

Information management tools are used to collect, distribute, and share explicit information with people. The project management information system (PMIS) is one type of information management tool. Others include document libraries, web searches, published articles, the lessons learned register, and document management systems. Document management systems can be used by team members to collaborate and share documents and often have built-in retention rules so that documents past a certain age are purged or archived. They may also have rules such as how documents and records must be named, they may provide version control on the documents, and they may be able to restrict access to documents or libraries to certain groups of people or departments.

As a result of knowledge and information gathering and sharing in the Manage Project Knowledge process, you'll update the project management plan, if required, to reflect this information, update the lessons learned register, and update organizational process assets. Remember that organizational process assets includes policies, and they may require updating as a result of the new knowledge created on the project.

Managing Project Artifacts

The purpose of this process is to manage project information. Project artifacts are a key component of project information. *Project artifacts* are typically explicit information and include everything from the project charter to lessons learned documents and everything in between.

Project artifacts should be managed according to the communications management plan. You'll recall that this plan is based on both the organizational structure (i.e., functional, matrix, projectized) and the requirements the stakeholders have in terms of managing and communicating project information. If your organization has a PMO, this office will often outline all the processes for managing project artifacts. If you don't have a PMO, you'll need to take a few steps to manage your project artifacts. They include the following:

- Determine the requirements for managing project artifacts, including what artifacts to store, where to store them, when to store them, and the person(s) responsible for storing and archiving the artifacts according to the policy.

- Establish version controls so that the responsible parties can ensure the project artifacts are kept up to date and validated.

- Make sure that project artifacts are accessible to all stakeholders so that they are engaged and informed. Use the communications plan to help manage the flow of information to the stakeholders.

- Continually monitor and assess the processes used to manage, post, distribute, share, and archive project artifacts.

> One of the project manager's main objectives is to keep the stakeholders engaged and informed. Ensuring the project artifacts are accessible to stakeholders will help you meet this objective.

Managing Project Information

The *Manage Communications* process is concerned with gathering, creating, storing, distributing, retrieving, and the disposition of project communications. One of the primary functions of this process is distributing information about the project to the stakeholders in a timely manner. This can come about in several ways: status reports, project meetings, review meetings, and so on. Status reports inform stakeholders about where the project is today in regard to project schedule and budget, for example. They also describe what the project team has accomplished to date. This might include milestones completed to date, the percentage of schedule completion, and what remains to be completed. The Manage Communications process describes how this status report and other information are distributed and to whom.

In the Manage Communications process, the communications management plan that was defined during the Plan Communications Management process is put into action. This plan will define how you manage the flow of information to the stakeholders and will also help keep them informed and engaged.

There is a subtle difference between the communications management plan and the stakeholder engagement plan in the Executing process group. The communications management plan is used to manage the flow of information to the stakeholders in order to keep them engaged and informed. The stakeholder engagement plan helps maintain relationships with the stakeholders, helps to ensure their support, and manages stakeholder expectations.

Manage Communications has some new tools and techniques, which we'll look at next.

Communication and Conflict Resolution Skills

This process has several tools and techniques related to communication skills, conflict management skills, and more that we'll examine in this section. One of the techniques we've already discussed is communication technology and methods in Chapter 6, "Developing the Project Budget and Engaging Stakeholders." Pop quiz: Do you remember what the difference is between interactive, push, and pull communication methods? If not, jump back to Chapter 6 for a quick refresher.

Communication models are not a tool and technique of this process, but we'll explore this topic in a little more depth in this section because it ties in closely with the communication skills and interpersonal and team skills listed in this process. Communication skills are probably the single most important skill in your project management toolbox. You can't employ communication methods or use communication tools effectively without some sound communication skills, so let's take a look at this topic next.

Developing Great Communication Skills

Every aspect of your job as a project manager will involve communications. It has been estimated that project managers spend as much as 90 percent of their time communicating in one form or another. Therefore, communication skills are arguably among the most important skills a project manager can have. They are even more important than technical skills. Good communication skills foster an open, trusting environment, and excellent communication skills are a project manager's best asset.

Throughout this book I've emphasized how important good communication skills are. Now I'll discuss communication, listening behaviors, and conflict resolution. You'll employ each of these techniques with your project team, stakeholders, customers, and management team.

Information Exchange

Communication is the process of exchanging information. All communication includes three elements:

Sender   The sender is the person responsible for putting the information together in a clear and concise manner. The information should be complete and presented in a way that the receiver will be able to correctly understand it. Make your messages relevant to the receiver. Junk mail is annoying, and information that doesn't pertain in any way whatsoever to the receiver is nothing more than that.

Message   The message is the information being sent and received. It might be written, verbal, nonverbal, formal, informal, internal, external, horizontal, or vertical. *Horizontal communications* are messages sent to and received by peers. *Vertical communications* are messages sent and received down to subordinates and up to executive management.

Make your messages as simple as you can to get your point across. Don't complicate messages with unnecessary detail and technical jargon that others might not understand. A simple trick that helps clarify your messages, especially verbal messages, is to repeat the key information periodically. Public speakers are taught that the best way to organize a speech is to first tell the audience what you're going to tell them; second, tell them; and third, tell them what you just told them.

Receiver   The receiver is the person for whom the message is intended. Receivers are responsible for understanding the information correctly and making sure they've received all the information.

Keep in mind that receivers filter the information they receive through their knowledge of the subject, cultural influences, language, emotions, attitudes, and geographic locations. The sender should take these filters into consideration when sending messages so that the receiver will clearly understand the message that was sent.

This book is an example of the sender-receiver communication model. I'm the sender of the information. The message concerns topics you need to know to pass the PMP® exam (and, if I've done my job correctly, is written in a clear and easily understood format). You, the reader, are the receiver.

ELEMENTS OF A COMMUNICATION MODEL

Senders, receivers, and messages are the elements of communication. The way the sender packages or encodes the information and transmits it and the way the receiver unpacks or decodes the message are the models of communication exchange. This is a basic sender/receiver model.

Senders encode messages. Encoding is a method of putting the information into a format the receiver will understand. Language, pictures, and symbols are used to encode messages. Encoding formats the message for transmitting.

Transmitting is the way the information gets from the sender to the receiver. Spoken words, written documentation, memos, email, and voicemail are all transmitting methods.

Decoding is what receivers do with the information when they get it. They convert it into an understandable format. Usually, this means they read the memo, listen to the speaker, read the book, and so on.

The interactive communication model goes a couple of steps further in that the message is acknowledged and feedback is provided. Acknowledging involves receivers informing the sender they have received the message. This is not, however, an indication of agreement or that the receiver understands the message. It serves only to let the sender know the message has been received.

Feedback/response implies that the message was not only received and decoded but that the message from the sender was understood. In response, the receiver crafts a message and sends it back to the sender, starting the cycle all over again.

FORMS AND METHODS OF COMMUNICATION

Communication occurs primarily in written or verbal form. Granted, you can point to something or indicate what you need with motions, but usually you use the spoken or written word to get your message across.

Verbal communication is easier and less complicated than written communication, and it's usually a fast method of communication. Written communication, on the other hand, is an excellent way to get across complex, detailed messages. Detailed instructions are better provided in written form because it gives readers the ability to go back over information about which they're not quite sure.

Both verbal and written communication might take a formal or an informal approach. Speeches, presentations, and lectures are examples of formal verbal communication. Most project status meetings take more of a formal approach, as do most written project status reports. Generally speaking, the project manager should take an informal approach when communicating with stakeholders and project team members outside of the status meetings. This makes you appear more open and friendly and easier to approach with questions and issues.

Formal and informal approaches may also include techniques such as fact finding, asking questions, motivating team members, providing performance feedback to team members, and setting and managing stakeholder (and team) expectations.

There are many methods and skills you can use to communicate, including in person in a one-on-one setting; in small groups; by using a speaker to address an audience of people; by using mass communication methods such as newsletters and other publications; or by using network methods such as instant messaging, email, videos, press releases, and social networks.

Communication Competence

Communication competence involves having the skills necessary to craft your message for the intended audience, appropriately sharing information, and presenting yourself in a way that denotes you are a leader. It is the ability to communicate clearly and effectively in a way that you are understood. For example, if you are working on a complex project involving advanced engineering techniques, you'll communicate differently with the engineers than you will with the business members of the project team. The engineers will understand a certain level of vocabulary and terms that the business users will not likely be familiar with—and vice versa. Sometimes a project manager serves as a translator between groups or teams of professionals, and you'll need to be adept at understanding effective communication, based on the audience's level of understanding, as well as knowing what is considered appropriate communication for each group.

Feedback

Feedback involves two-way communication. Project managers must be open to feedback from team members, stakeholders, sponsors, and others. No one person has all the answers, and one of the best actions project managers can take to gain the trust of their team is to solicit feedback, actively listen to the feedback, and implement solutions or process improvements offered through the feedback process when it is appropriate and makes sense for the project. Warning! If your first reaction is to dismiss the suggestions of others and not actively engage their ideas, you will not gain a lot of trust and will shut yourself off from the communication flow, miss out on opportunities to improve, and worst case, find yourself looking for a new position elsewhere.

Nonverbal Communication Skills

I was in a meeting recently that was, to be honest, a bit tense. A senior staff member was asked a question by a VP. The question was posed in a way that indicated the VP's displeasure about the slowness of the progress made to date. The staff member wanted to assure the VP that all was well and there was no cause for concern. However, when answering the question, the staffer slumped in their chair and talked down to the table (instead of to the VP). Their body language didn't match the message and the VP picked up on this instantly. Nonverbals are just as important as the words you use. Make certain your posture, tone of voice, expressions, and gestures match the message you're delivering.

Presentations

Presentations come in many forms. They may include giving status reports; performing and presenting research for decision-making purposes; and presenting updates on deliverable status, risks, contract performance, and much more.

Project Reporting

This tool and technique involves collecting and distributing the project information, which can include status reports, measurements regarding the progress of the project, and forecasts concerning future performance. I'm certain you're all familiar with status reports. These can take many forms, but most project status reports include the progress of the project to date, the expected activities for the next period, schedule and budget updates, risk and issues updates, and change requests. Most project management software systems can produce dashboards with information about the project schedule, budget, quality, and other criteria the management team might like to see. The dashboards typically grade the health of these areas of the project with numeric scores or colors such as red, yellow, and green to indicate whether an area is in trouble, is headed into trouble and should be watched, or is in good shape.

Active Listening

What did you say? Often we think we're listening when we really aren't. In all fairness, we can take in only so much information at one time. However, it is important to perform active listening when someone else is speaking. As a project manager, you will spend the majority of your time communicating with team members, stakeholders, customers, vendors, and others. This means you should be as good a listener as you are a communicator.

You can use several techniques to improve your listening skills. Many books are devoted to this topic, so I'll try to highlight some of the most common techniques here:

- Appear interested and be interested in what the speaker is saying. This will make the speaker feel at ease and will benefit you as well. By acting interested, you become interested and thereby retain more of the information being presented.

- Making eye contact with the speaker is another effective listening tool. This lets the speaker know you are paying attention to what they're saying and are interested.

- Put your speaker at ease by letting them know beforehand that you're interested in what they're going to talk about and that you're looking forward to hearing what they have to say. While they're speaking, nod your head, smile, or make comments when and if appropriate to let the speaker know you understand the message. If you don't understand something and are in the proper setting, ask clarifying questions.

- Another great trick that works well in lots of situations is to recap what the speaker said in your own words and tell it back to them. Start with something like this, "Let me make sure I understand you correctly; you're saying ... ," and ask the speaker to confirm that you did understand them correctly.

- Just as your mother always said, it's impolite to interrupt. Interrupting is a way of telling the speaker that you aren't really listening and you're more interested in telling them what you have to say than listening to them. Interrupting gets the other person off track, they might forget their point, and it might even make them angry.

Not to disagree with Mom, but there are some occasions where interrupting is appropriate. For example, if you're in a project status meeting and someone wants to take the meeting off course, sometimes the only way to get the meeting back on track is to interrupt them. You can do this politely. Start first by saying the person's name to get their attention. Then let them know that you'd be happy to talk with them about their topic outside of the meeting or add it to the agenda for the next status meeting if it's something everyone needs to hear. Sorry, Mom.

Managing and Resolving Conflict

I said earlier in this chapter that if you have more than one person working on your project, you have a team. Here's another fact: if you have more than one person working on your project, you'll have conflict.

Everyone has desires, needs, and goals. Conflict comes into the picture when the desires, needs, or goals of one party are incompatible with the desires, needs, or goals of another party (or parties). *Conflict*, simply put, is the incompatibility of goals, which often leads to one party resisting or blocking the other party from attaining their goals. Wait—this doesn't sound like a party!

Exam Spotlight

The *PMBOK® Guide* notes that conflict can be reduced by implementing team ground rules and group norms and utilizing well-grounded project management processes. Regular and effective communication, and clear definitions of the roles and responsibilities of team members, will also go a long way in keeping conflict to a minimum.

Interpreting the Source and Stage of Conflict

The first step in managing conflict is to interpret the source and stage of the conflict. Is the conflict coming from within the team, outside the team, or from the organization, or is it a personal issue with a team member? Sources of conflict are limitless. According to the *PMBOK® Guide*, sources of conflict on teams come about due to schedule issues, availability of resources, and personal work styles. In my experience, most of the conflict on projects comes from personal interactions with other team members, stakeholders, or business units. However, ever-changing requirements on the project, budget cuts, inconsistency among executive managers on the priority of projects, mergers and acquisitions, executive leadership turnover, and thousands of other reasons could be potential sources of conflict. Sources of conflict may manifest in one of four ways:

- Interpersonal conflict exists between two individuals.

- Intrapersonal conflict exists within the mind of an individual.

- Intragroup conflict exists among team members (more than two people).

- Intergroup conflict exists among different teams across or within the organization.

After determining the source of the conflict, the next step is to determine the stage of the conflict. There are five stages of conflict developed by Louis R. Pondy, a professor of business administration and author on organizational management and other topics. They are as follows:

- In the latent stage, team members may not be aware they are in conflict. There are underlying issues here, such as roles and responsibilities, resource assignments and availability, and inconsistency in project goals among stakeholders, that could lead to later conflict.

- The perceived stage occurs when team members become aware they are in conflict or perceive they are in conflict. For example, team members may think they have a conflict and, after discussing their differences, find they are on the same page. The conflict was perceived and then immediately resolved. Or a team member may perceive there is a conflict but they don't care. Therefore, the conflict will remain in the perceived stage because the team member doesn't care and chooses not to be impacted by it. The perceived stage can occur without having experienced the latent stage. This is typically due to poor communication.

- The felt stage is when team members begin to experience stress or anxiety. Felt conflict follows perceived conflict. A team member perceives there is conflict and this conflict impacts the team member in some way, causing it to be a felt conflict.

- The manifest stage is when team members engage one another in the conflict through their interactions. Team members interact in a way that produces responses from others. This may include exerting aggression, displaying apathy, withdrawing, violence, increased competitiveness, and more.

- The aftermath stage occurs when the conflict is resolved or no longer present. If the resolution is positive and based on the needs and interests of the team members involved, they can build on this outcome and improve collaboration and communication. If the resolution just serves to keep everyone calmed down for a time or is suppressed, it will likely resurface and could be more serious than the original instance. It's always best to resolve the conflict whenever possible.

Analyzing and Resolving Conflict

To resolve the conflict, you need to understand what is causing it, so the next step in managing conflict involves analyzing the context of the conflict. Any number of techniques can be used to help with this, including root cause analysis, 5 Whys, observation, interviews, and more. You could also ask some basic questions of yourself or the team members, such as:

- What is the social or political context of the conflict?

- What is causing the conflict?

- What triggers seem to spark the conflict?

- Who are the people involved in the conflict? What are their interests, needs, and goals?

Always start by recognizing that a conflict exists and do your best to not take sides; doing so would involve you directly in the conflict. If you are involved, it's best to find a mediator or a manager or someone who has no interest in the outcome to help determine the cause and assist in finding a resolution. All team members need to be committed to finding a resolution or the conflict will not be mitigated. It may lie dormant for a while, but it will resurface. You or the mediator should make an effort to understand the concerns and perspectives of all the parties involved. Work with all the parties to mutually find behaviors that produce positive rather than negative feelings. Finally, monitor the agreed-on changes to ensure that the conflict is resolved. Unfortunately, you may encounter team members who are unwilling to agree to resolve the conflict, and you may need to terminate employees who refuse to change their behaviors or collaboratively work together to find resolution.

There are five styles of resolving conflict. Some represent permanent solutions, and others may only cover up the problem. The five styles are as follows:

Force/Direct   *Force* or *direct* is just as it sounds. One person forces a solution on the other parties. This is where the boss puts on the "Because I'm the boss and I said so" hat. Although this is a permanent solution, it isn't necessarily the best solution. People will go along with it because, well, they're forced to go along with it. It doesn't mean they agree with the solution. This isn't the best technique to use when you're trying to build a team. This is an example of a win-lose conflict resolution technique. The forcing party wins, and the losers are those who are forced to go along with the decision.

Smooth/Accommodate   The *smooth* or *accommodate* technique does not lead to a permanent solution. It's a temporary way to resolve conflict, where the areas of agreement are emphasized over the areas of difference so that the real issue stays buried. Smoothing can also occur when someone attempts to make the conflict appear less important than it is. Everyone looks at each other and scratches their head and wonders why they thought the conflict was such a big deal anyway. As a result, a compromise is reached, and everyone feels good about the solution until they get back to their desk and start thinking about the issue again. When they realize that the conflict was smoothed over and is more important than they were led to believe, or that they never dealt with the issue at hand, they'll be back at it, and the conflict will resurface. This is an example of a lose-lose conflict resolution technique because neither side wins. Smoothing is also known as accommodating.

Compromise/Reconcile   Parties that *compromise* or *reconcile* each give up something to reach a solution. Everyone involved decides what they will give on and what they won't give on, and eventually through all the give and take, a solution is reached. Neither side wins or loses in this situation. As a result, neither side is gung ho about the decision that was reached. They will drag their feet and reluctantly trudge along. If, however, both parties make firm commitments to the resolution, then the solution becomes a permanent one.

Collaborate/Problem Solve   The *collaborate* technique is also called *problem solve* and is the best way to resolve conflict. One of the key actions you'll perform with this technique is a fact-finding mission. The thinking here is that one right solution to a problem exists and the facts will bear out the solution. Once the facts are uncovered, they're presented to the parties and the decision will be clear. Thus, the solution becomes a permanent one and the conflict expires. This is a win-win solution. Multiple viewpoints are discussed and shared using this technique, and team members have the opportunity to examine all the perspectives of the issue. Collaborating will lead to true consensus where team members commit to the decision. This is the conflict resolution approach project managers use most often and is an example of a win-win conflict resolution technique.

Withdraw/Avoid When parties *withdraw* or *avoid*, they never reach resolution. The withdraw or avoid technique occurs when one of the parties gets up and leaves and refuses to discuss the conflict. It is probably the worst of all the techniques because nothing gets resolved. This is an example of a lose-lose conflict resolution technique.

Exam Spotlight

Know each of the stages of conflict and each of the conflict resolution techniques for the exam. Also remember that these techniques will not necessarily yield long-term results. The smoothing and withdrawal techniques have temporary results and aren't always good techniques to use to resolve problems. Resolutions reached through forcing and compromise might not always be satisfying for all parties, but they tend to produce longer-lasting results. Collaborating techniques are often successful and produce commitment to the decision, provided all parties believe they had the opportunity to submit their opinions and ideas.

Conflict is not always bad. *Functional conflict* is a type of conflict that leads to positive results. It often produces lively discussions among team members and creative solutions. This act of banter can often produce ideas and alternatives the team hadn't thought about before the conflict started. Analytical thinking, alternatives analysis, and good-natured competition may be used to analyze and resolve functional conflict.

You'll want to deal with conflict as soon as it arises. According to the *PMBOK® Guide*, when you have successfully resolved conflict, it will result in increased productivity and better, more positive working relationships.

We talked about the Manage Team process in Chapter 9, "Developing the Project Team." It's important to note that according to the *PMBOK® Guide*, most conflicts come about in the Manage Team process as a result of schedule issues, availability of resources (usually the lack of availability), or personal work styles. When project team members are having a conflict, address them first in private with the person who has the issue. Work in a direct and collaborative manner, but be prepared to escalate the issue into a more formalized procedure (potentially even disciplinary action) if needed.

If conflicts exist between the team members, encourage resolution between them without intervention on your part. The best conflict resolution will come about when they can work out the issues between them. When that isn't possible, you'll have to step in and help resolve the matter.

Remember that a solid team charter that establishes common goals for the team, ground rules, and established policies and procedures will help mitigate conflict before it arises.

Keep in mind that group size makes a difference when you're trying to resolve conflicts or make decisions. Remember the channels of communication you learned about in Chapter 6? The larger the group, the more lines of communication and the more difficult it will be to reach a decision. Groups of 5 to 11 people have a manageable number of participants and have been shown to make the most accurate decisions.

Use communication, listening, and conflict resolution skills wisely. As a project manager, you'll find that your day-to-day activities encompass these three areas the majority of the time. Project managers with excellent communication skills can work wonders. Communication won't take the place of proper planning and management techniques, but a project manager who communicates well with their team and the stakeholders can make up for a lack of technical skills any day, hands down. If your team and your stakeholders trust you and you can communicate the vision and the project goals and report on project status accurately and honestly, the world is your oyster.

Project Communications and Elements of Communicating

The Manage Communications process has four outputs that involve several aspects of communicating, including reports and updates. We'll look at several of them next.

The project communications outputs involve all of the work required for gathering, creating, storing, distributing, retrieving, and disposing of project communications. A few key elements you should consider regarding this output are the urgency of the message, the impact the message may have on the receiver (and perhaps the project as well), the delivery method you'll use for the message, and whether the message is sensitive or confidential and how to ensure that the message goes only to the intended receivers. According to the *PMBOK® Guide*, project communications typically consist of performance reports, deliverables status, schedule progress, cost incurred, and presentations.

The project management plan updates may include updating the communications management plan and the stakeholder engagement plan. It's important to keep the stakeholder engagement plan up to date because this will help keep the stakeholders engaged and supporting the team, and it helps you as the project manager manage their expectations. The Manage Communications process could require updates to the project scope, schedule, and/or cost baselines. Remember that together, the approved scope, schedule, and cost baselines are known as the performance measurement baseline. It's essential that you as the project manager update the project sponsor and key stakeholders about any changes to the performance management baseline.

There are six elements of the organizational process assets updates you should know:

Stakeholder Notifications   Remember that the focus of this process is distributing information. Stakeholder notifications involve notifying stakeholders when you have implemented solutions and approved changes, updated project status, resolved issues, and so on.

Project Reports   *Project reports* include the project status reports and minutes from project meetings, lessons learned, closure reports, and other documents from all the process outputs throughout the project. If you're keeping an issue log, the issues should be included with the project reports as well.

Project Presentations   *Project presentations* involve presenting project information to the stakeholders and other appropriate parties when necessary. The presentations might be formal or informal, depending on the audience and the information being communicated.

Project Records   As you might guess, *project records* include memos, correspondence, and other documents concerning the project. The best place to keep information like this is on the project website or in the PMIS. If you're keeping the information electronically, make certain it's backed up regularly. Individual team members might keep their own project records as well, but it's important to keep all the project information and records in a single repository. Encourage team members to store important project information on the project website. These records serve as historical information once the project is closed.

Feedback from Stakeholders   Feedback you receive from the stakeholders that can improve future performance on this project or performance on future projects should be captured and documented. If the information has an impact on the current project, distribute it to the appropriate team members so that future project performance can be modified to improve results.

Lessons Learned Documentation   *Lessons learned* includes information that you gather and document throughout the course of the project that can be used to benefit the current project, future projects, or other projects currently being performed by the organization. Lessons learned might include positive as well as negative lessons.

During the Manage Communications process, you'll continue conducting lessons learned meetings focusing on many different areas, depending on the nature of your project. These areas might include project management processes, product development, technical processes, project team performance, stakeholder involvement, and so on. The lessons learned register should be updated with this information.

Lessons learned meetings should always be conducted at the end of project phases and at the end of the project at minimum. Team members, stakeholders, vendors, and others involved with the project should participate in these meetings. It's important to understand, and to make your team members understand, that this is not a finger-pointing meeting. The purpose of lessons learned is to understand what went well and why—so you can repeat it on future projects—and what didn't go so well and why—so you can perform differently on future projects. These meetings can make good team-building sessions because you're creating an atmosphere of trust and sharing and you're building on each other's strengths to improve performance.

You should document the reasons or causes for the issues, the corrective action taken and why, and any other information from which future projects might benefit.

Exam Spotlight

According to the *PMBOK® Guide*, it's a project manager's professional obligation to hold lessons learned meetings. Understanding lessons learned enables continuous improvement on the current and future projects.

Communicating on Agile Projects

Communicating on agile projects has the same importance as communicating on predictive projects. Remember that on agile projects, the teams are small, which helps facilitate effective communications. The retrospective can also be used to discuss the effectiveness of project communications and how they might be improved. The following is a simple outline to ensure communication needs are met on any project.

Establish a communications management plan that's based on the organizational structure and the stakeholder requirements and follow it. This involves analyzing the communication needs of all the stakeholders and understanding what information they want, when, and how it should be delivered.

Determine what communications methods you'll use on the project. This may include verbal and written communications, formal and informal communications, meetings, presentations, one-on-one updates, and more. Keep in mind that communication should occur in multiple formats so that it's effective and keeps stakeholders with multiple learning styles engaged. Some stakeholders will read everything you put in front of them, whereas others want verbal updates and will barely glance at a written report. Effective communication meets the needs of the individual stakeholders and keeps them engaged.

Next, determine the frequency and level of detail for all stakeholders. Some stakeholders may want to know everything about the project and receive all communications. Others may want updates that pertain only to certain milestones or their own functional areas such as procurement or accounting.

Be certain to manage the flow of information according to the communications management plan so that you keep the stakeholders engaged and informed. Engaged stakeholders help ensure successful projects. If your stakeholders are uninformed and not engaged, they won't be able to make good decisions or solve problems. Unengaged stakeholders are typically not interested in the project, which could lead to a failed project. I've also experienced unengaged stakeholders who don't show up to meetings and aren't around when decisions are made on the project, only to show up later after significant progress has been made. You'll hear them bluff and blow about how they weren't included in the decision-making process and now this project is off course and will adversely impact their business area. Trust me on this—do your best to keep unengaged stakeholders informed. You can't force them to go to meetings or even to log in to the project site, but you can meet with them informally occasionally and send them emails with status updates or individual emails asking for their participation in an upcoming decision-making meeting.

Along with managing the flow of information, it's important to ensure project information and project updates are posted regularly and that you inform stakeholders when updates are available.

Last but not least, ensure that the communication is understood. One way to verify understanding is to ask for feedback. Feedback is important to you as the project manager, because it ensures that your stakeholders are informed and engaged and that they understand the current status and issues on the project. If you are not receiving feedback, you may assume everyone has the same level of understanding that you do about the status, only to encounter that bluffing and blowing stakeholder at the next status meeting asking how and why the project went off track. Remember that their expectations are not always in line with reality. The "off track" project in their mind may simply be a lack of understanding because they haven't been engaged until now. However, you are not off the hook! It's your responsibility to ensure that all key stakeholders are up to date on the latest status so that expectations stay aligned with project progress.

 Hopefully you've picked up a theme as you progressed through this outline. Effective communications, information, and updates that follow the communications management plan will keep your stakeholders engaged and informed, which means your project has a greater chance of a successful outcome.

Speaking of stakeholder engagement, we'll look at the Manage Stakeholder Engagement process next.

Managing Stakeholder Engagement

The *Manage Stakeholder Engagement* process is about satisfying the needs of the stakeholders by managing communications with them, resolving issues, engaging them on the project, managing their expectations, improving project performance by implementing requested changes, and managing concerns in anticipation of potential problems.

In my experience, managing stakeholder expectations is much more difficult than managing project team members. Stakeholders are often managers or directors in the organization who might be higher in the food chain than the project manager and aren't afraid to let you know it. Having said that, you *can* manage stakeholder expectations, and you do so using communication. Stakeholders need lots of communication in every form you can provide. If you are actively engaged with your stakeholders and interacting with them, providing project status, and resolving issues, your chances of a successful project are much greater than if you don't do these things. We looked at the multistep approach in communicating with stakeholders in the previous section. There are a few more tools and techniques you can use to determine whether your stakeholders are engaging at the right levels. We'll look at those next.

Observing and Conversing

Observation and conversation are a set of tools and techniques that are self-evident. To understand performance or quality issues, for example, you have to observe. Observation can also come into play with stakeholders and team members when you observe their body language, performance, and/or actions to determine their attitudes about the project. Conversation is carried out through communications. I hope you've learned how important communication is to the success of the project. This includes communicating with your team members. I know project managers who are reticent to engage their teams in conversation unless it's official project business. I've even known project managers who've instructed their administrative assistants to give specific directions to other team members to avoid having to converse with a team member. It's difficult to understand a team member's attitude or viewpoint toward the project if you're communicating through someone else. I recommend establishing an open door policy with your team members and live up to it. The benefits are so great that it's worth a few minutes a day of chitchat to establish that feeling of trust and camaraderie. If your team perceives you as open, honest, and willing to listen, you'll be the first person they come to when issues arise.

Ground Rules

You'll recall that ground rules are defined in the team charter and are expectations set by the project manager and project team that describe acceptable team behavior. For example, one of my pet peeves is team members who interrupt each other. In this case, one of the ground rules is one person speaks at a time during a meeting. Another ground rule might be reporting potential issues as soon as the team member becomes aware of them. Outlining ground rules like this helps the team understand expectations regarding acceptable behavior and increases productivity. Ground rules should also include expectations for stakeholder engagement. Ground rules are imperative for a healthy productive team no matter what methodology you are using to manage the work of the project.

The primary outputs of the Manage Stakeholder Engagement process are change requests and updates to the communications management plan, the stakeholder engagement plan, the change log, the issue log, the lessons learned register, and the stakeholder register.

One thing to note about the issue log in this process is that it acts more like an action item log where you record the actions needed to resolve stakeholder concerns and project issues they raise. Issues should be ranked according to their urgency and potential impact on the project. As with the issue log in the Manage Team process, you'll assign a responsible party and a due date for resolution.

Exam Spotlight

The issue log (or action item log) can be used to promote communication with stakeholders and to ensure that stakeholders and the project team have the same understanding of the issues. These issues may also be included in the RAID log we discussed earlier in the book. RAID stands for risks, action items, and decisions.

Agile Frameworks

Before diving into the agile frameworks, let's have a brief refresher on some of the principles of agile.

Agile is a methodology that focuses on delivering small, frequent units of work and may encompass any number of practices, including Scrum, XP, Crystal, Agile Unified Process (AUP), and more. Agile teams are made up of cross-functional staff with diverse skills and personalities and have between three and nine members. Lean thinking focuses on value; on delivering small, frequent units of completed functionality; and on eliminating waste from the process. Kanban and agile are both subsets of Lean thinking and share its attributes.

We talked about the four foundational values in the *Agile Manifesto* in Chapter 5. As a refresher, the values are as follows:

- Value individuals and interactions over process and tools.

- Value working software over comprehensive documentation.

- Value customer collaboration over contract negotiation.

- Value responding to change over following a plan.

The *Agile Manifesto* also defines 12 principles that clarify these values. I have recapped the principles next. I recommend understanding these for the exam.

- Satisfying the customer using continuous delivery of software is our highest priority.

- Welcome change requests at any time during development.

- Deliver code frequently. This is usually anywhere from two weeks to a couple of months. The shortest time frame possible is preferred.

- The cross-functional team of businesspeople and developers must work together daily throughout the project.

- Build your projects using motivated individuals. Give the team the environment and support they need.

- Face-to-face conversation is the most effective and efficient method of communicating and conveying information.

- The primary measure of progress for the project is working software.

- Agile teams should be able to maintain a constant and steady pace, thereby creating a sustainable development environment.

- Paying continuous attention to technical excellence and the practice of good design will enhance the team's agile capabilities.

- Simplicity is the key to success. Work on only what is essential.

- Self-organizing teams will produce the best architectures, requirements, and designs.

- Agile teams must take time to reflect on what went well, and not so well, at regular intervals so that they become more effective and change their behavior as needed to bring about improvement.

Agile Methodologies or Frameworks

Framework is a term that describes an agile methodology. I have primarily used the word *methodology* throughout this book, but you could substitute the word *framework* as well. We've discussed several throughout this book, including these:

- Scrum

- Kanban

- Extreme Programming (XP)

- Crystal methods

- Scrumban

- Feature-driven development (FDD)

- Dynamic systems development method (DSDM)

- Agile Unified Process (AUP)

Agile frameworks consist of standardized processes. The team may want to tailor their own version of agile to meet their needs using parts of multiple agile methodologies. For example, they may want to use the team structure and roles associated with Scrum along with the daily stand-ups, reviews, and retrospectives, but use a Kanban board to display work in process and control workflow. While executing the work, they might use pair programming, an XP methodology, to ensure quality and speed of delivery. Infinite possibilities exist for the team to tailor agile processes. If the team is new to agile, remember the principles of Shu Ha Ri. Start with discipline, outline the rules, and insist the team follow the rules. Once they are proficient at this, loosen up a bit and allow them some flexibility to tailor the process. Ultimately, once the concepts are mastered, the team is free to tailor the processes to match their needs.

> **Exam Spotlight**
>
> The team does not need to use an established, predefined agile framework such as Scrum or Kanban. They may develop their own agile methodology as long as it follows the principles of the *Agile Manifesto*.

Scaling Frameworks

Agile practices work best with small teams. But what happens when you have a large project that can't be accomplished using only three to nine team members? This is where scaling agile frameworks comes into play. This approach involves simply taking agile and scaling it to the enterprise. For example, you may have a large project that requires a couple of dozen people to produce the software code needed by the project deadline. You could use several agile teams to accomplish this goal, where each team has their own set of objectives and requirements. Together, they are all working toward the overarching goal of completing the project. The teams may reside in different locations, but it's important that each individual team be colocated if possible. The *Agile Practice Guide* (PMI®, 2017) mentions several scaling frameworks that we'll look at next. These are the most popular scaling frameworks used in organizations today.

Scrum of Scrums

Scrum of Scrums is an agile framework in which two or more Scrum teams are assembled (each team consisting of three–nine members) who work on the same project and, together, make up one large Scrum team. Work is coordinated across the teams, rather than one large team. Each Scrum team focuses on a portion of the project work and uses the standard Scrum practices of defining their own backlog, conducting a sprint planning meeting, working the sprint, performing reviews, and holding retrospectives. Typically, one team member from each Scrum team meets daily with representatives of other Scrum teams to discuss and coordinate progress, work on removing obstacles from the project, and discuss any future impediments that may impair the work of the collective team. This is a Scrum of Scrums and is similar to a daily stand-ups. These meetings may not always occur daily but should occur at least two times per week. The goal is to optimize the team's efficiency and ensure there are no obstacles or impediments to prevent progress. It also ensures the Scrum team is coordinating the work across the individual teams.

Large projects consisting of multiple Scrum teams may also have multiple Scrum of Scrums teams. For example, perhaps there are 12 Scrum teams on the project. It's best to divide these Scrum teams into groups of three or four Scrum of Scrums. If you have three Scrum of Scrums, they will each roll up to one Scrum of Scrums of Scrums. Each Scrum team rolls up to a Scrum of Scrums team, and multiple Scrum of Scrums teams roll up to one Scrum of Scrums of Scrums. Scrum of Scrums is also known as Meta-Scrum.

Large-Scale Scrum (LeSS)

Large-scale Scrum (LeSS) also scales up the principles of Scrum but rather than dividing the work up into individual teams as in Scrum of Scrums, LeSS focuses on the whole product. LeSS may consist of up to eight teams, with eight team members on each team. LeSS Huge may consist of a few thousand people who are all focused on the same project. LeSS is performed by holding a sprint planning meeting that includes people from all the Scrum teams. After the initial sprint planning meeting, each team holds their own planning meeting to further coordinate and clarify the work of the sprint. All team members work from the same backlog and work on the same sprint. Backlog refinement is a one-team event. The sprint review is conducted with all team members at the end of the sprint. Consider using a large conference room where each team can set up a table to display their results. The retrospective in LeSS includes the product owner, Scrum master, and representatives from each team. Members from each team should have an opportunity to attend a retrospective, so use a round-robin approach to ensure each team member gets this opportunity. The idea with the LeSS retrospective is to improve the overall process and focus on cross-team improvements rather than individual teams.

Enterprise Scrum

Enterprise Scrum applies to the organization as a whole, rather than an individual project. The idea here is to extend Scrum practices to all aspects of the organization to enable disruptive innovation. Tailoring comes into the picture here so that the organization can use, modify, and apply the techniques that will help them innovate.

Scaled Agile Framework (SAFe)

Scaled Agile Framework (SAFe) also involves implementing agile practices at an organizational or enterprise level. SAFe is a type of interactive knowledge base consisting of technical guidance, knowledge, and information about agile. It also contains knowledge and information about the organization. LeSS and SAFe have some things in common. They both use a single Scrum team and they all share the same backlog, collaboration occurs across the teams, and the teams are self-organized and self-managed. SAFe has several principles, and they are all outlined in the *Agile Practice Guide* (PMI®, 2017). The principles are focused on using systems thinking, taking an economic view, decentralizing decision-making, tapping into the intrinsic motivation of workers, and more. I recommend you review the entire list of principles for the exam.

Disciplined Agile (DA)

Disciplined Agile (DA) is a framework that combines several agile best practices in order to help the organization transform to agile methodologies. It's a tailoring approach in a way and combines the best of the more popular agile methodologies to fit the organization's needs. It includes information from functional areas such as finance, HR, the PMO, and more to help you make decisions and organize around the goals of the organization. The guiding principles of DA all center on delighting the customer by meeting, or better yet, exceeding their expectations. According to the *Agile Practice Guide* (PMI®, 2017), the principles include putting people first, focusing on collaborating and learning, supporting fit-for-purpose life cycles, focusing on goals and tailoring where necessary to meet the goals, maintaining enterprise awareness, and keeping the processes scalable. As a refresher, *fit for purpose* means it's appropriate and suitable for its intended purpose.

You've successfully completed the Executing process group. Congratulations. Remember that the Executing processes and Monitoring and Controlling processes serve as inputs to each other and that Executing and Monitoring and Controlling are both iterative process groups. As your project progresses and it becomes evident that you need to exercise controls to get the project back on track, you'll come back through the Executing process group and then proceed through the Monitoring and Controlling processes again. You'll move on to the Monitoring and Controlling process group now and find out what it's all about.

 Real World Scenario

Project Case Study: New Kitchen Heaven Retail Store

Dirk Perrier logs on and finds the following status report addressed to all the stakeholders and project team members from you:

Project Progress Report

> **Project Name**   Kitchen Heaven Retail Store
>
> **Project Number**   081501-1910
>
> **Prepared By**   Project manager
>
> **Date**   October 8

Section 1: Action Items

- *Action Item 1*: Call cable vendor. Responsible party: Ricardo. Resolution date: 9/14

- *Action Item 2*: Check T1 connection status. Responsible party: Ricardo. Resolution date: Pending

- *Action Item 3*: Buildout begins. Responsible party: Jake. Resolution date: Pending

Section 2: Scheduled and Actual Completion Dates

- *Sign lease*: Scheduled: 8/21. Completed: 8/21

- *Gomez contract signed*: Scheduled: 9/12. Completed: 9/12

- *Ethernet cable run*: Scheduled: 9/18. Completed: 9/19

- *Buildout started*: Scheduled: 9/20. Completed: In progress

Section 3: Activity That Occurred in the Project This Week

The cable run was completed without a problem.

Section 4: Progress Expected This Reporting Period Not Completed

None. Project is on track to date.

Section 5: Progress Expected Next Reporting Period

Buildout will continue. Jake reports that Gomez expects to have electrical lines run and drywall started prior to the end of the next reporting period.

Ricardo should have a T1 update. There's a slim possibility that the T1 connection will have occurred by next reporting period.

Section 6: Issues

We had to start the buildout on the last day of the cable run (this is called *fast tracking*) to keep the project on schedule. Jake and Ricardo reported only minor problems with this arrangement in that the contractors got in each other's way a time or two. This did not impact the completion of the cable run because for most of day 2 this team was in the back room and Gomez's crew was in the storefront area.

A key member of the Gomez construction crew was out last week because of a family emergency. Gomez assures us that it will not impact the buildout schedule. They replaced the team member with someone from another project, so it appears so far that the build-out is on schedule.

Ricardo is somewhat concerned about the T1 connection because the phone company won't return his calls inquiring about status. We're still ahead of the curve on this one because hardware isn't scheduled to begin testing until January 21. Hardware testing depends on the T1 connection. This is a heads up at this point, and we'll carry this as an issue in the status report going forward until it's resolved.

One of the gourmet food item suppliers Jill uses regularly went out of business. She is in the process of tracking down a new supplier to pick up the slack for the existing stores and supply the gourmet food products for the new store.

Status Meeting in Early December

"Thank you all for coming," you begin. You note those stakeholders who are present and pass out the agendas. "First, we have a contract update. Jake, would you give us the update, please?"

"Gomez Construction has submitted a seller payment request for the work completed through November 30. Shelly in our contract management office manages the payment system and handles all payment requests. She'll get a check cut and out to Gomez by the end of this month. They are doing an outstanding job as always. As you know, I've also hired an independent inspector, aside from the city and county types, so that we make sure we're up to code before the city types get there. I don't want to get caught in that trap and end up delaying the project because we can't get the city inspector back out to reinspect quickly enough."

"Thank you, Jake. Any problems with those inspections so far?"

Jake clears his throat. "It turned out to be a good move because the contractor did find some things that we were able to correct before bringing out the official inspectors."

"Ricardo, do you have a contract update for us today?"

"Yes. The contract management office used a fixed-price contract on the hardware and IT supplies order. That contract is just now making its way through the sign-off processes. My group will manage the quality control and testing once all this equipment arrives."

"Jill, can you give us the update on the store?" you ask.

"I've ordered all the retail products, have ordered the cookware line, and have lined up the chef demos. The costs for the new gourmet supplier we're using are higher than our original vendor. This impacts ongoing operations, but the hit to the project budget is minimal. I should also mention that a change request was submitted."

You point everyone's attention to the issues list. "In the last meeting I reported that Gomez had an important crew member out on a family emergency. Gomez was able to replace the team member with no impact to the schedule. The next issue was the T1 connection—I reported that Ricardo was not receiving return phone calls. Good news—that issue has been cleared, the date has been set, and we can close this issue. Are there any new issues to be added this week?"

No additional issues were reported.

"What's the forecast?" Dirk asks. "Are we on track for meeting the grand opening date given all these issues?"

"I'll have performance figures for you at the next status meeting, and I have some ideas on how we can make up this time in other ways so that we still meet the date."

You thank everyone for coming and remind them of the next meeting time.

Project Case Study Checklist

- Conduct Procurements
- Procurement negotiations
- Agreements
- Manage Quality
- Quality audits

- Making certain the project will meet and satisfy quality standards

- Manage Communications

- Project information is delivered to stakeholders in a timely manner.

- Manage Communications method is status reports via email and project meetings

- Status report is part of the project reports filed in the project notebook or filed for future reference as historical information

- Manage Stakeholder Engagement

- Communications skills

- Issue log

- Resolved issues

Understanding How This Applies to Your Next Project

This chapter is jammed with information you need to know for the exam as well as on the job. Depending on the size of your organization, many of the processes I discussed in the Executing process group might be handled by another department in your organization. I've worked in small companies (fewer than 100 people) as well as very large companies and I've always had either a person or a department who was responsible for the vendor selection, contract negotiation, and contract administration. In all these situations and in my current role overseeing major projects, I have significant input to these processes, but the person or department responsible for procurement has the ultimate control. For example, we use RFPs (and occasionally RFIs) to solicit vendors. We typically use a weighted scoring model in combination with a screening and rating system to choose a vendor. For small projects, we have a list of prequalified vendors from which to choose.

Someone who is skilled at writing contracts can best handle the contracts, which ideally should be reviewed by the legal team. The legal team should also review changes proposed by the vendor before someone signs on the dotted line. Clear, concise, and specific contracts, in my experience, are critical success factors for any project. I've too often seen contract issues bring a project to a sudden halt. Another classic contract faux pas allows the vendor to think the work is complete while the buyer believes the vendor is weeks or months away from meeting the requirements. These types of disputes can almost always be traced back to an unclear, imprecise contract. It's important to be specific in your statement of work. Make sure that the project requirements are clear and broken down far enough to be measurable and that deliverables are defined with criteria that allow you to inspect for contract compliance prior to final acceptance.

I'll emphasize one last time the importance of communication, good communication skills, and getting that information into the appropriate hands at the right time. Communication is critical on all projects no matter what methodology you're using to manage the project. Enough said.

Engaging your stakeholders on the project and managing their expectations is as important as managing the project team. Stakeholders need plenty of communication, and issues need to be discussed and resolutions agreed on. Remember the old adage that most people need to hear the same information six times before it registers with them. Stakeholders, understandably, are notorious for hearing what they want to hear. Here's an example: If I asked you to picture an elephant, you would likely picture an elephant in your mind. It might be live or the stuffed variety and it could be gray, brown, or pink, but you'd clearly understand the concept. *Elephant* is a word that's pretty hard to misinterpret. But what if I asked you to picture a three-bedroom house? There's lots of room for interpretation there. When I said *three-bedroom house*, I meant ranch style with the master on one end of the house and the other two bedrooms at the other end of the house. What did you picture? Make certain you're using language your stakeholders understand, and repeat it often until you're certain they get it.

Summary

This chapter finished up the Executing process group. We looked at Implement Risk Responses, Conduct Procurements, Manage Quality, Manage Communications, Manage Project Knowledge, and Manage Stakeholder Engagement. We finished up the chapter by looking at scaling frameworks for agile.

Implement Risk Responses concerns implementing the agreed-on risk response plans and making certain risk owners are monitoring for risks so that they can take action when appropriate.

Conduct Procurements involves obtaining responses to bids and proposals from potential vendors, selecting a vendor, and awarding the contract. In this process, the selection committee will use proposal evaluation techniques to prioritize the bids and proposals, and the outcome is that a seller (or sellers) is selected and contracts awarded.

Contracts have cycles of their own, much like projects. Phases of the contracting life cycle include requirement, requisition, solicitation, and award. Changes to the contract are managed with change requests.

Changes that cannot be agreed on are contested changes. These take the form of disputes, claims, or appeals. They might be settled among the parties directly, through a court of law, or through arbitration.

During the Manage Quality process, quality audits are performed to ensure that the project will meet and satisfy the project's quality standards set out in the quality management plan. Quality criteria are part of the acceptance criteria on an agile project.

Quality is usually higher on an agile project due to the shortened time frames and limited work in process. Retrospective meetings are the most important practice on any agile project and help ensure quality criteria are met. Kanban, DSDM, XP, and LSD all use principles and practices that incorporate quality results, speed of delivery, and alignment with customer expectations.

The Manage Project Knowledge process involves sharing organizational as well as project knowledge and creating new knowledge that can be shared in the future. The purpose for sharing the knowledge is to increase the chance of success for your project and to improve project results. Project artifacts should be managed according to the communications management plan. The steps involve determining the requirements for managing artifacts, implementing version control, making artifacts accessible to all stakeholders, and monitoring and assessing processes used to manage artifacts.

The Manage Communications process is concerned with making sure project information is available to stakeholders at the right time and in the appropriate format. This process is performed throughout the life of the project. Good communication skills foster an open, trusting environment.

Managing and resolving conflict involves interpreting the source and stage of the conflict, understanding the context of the conflict, and resolving the conflict. The stages of conflict are latent, perceived, felt, manifest, and aftermath. Conflict resolution techniques include force/direct, smooth/accommodate, compromise/reconcile, collaborate/problem solve, and withdraw/avoid. Functional conflict is a type of conflict that leads to positive results.

The Manage Stakeholder Engagement process involves making certain communication needs and expectations of the stakeholders are met, managing any issues stakeholders might raise that could become future issues, and resolving previously identified issues. Be certain to establish ground rules.

There are several scaling frameworks you should be familiar with for the exam, including Scrum of Scrums, LeSS, Enterprise Agile, SAFe, and Disciplined Agile.

Exam Essentials

Be able to describe the purpose of the Implement Risk Responses process.   Implement Risk Responses concerns putting the agreed-on risk response plans into action when needed.

Be able to describe the purpose of the Conduct Procurements process.   Conduct Procurements involves obtaining bids and proposals from vendors, selecting a vendor, and awarding a contract.

Be able to name the contracting life cycle stages.   Contracting life cycles include requirement, requisition, solicitation, and award stages.

Be able to name the type of contract that's best for an agile project.   A master service agreement (MSA) is the best type of contract to use for an agile project. Agile projects involve frequent, small deliveries, and an MSA allows you to add work to the contract via an appendix without having to rewrite the contract.

Be able to describe the purpose of the Manage Quality process.   The Manage Quality process is concerned with making certain the project will meet and satisfy the quality standards of the project.

Be able to describe managing quality on an agile project.   Quality criteria are part of the acceptance criteria and define a completed work unit. Quality on agile teams is higher due to shortened time frames, small deliverables, and automated testing, and outside dependencies are eliminated.

Be able to describe automated code quality analysis.   A method of testing code using scripts that review the code line by line to identify bugs or vulnerabilities in the programming code.

Be able to name the steps involved in managing project artifacts.   Determine the requirements for managing project artifacts, apply version controls, make artifacts accessible to all stakeholders, and continually monitor and assess the processes involved in managing artifacts.

Be able to differentiate between senders and receivers of information.   Senders are responsible for clear, concise, complete messages, whereas receivers are responsible for understanding the message correctly.

Be able to describe the sender-receiver model and the interactive communication model.   The sender-receiver model involves a sender who encodes and transmits a message and a receiver who decodes and interprets the message. The interactive communication model includes the sender-receiver model as well as an acknowledgment that the message was received and feedback regarding the contents of the message.

Be able to describe the purpose of the Manage Communication process.   Manage Communication involves making sure project information is available to stakeholders at the right time and in the appropriate format.

Be able to identify the five stages of conflict defined by Louis Pondy.   The five stages are latent, perceived, felt, manifest, and aftermath.

Be able to identify the five styles of conflict resolution.   The five styles of conflict resolution are force/direct, smooth/accommodate, compromise/reconcile, collaborate/problem solve, and withdraw/avoid.

Be able to describe the purpose of the Manage Project Knowledge process.   Manage Project Knowledge involves sharing organizational as well as project knowledge and creating new knowledge that can be shared in the future in order to increase the chance for success for the project.

Be able to name the most important benefit of communicating with and providing project artifacts to stakeholders. This helps ensure stakeholders are engaged and informed. This is critical for project success.

Be able to describe the purpose of the Manage Stakeholder Engagement process. Manage Stakeholder Engagement involves satisfying the needs of the stakeholders and successfully meeting the goals of the project by managing communications with stakeholders, resolving issues, engaging them on the project, managing their expectations, improving project performance by implementing requested changes, and managing concerns in anticipation of potential problems.

Be able to name the purpose of an agile scaling framework. This allows for scaling agile from individual teams to the enterprise.

Be able to name some popular agile scaling frameworks. These include Scrum of Scrums, Large-Scale Scrum, Enterprise Scrum, Scaled Agile Framework, and Disciplined Agile.

Review Questions

You can find the answers to the review questions in Appendix A. Be sure to download the Bonus Exams and Bonus Questions so that you'll have a broader exposure and more experience answering questions related to the topics in this chapter.

1. You have been hired as a contract project manager for Grapevine Vineyards. Grapevine wants you to design an Internet wine club for its customers. Customers must register before being allowed to order wine over the Internet so that legal age can be established. You know that the module to verify registration must be written and tested using data from Grapevine's existing database. This new module cannot be tested until the data from the existing system is loaded. You are going to hire a vendor to perform the programming and testing tasks for this module to help speed up the project schedule. You decide to use an IFB and include a detailed SOW. This is an example of which of the following?

   A. Source selection criteria

   B. Seller proposals

   C. Advertising

   D. Procurement documentation

2. Receivers in the communication model filter their information through all of the following. (Choose all that apply.)

   A. Culture

   B. Knowledge of subject

   C. Habits

   D. Language

   E. Attitudes

3. You have accumulated project information throughout the project and regularly post it to the project site. You just received some important information that you need to ensure the stakeholders receive as soon as possible. Which of the following are true regarding this question?

   A. Keeping stakeholders informed with up-to-date project information will help keep them engaged.

   B. Engaged and informed stakeholders increase the chances of a successful project.

   C. You will use distribution tools to ensure the project information and artifacts are accessible by all stakeholders.

   D. Good communication skills foster an open and trusting environment.

   E. A, B, C, D

   F. A, B, D

4. You know that the next status meeting will require some discussion and a decision for a problem that has surfaced on the project. You need to ensure that the impediment on the project is resolved at this meeting. To make the most accurate decision, you know that the number of participants in the meeting should be limited to how many?

   A. 1 to 5

   B. 5 to 11

    C. 7 to 16

    D. 10 to 18

5. You are holding end-of-phase meetings with your team members and key stakeholders to learn what has hindered and helped the project team's performance of the work. Which of the following options are true regarding this question? (Choose three.)

    A. These meetings are called lessons learned meetings. Lessons learned documentation is part of the organizational process assets output of the Manage Communications process. These meetings are also a good team-building activity.

    B. The information learned from these meetings concerns processes and activities that have already occurred, so it will be most useful to document the lessons learned and use it for future projects.

    C. These meetings should be documented and could include the effectiveness of the project management processes, product development, technical processes, and the effectiveness of agile project management techniques.

    D. These meetings are similar to review meetings that occur at the end of an iteration on an agile project.

6. You are a project manager for Dakota Software Consulting Services. You're working with a major retailer that offers its products online. The company is interested in knowing customer characteristics, the amounts of first-time orders, and similar information. As a potential bidder for this project, you worked on the RFP response and submitted the proposal. When the selection committee received the RFP responses from all the vendors bidding on this project, they used a weighted system in combination with a seller rating system to make a selection. Which of the following are true? (Choose three.)

    A. The procurement officers for both seller and buyer determine how extensive the complexity of the contract negotiations might be.

    B. Weighted systems are a proposal evaluation technique in the Conduct Procurements process.

    C. Procurement officers typically handle the contract negotiations but project managers should participate in the negotiations. Procurement negotiations require sound negotiating skills. Both parties will discuss elements such as price, roles and responsibilities, and the overall approach to the project.

    D. Seller rating systems may be used as a sole evaluation technique when evaluating and selecting vendors.

    E. Procurement negotiations can be performed as a stand-alone process with its own inputs and outputs.

7. You have been asked to submit a proposal for a project that has been put out for bid. First you attend the bidder conference to ask questions of the buyers and to hear the questions some of the other bidders will ask. Which of the following statements are true?

    A. Bidder conferences are also known as contractor conferences.

    B. Bidder conferences are also known as vendor conferences.

    C. Bidder conferences are also known as prebid conferences.

    D. Bidder conferences are also known as procurement conferences.

    E. A, B, C, D

    F. A, B, C

8. You have been asked to submit a proposal for a project that has been put out for bid. Prior to submitting the proposal, your company must register so that its firm is on the qualified sellers list. Which of the following statements are true? (Choose two.)

   A. The qualified sellers list is a list of prospective sellers who have been preapproved or prequalified to provide goods or services to the organization.

   B. The qualified sellers list provides information about the project and the company that wrote the RFP and is an output of the Conduct Procurements process.

   C. The qualified sellers list provides information about the project and the company that wrote the RFP and is a tool and technique of the Conduct Procurements process.

   D. The qualified sellers list is useful on an agile project because the team has vendors that are preapproved and work orders can be issued quickly.

9. You are a project manager for Fountain of Youth Spring Water bottlers. Your project involves installing a new accounting system. You have identified several problems and want to brainstorm the causes of those problems with the project team because there isn't sufficient quantitative data to analyze the problems. Which of the following diagrams will you use to outline each problem, including the causes and effects of the problems?

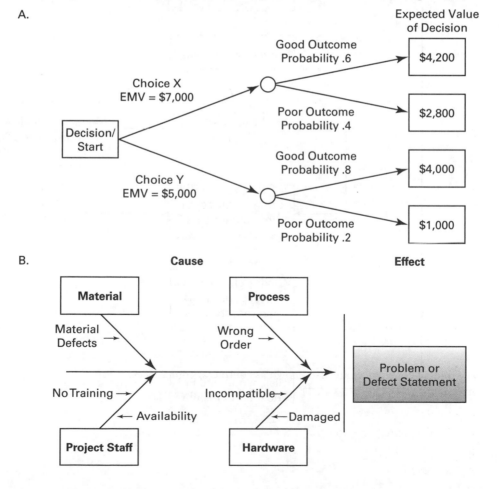

C.

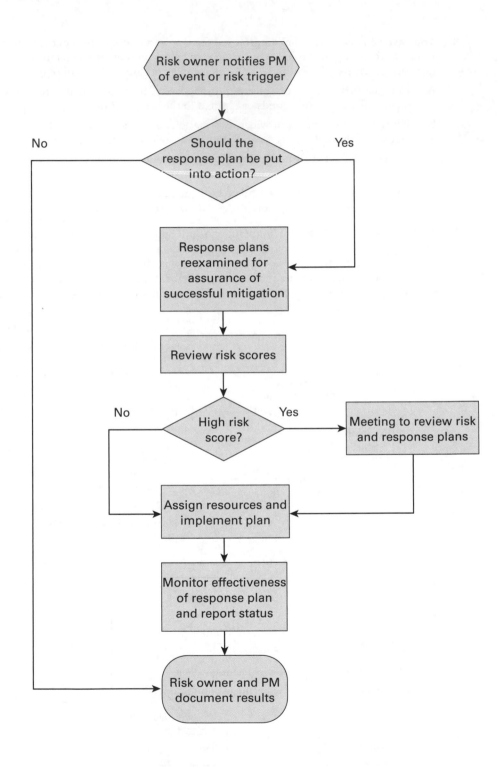

10. During the opening rounds of contract negotiation, the other party uses a fait accompli tactic. Which of the following statements is true about fait accompli tactics?

    A. One party agrees to accept the offer of the other party but secretly knows they will bring the issue back up at a later time.

    B. One party claims the issue was documented and accepted in the issue log but they're willing to reopen it for discussion.

    C. One party claims the issue under discussion has already been decided and can't be changed.

    D. One party claims to accept the offer of the other party, provided a contract change request is submitted describing the offer in detail.

11. According to the *Agile Manifesto,* which of the following should be more highly valued than contract negotiation?

    A. Satisfying the customer using continuous delivery

    B. Using motivated individuals to deliver business value

    C. Simplicity

    D. Customer collaboration

12. Which of the following describe the purpose of a quality audit?

    A. To determine which project processes or activities are inefficient or ineffective

    B. To examine the work of the project, or the results of a process, and accept the work results

    C. To improve processes and reduce the cost of quality

    D. To improve processes and increase the percentage of product or service acceptance

    E. A, C, D

    F. A, B, C, D

13. You are a contract project manager for a wholesale flower distribution company. Your project involves developing a website for the company to allow retailers to place their flower orders online. You will also provide a separate link for individual purchases that are ordered, packaged, and mailed to the consumer directly from the grower's site. This project involves coordinating the parent company, growers, and distributors. You've discovered a problem with one of the technical processes needed to perform this project. You examine elements such as constraints, problems, and activities that don't add value to the project in order to identify process improvements and reduce or eliminate the problems with the technical processes. Which of the following is true?

    A. You are using the process analysis technique.

    B. You are using the quality audit technique.

    C. You are using the Design for X technique.

    D. You are using a design of experiments tool and technique.

14. You are performing the Manage Quality process and establishing expected values for the dependent relationships in the hierarchy. Which technique are you using?

    A. Affinity diagrams

    B. PDPC

    C. Tree diagrams

    D. Interrelationship digraphs

15. You are working on a project and discover that one of the business users responsible for testing the product never completed this activity. She has written an email requesting that one of your team members drop everything to assist her with a problem that could have been avoided if she had performed the test. This employee reports to a stakeholder, not to the project team. All project team members and stakeholders are colocated. Because your team also needs her to participate in an upcoming test, you decide to do which of the following?

    A. You decide to record the issue in the issue log and bring it up at the next status meeting. Everyone can benefit from understanding the importance of stakeholders fulfilling their roles and responsibilities on the project.

    B. You decide to record the issue in the issue log and then phone the stakeholder to explain what happened. You know speaking with the stakeholder directly is the most effective means for resolving issues.

    C. You decide to have a face-to-face meeting with the stakeholder because this is the most effective means for resolving issues with them.

    D. You decide to email the stakeholder and explain what happened in a professional manner so you can document that you surfaced the issue and then incorporate their response in the next status report.

16. Your project is progressing as planned. The project team has come up with a demo that the sales team will use when making presentations to prospective clients. You will do which of the following at your next stakeholder project status meeting?

    A. Preview the demo for stakeholders and obtain their approval and sign-off.

    B. Report on the progress of the demo and note that it's a completed task.

    C. Review the technical documentation of the demo and obtain approval and sign-off.

    D. Report that the demo has been noted as a completed task in the project information system.

17. You are a contract project manager for a wholesale flower distribution company. Your project involves developing a website for the company to allow retailers to place their flower orders online. You will also provide a separate link for individual purchases that are ordered, packaged, and mailed to the consumer directly from the grower's site. This project involves coordinating the parent company, growers, and distributors. You are using an agile methodology to manage this project and have discovered that some stakeholders do not understand their role on the project, and this is causing conflict for the team. Which of the following should you do to help reduce conflict for the stakeholders and the team?

A. Clearly define the roles and responsibility of team members and stakeholders.

B. Establish team ground rules in the team charter including stakeholder expectations.

C. Communicate effectively and ensure project artifacts are accessible to stakeholders.

D. Address conflict as soon as it occurs to reduce or eliminate its impact.

E. A, C, D

F. A, B, C, D

18. According to the *PMBOK® Guide*, this party is responsible for managing stakeholder expectations.

A. Project manager, because actively managing stakeholder expectations will reduce project risk and help unresolved issues get resolved quickly.

B. Project sponsor, because the overall success or failure of the project rests on this individual.

C. All stakeholders have the responsibility to make sure their expectations are managed and that they receive the proper information at the right time.

D. Project manager and the project team members together have the responsibility because the project manager alone cannot manage all the stakeholders on a large, complex project.

19. These two conflict resolution techniques are lose-lose techniques.

A. Problem solving

B. Withdrawal

C. Reconcile

D. Forcing

E. Accommodating

20. Match the following scaling agile frameworks with their description.

Scaling Agile Frameworks

Framework	Description
A. Scrum of Scrums	1. Two or more Scrum teams work on the project together. Each Scrum team focuses on a portion of the work.
B. SAFe	2. This consists of up to eight Scrum teams with up to eight members each who all work on the project together.
C. DA	3. This extends Scrum practices to all aspects of the organization.

Framework	Description
D. LeSS	4. This combines several agile best practices and includes information from functional areas of the business.
E. Enterprise Scrum	5. This is an interactive knowledge base consisting of technical guidance, knowledge, and information on agile.

Chapter

11

Measuring and Controlling Project Performance

THE PMP® EXAM CONTENT FROM THE PEOPLE DOMAIN COVERED IN THIS CHAPTER INCLUDES THE FOLLOWING:

✓ **Task 1.4 Empower team members and stakeholders**
- 1.4.3 Evaluate demonstration of task accountability

✓ **Task 1.8 Negotiate project agreements**
- 1.8.2 Assess priorities and determine ultimate objectives
- 1.8.4 Participate in agreement negotiations

✓ **Task 1.9 Collaborate with stakeholders**
- 1.9.1 Evaluate engagement needs for stakeholders
- 1.9.2 Optimize alignment between stakeholder needs, expectations, and project objectives

THE PMP® EXAM CONTENT FROM THE PROCESS DOMAIN COVERED IN THIS CHAPTER INCLUDES THE FOLLOWING:

✓ **Task 2.2 Manage communications**
- 2.2.1 Analyze communication needs of stakeholders
- 2.2.2. Determine communication methods, channels, frequency, and level of detail for all stakeholders
- 2.2.3 Communicate project information and updates effectively

✓ **Task 2.3 Assess and manage risks**
- 2.3.2 Iteratively assess and prioritize risks

✓ **Task 2.4 Engage stakeholders**

- 2.4.1 Analyze stakeholders

- 2.4.4 Develop, execute, and validate a strategy for stakeholder engagement

✓ **Task 2.7 Plan and manage quality of products/ deliverables**

- 2.7.1 Determine quality standard required for project deliverables

- 2.7.2 Recommend options for improvement based on quality gaps

- 2.7.3 Continually survey project deliverable quality

✓ **Task 2.9 Integrate project planning activities**

- 2.9.1 Assess consolidated project plans for dependencies, gaps and continued business value

- 2.9.2 Analyze the data collected

- 2.9.3 Collect and analyze data to make informed project decisions

- 2.9.4 Determine critical information requirements

✓ **Task 2.10 Manage project changes**

- 2.10.1 Anticipate and embrace the need for change

- 2.10.2 Determine strategy to handle change

- 2.10.3 Execute change management strategy according to the methodology

- 2.10.4 Determine a change response to move the project forward

✓ **Task 2.11 Plan and manage procurement**

- 2.11.3 Manage suppliers/contracts

- 2.11.4 Plan and manage procurement strategy

✓ **Task 2.12 Manage project artifacts**

- 2.12.2 Validate that the project information is kept up to date

- 2.12.3 Continually assess the effectiveness of the management of the project artifacts

✓ **Task 2.15 Manage project issues**

- 2.15.2 Attack the issue with the optimal action to achieve project success

- 2.15.3 Collaborate with relevant stakeholders on the approach to resolve the issues

THE PMP® EXAM CONTENT FROM THE BUSINESS ENVIRONMENT DOMAIN COVERED IN THIS CHAPTER INCLUDES THE FOLLOWING:

✓ **Task 3.1 Plan and manage project compliance**

- 3.1.3 Determine potential threats to compliance

- 3.1.4 Use methods to support compliance

- 3.1.5 Analyze the consequences of noncompliance

- 3.1.7 Measure the extent to which the project is in compliance

✓ **Task 3.2 Evaluate and deliver project benefits and value**

- 3.2.1 Investigate that benefits are identified

- 3.2.3 Verify measurement system is in place to track benefits

- 3.2.4 Evaluate delivery options to demonstrate value

- 3.2.5 Appraise stakeholders of value gain progress

✓ **Task 3.3 Evaluate and address external business environment changes for impact on scope**

- 3.3.1 Survey changes to external business environment

- 3.3.2 Assess and prioritize impact on project scope/ backlog based on changes in external business environment

- 3.3.3 Recommend options for scope/backlog changes
- 3.3.4 Continually review external business environment for impacts on project scope/backlog

✓ **Task 3.4 Support organizational change**

- 3.4.2 Evaluate impact of organizational change to project and determine required actions
- 3.4.3 Evaluate impact of the project to the organization and determine required actions

This chapter introduces the Monitoring and Controlling process group. The Monitoring and Controlling process group involves taking measurements and performing inspections to find out whether variances exist in the plan. If you discover variances, you need to take corrective action to get the project back on track and repeat the affected project Planning processes to make adjustments to the plan as a result of resolving the variances.

We'll start with the Monitor and Control Project Work process. This process involves tracking, reviewing, adjusting, and controlling progress to meet the project performance objectives. We'll also look at Control Procurements, Monitor Communications, Perform Integrated Change Control, Monitor Stakeholder Engagement, Control Resources, Control Quality, and Monitor Risks. I'll conclude the chapter with some considerations for monitoring project management integrations.

Control Procurements is where contract performance is monitored, procurement relationships are managed, and corrections or changes are implemented where needed.

Work performance information is an output of the Monitor Communications process, and this output is used and maintained during the Monitoring and Controlling processes, along with tools and techniques, to monitor and control project communication throughout the life of the project. This information allows the project manager to take corrective action during the Monitoring and Controlling processes and to remedy wayward project results and realign future project results with the project management plan.

Perform Integrated Change Control lays the foundation for all the control processes that we'll talk about in this chapter. During this discussion, I'll take time to explain how change control and configuration management works. When you read these sections, remember that the information from the Perform Integrated Change Control process applies to the remaining control processes we'll talk about in Chapter 12, "Controlling Work Results and Closing Out the Project."

Monitor Stakeholder Engagement is about continuing your relationship with the stakeholders and maintaining their engagement on the project. Stakeholders sometimes lose interest once you get into the later stages of the project, and you don't want that to happen.

Control Resources concerns monitoring the physical resources on the project. Control Quality ensures the quality objectives in the final product meet the planned quality objectives. We'll look at some new tools and techniques in this process.

Monitor Risks concerns keeping an eye on the risk register and risk response plans and staying diligent about watching for new risks. The tailoring and project integration you performed earlier in the project should be analyzed to determine whether the processes you used were effective.

You have learned a lot about project management by this point and have just a few more topics to add to your study list. Keep up the good work. You're getting closer to exam day.

The process names, inputs, tools and techniques, outputs, and descriptions of the project management process groups and related materials and figures in this chapter are based on content from *A Guide to the Project Management Body of Knowledge (PMBOK® Guide), Sixth Edition* (PMI®, 2017). The references to adaptive and hybrid methodologies, related materials, and figures in this chapter are based on content from the *Agile Practice Guide* (PMI®, 2017).

Monitoring and Controlling Project Work

The processes in the Monitoring and Controlling process group concentrate on monitoring and measuring project performance at regular intervals to identify and quantify variances from the project management plan and get it back on track using corrective actions.

The *Monitor and Control Project Work* process is concerned with monitoring all the processes in the Initiating, Planning, Executing, and Closing process groups. Collecting data, measuring results, comparing results to what was planned, and reporting on performance information are some of the activities you'll perform during this process. Many key outputs from other Monitoring and Controlling processes are inputs to this process.

According to the *PMBOK® Guide*, the Monitor and Control Project Work process involves the following:

- Reporting and comparing actual project results against the project management plan

- Analyzing performance data and determining whether corrective or preventive action should be recommended

- Monitoring the project for risks to make certain they're identified and reported, their status is documented, and the appropriate risk response plans have been put into action

- Documenting all appropriate product information throughout the life of the project

- Gathering, recording, and documenting project information that provides project status, measurements of progress, and forecasting to update cost and schedule information that is reported to stakeholders, project team members, management, and others

- Monitoring approved change requests

- Providing reports on the project processes and program management status when the project is part of an overall program

Forecasting Methods

Most of the project information you are gathering and reporting on in this process involves measurements, metrics, and formulas. Forecasting is one way of measuring project performance. We'll look at that, as well as work performance information, next.

Schedule Forecasts

Schedule forecasts are an output of the Control Schedule process. Schedule forecasts help determine the progress of the project schedule as it compares to the schedule baseline. This can involve a series of formulas to determine elements such as when a milestone was scheduled to be completed versus when it was actually completed. We will look at all the schedule forecast formulas in the next chapter when we talk about controlling the project schedule.

Cost Forecasts

Cost forecasts are an output of the Control Costs process. Much like the schedule forecasts, cost forecasts help us determine project progress as far as costs are concerned. This includes formulas to determine factors such as what we anticipated spending at this point in time versus what was actually spent. These formulas are known as earned value management formulas, and we will look at them as well as the cost forecasts formulas, in the next chapter when we talk about controlling project costs.

Cost and Schedule Forecasting Methods

There are several forecasting methods for both schedule and cost forecasting that you may need to know for the exam. Let's take a quick look at each of them.

> *Forecasting* involves examining the project performance data to date and making predictions about future project performance based on this data. According to the *PMBOK® Guide*, cost forecasts are determined using earned value management formulas such as estimate to complete and estimate at completion. Forecasting can also include methods such as time series, scenario building, and simulation. These forecasting methods fall into four different categories, and each category has several types of forecasting methods. We'll look at a brief description of each of the categories next, but it is beyond the scope of this book to go into detail on each one. Keep in mind these are not discussed in the *PMBOK® Guide* but they are methods you should know, since you may see them in practice.

Time-Series Methods   These forecasting methods use historical data to predict future performance. Earned value, moving average, extrapolation, trend estimation, linear prediction, and growth curve are some of the methods that fall into this category. We'll cover earned value in Chapter 12.

Causal/Econometric Methods   Causal methods are based on the ability to identify variables that may cause or influence the forecast. A drop in interest rates, for example, may spur an increase in home sales. The methods included in this category are regression analysis, autoregressive moving average (ARMA), and econometrics.

Judgmental Methods   If your mom is anything like mine, you were taught not to be judgmental. This category of forecasting does just that, only it doesn't judge people. It uses opinions, intuitive judgments, and probability estimates to determine possible future results. Methods within this category include composite forecasts, the Delphi method, surveys, technology forecasting, scenario building, and forecast by analogy.

Other Methods   Other types of forecasting methods include simulation (like Monte Carlo analysis), probabilistic forecasting, and ensemble forecasting.

Work Performance Information

Work performance information concerns gathering the performance data, or work results, from the Executing processes and then reporting and analyzing the information in the Monitoring and Controlling processes. This might include performance information such as the status of deliverables, schedule forecast data, cost forecast data, the status of change requests, costs incurred to date, and more. This information is analyzed against the project management plan, project documents, and other project information. For example, you'll recall that quality metrics were defined in the Plan Quality Management process. In order to know if the quality metrics you've gathered are on target with the plan, you have to compare the results to the quality management plan. Work performance information is reported to the stakeholders according to the requirements laid out in the communications management plan and the stakeholder management plan.

Work Performance Reports

Work performance reports are the primary output of this process. It's here that the work performance data and work performance information gathered and analyzed throughout the Monitoring and Controlling processes are documented and reported to the stakeholders as outlined in the communications management plan. Work performance reports might take many forms, including status reports, issues or action item logs, project documents, and so on. They may be represented in hardcopy form or electronic form or both.

Work performance reports may range from simply stated status reports to highly detailed reports. Dashboards are an example of a simple report that may use stoplight-type indicators like red-yellow-green to show the status of each area of the project at a glance. Red means the item being reported is in trouble or behind schedule, yellow means it's in danger of falling behind schedule and corrective action should be taken, and green means all is well.

Almost any type of report or chart or graph can be included as part of the work performance report output. Other examples might include defect histograms, burndown charts, trend lines, and more. The information in these reports pertains to project work results (or performance metrics) but can also be used to report on contract performance.

Exam Spotlight

According to the *PMBOK® Guide*, the work performance reports produced as a result of this process are used by the project management team and stakeholders to create awareness of issues, make decisions about those issues, and determine actions to take to resolve the issues. They are a subset of project documents.

More detailed reports may include the following elements:

- Analysis of project performance for previous periods
- Risk and issue status
- Work completed in the current reporting period
- Work expected to be completed during the next reporting period
- Changes approved in the current reporting period (or a summary if there are numerous changes)
- Results of variance analysis
- Time completion forecasts and cost forecasts
- Other information stakeholders want or need to know

The information contained in these reports is used at status meetings or in meetings with the project sponsor to generate action, to make decisions, or to inform. This information can also be shared in a daily stand-up on agile projects. This information should be in written format (such as a status report or memo) and stored and version controlled with other project documents. Work performance reports are considered a subset of project documents and should be shared and distributed according to your communications plan.

Controlling Procurements

We'll now take another look at the procurement arena. You learned about Conduct Procurements in Chapter 10, "Sharing Information." Now that the contract has been awarded, you need to administer it.

The *Control Procurements* process involves monitoring the vendor's performance and ensuring that all the requirements of the procurement agreement (this is usually a work order or contract) are met. The monitoring you conduct when using vendors to perform work must align with the elements detailed in the procurement management plan. This includes any details about the human resources that will be working on the project. You will verify compliance with the project objectives by measuring and monitoring vendor performance. We will cover several measurement and monitoring techniques in this chapter and the next.

As the project manager, you need to manage stakeholder expectations regarding project goals for all projects, whether you are using vendors or performing the work in-house. This includes identifying key deliverables (from the business requirements) and measuring and monitoring them to ensure compliance with objectives.

The last activity in this process is closing out the procurement when the work is complete. When multiple vendors are providing goods and services to the project, Control Procurements entails coordinating the interfaces among all the vendors as well as administering each of the contracts or procurement agreements. If vendor A has a due date that will impact whether vendor B can perform their service, the management and coordination of the two vendors become important. Vendor A's contract and due dates must be monitored closely because failure to perform could affect another vendor's ability to perform, not to mention the project schedule. You can see how this type of situation could multiply quickly when you have six or seven or more vendors involved.

It's imperative that the project manager and project team be aware of any contract agreements that might impact the project so that the team does not inadvertently take action that violates the terms of the contract.

Depending on the size of the organization, administering the contract or other procurement agreement might fall to someone in the procurement department. This doesn't mean you're off the hook as the project manager. It's still your responsibility to oversee the process and make sure the project objectives are being met, regardless of whether a vendor or your project team members are performing the activities. You'll be the one monitoring the performance of the vendor and informing them when and if performance is lacking. You'll also monitor the procurement agreement's financial conditions. For example, the seller should be paid in a timely manner when they've satisfactorily met the conditions of the agreement, and it will be up to you to let the procurement department know it's okay to pay the vendor. You might have to terminate the contract when the vendor violates the terms or doesn't meet the agreed-on deliverables. If the procurement department has this responsibility, you'll have to document the situation and provide this to the procurement department so that they can enforce or terminate the contract.

Control Procurements involves monitoring procurement activities to verify that you are meeting the project objectives, and it is closely linked with other project management processes. You'll manage vendor relationships, monitor the progress of the contract, execute plans, track costs, measure outputs, approve changes, take corrective action, and report on status, just as you do for the project itself. As the project manager, you are responsible for monitoring procurement activities no matter whether you are using predictive or agile methodologies to manage the work of the project.

Exam Spotlight

Buyers and sellers are equally responsible for monitoring the contract. Each has a vested interest in ensuring that the other party is living up to their contractual obligations and that their own legal rights are protected.

Procurement Documents and Approved Change Requests

Control Procurements has many inputs you've seen before. We'll look at approved procurement documentation, change requests, and some additional content for work performance data as it pertains to this process next.

Procurement Documentation

It has been a while since I've talked about documentation, but discussing procurement documentation reminds me of the importance of having an organized system for your procurement documents. These may include the contract itself, the statement of work, payment information, drawings and plans, policies and procedures, and so on. A *records management system* provides a way to organize, retrieve, and archive these documents when needed. Records management might be part of your project management information system or a stand-alone system. In my organization, the PMIS is separate from the records management system. Records are clearly defined, have specific naming conventions, are indexed for easy filing and retrieval, and have retention policies associated with them. Both systems may provide control functions, workflows, automated tools, and a way to link to documents in the PMIS (or vice versa).

Approved Change Requests

Sometimes as you get into the work of the project, you discover that changes need to be made. This could entail changes to the contract as well. Approved change requests are used to process the project or contract changes and might include things such as modifications to deliverables, changes to the product or service of the project, changes in contract terms, changes to the schedule, changes in vendor personnel, termination for poor performance, and more.

Generally speaking, contracts end for three reasons: cause, convenience, or default. When a contract ends for cause, it's because one of the parties has violated the terms of the contract. A contract ends for convenience when one of the parties determines they no longer wish to participate in the contract. And contracts that end by default are those where one of the parties has failed to perform according to the terms of the contract.

Contracts can be amended at any time prior to contract completion, provided the changes are agreed to by all parties and conform to the change control processes outlined in the contract.

Contract Change Control System

The purpose of the contract change control system is to establish a formal process for submitting change requests. It documents how to submit changes, establishes the approval process, and outlines authorization levels. The contract change control system is part of the enterprise environmental factors input to the Plan Procurement Management process. Why am I mentioning it here? Because changes to the contract generally occur during the Executing or Monitoring and Controlling processes after the work of the project has started and when results are being monitored. Approved change requests for contracts are processed through the contract change control system.

Much like the management plans found in the Planning processes, the *contract change control system* describes the processes needed to make contract changes. Because the contract is a legal document, changes to it require the agreement of all parties. A formal process must be established to process and authorize (or deny) changes. (Authorization levels are defined in the organizational policies.) It includes a tracking system to number the change requests and record their status. The procedures for dispute resolution are spelled out in the contract change control system as well.

The change control system, along with all the management plan outputs, becomes part of the Perform Integrated Change Control process that I'll discuss later in this chapter.

Exam Spotlight

Work performance data is an output of the Direct and Manage Project Work process. The results you gather in the Control Procurements process are actually collected as part of the Direct and Manage Project Work process output.

Work Performance Data

Work performance data include technical documentation that is developed by the seller according to the terms of the contract as well as work performance information as it relates to the seller (or vendor).

Work performance information in the Control Procurements process concerns monitoring vendor work results and examining the vendor's deliverables. This includes monitoring their work results against the contract statement of work and making sure activities are performed correctly and in sequence. You'll need to determine which deliverables are complete and which ones have not been completed to date. You'll also need to consider the quality of the deliverables and the costs that have been incurred to date.

Vendors request payment for the goods or services delivered in the form of seller invoices. *Seller invoices* should describe the work that was completed or the materials that were delivered and should include any supporting documentation necessary to describe what was delivered. The contract should state what type of supporting documentation is needed with the invoice. The *payment system* is used to issue payments for the invoices. The organization might have a dedicated department, such as accounts payable, that handles vendor payments, or it might fall to the project manager. In either case, follow the policies and procedures the organization has established regarding vendor payments and make sure the payments themselves adhere to the contract terms.

Monitoring Vendor Performance

The Control Procurements process has several new tools and techniques you will use to monitor vendor performance, including claims administration, performance reviews, inspections, and audits. We'll look at these next.

Claims Administration

Claims administration involves documenting, monitoring, and managing contested changes to the contract. Changes that cannot be agreed on are called *contested changes*. Contested changes usually involve a disagreement about the compensation to the vendor for implementing the change. You might believe the change is not significant enough to justify additional compensation, whereas the vendor believes they'll lose money by implementing the change free of charge. Contested changes are also known as *disputes*, *claims*, or *appeals*. These can be settled directly between the parties themselves, through the court system, or by a process called arbitration. *Arbitration* involves bringing all parties to the table with a disinterested third party who is not a participant in the contract to try to reach an agreement. The purpose of arbitration is to reach an agreement without having to go to court. Arbitration is a form of alternative dispute resolution (ADR) and could be a named tool in the contract for resolving disputes. Reaching a negotiated settlement through the aid of ADR techniques is the most favorable way to resolve the dispute.

Procurement Performance Reviews

Procurement performance reviews examine the seller's performance on the contract to date. These reviews can be conducted at the end of the contract or at intervals during the contract period. Procurement performance reviews can be performed at the end of an iteration, a work interval, or release on agile projects, or any other predetermined time periods.

Procurement reviews examine the contract terms and seller performance for elements such as these:

- Meeting project scope
- Meeting project quality
- Staying within project budgets
- Meeting the project schedule

The performance reviews themselves might take the form of quality audits or inspections of documents as well as the work of the product itself, or they may be part of a status review. The point of the review is to determine where the seller is succeeding at meeting scope, quality, cost, and schedule issues, for example, or where they're not measuring up. If the seller is not in compliance, action must be taken to either get them back into compliance or terminate the contract. The yardstick you're using to measure their performance is the contract SOW and the terms of the contract. When the RFP is included as part of the contract, you might also use it to determine contract compliance.

Inspections

Inspections involve physically examining or inspecting the work. This may be performed by the buyer or a designated third party. Inspections may occur during the work of the project as well as at the completion of the work.

Audits

Audits in this process consist of reviews of procurement activities to determine whether they are meeting the right needs and are being performed correctly and according to standards. Procurement audits also examine the procurement processes to determine areas of improvement and to identify flawed processes or procedures. This allows you to reuse the successful processes on other procurement items for this project, on future projects, or elsewhere in the organization. It also alerts you to problems in the process so that you don't repeat them.

Procurement audits might be used by either the buyer or the vendor, or by both, as an opportunity for improvement. It's a good idea to document what you learned during the audits as lessons learned—including the successes and failures that occurred. Doing so allows you to improve other procurement processes currently underway on this project or other projects. It also gives you the opportunity to improve the process for future projects.

Closing Out Procurements

The primary output of the Control Procurements process is closed procurements, and we'll look at this next. I'll also cover some new information in the procurement documentation updates output and the organizational process assets updates. These outputs relate to the tools and techniques we just talked about and, in practice, work hand in hand with them.

Closed Procurements

When the work of the contract is complete, you'll provide formal notice to the seller—in written form—that the procurement is complete. This is a formal acceptance and closure of the procurement. It's your responsibility as project manager to document the formal acceptance of the procurement. Many times the provisions for formalizing acceptance and closing the procurement are spelled out in the procurement documents and should be included in the procurement management plan.

If you have a procurement department that handles contract administration, they will expect you to inform them when the procurement is completed and will in turn follow the formal procedures to let the seller know it's complete. However, you'll still note the procurement completion in your copy of the project records.

Depending on the terms of the procurement, early termination (whether by agreement, via default, or for cause) might result in additional charges to the buyer. Be certain to note the reasons for early termination in your procurement documentation.

Closing procurements should occur after receiving all the deliverables, ensuring that they meet the technical and quality standards set out in the contract, that there is no remaining work, and that all payments have been made to the vendor. If the product or service does not meet expectations, the vendor will need to correct the problems before you issue a formal acceptance notice. Ideally, quality inspections have been performed during the course of the project, and the vendor was given the opportunity to make corrections earlier in the process, before the closing stage. It's not a good idea to wait until the very end of the project and then spring all the problems and issues on the vendor at once. It's much more efficient to discuss problems with your vendor as the project progresses because it provides the opportunity for correction when the problems occur.

Some of the administrative functions of closing out the procurement documents, such as confirming formal acceptance, closing out open claims, and updating and archiving records, will happen in the Closing process group. We'll talk more about these procedures in Chapter 12.

Procurement Documentation Updates

Documentation should be one of your favorite topics by now. Document updates are part of managing your project artifacts. As we've discussed, these artifacts need to be version controlled and updated, and must remain accessible to stakeholders and team members. Updating procurement documents includes (but isn't limited to) updating all of the following:

- Contract or other procurement documents
- Deliverables
- Performance reports

- Technical documentation
- Warranties
- Financial information (such as invoices and payment records)
- Inspection and audit results
- Supporting schedules
- Approved and unapproved changes

The records management system I talked about earlier is the perfect place to keep all these documents.

Change Requests

Requested contract changes are coordinated with the Direct and Manage Project Work and Perform Integrated Change Control processes so that any changes impacting the project are communicated to the project team and appropriate actions are put into place to realign the objectives. This might also mean you'll have to change the project management plan, including the cost baseline and/or the schedule baseline. That requires you to jump back to the Planning process group to bring the project management plan, or other documents, up to date. Once the plan is updated, the Direct and Manage Project Work process might require changes as well to get the work of the project in line with the new plan. Remember that project management is an iterative process, and it's not unusual to revisit the Planning or Executing processes, particularly as changes are made or corrective actions are put into place.

Contract changes will not always impact the project management plan, however. For example, late delivery of key equipment probably would impact the project management plan, but changes in the vendor payment schedule probably would not. It's important that you are kept abreast of any changes to the contract so that you can evaluate whether the project management plan needs adjusting.

Organizational Process Assets Updates

Organizational process assets updates consist of organizational policies, procedures, and so on. Three elements of this output relate to contracts:

Procurement File   The procurement file includes contract documents, the closed contract documents, and correspondence related to the contract and contract performance. Correspondence is information that needs to be communicated in writing to either the seller or the buyer. Examples include changes to the contract, clarification of contract terms, results of buyer audits and inspections, and notification of performance issues. You'll use correspondence to inform the vendor if their performance is unsatisfactory. If they don't correct it, you'll use correspondence again to notify a vendor that you're terminating the contract because performance is below expectations and is not satisfying the requirements of the contract.

Payment Schedules and Requests   Many times, contracts are written such that payment is made based on a predefined performance schedule. For example, perhaps the first payment is made after 25 percent of the product or service is completed, or maybe the payment schedule is based on milestone completion. In any case, as the project manager, you will verify that the vendor's work (or delivery) meets expectations before the payment is authorized. It's almost always your responsibility as the project manager to verify that the terms of the contract to date have or have not been satisfied. Depending on your organizational policies, someone from the accounting or procurement department might request written notification from you that the vendor has completed a milestone or made a delivery. Monitoring the work of the vendor is as important as monitoring the work of your team members.

If the procurement department is responsible for paying the contractor, then that department manages the payment schedule (which is one of the terms of the contract). The seller submits a payment request, an inspection or review of their performance is conducted to make certain the terms of the contract were fulfilled, and the payment is made.

Seller Performance Evaluation   Seller performance evaluation is a written record of the seller's performance on the contract and is prepared by the buyer. It should include information about whether the seller successfully met contract dates, whether the seller fulfilled the requirements of the contract and/or contract statement of work, whether the work was satisfactory, and so on. Seller performance evaluations can be used as a basis for terminating the existing contract if performance is not satisfactory. They can also be used to determine penalty fees in the case of unsatisfactory performance or incentives that are due according to the terms of the contract. They should also indicate whether the vendor should be allowed to bid on future work. Seller performance evaluations can also be included as part of the qualified sellers lists.

Prequalified Seller List Updates   Prequalified sellers are sellers who are preapproved to bid on contract work. If you have an underperforming vendor, you'll want to remove them from this list.

Lessons Learned Repository   Everything you learned about the procurement process, successes *and* failures, should be documented in the lessons learned repository for future reference. This should also include an assessment of the procurement management plan and how it compared to the actual procurement process.

Exam Spotlight

I recommend that you understand the purposes of the procurement processes and don't simply memorize their components. Here's a brief recap:

- Plan Procurement Management: Preparing the SOW and procurement documents and determining source selection criteria

- Conduct Procurements: Obtaining bids and proposals from potential vendors, evaluating proposals against predetermined evaluation criteria, selecting vendors, and awarding the contract

- Control Procurements: Monitoring vendor performance to ensure that contract requirements are met

Monitoring Communications

The Monitoring and Controlling process group concentrates on monitoring and measuring project performance to identify variances from the project management plan. *Monitor Communications* is the process that monitors and controls communications throughout the life of the project, and it is part of the Project Communications Management Knowledge Area. You will recall that this Knowledge Area is concerned with collecting, distributing, storing, archiving, and organizing project communications. Therefore, it plays a key role in collecting information regarding project progress and project accomplishments and reporting it to the stakeholders. This information might also be reported to project team members, the management team, and other interested parties through the Monitor and Control Project Work process. When reporting information outside of the organization, you'll want people available with expertise in media communications, public outreach, and international communications. Remember also that communication among and between virtual teams is important in the communications processes.

Reporting might include information concerning project quality, costs, scope, project schedules, procurement, and risk, and it can be presented in the form of status reports, progress measurements, or forecasts. As with other processes in the Monitoring and Controlling process group, Monitor Communications may produce changes that will require a trip back through the Plan Communications Management process and/or the Manage Communications process.

Exam Spotlight

According to the *PMBOK® Guide*, the Monitor Communications process is concerned with evaluating and controlling the impact messages may carry and delivering the right message to the right people at the right time. This process is also about making certain that the project information needs and, most importantly, the information needs of the stakeholders are met.

Documents to Help Monitor Communications

You've examined most of the inputs to the Monitor Communications process previously, but we'll take another look at a few of them from the perspective of monitoring communications.

Project Management Plan   The project management plan contains the project management baseline data (typically cost, schedule, and scope factors), which you'll use to monitor and compare results. Deviations from this data are reported to management. The resource management plan, communications management plan, and stakeholder engagement plan are useful in keeping up with the communication needs of the stakeholders. Following these plans will help you manage the flow of information and help keep stakeholders engaged and informed.

Issue Log   The issue log documents issues, assigns owners, tracks due dates, and more. The issue log is useful in status meetings and in the Monitor Communications process as a way to determine what issues remain that are potentially blocking progress and what's been resolved. I have found that reviewing issue logs at every status meeting helps to maintain stakeholder expectations and also holds them accountable for actions they are required to take to resolve the issue. Remember that issues might also be logged in the RAID log (Risk, Action Items, and Decision Items).

Lessons Learned Register   Lessons learned are helpful in improving the effectiveness of your communications. Review lessons learned from earlier in the project to see if there are areas where you can improve communications (or that have been improved). Lessons learned from similar projects completed in the past are helpful as well.

Project Communications   Project communications include information such as status, progress on performance, cost and budget information, schedule progress, and more. This information is used to make project decisions and take action where needed.

Monitoring Communications with Meetings

Who doesn't love meetings? Meetings are the primary tool and technique of this process and are key in presenting project information to the stakeholders. Before diving into all those project meetings, it might be helpful to review the stakeholder engagement assessment matrix, another tool and technique of this process (as part of the data representation tool and technique). You'll recall that this matrix documents the communication needs of the stakeholders. Hopefully, you've been keeping this updated as the project progresses and periodically evaluating whether stakeholders' communication needs are being met. You'll also want to ensure that the communications have been effective in informing stakeholders about all aspects of the project. You should also check in with the stakeholders occasionally to ask about the effectiveness of the communication, the methods of communication, and whether the project meetings provide them with information they find useful. You could do that using observation and conversation, another tool and technique of this process (as part of the data representation tool and technique). Observation and conversation is used to gather information from team members and stakeholders on project performance, to communicate project updates, to determine whether conflict exists among or between teams, and to address team member performance issues.

Meetings in this process include meetings involving the project team, vendors, functional managers, stakeholders, the project sponsor, and others you will be working with on the project. Meetings may be formal, informal, in person, or online. The important point here is that you are meeting with all those involved in your project and exchanging information. Meetings with the project team are concerned with updating project performance information and communicating those updates to the appropriate parties and working with stakeholders to provide them with requested project information.

One of the important meetings you will conduct throughout the life of the project is the *status meeting*. Status meetings are a type of interactive communication and are important functions during the course of the project. The purpose of the status meeting is to provide updated information regarding the progress of the project. These are not show-and-tell meetings. If you have a prototype to demo, set up a different time to do that. Status meetings are meant to exchange information and provide project updates. They are a way to formally exchange project information. It's not unusual for projects to have three or four status meetings conducted for different audiences. They can occur between the project team and project manager, between the project manager and stakeholders, between the project manager and users or customers, between the project manager and the management team, and so on.

The project manager should always be included in status meetings. However, take care that you don't overburden yourself with meetings that aren't necessary or meetings that could be combined with other meetings. Having any more than three or four status meetings per month is unwieldy.

Regular, timely status meetings prevent surprises down the road because you are keeping stakeholders and customers informed of what's happening. Team status meetings alert the project manager to potential risk events and provide the opportunity to discover and manage problems before they get to the uncontrollable stage.

The project manager typically hosts the status meeting. As such, it's your job to use status meetings wisely. Make certain that the information you'll share is relevant and tailored to the audience. Don't waste your team's time or the stakeholders' time by holding a meeting for the sake of a meeting or droning on about information no one cares about. Notify attendees in writing of the meeting time and place. Publish an agenda prior to the meeting, and stick to the agenda during the meeting. Every so often, summarize what has been discussed during the meeting. Don't let side discussions lead you down rabbit trails, and keep irrelevant conversations to a minimum. It's also good to publish status meeting notes at the conclusion of the meeting, especially if any action items resulted from the meeting. This will give you a document trail and serve as a reminder to the meeting participants of what actions need to be resolved and who is responsible for each action item. You should also log action items in the RAID log or the action item log. If you prefer, you could reference the link to the RAID log in the status report. Be advised that I've found that it's best to list action items in the status report itself (yes I know it's extra work) and also record them in the RAID log. Stakeholders won't always take the time to review the RAID log and they will forget that they have items to resolve. If their item is listed on the status report, they are more likely to remember it. This is because they reference the status report frequently.

It's important that project team members are honest with the project manager and that the project manager is in turn honest about what they report. A few years ago, a department in my agency took on a project of gargantuan proportions and unfortunately didn't employ good project management techniques. One of the biggest problems with this project was that the project manager did not listen to the highly skilled project team members. The team members warned of problems and setbacks, but the project manager didn't want to hear about it. The project manager took their reports to be of the "Chicken Little" ilk and refused to believe the sky was falling. Unfortunately, the sky *was* falling! Because the project manager didn't believe the reports, she refused to report the true status of the project to the stakeholders and oversight committees. Millions of dollars were wasted on a project that was doomed to failure while the project manager continued to report that the project was on time and activities were completed when in fact they were not.

There are hundreds of project stories like this, and I'll bet you've got one or two from your experiences as well. Don't let your project become the next bad example. Above all, be honest in your reporting. No one likes bad news, but delivering bad news too late along with wasting millions of dollars is a guaranteed career-limiting move.

Work Performance Information

The primary output of the Monitor Communications process is work performance information and change requests. You'll recall that work performance information includes information you've gathered and analyzed about the progress of the project to date. This information is typically documented in the work performance reports (an output of the Monitor and Control Project Work process we discussed earlier in this chapter) and is distributed to the stakeholders. Change requests come about as a result of actions taken to correct issues; actions taken to modify scope, schedule, or budget; or actions taken to address other items. These actions and decisions are a result of reviewing work performance information. Change requests are generated from this process (and others) but are processed through the Integrated Change Control process that we'll discuss in the next section of this chapter. Change requests may require updates to the project management plan, project documents, project baselines, and more. Remember to keep all this information up to date, version controlled, and accessible to the stakeholders, and continually assess the effectiveness of the management of this information and other project artifacts.

Exam Spotlight

For the exam, remember that one of the purposes of the Monitoring and Controlling process group is to gather performance metrics, including work efforts, costs (both expended and remaining forecasted costs), milestone measures, and other work performance measures to determine project progress. The Monitor Communications process brings much of this information together into one format.

Exam Spotlight

Work performance data, work performance information, and work performance reports are all either inputs or outputs of various processes in the Executing and Monitoring and Controlling process groups. They are similar but do have differences. I recommend you understand the differences for the exam. Here is a recap of each for your study time. According to the *PMBOK® Guide*, the definitions are as follows:

Work Performance Data

This data is the result of measuring and observing project activities. It may include data such as reports on the percentage of work complete, status of deliverables, technical performance measurements, the actual start and finish dates of schedule activities (schedule progress to date), the number of change requests or defects, costs incurred to date, and so on. Work performance data is an input of the following processes:

- Validate Scope

- Control Scope

- Control Schedule

- Control Costs

- Control Quality

- Monitor Communications

- Monitor Risks

- Control Procurements

- Control Resources

- Monitor Stakeholder Engagement

Work performance data is an output of the following process:

- Direct and Manage Project Work

Work Performance Information

Work performance information is used in the Monitoring and Controlling processes to analyze information such as the status of deliverables, the status of change requests, and forecasts such as estimate to complete (i.e., the work performance data).

Work performance information is an input of the following process:

- Monitor and Control Project Work

Work performance information is an output of the following processes:

- Validate Scope
- Control Scope
- Control Schedule
- Control Costs
- Control Quality
- Control Resources
- Monitor Communications
- Monitor Risks
- Control Procurements
- Monitor Stakeholder Engagement

Work Performance Reports

Work performance reports are the physical manifestation of the work performance information you've gathered and analyzed using variance analysis, earned value calculations, or forecasts. Work performance reports may include status reports, memos, updates, and more. Work performance reports are intended to aid decision-making and action plans.

Work performance reports are an input of the following processes:

- Manage Team
- Manage Communications
- Perform Integrated Change Control
- Monitor Risks

Work performance data is an output of the following process:

- Direct and Manage Project Work

Performing Integrated Change Control

The *Perform Integrated Change Control* process is where you implement the project's change control process in accordance with the change management plan. This process is performed throughout the life of the project from the point the project baselines are established to the end of the project. Changes, which encompass corrective actions, preventive actions, and defect repair, are common across all the Monitoring and Controlling processes. (Don't forget that change requests are also an output of the Direct and Manage Project Work process, which is an Executing process.) These processes can generate the change requests that are managed through the Perform Integrated Change Control process. Changes on predictive projects need to be carefully managed to ensure that the project objectives are not compromised and that the project remains on budget and on schedule. Remember that change is welcomed on agile projects and is easily accommodated. The feedback received during the retrospective and the backlog grooming that occurs at the beginning of the iteration allows change to be a regular part of the project. Rather than log, review, and go through an extensive change approval process that you'll learn about shortly, an agile project reviews, approves, and incorporates change in the upcoming iteration.

You may see exam questions on several topics that involve change and the Perform Integrated Change Control process. Configuration management, for example, isn't listed as a tool and technique of this process, but you really can't perform the Perform Integrated Change Control process without it. There are several elements of configuration management listed as part of the change control tool and technique of this process, but configuration management itself needs some explanation. I'll discuss this shortly. First, though, you'll look at change, what it is, and how it comes about.

Exam Spotlight

The project manager is responsible for performing and overseeing the Perform Integrated Control process.

Changes come about on projects for many reasons. It's the project manager's responsibility to manage these changes and see to it that organizational policies regarding changes are implemented. Changes don't necessarily mean negative consequences. Changes can produce positive results as well. It's important that you manage this process carefully, because too many changes—even one significant change—will impact cost, schedule, scope, and/or quality. Once a change request has been submitted, you have some decisions to make. Ask yourself questions such as these:

- Should the change be implemented?
- If so, what's the cost to the project in terms of project constraints: cost, time, scope, and quality?
- Will the benefits gained by making the change increase or decrease the chances of project completion?
- What are the risks associated with this change?
- Do the benefits of the change outweigh the risks?

Just because a change is requested doesn't mean you have to implement it. You'll always want to discover the reasons for the change to determine whether they're justifiable and aligned with business needs, and you'll want to know the cost of the change. Remember that cost can also take the form of increased time. Let's say the change you're considering will result in a later schedule completion date. That means you'll need human resources longer than expected. If you've leased equipment or project resources for the team members to use during the course of the project, a later completion date means your team needs the leased equipment for a longer period of time. All this translates to increased costs. Time equals money, as the saying goes, so manage time changes wisely, and dig deep to find the impacts that time changes might have on the budget.

How Change Occurs

As the project progresses, the stakeholders or customers might request a change directly. Team members might also recommend changes as the project progresses. For example, once the project is under way, they might discover more efficient ways of performing tasks or producing the product of the project and recommend changes to accommodate the new efficiencies. Changes might also come about as a result of mistakes that were made earlier in the project in the Planning or Executing processes. (However, I hope you've applied all the great practices and techniques I've talked about to date and you didn't experience many of these.)

Changes to the project might occur indirectly as a result of contingency plans, other changes, or team members performing favors for the stakeholders by making that one little change without telling anyone about it. Many times, the project manager is the last to know about changes such as these. There's a fine line here because you don't want to discourage good working relationships between team members and stakeholders, yet at the same time, you want to ensure that all changes come through the change control process. If a dozen little changes slip through like this, your project scope suddenly exits stage left.

Change Control Concerns

Perform Integrated Change Control, according to the *PMBOK® Guide*, is primarily concerned with the following:

- Influencing the factors that cause change control processes to be circumvented
- Promptly reviewing and analyzing change requests
- Managing approved changes
- Maintaining the integrity of the project baselines (including scope, quality, schedule, cost, and performance measurement baselines) and incorporating approved changes into the project management plan and other project documents
- Promptly reviewing and analyzing corrective and preventive actions
- Coordinating and managing changes across the project
- Documenting requested changes and their impacts

Factors that might cause change include project constraints, stakeholder requests, team member recommendations, vendor issues, and many others. You'll want to understand the factors that are influencing or bringing about change and how a proposed change might impact the project if implemented. Performance measures and corrective actions might dictate that a project change is needed as well.

Modifications to the project are submitted in the form of change requests and managed through the change control process. Obviously, you'll want to implement those changes that are most beneficial to the project. I'll talk more about change requests later in this chapter.

Managing changes might involve making changes to the project scope, schedule, or cost baseline, also known as the *performance measurement baseline*. This baseline might also involve quality or technical elements. The performance measurement baseline is the approved project management plan that describes the work of the project. This is used through Executing and Monitoring and Controlling to measure project performance and determine deviations from the plan. It's your responsibility to maintain the reliability of the performance measurement baselines. Changes that impact an existing or completed project management process will require updates to those processes, which might mean additional passes through the appropriate Planning and Executing processes.

The management plans created during the Planning process group should reflect the changes as well, which might require updates to the project management plan or the project scope statement. This demands you keep a close eye on coordination among all the processes that are impacted. For example, changes might require updates to risk response alternatives, schedule, cost, resource requirements, or other elements. Changes that affect product scope always require an update to the project scope.

I caution you to not change baselines at the drop of a hat. Examine the changes, their justification, and their impacts thoroughly before making changes to the baselines. Make certain your project sponsor approves baseline changes and that the project sponsor understands why the change occurred and how it will impact the project. Be sure to keep a copy of the original baseline for comparison purposes and for lessons learned.

Configuration Control

Configuration control is concerned with changes to the specifications of the deliverables or project management processes. Configuration control is managed through the configuration management system. The *configuration management system* is a subsystem of the project management information system. It documents the procedures and authorized approval levels for managing and controlling changes to the physical characteristics of the deliverables.

The following items are part of the change control tool and technique of this process and are activities associated with configuration control, so we'll look at them here:

Configuration Identification Configuration identification describes the characteristics of the product, service, or result of the project. This description is the basis that's used to verify when changes are made and how they're managed. Configuration identification is also used to label products or documents, manage changes, and verify changes.

Configuration Status Accounting   This activity doesn't have to do with financials as you might assume from its title. It's about accounting for the status of the changes by documenting and storing the configuration information needed to effectively manage the product information. This includes the approved configuration identification, the status of proposed changes, and the status of changes currently being implemented.

Configuration Verification and Auditing   Verification and audits are performed to determine whether the configuration item is accurate and correct and to make sure the performance requirements have been met. It ensures that changes are registered in the configuration management system and that they are assessed, approved, tracked, and implemented correctly.

Change Control System

If you were to allow changes to occur to a project using a predictive methodology whenever requested, you would probably never complete the project. Stakeholders, the customer, and end users would continually change the project requirements if given the opportunity to do so. That's why careful planning and scope definition are important in the beginning of the project. It's your job as project manager to drive out all the compelling needs and requirements of the project during the Planning process so that important requirements aren't suddenly "remembered" halfway through the project. However, we're all human, and sometimes things are not known, weren't thought about, or simply weren't discovered until a certain point during the project. Stakeholders will probably start thinking in a direction they weren't considering during the Planning process, and new requirements will come to light. This is where the *change control system* comes into play.

If you know early on that your project will have a lot of change, or that your stakeholders are unable to clearly articulate their goals for the end product, consider using a hybrid or adaptive methodology. These methodologies can accommodate change more efficiently than a predictive methodology.

Exam Spotlight

Understand for the exam that configuration management involves identifying the physical characteristics of the product, service, or result of the project (or its individual components); controlling changes to those characteristics; and documenting changes to verify that requirements are met. It also includes the change management system and documents the process for requesting, tracking, and determining whether change requests should be approved or denied.

The Purpose of the Change Control System

Change control systems are documented procedures that describe how the deliverables of the project and associated project documentation are controlled, changed, and approved. They also often describe how to submit change requests and how to manage change requests. They may include preprinted change request forms that provide a place to record general project information such as the name and project number, the date, and the details regarding the change request. Change control systems are usually subsystems of the configuration management system.

The change control system also tracks the status of change requests, including their approval status. Not all change requests will receive approval. Those changes that are not approved are also tracked, communicated to the appropriate stakeholders, and filed in the change control log for future reference.

The change control system might define the level of authority needed to approve changes, if it wasn't previously defined in the configuration management system. Some change requests could receive approval based on the project manager's decision; others might need to be reviewed and formally approved by the project sponsor, executive management, and so on.

The change control system is a subset of the project management system. It includes the processes for identifying the characteristics of the product, service, or result and includes the process for documenting, tracking, and approving changes.

Procedures that detail how emergency changes are approved should be defined as well. For example, you and your team might be putting in some weekend hours and be close to the completion of a deliverable when you discover that thing 1 will not talk to thing 2 no matter what you do. The team brainstorms and comes up with a brilliant solution that requires a change request. Do you stop work right then and wait until the change control board or committee can meet sometime next week and make a decision, or do you—the project manager—make the decision to go forward with this solution and explain the change to the appropriate parties later? That answer depends on the change procedures you have in place to handle situations like this and the authority you have to make emergency changes as outlined in the change control system.

Many organizations have formal change control or change request systems in place. If that's the case, you can easily adopt those procedures and use the existing system to manage project change. But if no procedures exist, you'll have to define them.

There are three objectives you should know about for implementing and using configuration management systems and change control processes:

- Establish a method to consistently identify changes, request changes to project baselines, and analyze and determine the value and effectiveness of the changes.

- Continuously authenticate and improve project performance by evaluating the impact of each change.

- Communicate all change requests—whether approved, rejected, or delayed—to all the stakeholders.

🌐 **Real World Scenario**

But I Thought You Said...

Marcus was working on a web redesign project for the marketing department in his organization. The project started out as a simple redesign of the look and feel of the site. Marcus made the mistake of not defining change control procedures for the project from the beginning because he reasoned that the project was small, the design changes were well understood by all, and the project could be finished in a matter of weeks.

After the work of the project started, Marcus's team showed the initial design results to the business lead, Kendra. Kendra asked for a few modifications that seemed minor, and Marcus was happy to accommodate. When his team finished the work and turned the site over for initial testing, Kendra created a list of changes that extended beyond the initial scope of the project. Marcus thought everyone had agreed that the project involved only an update to the look and feel of the site, but Kendra was requesting changes to the applications people used to order products from the website. Marcus knew that application changes needed their own set of requirements and testing. Changing the look of the site was a lot different from changing the way an application worked and the results it produced. Kendra and Marcus were in disagreement over what the changes entailed. Kendra thought she had carte blanche to make changes until she was satisfied with the project. Marcus knew that the projects waiting in the queue were going to suffer because of the never-ending stream of changes to Kendra's project. The additional time her project was taking already had pushed out the deliverable of the next project on their list by two weeks.

To resolve the dilemma, Marcus had to negotiate with Kendra on the list of changes she had requested. He agreed to all changes that required less than four hours to complete, and the remaining changes were moved to a new project request. Marcus also vowed to implement change control procedures for all projects from this point forward, no matter what the size or complexity of the project. Another important lesson Marcus learned was that this project should have been conducted as an agile project. The change requests Kendra asked for could have been incorporated into the iterations and functionality could have been rolled out at the end of each iteration, rather than waiting until the end of the project to incorporate a host of new requirements.

Requirements for Change

You should require two things regarding change for all projects. First, require that all change requests be submitted in writing. This is to clarify the change and make sure there's no confusion about what's requested. It also allows the project team to accurately estimate the time it will take to incorporate the change.

Second, require that all change requests go through the formal change control system. Make sure no one is allowed to go directly to team members and request changes without the project manager knowing about them. Also make certain your stakeholders understand that going around the project manager can cause schedule delays, cost overruns, and sacrifices to quality and that it isn't good change management practice. From the beginning of the project, encourage the stakeholders to use the formal procedures laid out in the change control system to request changes.

It's good practice to require all change requests in writing. This should be a documented procedure outlined in your change control system. Beware! Stakeholders are notorious for asking for changes verbally even when there is a detailed process in place. If they don't want to follow the process, someone on the project team should have responsibility for documenting and logging the change requests for future reference.

Change Control Board

In some organizations, a *change control board (CCB)* is established to review all change requests. The board is officially chartered and given the authority to approve or deny change requests as defined by the organization. It's important that its authority be clearly defined and that separate procedures exist for emergency changes. The CCB might meet only once a week, once every other week, or even once a month, depending on the project. When emergencies arise, the established procedures allow the project manager to implement the change on the spot. This always requires follow-up with the CCB and completion of a formal change request, even though it's after the fact.

CCB members might include stakeholders, managers, project team members, and others who might not have any connection to the project at hand. Some organizations have permanent CCBs that are staffed by full-time employees dedicated to managing change for the entire organization, not just project change. You might want to consider establishing a CCB for your project if the organization does not have one.

Exam Spotlight

For purposes of the exam, note that the *PMBOK® Guide* states that the CCB is made up of a group of stakeholders who review and then approve, delay, or reject change requests. Agile projects incorporate change in each iteration and do not require a CCB. The product owner and Scrum master will decide what changes will be incorporated into the next iteration.

Some organizations use other types of review boards that have the same responsibilities as the CCB. Some other names you might see are technical assessment board (TAB), technical review board (TRB), engineering review board (ERB), and configuration control board (CCB).

Approved Change Requests

The primary purpose of the Perform Integrated Change Control process is to create approved change requests. On a predictive project, this means you'll need to consider the project management plan, the scope baseline (consisting of the schedule, cost, and scope baseline), and work performance reports. All of this information will be analyzed to determine the impact of the change if it were implemented. Obviously, the most important input to this process is change requests.

Change control tools are employed in this process to track and record the change requests, record the decisions by the change control board, and help manage the distribution of the change request decisions and more. The configuration tools I mentioned earlier—configuration item description, configuration status accounting, and configuration item verification—are all considered change control tools. They help support identifying and documenting changes, tracking changes, and making decisions regarding the disposition of the change.

Approved change requests are the primary output of this process; however, they are implemented in the Direct and Manage Project Work process. This is a key interaction between the process groups. An approved change request (approved in the Monitoring and Controlling process group) has no effect if you don't implement it. Implementation activities occur in the Executing process group. Therefore, once the change request is approved, you will repeat the Executing process group in order to bring about the change. Then you will again monitor and verify that the change had the desired effect by going back through the Monitoring and Controlling process group.

Exam Spotlight

Changes should be managed according to the change management plan. They are analyzed, implemented, and monitored according to the Perform Integrated Change process. The key purposes of the change processes are to verify that the changes do not modify the goals of the project and that the goals remain aligned with business needs.

Change requests are documented in the change log (typically incorporated into a change control tool) along with other items such as costs, risks associated with the changes, the timing of the change requests, and more. Project management plan updates are typically required as a result of an approved change or corrective action, especially those changes that impact a project baseline. It may sound obvious, but I'll tell you anyway: Changes made to baselines should reflect changes from the current point in time forward. You cannot change past performance. These changes are noted in the change control system or the configuration management system, and stakeholders are informed at the status meetings of the changes that have occurred, their impacts, and where the description of the changes can be found.

You should document all the actions taken in the Perform Integrated Change Control process (whether implemented or not). You should also record the reason for the change request. In other words, how did this particular change request come about? How did it change the original project management plan? Is this something you could or should have known about in the Planning processes? You should also note the corrective action taken and the justification for choosing that particular corrective action as part of lessons learned. You can use the information captured in your configuration management system as lessons learned for future projects. When you take on a new project, it's a good idea to review the lessons learned from similar projects so that you can plan appropriately and avoid, where possible, the variances that occurred in those projects.

Exam Spotlight

Remember that approved change requests are an output of the Perform Integrated Change Control process but that they are implemented in the Direct and Manage Project Work process (an Executing process).

In the next chapter, you'll explore the individual change control processes (like Control Costs and Control Schedule) and the measurement tools you'll use to provide the variance measurements that are gathered and reported to the stakeholder via the Monitor Communications process that we discussed earlier in this chapter.

Changes in the Business Environment

We've discussed changes to the project throughout the book. In this section, we'll change focus a bit and examine changes to the business environment itself. Organizations undergo changes, just as projects do. For example, the chief executive officer retires, so the organization brings on a new CEO. This will undoubtedly change the organizational culture, not to mention bring about a new vision and mission that will impact the type of projects you are conducting. Board member changes can have similar impacts.

Organizations that are new to agile methodologies will also undergo changes. Because agile works with cross-functional team members who deliver business value frequently, the human resources department may need to examine hiring practices and work assignment policies that will enable teams to form quickly and to self-manage. They may also need to change the performance review process to reflect the self-managed, cross-functional style that agile promotes. The finance department may have to reexamine how the deliverables an agile team produces are capitalized or expensed. The procurement team may have to examine contract templates and processes for engaging vendors on agile projects so that business value can be realized quickly. Other functions such as internal audit, risk management, equal employment opportunity office, and corporate management policies and directives may need to adjust their processes to align with the agile practices.

Changes may occur to the external business environment that could impact your organization and your project, no matter the methodology you are using to conduct the project. We'll look at examples of external changes next. Some of these may look familiar. That's because they may also bring about new projects. You can review the needs and demands that bring about projects in Chapter 1, "Building the Foundation," if you need a refresher.

Mergers and Acquisitions   Mergers and acquisitions are commonplace in today's business environment. *Mergers* occur when two organizations agree to come together to form a new business entity. If the organizations are publicly held, both companies must first obtain approval from their shareholders. Once approved, the two companies become one and are managed and operated as one company.

Acquisitions occur when one organization acquires the assets, stock, or equity interests in another firm. This can occur with the agreement of the other firm (considered a friendly acquisition) or without their consent (considered a hostile takeover). Acquisitions usually involve large companies purchasing or acquiring smaller companies. The acquired company ceases to exist.

Many start-up companies invent a new product or service in the hope of being acquired by a bigger company so that the founders of the start-up can make big money and perhaps go on to launch another company and repeat the process.

Regulatory   Regulatory factors can include any number of factors that can bring about change, including laws and regulations in the country where you're doing business, government policies, taxes, tariffs, prohibitions on imports/exports, healthcare regulations, specific industry requirements and standards, and more.

Technology and Innovation   Technology and innovation is another area that can bring about any number of changes. For example, technology has enabled employees to work from almost anywhere on the planet, and virtual teams can work across the globe. Cloud technologies have changed the way organizations manage their systems, and employee access is no longer the issue it was when systems resided on premises. Don't forget to consider the infrastructure that supports technology when examining this category of change. The power grid and reliability of electricity can be an issue in some countries. There are so many innovations and advances in technology, it is difficult to keep up. Artificial intelligence, machine learning, autonomous vehicles, disease detection, and wearable technology don't even scratch the surface.

Geopolitical   Change can occur rather quickly in this category. Examine such factors as political stability, geographic factors, foreign policy, trade agreements, sanctions, national power, and national policies.

Market   The marketplace can encompass a host of changes. Consumers can be fickle, and their tastes may change quickly. Political, social, and economic factors can also influence market changes. For example, a recession may drive consumers to save more than they spend, thereby decreasing demand for "nice to have" items they might have purchased in good economic times. Globalization impacts the marketplace and may increase competition, increase demand, or have the opposite effect due to saturation in the market.

Social   Social factors can cover a lot of ground. Consumers and employees may value whether an organization they do business with has a charitable mindset or if they are environmentally responsible. Inclusivity and diversity are important social factors for many employees. Also consider the customs and cultural values of the people you'll be working with (or selling to). Can your organization embrace these customs? Will your product align with the cultural values of the geographic area you're selling to? Some work cultures value 32-hour workweeks; others provide extended paid time off in the summer for all employees; others give employees extended midday breaks to run errands, visit family, or rest.

Economic   Economic factors may include the economic system of the country, economic policies, recessions, economic growth, debt levels, lending practices, interest rates, taxes, exchange rates, and more.

After gaining understanding of the external changes that can impact your organization, you should assess and prioritize the impact of these changes. Perhaps the organization will need to change policies or processes as a result of these changes. For example, maybe the manufacturing process needs to change to meet environmental standards. If so, you'll need to address questions such as what those changes will entail, what the cost is, and whether your organization is prepared to undertake the risks associated with the change.

External changes will likely impact your projects as well. You'll need to analyze and evaluate the impact of these changes on your project. Be certain to examine the impacts on scope (or backlog on an agile project), risk, quality, schedule, costs, and resources. Develop recommendations and action items to accommodate these changes, and communicate them to your executive sponsor.

You should also assess the cultural impacts to the organization due to internal or external changes. Organizational culture may shift due to a change such as a new CEO coming on board, as I mentioned earlier. Remember also that the organization's culture will always influence the development life cycle methodology you'll choose to manage projects. You may be well under way on your agile project only to find that an external change forces you to reconsider the development life cycle methodology for this or future projects.

Implementing successful organizational change is not an easy feat. Can you guess what the critical success factor is in making it successful? Congratulations. You guessed it:

communication. As soon as practical, you need to get the message out that change is coming. Then craft your message in a way that introduces elements of the change in digestible bites over time. Communicate multiple times in multiple ways. Executive management should also participate in the communication activities. This demonstrates their buy-in and helps reinforce the benefits of the change. Employees must understand the implications of the change in a way that answers this question: "How will this change impact me and my job?" If you can communicate the change and its benefits in a way that employees can embrace and support it, and they understand how the change impacts their jobs, you'll go a long way toward implementing successful change.

Managing change is an expansive topic. We can't cover all aspects of change in this book. The key is to evaluate the impact of the change and determine what actions need to be taken to accommodate it. Equally important is continually monitoring and reviewing the internal and external environment for changes that can impact the project.

Monitoring Stakeholder Engagement

The purpose of the *Monitor Stakeholder Engagement* process is to monitor your stakeholder relationships and ensure the stakeholders' continued engagement in the project. It's easy for stakeholders to become distracted on long projects. As more time goes by, the current project fades in importance and other issues can take its place. The primary focus of this process (and all the stakeholder processes) is keeping stakeholders informed and engaged. This will help overall project success, it will help you efficiently manage the processes as the project progresses, and it will help your stakeholders stay up to date on issues and status as the project evolves so that they can make informed decisions. Many of the tools and techniques you've seen before are used here as well—for example, stakeholder analysis, root cause analysis, decision-making, stakeholder engagement assessments, communication, and interpersonal skills. As you monitor your stakeholders' involvement and engagement, be certain to keep the stakeholder engagement plan and the communications management plan up to date. Stakeholders will come and go on a project, and you'll need to keep track of the new stakeholders' communication needs. Also, record any findings about stakeholders from this process in the lessons learned register for future reference. For example, it might be important to know that a key stakeholder on past projects has a consistent habit of dropping out midway through the project. You can address this during the Planning phases and replace this stakeholder with someone who will be there for the long haul.

Controlling Resources

The *Control Resources* process concerns making certain the right physical resources are available at the time they are needed and at the location they are needed. Resources are assigned earlier in the Planning process group, and when they are finished performing their work, this process ensures their release. This process also concerns taking corrective action when the resources are not being utilized as planned. You will perform this process throughout the project as resources are deployed and released.

Problem solving is used in this process as a tool and technique to monitor and analyze resource usage. It's used for solving issues and finding answers to problems and challenges surrounding the physical resources and their utilization during the project. Problems can come from inside or outside the organization. Typically, the problems you'll encounter outside of the organization will have more to do with the vendor providing the equipment or resource than the resource itself. As you'll recall, according to the *PMBOK® Guide*, problem solving identifies and defines the problem, investigates and analyzes the data, determines possible solutions, implements the best solution, and then checks the solution to ensure the problem is fixed.

The primary outputs of this process are work performance information, change requests, and physical resource assignment updates. Work performance information concerns gathering the performance data and work results about the resources. This will include documenting a performance review of the physical resource. You will use this review to determine whether the resource utilization is on track with what was planned. This may involve analyzing schedule performance and cost performance as well to learn how and why the resources are not being used (or performing) as planned. Change requests may come about when new physical resources are needed for the project or existing resources don't work as planned.

Resource availability can change. Hopefully, during the Planning and Procurement processes you were able to secure some type of assurance from your vendors that the resources will be available at the time and the location agreed on. However, many things that are outside of anyone's control, such as weather, scarcity, and transportation issues, may impact resource assignments. The resulting changes may require an update to the physical resources assignments as well as the project schedule, cost baseline, and other project management plans and documents.

Utilizing Control Quality Techniques

Control Quality is specifically concerned with monitoring work results and project deliverables to see whether they comply with the stakeholders' expectations. You should practice Control Quality throughout the project to identify and remove the causes of unacceptable results. Remember that Control Quality is concerned with project results, both from a management perspective, such as schedule and cost performance, and from a product perspective. It also ensures that quality activity results meet the quality requirements laid out in the plan. In other words, the end product should conform to the requirements and product description defined during the Planning processes.

Control Quality Tools and Techniques

The primary purpose of the tools and techniques in this process is to examine the product, service, or result of the project, as well as the project processes, for conformity to standards. They are used with the plan-do-check-act cycle we talked about in Chapter 1 to help identify and resolve problems related to quality defects. If the results fall within the tolerance range specified, the results are acceptable. Alternatively, if the results fall within the control limits set for the product (as defined by the various tools and techniques I'll discuss in the following sections), the process you are examining is said to be *in control*. Spend time understanding these tools and their individual uses because you might see exam questions about each of them.

In this section, we'll look at checklists, check sheets, and statistical sampling, which are all forms of data gathering. We'll look at inspections and testing and product evaluations and how they're used to monitor quality. We'll also explore how to represent the data we gathered in the form of cause-and-effect diagrams, control charts, histograms, and scatter diagrams. Let's get to it.

Check Sheets

Check sheets, also known as tally sheets, are used to help organize data when you're performing inspections and gathering information on defects. They are similar to checklists. The frequencies or impacts of defects that are recorded on a check sheet can be displayed as a Pareto diagram.

Statistical Sampling

Statistical sampling involves taking a sample number of parts from the whole population and examining them to determine whether they fall within acceptable variances. The formula for calculating the correct sample size is beyond the scope of this book. However, Creative Research Systems has an online calculator and an explanation of statistical sampling that you might find useful; visit www.surveysystem.com/sscalc.htm.

Perhaps you determine to statistically sample 25 parts out of a lot or run. The quality plan outlines that the lot will pass if four parts or fewer fall outside the allowable variance.

Statistical sampling might also involve determining the standard deviation for a process. The quality management plan determines whether plus or minus two standard deviations—95.44 percent of the population—is adequate or whether plus or minus three standard deviations—99.73 percent—is adequate.

Inspection

Inspection involves physically looking at, measuring, or testing results to determine whether they conform to the requirements or quality standards. It's a tool used to gather information and improve results. Inspections might occur after the final product is produced or at intervals during the development of the product to examine individual components. Acceptance decisions are made when the work is inspected and is either accepted or rejected. When work is rejected, it might have to go back through the process for rework. According to the *PMBOK® Guide*, inspection is also known as *reviews, peer reviews, audits,* or *walkthroughs.*

Inspection might take actual measurements of components to determine whether they meet requirements. Maybe a component part for your product must be exactly 5 mm in length. To pass inspection, the parts are measured and must meet the 5 mm length requirement. If they measure 5 mm, they pass; if they do not, they fail.

You can use check sheets, described earlier in this section, to help track and record inspection data.

Exam Spotlight

Don't confuse inspection with prevention; they're two different tools. Inspection keeps errors in the product from reaching the customer. *Prevention* keeps errors from occurring in the process. It always costs less to prevent problems in the first place than it does to fix problems built into the product after the fact. Rework, labor costs, material costs, and potential loss of customers are all factors to consider when weighing prevention costs versus the cost of rework. Philip Crosby developed the theory of Zero Defects, which deals with prevention costs. Loosely translated, Zero Defects means doing it right the first time.

Measurements can vary even if the variances are not noticeable. Machines wear down, people make mistakes, the environment might cause variances, and so on. Measurements that fall within a specified range are called *tolerable results*. So, instead of 5 mm exactly, maybe a range between 4.98 mm and 5.02 mm is an acceptable or tolerable measurement for the component. If the samples that are measured fall within the tolerable range, they pass; otherwise, they fail inspection.

One inspection technique uses measurements called *attributes*. The measurements taken during attribute sampling determine whether they meet one of two options, conforming or nonconforming. In other words, the measurements conform or meet the requirement or they do not conform. This can also be considered a pass/fail or go/no-go decision.

Attribute conformity and inspections are not necessarily performed on every component part or every end product that's produced. That's time-consuming and inefficient when you're producing numerous components. Inspection in cases like this is usually performed on a sampling of parts or products where a specified number of parts are tested for conformity or measurement specifics.

Inspection will tell you where problems exist and will give you the opportunity to correct them, thereby leading to quality improvements. The other tools and techniques I'll talk about in this section also lead to quality improvements in the product or process or in both.

Real World Scenario

An Ounce of Prevention

One of the main thoroughfares into your city requires a bridge replacement. You were appointed the project manager for the city and have managed this project since its initiation 15 months ago. The project entailed hiring a contractor to build the new bridge and manage the contract.

Approximately 28,000 vehicles travel across this bridge daily, carrying commuters and college students back and forth to the downtown area. One of the requirements was no more than three of the six lanes of traffic could be closed at one time during construction. Another requirement was that each piece of steel had to be painted with two coats of paint before it was brought on site. A third coat of paint was to be applied at the site after construction. The paint must be guaranteed to last 25 to 30 years.

An onsite quality control inspection revealed that some of the paint was peeling. After further investigation, you discovered that the contractor did not allow the first coat of paint to cure properly, so when the second coat was applied, it peeled and flaked.

You informed the contractor that, according to the terms of the contract and the SOW specifications, they were required to apply three coats of paint to the bridge, and the paint was required to last 25 to 30 years. Paint that peeled before construction was completed did not comply with specifications. Corrective action was needed. As a result, the contract company decided to subcontract the painting work to another company while they finished their remaining tasks on the project.

Unfortunately, the subcontractor they hired was not up to the task and was unable to complete the paint job. Several months passed, and the original project completion date was missed. Obviously, revisions to the project schedule were required when it became clear that the subcontractor wasn't going to make the deadline to complete the painting task.

The original contractor found another subcontractor capable of completing the paint job. Because it was the middle of winter and temperatures were cold, the painting crew had to hang insulated tarps between the bays on the bridge and use heaters to warm up small areas of steel to the proper temperature to apply the paint. This process, combined with the previous mistakes, extended the completion date by more than three times its original estimate and ultimately delayed the completion of the project by two years. Additional costs were incurred to hire the subcontractor and rent the heaters.

Corrective action was taken as a result of the inspection, and eventually the project was completed, but not without schedule delays, schedule changes, scope changes, and rework—not to mention the increased cost to the original contractor. Because the contract was a fixed-price contract, the contractor's profit was eaten away paying for the painting job. The cost to correct the quality issue did not impact the city, but it did impact the contractor. This is a case where an ounce of prevention would have been worth several gallons of cure, as the old saying goes.

Testing and Product Evaluations

Testing is a structured activity that measures or observes whether the product, service, or result meets the project requirements and the quality objectives. Testing can take many forms depending on the type of deliverables produced by the project. Product evaluations are often conducted in the same manner as testing. Visual observation, measurements, and so on will determine whether the product meets the quality objectives.

The quality management plan outlines the type and amount of testing needed to ensure that the quality of the product, service, or result of the project conforms to the project requirements. Testing ensures that the product is free from errors and defects or any other nonconformance issues. Conversely, testing and product evaluation will alert you that corrections need to be made to get the product, service, or result into an error-free state that meets the quality standards for the project.

We talked about the fishbone diagram (a cause-and-effect diagram), also known as the Ishikawa diagram, in Chapter 10. The fishbone diagram is used to help identify root causes. Kaoru Ishikawa is known for the fishbone diagram, but he was also a significant contributor in the realm of quality.

Control Charts

Control charts measure the results of processes over time and display the results in graph form. Control charts are a way to measure variances to determine whether process variances are in control or out of control.

A control chart is based on sample variance measurements. From the samples chosen and measured, the mean and standard deviation are determined. Control Quality is usually maintained—or said to be in control—within plus or minus three standard deviations. In other words, Control Quality says that if the process is in control (that is, the measurements fall within the control limits), you know that 99.73 percent of the parts going through the process will fall within an acceptable range of the mean. If you discover a part outside of this range, you should investigate and determine whether corrective action is needed. Figure 11.1 illustrates an example of a control chart.

FIGURE 11.1   Control chart

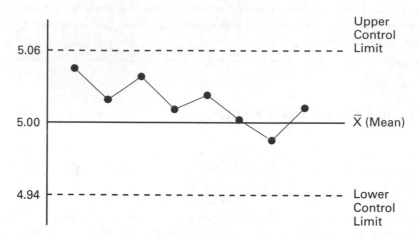

Let's assume you've determined from your sample measurements that 5 mm is the mean in the example control chart. One standard deviation equals 0.02. Three standard deviations on either side of the mean become your upper and lower control points on this chart. Therefore, if all control points fall within plus or minus three standard deviations on either side of the mean, the process is in control. If points fall outside the acceptable limits, the process is not in control and corrective action is needed.

Differences in results will occur in processes because there is no such thing as a perfect process. When processes are considered in control, differences in results might occur because of common causes of variances or special-cause variances.

Common causes of variances come about as a result of circumstances or situations that are relatively common to the process you're using and are easily controlled at the operational level. *Special-cause variances* are variances that are not common to the process. For example, perhaps you have very detailed processes with specific procedures that must be followed in order to produce the output, and a process gets missed. Or maybe your project requires the manufacturing of a certain part, and a machine on the line has a problem and requires a special calibration. This is an easy set of terms to remember because their names logically imply their definitions.

Three types of variances make up common causes of variances:

Random Variances   Random variations might be normal, depending on the processes you're using to produce the product or service of the project, but they occur, as the name implies, at random.

Known or Predictable Variances   Known or predictable variances are variances that you know exist in the process because of particular characteristics of the product, service, or result you are processing. These are generally unique to a particular application.

Variances That Are Always Present in the Process   The process itself will have inherent variability that is perhaps caused by human mistakes, machine variations or malfunctions, the environment, and so on, which are known as variances always present in the process. These variances generally exist across all applications of the process.

Common-cause variances that do not fall within the acceptable range are difficult to correct and usually require a reorganization of the process. This has the potential for significant impact, and decisions to change the process always require management approval.

Exam Spotlight

When a process is in control, it should not be adjusted. When a process falls outside the acceptable limits, it should be adjusted.

The *Rule of 7* is another way for the project team to use control charts and determine whether the process is in control. The Rule of 7 works like this: If seven consecutive points or more fall on one side of the mean, this may indicate there are factors influencing the result and they should be investigated. So, while the overall results are within the control limits, the process may not be in control and those factors should be examined more closely.

Control charts are used most often in manufacturing settings where repetitive activities are easily monitored. For example, the process that produces widgets by the case lot must meet certain specifications and fall within certain variances to be considered in control. However, you aren't limited to using control charts only in the manufacturing industry. You can use them to monitor any output. You might consider using control charts to track and monitor project management processes. You could plot cost variances, schedule variances, frequency or number of scope changes, and so on to help monitor variances.

Pareto Diagrams (Histograms)

You have probably heard of the 80/20 rule. Vilfredo Pareto, an Italian economist and sociologist, is credited with discovering this rule. He observed that 80 percent of the wealth and land ownership in Italy was held by 20 percent of the population. Over the years, others have shown that the 80/20 rule applies across many disciplines and areas. As an example, generally speaking, 80 percent of the deposits of any given financial institution are held by 20 percent of its customer base. Let's hope that rule doesn't apply to project managers, though, with 20 percent of the project managers out there doing 80 percent of the work!

The 80/20 rule as it applies to quality says that a small number of causes (20 percent) create the majority of the problems (80 percent). Have you ever noticed this with your project or department staff? It always seems that just a few people cause the biggest headaches. But I'm getting off track.

Pareto diagrams are displayed as histograms that rank the most important factors—such as delays, costs, and defects—by their frequency over time. Pareto's theory is that you get the most benefit if you spend the majority of your time fixing the most important problems. The information shown in Table 11.1 is plotted on an example Pareto chart shown in Figure 11.2.

TABLE 11.1   Frequency of failures

Item	Defect frequency	Percent of defects	Cumulative percent
A	800	33	33
B	700	29	62
C	400	17	79
D	300	13	92
E	200	8	100

The problems are rank according to their frequency and percentage of defects. The defect frequencies in this figure appear as bars, and the cumulative percentages of defects are plotted as circles. The rank of these problems shows you where corrective action should be taken first. You can see in Figure 11.2 that problem A should receive priority attention because the most benefit will come from fixing this problem.

Scatter Diagrams

Scatter diagrams use two variables: an *independent* variable, which is an input, and a *dependent* variable, which is an output. Scatter diagrams, also known as correlation charts, display the relationship between these two elements as points on a graph. The independent variable is shown on the horizontal axis on the graph, and the dependent variable is shown on the vertical axis. This relationship is typically analyzed to prove or disprove cause-and-effect relationships. As an example, maybe your scatter diagram plots the ability of your employees to perform a certain task. The length of time (in months) they have performed this task is plotted as the independent variable on the x-axis, and the accuracy they achieve in performing this task, which is expressed as a score—the dependent variable—is plotted on the y-axis. The scatter diagram can help you determine whether cause and effect (in this case, increased experience over time versus accuracy) can be proved. Scatter diagrams can also help you look for and analyze root causes of problems.

FIGURE 11.2   Pareto chart

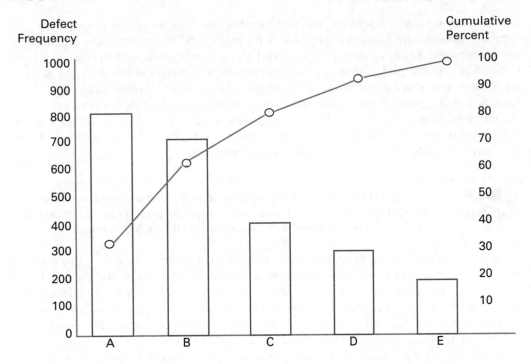

The important point to remember about scatter diagrams is that they plot the dependent and independent variables, and the more the points resemble a diagonal line, the more closely these variables are related. Figure 11.3 shows a sample scatter diagram.

FIGURE 11.3   Scatter diagram

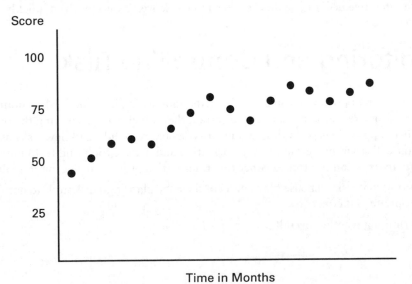

Verifying Deliverables

Quality improvement is a primary goal of the quality processes. Failure to meet quality requirements can have a significant impact on the project and the project team and might result in rework. *Rework* causes a project to take longer and cost more than originally planned because the project team has to repeat processes to correct the work. You should try to keep rework to a minimum so as not to impact the project schedule and budget. Rework has the potential to cause morale issues as well, especially if the team members thought they were doing a good job all along. Rework might require the project team to put in long hours, which in turn might cause more errors or other negative consequences. Monitor quality periodically so that rework is kept to a minimum.

 Manage Quality is concerned with ensuring that the project is using the correct and most efficient processes to meet the project requirements; Control Quality is concerned with the accuracy of the project results.

Verifying deliverables is the primary output of Control Quality. It involves using the tools and techniques of this process to determine whether the deliverable is correct and accurate and meets the users' needs. Don't forget to update your lessons learned document to include the causes of variances found during this process and why the corrective actions were recommended. It should also include how quality defects could have been avoided.

The results of changes, defect repairs, or variances that have been inspected and corrected are called *validated changes*. Validated changes, particularly corrective and preventive actions, can contribute to overall quality improvements and should be noted in the lessons learned documentation. Remember that processes that are in control should not be adjusted. Processes that are out of control might require adjusting, but this should occur only as a result of a management decision.

Quality control is built into agile projects. Because agile projects produce small units of work in an iterative fashion, quality issues can be discovered and resolved quickly.

Monitoring and Controlling Risk

The *Monitor Risks* process involves tracking and monitoring identified risks, monitoring implemented risk response plans to make certain they are effective, ensuring the overall risk management approach is still valid, and identifying and responding to new risks as they occur. You will examine performance information and use the tools and techniques of this process to analyze and implement other functions of this process, including the following:

- Ensuring that risk response plans and contingency plans that are put into action are appropriate and effective
- Determining whether overall project risk has changed

- Reexamining existing risks to learn if they have changed or should be closed out
- Monitoring for new individual risks
- Evaluating the overall appropriateness of the risk management approach
- Reassessing project assumptions and determining validity
- Ensuring that risk management policies and procedures are followed
- Ensuring that contingency reserves (for schedule and cost) are updated according to the updated risk assessment
- Ensuring the overall project strategy is valid

Monitor Risks is a busy process. During the course of the project, risk responses, which were developed during the Planning process group and implemented in the Executing process group, are still effective and have reduced or averted the impact of negative risk events (or you hope they did).

 Implement Risk Responses and Monitor Risks go hand in hand, but they each have their distinct function. Implement Risk Responses is performed by the risk owners (who are responsible for monitoring risks to determine whether response plans are needed). Risk owners put the response plan into action in this process. Monitor Risks monitors the response plans for effectiveness along with all the other functions listed earlier in this section.

This process requires you to review the risk register often. You'll recall that the risk register tracks and ranks individual risks, identifies the risk owner, describes risk triggers and residual risks, and lists the response plans and strategies you should implement if an actual risk event occurs. Keep in mind that some risk events identified in risk planning will happen and some will not. You will have to stay alert for risk events and be prepared to respond to them when they do occur. This means you should monitor the risk register regularly.

Another input that will assist you in this process is work performance data. This includes information that may help you determine that a new risk event is about to occur, or it may assist you in monitoring previously identified risks. Remember that this entails elements such as the status of deliverables, costs to date, cost changes, schedule progress to date, schedule changes, and scope changes. You will use variance analysis, forecasting, and earned value analysis on the information you've gathered and report the results in the work performance report (usually the status report). This information will help aid in decision-making, implementing risk responses, and taking corrective actions. These results should also be examined from the perspective of risks that might need close monitoring or response plans that could require changes to coincide with the data in the work performance reports.

Additional risk response planning might be needed to deal with the new risks or with expected risks whose impact might be greater than expected. This might require repeating the Plan Risk Responses process to create new contingency plans or alternative plans to deal with the risk, or it might require modification to existing plans.

Monitor Risks Analysis and Meetings

The tools and techniques of Monitor Risks are used to monitor risks throughout the life of the project. You should perform periodic reviews, audits, and new earned value analyses to check the pulse of risk activity and to make certain risk management is enacted effectively. We will look at a few new techniques here and revisit meetings to understand some key elements needed for the Monitor Risks process.

Technical Performance Analysis

The *technical performance analysis* tool and technique compares the technical accomplishments of the project objectives completed during the Executing and Monitoring and Controlling processes to the technical objectives defined in the project management Planning processes. Variances might indicate that a project risk is looming, and you'll want to analyze and prepare a response to it if appropriate. For example, a technical objective for a new computer software project might require that the forms printed from a particular module include a barcode at the bottom of the page. If the barcode functionality does not work once the module is coded, a technical deviation exists, which means you should reexamine project risks. In this particular example, project scope is likely at risk. Any technical performance measurement or objective that was defined in the Planning process group should be examined here.

Reserve Analysis

You'll recall that contingency reserves hold project funds, time, or resources in reserve to offset any unavoidable threats or opportunities that might impact the project scope, schedule, budget, or quality. This tool and technique examines the remaining reserves and compares them to the remaining risks to ensure there are enough reserves to cover any potential risks that may occur from now through the end of the project.

Risk Audits

Risk audits are carried out during the entire life of the project by risk auditors. Risk auditors are not typically project team members and are expertly trained in audit techniques and risk assessment. Risk audits are specifically interested in examining the risk management process. Risk audits also evaluate the implementation of response plans and their effectiveness at dealing with risks and their root causes. According to the *PMBOK® Guide,* the project manager is responsible for making certain audits that are performed throughout the project as appropriate and/or according to plan.

Meetings

Periodic, scheduled risk reviews of identified risks, risk responses, and risk priorities should occur during the project. The idea here is to monitor risks and their status and determine whether their consequences still have the same impact on the project objectives as when they were originally planned. If the risk is no longer viable, it should be closed. Every status meeting should have a time set aside to discuss and review risks and response plans, and risk review meetings should be scheduled periodically for the sole purpose of reviewing risk responses.

Risk identification and monitoring is an ongoing process throughout the life of the project. Risks can change, and previously identified risks might have greater impacts than originally thought as more facts are discovered. Reassessment of risks should be a regular activity performed by everyone involved on the project. Monitor the risk register (including those risks that have low scores) and risk triggers. You should also monitor the risk responses that have been implemented for their effectiveness in dealing with risk. You might have to revisit the Perform Qualitative and Perform Quantitative Risk Analysis processes when new risk consequences are discovered or risk impacts are found to be greater than what was originally planned. Remember that risk identification, analysis, and monitoring occur during every iteration of an agile project.

Monitor Risks Updates

Work performance information is a key output of this process, as it is for most of the Monitoring and Controlling processes. This update includes how the project risk management approach and performance compares to the planning processes where you identified anticipated risks, documented their expected outcomes, and documented response plans. In other words, did you do a good job of identifying risks, did their impacts manifest as you expected, and were the response plans effective in reducing threats or gaining opportunities?

Change requests may also result from this process. They might take the form of corrective actions or preventive actions. Or implementing a contingency plan or a workaround may bring about the need for a change request. A *workaround* is an unplanned response to a negative risk event. It attempts to deal with the risk in a productive, efficient manner. If no risk response plan exists (this might be the case when you accept a risk event during the Planning processes) or an unplanned risk occurs, workarounds are implemented to deal with the consequences of the risk.

The risk register needs to be updated with new risks whenever they occur on the project. You'll also update the risk response plans and close out risks that are no longer relevant.

The risk report is updated with the status of current individual risks and the up-to-date level of overall project risk. It may also include the results of the risk events that have occurred, the effectiveness of the risk response plans, and updates regarding risk owners. The risk report should be updated when a risk audit or risk reassessment concludes that some element of the original risk information has changed—for example, the impact or probability scores are updated to reflect new conditions, the priority of the risk has changed, the owner of the risk has changed, or the response plan has been updated.

According to the *PMBOK® Guide*, this process is performed throughout the life of the project and ensures that (1) stakeholders understand and are aware of the up-to-date risk exposure, and (2) the work of the project is continually monitored to identify new risks, to review existing risks and determine whether they've changed, and to identify risks that are no longer relevant.

Monitoring Project Management Integrations

We talked earlier in this book about integration management and tailoring project processes. In the Monitoring and Controlling process group, you should review the tailoring processes and determine whether they are delivering the efficiencies you expected. You should document your findings in the lessons learned document for future reference. According to the *PMBOK Guide®*, the key elements you should consider when tailoring projects and reviewing the effectiveness of project integration management include the following:

Project Life Cycles   Consider whether the project life cycle was appropriate and assisted in delivering the business value expected by the stakeholders. We discussed life cycles in Chapter 1 if you need a refresher.

Development Life Cycles   You will have used a predictive, adaptive, or hybrid style to develop the project. Analyze the success of the development life cycle you chose and determine whether it was successful in delivering business value according to stakeholder expectations.

Management Approaches   Elements to consider here include management policies, processes, culture, and project complexity. For example, did the human resources department assist with establishing processes for creating agile teams?

Project Knowledge Management   I talked about this topic in Chapter 10. Consider whether the repositories and other information management tools you used were effective. Also, consider whether the project artifacts were managed according to the communications management plan and whether the stakeholders could easily access them.

Change management   We covered change management earlier in this chapter. Consider whether these processes were effective and if there were changes that had significant impacts on the project.

Governance Committees   There are several committees we've discussed throughout the book that you should monitor for effectiveness. Examples include the project steering committee, the project request and approval committee, the risk strategy committee, and the change control board.

Lessons Learned   Consider whether you are capturing information that will articulate what worked well—and not so well—on the project. Most importantly, think about the information from the perspective of a future project manager reviewing the lessons learned. Will they find this information useful on their future projects? Will it help prevent delays, conflict, schedule issues, and more?

Reporting on Expected Benefits   We talked about the benefits management plan in Chapter 2, "Assessing Project Needs." This plan outlines the intended benefits and business value the project will bring about, how the benefits will be measured, and how they will be obtained. Analyze whether your benefit measurements and goals were met. Were they reported on throughout the project? Did the assumptions you made about the project come to pass? Did the finished project align with the strategic organization's goal you identified early in the project?

 **Real World Scenario**

Project Case Study: New Kitchen Heaven Retail Store

Your regularly scheduled status meeting is in progress. Let's see how it's progressing.

Your next agenda item is an update on change requests. Ricardo submitted a change request regarding the hardware installation at the new store site. A new, much anticipated operating system was just released, and Ricardo has plans to upgrade the entire company to the new operating system. Since he must purchase new equipment for this store anyway, he contends that it makes sense to go ahead and purchase the hardware with the newest operating system already loaded. His staff won't have to upgrade this store as part of the upgrade project because the store will already have the new operating system.

Ricardo's change request was submitted in writing through the change control system. A CCB was set up during the Planning stages of this project to handle change requests. At the CCB meeting, the following questions come up regarding Ricardo's request: Has the new operating system been tested with the existing system? Are there compatibility problems? If so, what risks are associated with getting the problems resolved and the equipment installed by opening day? Is the vendor ready to ship the new operating system?

The CCB defers their decision for this request until Ricardo answers these questions.

During the next CCB meeting, the board reviews a change request from Jill. The gourmet food supplier she used went out of business, and Jill contracted with a new vendor. She received a sample shipment from the new vendor and was very unhappy with the quality. Upon inspecting the products, she found broken containers and damaged packaging. Meanwhile, Jill found another vendor, who has sent her a sample shipment. She is very pleased with the new vendor's products and service. However, this vendor's prices are even higher than the first replacement vendor with whom she contracted. Jill submitted a cost change request to the CCB because of the increased cost of the gourmet food product shipment. The change in cost does not have a significant impact on the budget.

Jake reported that payments have been made to the vendor supplying the store's display cases and shelving and also gave an update on the vendor supplying the lighting for the store. Both vendors met or exceeded performance requirements and he's happy with their service.

You hold a seat on the CCB and are aware of the change requests and their impacts on the project. Ricardo satisfactorily answered all the questions the CCB had, so his request and Jill's request were both approved during the meeting.

You inform the group that you will be setting up some one-on-one meetings with each of the key stakeholders to check in with them on how the project is progressing from their viewpoint and to listen to their input and/or any issues they may have. You also tell them it's time for a risk update meeting and you'll schedule that for next week.

"Now I'd like to give a brief update on the project forecast," you tell the group. "Based on the preliminary data I've gathered, we are somewhat behind schedule but the budget appears to be on track. I will have some forecast numbers for you at the next meeting."

Project Case Study Checklist

- Control Procurements
  - Monitor vendor performance
  - Document vendor performance
  - Monitor payments to seller
- Perform Integrated Change Control
  - Review and approve changes
  - Document changes in change log

- Configuration management systems
  - Configuration identification
  - Configuration status accounting
  - Configuration verification and auditing
  - Change control system
- Monitor Communications
  - Work performance information (status reports and meetings)
  - Forecasts
- Monitor Stakeholder Engagement
  - Meetings
  - Work performance information
- Monitor Risk
- Control Quality

Understanding How This Applies to Your Next Project

I've learned from experience the value of having a change control process in place for all projects. I've never managed a project that didn't encounter change—and there are hundreds of reasons that bring about change. One of the ways to help reduce the amount of change you might experience is to make certain you've documented the requirements of the project accurately and have obtained sign-off from the stakeholders. Beware! Just because the stakeholders have agreed to the requirements doesn't mean they won't want change. As you elaborate the deliverables and requirements, the product or end result becomes clearer, and that means some elements not previously known or at least not known in their entirety early in the planning process will require change.

If you don't already have a change control process in place, I recommend setting one up before you begin your next project. Document the procedures for requesting, tracking, and approving or denying changes. Make change control procedures one of the agenda items for discussion at your project kickoff meeting. It's easier to enforce change procedures (and deny changes that are out of scope) if the process is discussed with the stakeholders early in the project. You will likely want to include important stakeholders on the change control board, and that will give you another great opportunity to discuss and reinforce the process.

Administering contracts and procurements, as I mentioned earlier, may be managed by someone in your procurement department. In my experience, there is always someone from this department involved from the request process through contract administration. However, it will likely be up to you to monitor vendor performance, make sure deliverable dates are met, and verify time cards against the submitted invoice.

Keeping your stakeholders engaged throughout the entire project is key to its success. Don't let stakeholder interests drift. Make an effort to reach out to them and meet with them on a regular basis. Your enthusiasm for the project is contagious and will keep the stakeholders engaged.

Control Quality concerns monitoring work results and project deliverables to see whether they comply with the stakeholders' expectations. This process is performed throughout the project to identify and remove the causes of unacceptable results. You will determine whether deliverables are within tolerable levels and whether the process is in control.

Monitoring Risks is a process you'll perform once the work of the project begins and throughout the remainder of the project. Just like change, risk is something that will occur on most projects you undertake. I've never managed a project (except for projects that were started and finished within a matter of days) that didn't encounter risk. My experience has been that most risks are known-unknown, which makes contingency reserves (both time and money) essential on any project. Unknown-unknown risks are also common and require adequate management reserves. Taking on a project without knowing that risks will occur and not having some contingency set aside is a huge gamble because even the smallest projects have risks.

Change control, quality control, and monitoring for risks are inherent in agile projects. Each of these can be monitored and adjusted during each iteration. As feedback is received, changes can be incorporated, risks can be mitigated, and quality can be adjusted to meet stakeholder expectations.

Summary

This chapter examined several change control processes, starting with the Monitor and Control Project Work process. This process is responsible for reviewing, tracking, and controlling project progress. It ensures that the performance objectives outlined in the project management plan are met.

The Control Procurements process is concerned with managing vendor relationships, monitoring performance, and implementing changes or corrections when necessary. Both buyer and seller are responsible for administering the contract (or other procurement vehicle) to ensure that contractual obligations are being met.

The Monitor Communications process involves monitoring and measuring project performance to identify variances from the project management plan. This process monitors and controls communications throughout the life of the project.

Perform Integrated Change Control is an important part of the project process. It's your responsibility as project manager to manage change and implement corrective action where needed to keep the project on track with the plan. Perform Integrated Change Control concerns influencing the things that cause change and managing the change once it has occurred. One or more of the project baselines may be affected when change occurs. Managing change might involve changes to the project plan, the project schedule, the project budget, project scope, and so on. Changes that impact processes you've already completed require updates to those processes. Corrective action is often a result of a change and ensures that the future performance of the project lines up with the project management plan.

Configuration management systems typically include change control systems that document the procedures to manage change and how change requests are implemented. Change requests might come in written or verbal forms, but ideally you should ask for all your change requests in writing. Change requests are processed through a formal change control system, and configuration control boards have the authority to approve or deny change requests.

Monitor Stakeholder Engagement concerns monitoring your relationships with the stakeholders and keeping them engaged throughout the project.

Control Resources monitors the availability and performance of the physical resources for the project. Control Quality concerns monitoring work results and project deliverables to see whether they comply with the stakeholders' expectations. This process is performed throughout the project to identify and remove the causes of unacceptable results. You will determine whether deliverables are within tolerable levels and whether the process is in control.

Monitor Risks is performed throughout the life of the project. You should always be on the lookout for new risks or for changes that could impact identified risks. The risk owner is responsible for implementing the response plans, and you'll want to document everything you experienced about risk, including the effectiveness of the response plans, in the lessons learned document.

Tailoring processes and integrating project management processes were likely performed during the Planning process group and may have been adjusted some during the Executing process. In the Monitoring and Controlling process, you want to ensure the tailoring practices you put into place are working according to plan and document results in the lessons learned document.

Exam Essentials

Describe the purpose of the Monitor and Control Project Work process.   This process is responsible for reviewing, tracking, and controlling project progress.

Name the processes that integrate with the Control Procurements process.   These processes are Direct and Manage Project Work, Monitor Communications, Control Quality, Perform Integrated Change Control, and Monitor Risks.

Describe the purpose of the Monitor Communications process.   Monitor Communications concerns monitoring and controlling communications throughout the life of the project.

Name the purpose of the Perform Integrated Change Control process.   Perform Integrated Change Control is performed throughout the life of the project and involves reviewing all the project change requests, establishing a configuration management and change control process, and approving or denying changes.

Be able to define the purpose of a configuration management system.   Configuration management systems are documented procedures that describe the process for submitting change requests, the processes for tracking changes and their disposition, and the processes for defining the approval levels for approving and denying changes. The configuration management system also includes a process for authorizing the changes. Change control systems are generally a subset of the configuration management system. Configuration management also describes the characteristics of the product of the project and ensures the accuracy and completeness of the description.

Be able to describe a CCB.   The change control board (CCB) has the authority to approve or deny change requests. Their authority is defined and outlined by the organization. A CCB is made up of stakeholders.

Be able to describe the purpose of the Monitor Stakeholder Engagement process.   The purpose is to monitor stakeholder relationships and ensure their continued engagement in the project.

Be able to name the purpose of the Control Resources process.   The purpose of Control Resources is to make certain the right physical resources are available at the time they are needed and at the location they are needed. This process also ensures that the resources are released when their job is completed and that corrective actions are taken when resources are not utilized as planned.

Be able to name the purpose of the Control Quality process.   The purpose of the Control Quality process is to monitor work results to see whether they comply with the standards set out in the quality management plan.

Describe the purpose of Monitor Risks.   Monitor Risks involves identifying and responding to new risks as they occur. Risk monitoring and reassessment should occur throughout the life of the project.

Describe the purpose of monitoring project integrations.   You should review the tailoring processes and determine whether they are delivering the efficiencies you expected. You should document your findings in the lessons learned document for future reference. Some elements to examine include project and development life cycles, change management, lessons learned, and reporting on expected benefits.

Review Questions

You can find the answers to the review questions in Appendix A. Be sure to download the Bonus Exams and Bonus Questions so that you'll have a broader exposure and more experience answering questions related to the topics in this chapter.

1. Which of the following processes is responsible for reporting and comparing actual project results against the project management plan, analyzing performance data to learn if action needs to be taken, monitoring for risks, and monitoring approved change requests, among other functions?

    A. Perform Integrated Change Control

    B. Monitor Stakeholder Engagement

    C. Monitor Communications

    D. Monitor and Control Project Work

2. You are working on a project and discover that one of the business users responsible for testing the product never completed this activity. She has written an email requesting that one of your team members drop everything to assist her with a problem that could have been avoided if she had performed the test. This employee reports to a stakeholder, not to the project team. You estimate that the project might not be completed on time as a result of this missed activity. Which of the following options are true? (Choose three.)

    A. You should recommend a corrective action to bring the expected future project performance back into line with the project management plan because of this employee's failure to perform this activity.

    B. You should recommend a preventive action to reduce the possibility of future project performance veering off track because of this employee's failure to perform this activity.

    C. This situation was likely discovered while performing the Monitor and Control Project Work process.

    D. You might have to request a change to the project schedule as a result of this missed activity.

    E. You should document the performance issues with this employee in the lessons learned document for future project reference.

3. Which one of the following is the most preferred method of settling claims and disputes according to the *PMBOK® Guide*?

    A. Arbitration

    B. Collaboration

    C. Claims administration

    D. Negotiation

4. You have contracted with a vendor to perform your project. Your internal stakeholders seem to have a memory lapse on the project objectives. Your vendor has complained to you that your stakeholders are difficult to work with and their expectations are unreasonable.

Your vendor has also missed some requirements, which you discovered when you examined one portion of the finished product. Which of the following are true regarding this question? (Choose three.)

A. You will speak with the vendor in the upcoming performance review about their poor management of stakeholder expectations.

B. Inspections are a structured review of the procurement process, and audits are structured reviews of the work performed by the contractor.

C. Procurement performance reviews examine the vendor's performance to date to determine whether elements such as project scope, quality, budget, and schedule are aligned with the contract.

D. Procurement performance reviews might take the form of quality audits, inspections, or status review.

E. Both you and the vendor are equally responsible for monitoring the contract to ensure that the other party is meeting their obligations.

5. You are a project manager for an engineering company. Your company won the bid to add ramp-metering lights to several on-ramps along a stretch of highway at the south end of the city. You subcontracted a portion of the project to another company. The subcontractor's work involves digging the holes and setting the lamp poles in concrete. The subcontractor's performance to date does not meet the contract requirements. Which of the following is not a valid option?

A. You document the poor performance in written form and send the correspondence to the subcontractor.

B. You terminate the contract for cause and submit a change request through Control Procurements.

C. You terminate the contract due to default and submit a change request through Control Procurements.

D. You agree to meet with the subcontractor to see whether a satisfactory solution can be reached.

6. The Control Procurements process is closely integrated with all of the following processes except for which one?

A. Direct and Manage Project Work

B. Monitor Communications

C. Perform Integrated Change Control

D. Monitor Risks

7. You are holding a regularly scheduled status meeting for your project. You know all of the following are true regarding status meetings except which one?

A. Observation and conversation may be used to gather information on project performance and reported at the status meeting.

B. Status meetings are a type of interactive communication.

    C. Status meetings are a way to formally exchange information and update the stakeholders regarding project status.

    D. Status meetings should be held throughout the project and at regularly scheduled intervals.

    E. Information shared at status meetings always includes updates to elements such as scope, budget, and schedule and does not need to be tailored to the audience.

8. Your project was kicked off more than two years ago. Your stakeholders are not as engaged as they once were on this project. There are only four months to go until the project is complete. Which of the following are true regarding this scenario?

    A. Use the Monitor Stakeholder Engagement process to ensure that your stakeholder relationships are in good standing and to encourage continued engagement in the project.

    B. Use the Monitor Communications process to ensure that you are reporting information regarding the progress of the project and that the right information is delivered to the right people at the right time.

    C. One result of both the Monitor Stakeholder Engagement process and the Monitor Communications process is to help stakeholders make informed decisions.

    D. Ask your stakeholders if the meetings and information they are receiving are useful or if there are other ways or different communication techniques that will help keep them informed and engaged.

    E. A, B, C, D

    F. A, B, D

9. Your project is in a bit of trouble. You are in the Control Resources process and the specialized equipment you were expecting for the next set of project tasks is not available. The truck with the equipment was stolen and the vendor providing the machine you need has only two of them. The other is already deployed on another job. Which of the following are true?

    A. Control Resources may have impacts on resource assignments that are beyond the control of the project manager.

    B. You decide to engage the vendor and some key project team members in problem solving to try to find a workaround for this situation. You will identify the problem, investigate and analyze the data, and determine and implement the best possible solution. Later, you'll check again on the solution to ensure that it fixed the problem.

    C. You may need to engage in some preventive action since the resources are not being utilized as planned. You will document all of this in the lessons learned document.

    D. Problems encountered outside the organization in the Control Resources process are usually related to the vendor providing the resources.

    E. A, B, C, D

    F. A, B, D

10. The Delphi method, technology forecasting, and forecast by analogy are examples of what category of forecasting methods?

    A. Time series

    B. Judgmental

    C. Causal

    D. Econometric

11. You are working on a project that was proceeding well until a manufacturing glitch occurred that requires corrective action. It turns out the glitch was an unintentional enhancement to the product, and the marketing people are absolutely crazy about its potential. The corrective action is canceled, and you continue to produce the product with the newly discovered enhancement. As the project manager, you know that a variance has occurred. Which of the following options is true?

    A. Common causes of variance, also known as special-cause variances, are situations that are unique and not easily controlled at the operational level.

    B. Random variances, known or predictable variances, and variances that are always present in the process are known as common causes of variance.

    C. Attribute inspection determines whether measurements fall within tolerable results.

    D. Scatter diagrams display the relationships between an independent and a dependent variable to show variations in the process over time.

12. Your project progressed as planned until yesterday. Suddenly, an unexpected risk event occurred. You quickly devised a response to deal with this negative risk event using which of the following?

    A. Risk management plan updates

    B. Workarounds

    C. Corrective action

    D. Additional risk identification

13. You are a project manager for Bluebird Technologies. Bluebird writes custom billing applications for several industries. A schedule change has been requested. From the perspective of the Perform Integrated Change Control process, change is concerned with which of the following?

    A. Influencing factors that circumvent the change control process

    B. Initiating change requests

    C. Reviewing change

    D. Maintaining the integrity of baselines

    E. Coordinating and managing changes across the project

    F. A, C, D, E

    G. A, B, C, D, E

14. You are a project manager for Bluebird Technologies. Bluebird writes custom billing applications for several industries. One of your users asks for some changes to the invoice functionality during a status meeting. You explain to her that the change needs to go through the change control system, which is a subset of the configuration management system. You explain that a change control system does which of the following?

A. Documents procedures for managing change requests

B. Tracks the status of change requests

C. Describes the management impacts of change

D. Documents the process for approving changes

E. Defines the level of authority needed to approve emergency change requests

F. A, B, D, E

G. A, B, C, D, E

15. You are a project manager for Star Light Strings. Star Light manufactures strings of lights for outdoor display. Its products range from simple light strings to elaborate lights with animal designs, bug designs, memorabilia, and so on. Your newest project requires a change. You are performing the Control Quality process and are looking at and testing a sampling of the product for conformance to quality standards. The tolerable result for the bulbs used on the string is 995–1,000 hours. All of the light bulbs must be LED bulbs. Which of the following are true regarding this process?

A. Tolerable results are assessed during an inspection and are expressed within a specified range.

B. The LED bulbs are an attribute. Attributes have two measures, conforming or nonconforming.

C. Prevention cost should be considered against the cost of rework, labor, material, and potential loss of customers.

D. You are performing an inspection. Inspections sometime use check sheets to help track and record data.

E. A, B, C, D

F. A, B, C

16. You are a project manager for Star Light Strings. Star Light manufactures strings of lights for outdoor display. Its products range from simple light strings to elaborate lights with animal designs, bug designs, memorabilia, and so on. Your newest project requires a change. One of the business unit managers submitted a change through the change control system, which utilizes a CCB. Which of the following is true regarding the CCB?

A. The CCB describes how change requests are managed.

B. The CCB requires all change requests in writing.

C. The CCB approves or denies change requests.

D. The CCB requires updates to the appropriate management plan.

17. All of the following are activities of the configuration management system except for which one?

    A. Variance analysis

    B. Identification

    C. Status accounting

    D. Verification and auditing

18. You are using a process to measure the quality results over time. You are reviewing this graphic at your status meeting and explain that this information is based on sample variance measurements. Your measurements are within two standard deviations. Which of the following options are true? (Choose two.)

    A. You are using a scatter diagram.

    B. The process is in control and should not be adjusted.

    C. You are using a cause-and-effect diagram.

    D. You are using a control chart.

19. You are in the Monitor Risks process and are reviewing the risk register and notice something strange. A risk that was designated as low-low in the Planning phase of the project has suddenly become a high risk. There is no response plan for this risk because of its original ranking. Which of the following should be the first action you take out of all the options listed?

    A. You should inform the risk auditors what happened with this risk so they can include an evaluation in their next risk audit.

    B. You should conduct testing and a product evaluation to determine whether the objectives will be met and whether variances were caused as part of this risk.

    C. You should perform a reserve analysis to ensure there are enough reserves to cover this risk.

    D. You should calculate a forecast to determine whether the future project performance will be impacted by this risk.

20. You work for a small software development firm and use agile methodologies on all your projects. You learned that your company is being bought out by a larger company. The good news is that everyone on staff will retain their jobs with the new company, but your existing company will cease to exist. You are managing several projects that will be impacted by this. Which of the following are true regarding this?

    A. Your organization is undergoing an acquisition.

    B. You should assess and prioritize the impact of these changes to the projects and develop recommendations and action items and communicate them to the new management team.

    C. Your organization is undergoing a merger.

    D. This is a type of external business environment change.

    E. A, B, D

    F. B, C, D

Chapter 12

Controlling Work Results and Closing Out the Project

THE PMP® EXAM CONTENT FROM THE PEOPLE DOMAIN COVERED IN THIS CHAPTER INCLUDES THE FOLLOWING:

✓ **Task 1.4 Empower team members and stakeholders**

- 1.4.4 Determine and bestow level(s) of decision-making authority

✓ **Task 1.5 Ensure team members/stakeholders are adequately trained**

- 1.5.2 Determine training options based on training needs
- 1.5.4 Measure training outcomes

✓ **Task 1.7 Address and remove impediments, obstacles, and blockers for the team**

- 1.7.4 Reassess continually to ensure impediments, obstacles, and blockers for the team are being addressed

THE PMP® EXAM CONTENT FROM THE PROCESS DOMAIN COVERED IN THIS CHAPTER INCLUDES THE FOLLOWING:

✓ **Task 2.2 Manage communications**

- 2.2.3 Communicate project information and updates effectively

✓ **Task 2.5 Plan and manage budget and resources**

- 2.5.3 Monitor budget variations and work with governance process to adjust as necessary

✓ **Task 2.6 Plan and manage schedule**

- 2.6.4 Measure ongoing progress based on methodology

- 2.6.5 Modify schedule, as needed, based on methodology

✓ **Task 2.8 Plan and manage scope**

- 2.8.3 Monitor and validate scope

✓ **Task 2.9 Integrate project planning activities**

- 2.9.2 Analyze the data collected

✓ **Task 2.12 Manage project artifacts**

- 2.12.2 Validate that the project information is kept up to date

✓ **Task 2.16 Ensure knowledge transfer for project continuity**

- 2.16.1 Discuss project responsibilities within team

- 2.16.2 Outline expectations for working environment

- 2.16.3 Confirm approach for knowledge transfers

✓ **Task 2.17 Plan and manage project/phase closure or transitions**

- 2.17.1 Determine criteria to successfully close the project or phase

- 2.17.2 Validate readiness for transition

- 2.17.3 Conclude activities to close out project or phase

Congratulations! You've made terrific progress, and after finishing this chapter you will be well equipped to take the Project Management Professional (PMP)® exam.

This chapter covers the last group of project management processes in the Monitoring and Controlling process group. I'll cover the Control Costs, Control Schedule, Validate Scope, and Control Scope processes in this chapter. A significant amount of information is packed into this chapter, and I recommend you memorize all the formulas presented here for the exam.

The Control Costs and Control Schedule processes are similar to Perform Integrated Change Control, which we discussed in the previous chapter. When you're reading these sections, remember that the information from the Perform Integrated Change Control process applies to these areas as well.

The Validate Scope process involves validating and accepting work results. Control Scope is like the change control processes I discussed in Chapter 11, "Measuring and Controlling Project Performance," and is concerned with controlling changes to project scope.

All of the Monitoring and Controlling processes are incorporated into agile projects, and we'll take a look at how that happens in each of these sections.

The Closing process group, which is the last process group we'll cover, has one process: Close Project or Phase. The Close Project or Phase process is concerned with verifying that the work of the project was completed correctly and to the stakeholders' satisfaction. The Close Project or Phase process also verifies that the work of the project was completed correctly and that the deliverables were accepted.

Once you've obtained the PMP® designation, you have an obligation to maintain integrity, apply your subject matter and project management knowledge, and maintain the code of conduct published by the PMI®. You'll also be required to balance the interests and needs of stakeholders with the organization's needs. The exam might include questions on any of these topics.

As a project manager, you'll find yourself in many unique situations, different organizations, and possibly even different countries. Even if you never get involved in international project management, you will still come in contact with people from cultures and backgrounds different from yours. If you work as a contract project manager, you'll be exposed to many different organizations; each will have its own culture and ways of doing things. You should always strive to act in a professional, courteous manner in these situations.

The process names, inputs, tools and techniques, outputs, and descriptions of the project management process groups and related materials and figures in this chapter are based on content from *A Guide to the Project Management Body of Knowledge (PMBOK® Guide), Sixth Edition* (PMI®, 2017). The references to adaptive and hybrid methodologies, related materials, and figures in this chapter are based on content from the *Agile Practice Guide* (PMI®, 2017).

Controlling Cost Changes

The *Control Costs* process monitors the project budget and manages changes to the cost baseline. It's concerned with monitoring project costs to prevent unauthorized or incorrect costs from being included in the cost baseline. This means you'll also use Control Costs to ensure that the project budget isn't exceeded (resulting in cost overruns). If a change is implemented, you'll have to make certain the budget for the changed item stays within acceptable limits. All budget changes should be agreed to and approved by the project sponsor where applicable (the criteria for approvals should be outlined in the change control system documentation). Increases to the authorized budget should be submitted as change requests and approved through the Perform Integrated Change Control process. Stakeholders should be made aware of all budget changes.

This process involves examining the cost baseline by using its tools and techniques to compare actual expenditures to date to the baseline. You'll also refer to the cost management plan because it details how costs should be monitored and controlled throughout the life of the project.

The following list includes some of the activities you'll be involved in during this process:

- Monitoring changes to costs or the cost baseline and understanding variances from the baseline

- Monitoring change requests that affect cost and resolving them in a timely manner

- Informing stakeholders of approved changes and their costs

- Ensuring that the project budget does not exceed acceptable limits by taking action when overruns are imminent

- Ensuring that the project budget does not exceed the total funding authorized for the project or for the project phase

There are several new tools and techniques of the Control Costs process we will look at, including earned value analysis (EVA), variance analysis, trend analysis, and to-complete performance index (TCPI). All of these contribute to the primary output of this process, which is cost forecasts.

Earned Value Analysis

Earned value analysis (EVA) is an analysis technique that compares the actual schedule and cost to the performance measurement baseline. You can accomplish performance measurement analysis using a technique called *earned value management (EVM)*. Simply stated, EVM compares what you've received or produced to what you've spent. The EVM continuously monitors the planned value, earned value, and actual costs expended to produce the work of the project (I'll cover the definition of these terms shortly). When variances that result in cost changes are discovered (including schedule variances and cost variances), those changes are managed using the project change control system. The primary function of this analysis technique is to determine and document the cause of the variance, to determine the impact of the variance, and to determine whether a corrective action should be implemented as a result. We'll walk through various examples that illustrate how to determine these variances later in this chapter.

EVM looks at schedule, cost, and scope baselines and compares them to the actual work completed to date. You may recall that the schedule, cost, and scope baselines together make up the performance measurement baseline (PMB). The PMB and EVM will be used throughout the project to measure the progress and performance of the project. Remember that it does not include management reserves but that it does include contingencies.

EVM is the most often used performance measurement method. EVM is performed on the work packages and the control accounts of the WBS. To perform the EVM calculations, you need to first gather the three measurements known as the planned value (PV), actual cost (AC), and earned value (EV).

If you do any research on your own regarding these values, you might come across acronyms that are different from what you see here. I've included their alternative names and acronyms at the end of each description. I recommend you memorize planned value (PV), actual cost (AC), and earned value (EV) and make certain you understand the meaning of each before you continue.

Let's take a look at some definitions of these key measurements before diving into the actual calculations:

Planned Value The *planned value (PV)* is the cost of work that has been authorized and budgeted for a schedule activity or WBS component during a given time period or phase. These budgets are established during the Planning processes. For any given day, PV equals the planned cost of work that is scheduled to be completed on that day, whether or not the work is actually completed. PV is also called budgeted cost of work scheduled (BCWS).

Exam Spotlight

Remember to read exam questions carefully. PV might mean present value (as I talked about in Chapter 2) or planned value (as defined here).

Actual Cost   *Actual cost (AC)* is the actual cost of completing the work component in a given time period. Actual costs might include direct and indirect costs but must correspond to what was budgeted for the activity. If the budgeted amount did not include indirect costs, do not include them here. Later you'll see how to compare this to PV to come up with variance calculation results. Actual costs include whatever is spent to complete the work regardless of what was budgeted. AC is also called actual cost of work performed (ACWP).

Earned Value   *Earned value (EV)* is the value of the work completed to date as it compares to the authorized budgeted amount assigned to the work component. EV is typically expressed as a percentage of the work completed compared to the budget. For example, if the budgeted amount for our work component is $1,000 and we have completed 30 percent of the work so far, our EV is $300. Therefore, EV cannot exceed the PV budget for the activity. EV is also called budgeted cost of work performed (BCWP).

Exam Spotlight

PV, AC, and EV are easy to mix up. In their simplest forms, here's what each means:

PV—The approved budget assigned to the work to be completed during a given time period

AC—Money that's actually been expended during a given time period for completed work

EV—The value of the work completed to date compared to the budget

---

According to the earlier definition, EV is the sum of the cumulative budgeted costs for completed work for all activities that have been accomplished as of the measurement date. For example, if your total budget is $1,000 and 50 percent of the work has been completed as of the measurement date, your EV would equal $500. You can plot all the PV, AC, and EV measurements graphically to show the variances between them. If there are no variances in the measurements, all the lines on the graph remain the same, which means the project is progressing as planned. Figure 12.1 shows an example that plots these three measurements.

FIGURE 12.1 Earned value

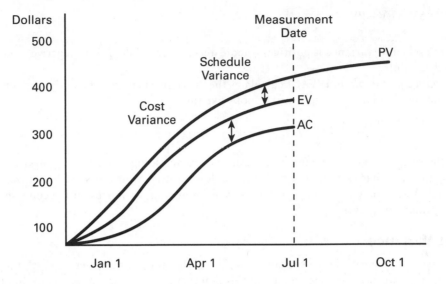

All of these measurements include a cost component. Costs are displayed in an S curve because spending is minimal in the beginning of the project, picks up steam toward the middle, and then tapers off at the end of the project. This means your earned value measurements will also take on the S curve shape.

Now you can calculate whether the project is progressing as planned or if variances exist in the approved baseline by using a variety of formulas discussed in the following sections. Use Figure 12.1 as your example for the formulas that follow. The Figure 12.1 totals are as follows:

PV = 400, EV = 375, AC = 325

Variance Analysis

Variance analysis in the Control Costs process examines the difference between the baseline cost or baseline schedule as they compare to actual performance, and/or the variance at completion of the project. Variances in this process are usually due to cost or schedule impacts, making cost and schedule the most commonly analyzed variances.

Cost variances are determined by subtracting AC from EV:

CV = EV – AC

If the cost variance is positive, your costs are under what was planned and you are doing better than expected. If they are negative, you are over what was planned.

Schedule variances are determined by subtracting PV from EV:

SV = EV – PV

If the schedule variance is positive, you are ahead of schedule; if it is negative, you are behind schedule.

Variance at completion (VAC) calculates the difference between the budget at completion and the estimate at completion. It looks like this:

VAC = BAC – EAC

If the result is a negative number, it means you're not doing as well with costs as you anticipated and that variance exists. If the result is positive, your costs are better than you planned. Assuming your project performance is improving, variances will become smaller as the project progresses.

Cost Variance

Cost variance is one of the most popular variances that project managers use. It's the difference between EV (where you are at this point) and AC (what you've spent). In other words, it tells you whether your actual costs are higher than expected (with a resulting negative number) or lower than expected (with a resulting positive number) at a certain point in time. It measures the actual performance to date (or during the period) against what's been spent.

The *cost variance (CV)* is calculated as follows:

CV = EV – AC

Let's calculate the CV using the numbers from Figure 12.1:

375 – 325 = 50

The CV is positive, which means you're spending less than what you planned for the work that you have completed as of July 1 (which Figure 12.1 shows because AC is less than EV).

If you come up with a negative number as the answer to this formula, it means that costs are higher than what you had planned for the work that was completed as of July 1. These costs are usually not recoverable.

Schedule Variance

Schedule variance, another popular variance, tells you whether the schedule is ahead of or behind what was planned for this period. It's the difference between where you are at this point (EV) and what was planned for this point (PV). This formula is most helpful when you've used the critical path methodology to build the project schedule. The *schedule variance (SV)* is calculated as follows:

$$SV = EV - PV$$

Let's plug in the numbers:

$$375 - 400 = -25$$

The resulting schedule variance is negative, which means you are behind schedule, or behind where you planned to be as of July 1.

Together, the CV and SV can be converted to *efficiency indicators* for the project and can be used to compare the performance of all the projects in a portfolio.

Performance Indexes

Cost and schedule performance indexes are primarily used to calculate performance efficiencies, and they're often used to help predict future project performance.

The *cost performance index (CPI)* measures the cost efficiency of the work completed against actual cost. Generally speaking, it is the most critical of all the EVM measurements because it tells you the cost efficiency for the work completed to date, or at the completion of the project. If CPI is greater than 1, you're spending less than anticipated to date. If CPI is less than 1, you are spending more than anticipated for the work completed and have a cost overrun on your hands.

The cost performance index (CPI) is calculated this way:

$$CPI = EV / AC$$

Let's plug in the numbers and see where you stand:

$$375 / 325 = 1.15$$

This means cost performance is better than expected. You get an A+ on this assignment!

The *schedule performance index (SPI)* measures the efficiency of the project team to date in completing work tasks against the progress that was planned. This formula should be used in conjunction with an analysis of the critical path activities to determine if the project will finish ahead of or behind schedule. If SPI is greater than 1, you are ahead of schedule and have completed more work than was planned. If SPI is less than 1, you are behind schedule and have not completed as much work as you planned to complete by the measurement date.

The schedule performance index (SPI) is calculated this way:

$$SPI = EV / PV$$

Again, let's see where you stand with this example:

$375 / 400 = 0.94$

Uh-oh, not so good. Schedule performance is not what you expected. Let's not grade this one.

There are two more indexes associated with cost and schedule that you should understand for the exam: cumulative CPI and cumulative SPI.

Cumulative CPI is a commonly used calculation to predict project costs at the completion of the project. It also represents the cumulative CPI of the project at the point at which the measurement is taken. First, you need to sum the earned value calculations taken to date, or cumulative EV, and the actual costs to date, or cumulative AC. The formula looks just like the CPI formula except that it uses the cumulative sums as follows:

Cumulative CPI = cumulative EV / cumulative AC

The difference between this and the CPI formula earlier is that the CPI formula is used for a single work period, whereas the cumulative CPI is calculated using the sum of all the costs of every work component for the project. Additionally, you might also use cumulative CPI to calculate the total cost of a work component such as a deliverable, for example. Let's say you have a deliverable that has five work packages. You would total the EV and AC at the measurement date for all five work packages to determine the cost performance index for the deliverable.

Cumulative SPI predicts schedule performance at the completion of the project. Like cumulative CPI, it also represents the cumulative SPI of the project at the point at which the measurement is taken. The formula is as follows:

Cumulative SPI = cumulative EV / cumulative PV

Trend Analysis

According to the *PMBOK® Guide*, trend analysis determines whether project performance is improving or worsening over time by periodically analyzing project results. These results are measured with mathematical formulas that attempt to forecast project outcomes based on historical information and results. You can use several formulas to predict project trends, but it's outside the scope of this book to go into all of them. For the exam, you're expected to understand the concept behind trend analysis, and to know that you can use charts (such as the S curves we discussed earlier) or forecasting to display trends. You can use the results you've analyzed to predict future project behavior or trends. Let's look at some of the formulas associated with forecasting next. Forecasting uses the information you've gathered to date and estimates the future conditions or future performance of the project based on what you know when the calculation is performed. Forecasts are based on work performance data (an output from the Executing process group) and your predictions of future performance.

The forecasting formulas you'll see later in this section are used to determine an *estimate at completion (EAC)* and an *estimate to complete (ETC)*. The EAC estimates (or forecasts) the expected total cost of a work component, a schedule activity, or the project at its completion by calculating the actual costs to date and then adding an estimate of what the remaining work will cost. The ETC is the anticipated cost estimate to finish the work of the project.

EAC is most often calculated by using actual costs incurred to date plus a bottom-up ETC estimate. The formula for the most typical EAC looks like this:

$$EAC = AC + \text{bottom-up ETC}$$

The bottom-up ETC estimate is usually provided by the members of the project team who are actually working on the project activities. They provide the project manager with an estimate of the amount of effort remaining (and, therefore, the cost of the effort) based on the activities they have completed to date and what they believe will occur in the future. Their estimates are summed to come up with a total ETC, also known as a bottom-up ETC.

There are three other EAC forecasting formulas outlined in the *PMBOK® Guide* that we'll look at next. A new term you'll need to know before we look at these formulas is *budget at completion (BAC)*. BAC is the total amount of PV (approved budgeted costs) for all the work of a work component or all the work of the project. It is the sum of all the budgets established for all the work in the work package, control account, schedule activity, or project.

 You may find that the EAC (estimate at completion) differs from the BAC (budget at completion). This could be due to changes in performance, risks, project changes, or any number of reasons. That means the BAC may no longer be a reasonable estimate given the changes in performance. When that occurs, use the EAC to project the cost of the project at completion.

The first EAC formula is called "EAC forecast for ETC work performed at the budgeted rate." I know that's a mouthful. Here's what you should know. This formula calculates EAC based on the actual costs to date and the assumption that ETC work will be completed at the budgeted rate. The formula looks like this:

$$EAC = AC + (BAC - EV)$$

Let's assume your AC to date is $800, BAC is $1,200, and EV is $600. EAC, assuming ETC work will be completed at the budgeted rate, is as follows:

$$\$800 + (\$1,200 - \$600) = \$1,400$$

In English, you'll spend $1,400 to complete this work component, assuming the remaining work is performed at the budgeted rate. That is $200 more than what you have budgeted because your EV is less than the actual cost to date. I recommend examining the quality of the work to date. You aren't getting what you're paying for.

The next EAC formula is called "EAC forecast for ETC work performed at the present CPI." (I didn't make up these titles!) Here's what you need to know. This forecast assumes that future performance will be just like the past performance for the project. The formula looks like this:

$$EAC = BAC / CPI$$

For this example, let's assume that BAC is $2,200 and CPI is 1.2. The formula looks like this:

$$\$2,200 / 1.2 = \$1,833.33$$

This result predicts you will spend less than the originally budgeted amount for the project. In this case, you are getting more work or goods for the dollars you're spending. Good for you.

The last formula is called "EAC forecast for ETC work considering both SPI and CPI factors." This formula assumes two things: there is a negative cost performance to date and the project schedule dates must be met. The formula looks like this:

$$EAC = AC + \left[(BAC - EV) / (CPI \times SPI) \right]$$

Let's assume AC is $1,000, BAC is $1,500, EV is $900, CPI is 0.97, and SPI is 1.05. Here's the resulting EAC:

$$\$1,000 + \left[(\$1,500 - \$900) / (0.97 \times 1.05) \right] = \$1,589.10$$

Based on the assumptions that cost performance to date is negative (AC is higher than EV) and that we must meet the project schedule date, EAC is $1,589.10. We will have a slight cost overrun at the end of the project in order to meet the schedule date.

> ### Exam Spotlight
>
> For study purposes, the EAC formula and the three EAC calculations are shown here. Remember that if you monitor EAC regularly, you'll know if the project is within acceptable tolerances. The formulas are as follows:
>
> The EAC formula:
>
> $$EAC = AC + bottom - up\,ETC$$
>
> EAC using actual costs to date and assuming ETC uses budgeted rate:
>
> $$EAC = AC + (BAC - EV)$$
>
> EAC assuming future performance will behave like past performance:
>
> $$EAC = BAC\,/\,CPI$$
>
> EAC when cost performance is negative and schedule dates must be met:
>
> $$EAC = AC + \left[(BAC - EV)\,/\,(CPI \times SPI)\right]$$

In addition to the bottom-up ETC provided by the project team, there are four other formulas for calculating ETC that you should be aware of for the exam. They are discussed next.

When you believe that the work will continue to proceed as planned, use this formula to calculate ETC:

$$ETC = EAC - AC$$

When you anticipate significant changes in the work and believe it will not continue as planned, you need to reestimate ETC from the bottom up. The formula, oddly enough, looks like this:

$$ETC = reestimate$$

When you believe that future cost variances will be similar to the types of variances you've seen to date, you'll use this formula to calculate ETC:

$$ETC = (BAC - EV) / CPI$$

Assuming your earned value is 725, CPI is 1.12, and BAC is 1,000, plug in the numbers:

$$(1,000 - 725) / 1.12 = 245.54$$

Therefore, at the measurement date, you need $245.54 to complete all the remaining work of this work component (or project if you're using project totals), assuming variances in the future will be the same as they have been to date. That's a little less than the BAC, so this is good news.

When you believe that future cost variances will *not* be similar to the types of variances you've seen to date, you'll use this formula to calculate ETC:

$$ETC = (BAC - \text{cumulative EV})$$

Now calculate your value:

$$(1,000 - 725) = 275$$

In this case, you need $275 to complete all the remaining work of this work component, assuming variances in the future are different than they have been to date. In this case, your project is on track and won't need any measure to correct performance.

Exam Spotlight

For study purposes, the ETC formulas are shown here.

Bottom-up ETC:

Manual summation of the costs of the remaining work based on estimates from the project team members working on these activities.

ETC when work is anticipated to proceed as planned:

$$ETC = EAC - AC$$

ETC when work is not anticipated to proceed as planned:

$$ETC = Reestimate$$

ETC when future cost variances will be similar to past variances:

$$ETC = (BAC - EV) / CPI$$

ETC when future cost variances are expected to be atypical:

$$ETC = (BAC - cumulative\,EV)$$

To-Complete Performance Index

To-complete performance index (TCPI) is the projected cost performance the remaining work of the project must achieve in order to meet the BAC or EAC. It's calculated by dividing the work that's remaining by the funds that are remaining.

The formula for TCPI when using the BAC is as follows:

$$TCPI = (BAC - EV) / (BAC - AC)$$

Assume for this example that BAC is \$1,000, EV is \$700, and AC is \$800:

$$(\$1,000 - \$700) / (\$1,000 - \$800) = 1.5$$

This means you'll need to reach a CPI rate that's 1.5 times what you've experienced to date in order to meet the BAC goal. You will have to improve the level of performance in this scenario in order to bring costs back into alignment with the authorized budget. This may or may not be possible given other project factors such as risk, schedule, and other performance factors. If the result is less than 1, future work does not have to be performed as efficiently as past performance.

When the BAC is no longer attainable, the project manager should calculate a new EAC. This new estimate becomes the goal you'll work toward once it's approved by management. The TCPI formula when EAC is the goal you're aiming for is as follows:

$$TCPI = (BAC - EV) / (EAC - AC)$$

We'll use the same assumptions we used in the formula earlier and note that EAC is $1,200. The formula looks like this:

$$(\$1,000 - \$700) / (\$1,200 - \$800) = 0.75$$

This result means that in order to complete the work within the EAC target, the project team needs to continue performing at an efficiency of 0.75. However, also remember that you revised the original BAC and are now using EAC as your estimate to complete, so additional costs were incurred.

There is one last thing to note regarding these formulas: if cumulative CPI falls below 1, all future project work must be performed at the TCPI in order to stay within your authorized project budget (BAC). That may or may not be possible given the risks, resources, schedule, and other considerations. If it is not possible, you as the project manager should calculate a new EAC and use this as the new goal.

 You'll be given some scratch paper when you go into the exam. I recommend that you write these formulas down on a piece of your scratch paper right after you start the test but before you start answering questions. Keep your list handy. That way, the formulas are off your mind and you've got them in front of you to reference when you get to the portion of the exam where these questions appear. You might want to use this tip for other items you've memorized as well. If you write them down before you begin, you don't have to jog your memory on every question. If you forget something, leave a blank space where it goes and as soon as you remember it or see a question that reminds you what it is, fill in the blank.

Recap of Formulas

You have a lot of formulas to memorize. Keep in mind that you'll be given a calculator when you take the exam, so you don't have to do the math manually. Here are the formulas I've covered in this chapter:

Performance Indexes

Cost performance index: CPI = EV / AC

Results > 1 are under planned cost.

Results < 1 are over planned cost.

Cumulative cost performance index: cumulative CPI = cumulative EV / cumulative AC

Results > 1 are under planned cost.

Results < 1 are over planned cost.

Schedule performance index: SPI = EV / PV

Results > 1 are ahead of schedule.

Results < 1 are behind schedule.

Cumulative schedule performance index: cumulative SPI = cumulative EV / cumulative PV

Results > 1 are ahead of schedule.

Results < 1 are behind schedule.

Variance Analysis

Cost variance: CV = EV – AC

Positive results mean the costs are under what was planned.

Negative results mean the costs are over what was planned.

Schedule variance: SV = EV – PV

Positive results mean the schedule is ahead of what was planned.

Negative results mean the schedule is behind what was planned.

Variance at completion: VAC = BAC – EAC

Forecasting

EAC formula: EAC = AC + bottom-up ETC

EAC using actual costs to date and assuming ETC uses budgeted rate:

$$EAC = AC + (BAC - EV)$$

EAC assuming future performance will behave like past performance:

$$EAC = BAC / CPI$$

EAC when cost performance is negative and schedule dates must be met:

$$EAC = AC + \left[(BAC - EV)/(CPI \times SPI)\right]$$

Bottom-up ETC: Summation of the costs of the remaining work based on estimates from the project team members working on these activities

ETC when work is anticipated to proceed as planned: ETC = EAC – AC

ETC when work is not anticipated to proceed as planned: ETC = reestimate

ETC when future cost variances will be similar to past variances: ETC = (BAC – EV) / CPI

ETC when future cost variances are expected to be atypical: ETC = (BAC – EV)

To Complete Performance Index

TCPI using BAC: TCPI = (BAC – EV) / (BAC – AC)

Results > 1 are more difficult to complete.

Results < 1 are easier to complete.

TCPI using EAC: TCPI = (BAC – EV) / (EAC – AC)

Results > 1 are more difficult to complete.

Results < 1 are easier to complete.

Problems with costs come about for many reasons, including incorrect estimating techniques, predetermined or fixed budgets with no flexibility, schedule overruns, inadequate WBS development, and so on. Good project management planning techniques during the Planning processes might prevent cost problems later in the project. At a minimum, proper planning will reduce the impact of these problems if they do occur.

Always inform appropriate stakeholders of revised budget or cost estimates and any changes of significant impact to the project. Keep them updated on changes, status, and risk conditions during regularly scheduled project meetings.

Earned Value Measures on Agile Projects

Controlling costs on an agile project does not usually require the rigor of the predictive type measures we calculated in the last section. Because agile produces frequent deliverables and you're in a continuous process of planning, working, reviewing, and revising, you are well aware of the costs and budget implications as the work progresses. This allows you to adjust elements that impact cost quickly and as needed.

However, you could easily use schedule performance index and cost performance index to measure story points on an agile project. Let's say your team planned on completing 42 story points at the beginning of the iteration but they only completed 36. SPI in this case would be the completed work, 36 story points, divided by the planned work, or 42 story points: 36 / 42 = .85. In other words, the team is working at 85 percent of the planned rate. When this result is less than 1, the team is behind schedule. When it is 1 or greater than 1, they are on schedule or ahead of schedule, respectively.

CPI is calculated by dividing the value of the features completed to date (earned value) by the actual costs to date. For example, perhaps the earned value expected for the iteration your team just completed is \$230,000. The actual costs at the end of the iteration are \$250,000. Divide the earned value by the actual costs to determine CPI. In this case, \$230,000 / \$250,000 = \$0.92. In this example, the team is spending \$0.08 cents per dollar more than expected at this point on the project. I derived that by taking \$1.00 − \$0.92 (the overage spent to date calculated using CPI) = \$0.08. When the CPI result is less than 1, the team is spending more than planned to date. When this result is 1 or greater than 1, the team is spending as planned or less than planned to date, respectively.

For quick reference, the formulas look like this:

SPI = Completed features / planned features or EV / PV

CPI = Earned value (value of the features expected to date) / actual costs or EV / AC

 **Real World Scenario**

New Accounting System for Mustang Enterprises

You are a stakeholder of the New Accounting System project for Mustang Enterprises. The existing accounting system is more than 12 years old and is at its end of life. Your company decided to hire a software services firm to implement a cloud-based accounting system so that the legacy program can be retired. You've also assigned a senior accountant to act as the project manager on behalf of your organization.

The project is in the Monitoring and Controlling process group, and the project manager keeps reporting that everything is okay and on schedule. When you asked him detailed questions and requested performance data, he patted you on the back and said, "Don't worry, I've got everything under control."

You are a little worried because some of the key project team members have come to you confidentially to inform you of the progress of the project.

After further investigation, you discover that the project manager instructed the vendor to change some of the system configurations without letting the stakeholders know. He also shortened the training phase and eliminated one of the testing phases so that only two payroll testing cycles will be performed. All of these changes could lead to project failure. The stakeholders need adequate time to train on this new system. No one in the business department has experience using this or similar systems. The payroll rules for the organization are complex and all stakeholders agreed at the beginning of the project that testing needs to include three testing cycles.

Because you're a key stakeholder, you decide to bring this information out into the open at the next project status meeting. Additionally, you plan to meet with the project sponsor and the procurement department to determine what alternatives you have to request that the vendor realign the project to meet the original contractual requirements. However, you fear that because the project manager is the one who gave the orders to make the changes, your organization might not have a lot of recourse. You will also make the project sponsor aware that the project manager doesn't have the skills needed to conduct this project and a new project manager should be hired as soon as possible. The project manager is invaluable to the organization as an accountant, but he doesn't have the project management experience needed to conduct a project of this size and complexity. This might cause a further setback to the project, but the project management plan and project schedule will require updates as a result of the existing project manager's decisions. You also decide to document all that has happened as a lesson learned and to set up a change control process to prevent this from happening in the future.

Monitoring and Controlling Schedule Changes

The *Control Schedule* process involves determining the status of the project schedule, determining whether changes have occurred or should occur, taking corrective or preventive action when needed, and influencing and managing schedule changes, including changes to the schedule baseline.

Keep in mind that the Control Schedule process works hand in hand with the Perform Integrated Change Control process we covered in Chapter 11, and that means any changes to the schedule baseline must be processed through Perform Integrated Change Control.

We've covered all of the earned value formulas that are involved in this process in the previous section. Next we'll look at burndown and burnup charts, which are used for agile projects, and performance reviews.

Keeping the schedule on track means you're monitoring and controlling time—one of the classic triple constraints.

Burndown and Burnup Charts

Burndown charts were first discussed in Chapter 5, "Creating the Project Schedule." They are used on agile projects and are a way to monitor and control the project schedule. You'll recall that burndown charts are used in agile projects to show the remaining work effort for the iteration. The y-axis shows the story points planned and completed for the iteration (or other units of work the team agrees on), and the x-axis shows the time period of the iteration. At the end of each day, team members update their progress on work completed and update their estimates for the remaining amount of work, which in turn updates the burndown chart. The burndown chart should be kept in a prominent place where all stakeholders and team members can see the amount of work remaining in the iteration. This allows the team to see at a glance if issues are cropping up so that they can address them before they get out of hand. If there is a variance between the actual work and ideal work remaining, you'll see a trend line emerging that forecasts the likely variance at the end of the iteration. If a large variance exists, you'll want to take action to get the remaining work completed on time or reprioritize it into a later iteration.

The accuracy of the burndown chart relies on the accuracy of the estimates the team provides for the units of work. Keep in mind that it only shows the work of the iteration and does not show the entire backlog. It is a good indicator of trends, but a burndown chart can't tell you if you are working on the right things. Figure 12.2 is similar to the burndown chart you saw in Chapter 5. This figure shows story points on the y-axis rather than hours.

FIGURE 12.2   Burndown chart

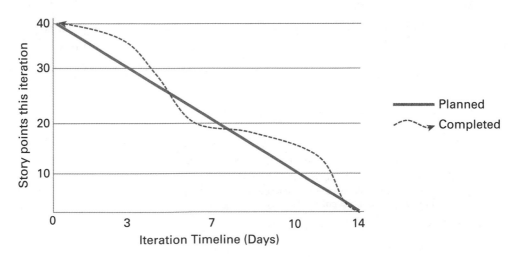

A product burndown chart is different than a burndown chart. It shows the work remaining for the entire project. It is constructed the same way as the burndown chart except that it shows all of the work effort on the y-axis and the timeline on the x-axis is the expected length of the whole project. Both charts show the number of work units completed but do not have a way to reflect changes or new requirements. The burndown chart is more commonly used in practice.

Burnup charts are similar to burndown charts and show much of the same information as a burndown chart. They can be used for the work of the iteration or for the entire project. The y-axis shows units of work such as story points, work hours, or work days, or other work units the team deems appropriate. The x-axis shows the amount of time in days, weeks, or iterations. A burnup chart has the ability to show changes in scope. If you look at Figure 12.3, you'll see that the original plan for this project was to complete 90 user stories in eight iterations. The planned number of user stories are shown above the planned line on the graph. At iteration 4, 10 more user stories were added to the scope. Up until iteration 4, the team had fallen behind their estimates. By iteration 4, the planned number of iterations seemed to be on track with the completed work. Adding 10 more user stories at week 4 means the team will not likely finish in eight iterations unless they can manage to complete more user stories per iteration than planned. Looking at the planned and trend in user story completions, they will complete 90 user stories in eight iterations and will need one more iteration to complete 100 user stories.

FIGURE 12.3   Burnup chart

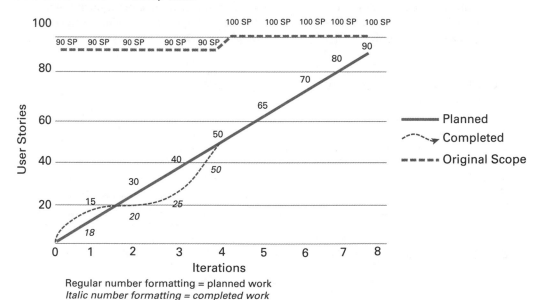

Regular number formatting = planned work
Italic number formatting = completed work

Exam Spotlight

Burndown charts are used to display the work remaining for an iteration and go to zero over time. They do not show changes in scope. Burnup charts are used to display the completed work of the project and will show changes in scope. Burnup charts start at zero and trend up to completion over time.

Performance Reviews

Performance reviews examine elements such as actual start and end dates for schedule activities and the remaining time to finish uncompleted activities. There are several techniques you can use in this process, including trend analysis, critical path method, critical chain method, and earned value management. If you've taken earned value measurements, the SV and SPI will be helpful in determining the impact of the schedule variations and in determining whether corrective actions are necessary.

You will recall that the critical path is the longest path on the project schedule with zero or negative float. If there are variances in critical path tasks, your schedule is likely at risk. Examining critical path tasks, or those near a critical path task, can help alert you to schedule risk.

If you're using the critical chain method to construct the schedule, you should compare the amount of buffer needed to the amount of buffer remaining to help determine if the schedule is on track. This will also indicate whether corrective actions are necessary to adjust the schedule.

In the Control Schedule process, because you're dealing with time issues, it's imperative that you act as quickly as possible to implement corrective actions so that the schedule is brought back in line with the plan and the least amount of schedule delay possible is experienced.

Schedule changes might be potential hot buttons with certain stakeholders and can burn you if you don't handle them correctly. No one likes to hear that the project is going to take longer than originally planned. That doesn't mean you should withhold this information, however. Always report the truth. If you've been keeping your stakeholders abreast of project status, they should already know that the potential for schedule changes exists. Nevertheless, be prepared to justify the reason for the schedule change or start dusting off your résumé—maybe both, depending on the company.

Be sure to examine the float variance of the critical path activities when monitoring the schedule. Thinking back to the Develop Schedule process, you'll recall that float is the amount of time you can delay starting an activity without increasing the amount of time it takes to complete the project. Because the activities with the least amount of float have the potential to cause the biggest schedule delays, examine float variance in ascending order of critical activities.

Keep in mind that not all schedule variances will impact the schedule. For example, a delay to a noncritical path task will not delay the overall schedule and might not need corrective action. Use caution here, though—if a delay occurs on a noncritical path task or its duration is increased for some reason, that task can actually become part of the critical path. Delays to critical path tasks will *always* cause delays to the project completion date and require corrective action. Careful watch of the variances in schedule start and end dates will help you control the total time element of the project.

Changes to the Schedule

Schedule changes will require updates to the schedule management plan, the schedule baseline, the cost baseline, and/or the performance measurement baseline. Changes to the cost baseline may be necessary when you've used a schedule compression or crashing technique. Changes to approved schedule start and end dates in the schedule baseline are called *revisions*. They generally occur as a result of a project scope change, or changes to activity estimates, and might result in a schedule baseline update. Schedule baseline updates occur when significant changes to the project schedule, such as the changes just mentioned, are made. This means a new schedule baseline is established that reflects the changed project activity dates. Once the new baseline is established, it is used as the basis for future performance measurements. Never re-baseline a schedule without first having it approved by the project sponsor and archiving a copy of the original baseline and schedule.

Take care when re-baselining a project schedule. Don't lose the original baseline information. Why do you care? Because the original baseline serves as historical information to reference for future projects. Make a backup copy of the original schedule so that you have a record of the original baseline as a reference. Even though some project management software allows you to save several baselines plus the original, it's still good practice to make a backup copy of the original.

Changes to the project schedule might or might not require updates to other elements of the project management plan as well. For example, extending a schedule activity involving a contractor might impact the costs associated with that activity.

The project documents updates output may require updates to the assumption log, basis of estimates, lessons learned register, schedule data, project schedule, resource calendars, and/or the risk register. For example, project schedule network diagrams require updates as a result of schedule model data updates. Don't forget to document these changes and inform your stakeholders.

Validating Project Scope

Managing and reporting on project progress make up the primary focus of the Monitoring and Controlling processes. The primary purpose of the *Validate Scope* process, which is one of those processes, is to formally accept completed deliverables and obtain sign-off that the deliverables are satisfactory and meet stakeholders' expectations and the documented requirements.

This process involves evaluating the deliverables to determine whether the work is complete and whether it satisfies the project objectives. Remember way back to the beginning of the project when success criteria were defined for the project? You will evaluate those criteria during this process. Evaluation is performed using inspection and decision-making techniques (voting), the only tools and techniques of this process. Even if the project is canceled, you should perform Validate Scope to document the degree to which the project was completed. This will serve as historical information, and if the project is ever started up again, you will have documentation that tells you what was completed and how far the project progressed.

The primary output of Validate Scope is accepted deliverables. Accepted deliverables are concerned with the formal acceptance of the work by the stakeholders. Remember that stakeholders include customers, the project sponsor, the project team, the management team, and so on. Document their acceptance with formal sign-off, and keep this with your project documents. Acceptance of deliverables on agile projects occurs at the end of the iteration and does not require a formal sign-off. If something is not right about the deliverable, it can be corrected and then accepted in the next iteration.

Validate Scope occurs at the end of every iteration on an agile project. The deliverable is examined by the product owner and compared to the acceptance criteria. This is usually a straightforward process because agile delivers working functionality (or a deliverable) at the end of the iteration or flow. If the functionality doesn't work, it isn't accepted. If it doesn't work as desired, it isn't accepted. Sometimes, agile projects end before they are completed, just as predictive projects do. If this happens, it's a good idea to perform the Validate Scope project and document the level and degree of completion.

Controlling Scope

The *Control Scope* process involves monitoring the status of both the project and the product scope, monitoring changes to the project and product scope, and managing changes to the scope baseline. As with all the change processes, change requests and preventive and corrective actions that come about during this process are managed through the Perform Integrated Change Control process. This process primarily concerns projects using a predictive development life cycle where the requirements have been documented and planned out in detail. Agile projects don't require this rigor, and agile welcomes change on a project. The purpose of agile is to allow for frequent deliverables and frequent feedback so that changes can be incorporated into the upcoming iterations. The remainder of this section will focus on projects conducted using the predictive development methodology.

Any modification to the agreed-on WBS is considered a scope change. (It has been eons ago that you looked at this, so remember that the work breakdown structure is a deliverables-oriented hierarchy that defines the total work of the project.) This means the addition or deletion of activities or modifications to the existing activities on the WBS constitutes a project scope change.

Changes in product scope require changes to the project scope as well. Let's say one of your project deliverables is the design of a piece of specialized equipment that's integrated into your final product. Now let's say that because of engineering setbacks and some miscalculations, the specialized equipment requires design modifications. The redesign of this equipment impacts the end product or product scope. Because changes to the product scope impact the project requirements, which are detailed in the scope document, changes to project scope are also required. This change, along with recommended corrective actions, should be processed through the Perform Integrated Change Control process.

Unapproved or undocumented changes that sometimes make their way into the project are referred to as *scope creep*. How often have you overheard a stakeholder speaking directly with a project team member and asking them to make "this one little change that doesn't impact anybody ... really, no one will notice"? Make certain your project team members are well versed in the change control process and insist that they inform you of shenanigans like this. Scope creep can kill an otherwise viable project. Little changes add up and eventually impact budget, schedule, and quality.

If you are using a configuration management system to control product scope, the change control system must also integrate with it. The configuration management system manages changes to product and project scope and ensures that these changes are reasonable and make sense before they're processed through the Perform Integrated Change Control process.

The primary output of Control Scope is change requests. Changes to scope will likely require that you repeat some of the project Planning processes and make any needed adjustments, including updating the project documents. Scope changes require an update to the project scope statement. This may require an update to the WBS and WBS dictionary as well. Here's a pop quiz: The project scope statement, WBS, and WBS dictionary are collectively known as what? The answer is the scope baseline. Scope baseline updates are part of the project management plan updates output of this process.

Scope changes include any changes to the project scope as defined by the agreed-on WBS. This in turn might require changes or updates to project objectives, costs, quality measures or controls, performance measurements baselines, or time in the form of schedule revisions. Scope changes almost always affect project costs and/or require schedule revisions.

Schedule revisions are almost always needed as a result of scope changes, but not all scope changes lengthen the project schedule. Some scope changes (a reduction in overall project requirements, for example) might reduce the number of hours needed to complete the project, which in turn might reduce the project budget. This most often occurs when the schedule is the primary constraint on the project and the start or end dates cannot be changed.

When scope changes are requested, all areas of the project should be investigated to determine what the changes will impact. The project team should perform estimates of the impact and of the amount of time needed to make the changes. Sometimes, however, the change request is so extensive that even the time to perform an estimate should be evaluated before proceeding. In other words, if the project team is busy working on estimates, they aren't working on the project. That means extensive change requests could impact the existing schedule because of the time and effort needed just to evaluate the change. Cases like these require you to make a determination or ask the change control board (CCB) to decide whether the change is important enough to allow the project team time to work on the estimates.

Always remember to update your stakeholders regarding the changes you're implementing and their impacts. They'll want to know how the changes affect the performance baselines, including the project costs, project schedule, project scope, and quality.

Measuring Work Results on Agile Projects

Agile projects focus on customer value, and the measurements and metrics you use on agile projects should as well. Agile projects deliver finished work at the end of each iteration. If you get to the end of the work period and the deliverable is not complete, you have a rock-solid measurement. There is no need to make educated guesses or conduct estimates on the progress of the work. It's complete, or it's not complete.

One of the measurements used on agile projects is the definition of done. *Definition of done (DoD)* is a checklist of elements needed to ensure that the deliverable or functionality produced at the end of the iteration is ready for the customer to use. DoD is not acceptance criteria. It measures the iteration, not the deliverable. The DoD can be defined by the team, the stakeholders, or a combination of the two. Some examples are integration testing completed, code review completed, and coding errors have been resolved.

The *definition of ready (DoR)* is another measurement used by agile teams. This determines the specifics of the tasks planned for the iteration before the team begins working on the iteration. Some examples of DoR might include the team has reviewed and estimated the work for the user story, the user story describes the functionality in terms of purpose and how it performs, acceptance criteria are defined and agreed on for the user story, and the product owner has approved the user story.

Exam Spotlight

DoD describes the elements needed to ensure the iteration is finished and the deliverable is ready for the customer to use. DoR describes the elements needed to ensure the iteration is ready to start. DoD and DoR measure the iteration itself, not the deliverable.

Demos and prototypes on agile projects are a visual way of monitoring project progress. The working demo is empirical evidence that the work is complete and satisfies the acceptance criteria. *Empirical measures* are typically expressed as deliverables, functionality, or features. Surrogate measures include elements like the percentages supplied in some of the cost and schedule estimates we discussed earlier, along with the classic percent complete measure. It's a standing joke in the IT world that all projects are 90 percent complete at almost any stage of the project and the remaining 10 percent of the work will take more time than the first 90 percent. Percent complete can be subjective and is reliant on the accuracy of the estimates. Surrogate measures are not as useful on an agile project as empirical measures.

One measure that is common on predictive development life cycle projects is the standard red-yellow-green stoplight status known the world over. We talked about this in the previous chapter. This is a predictive type of measurement (not to be confused with a predictive development life cycle) and is not necessary on an agile project. That's because you have the immediacy of knowing whether the planned work was completed at the end of the iteration. Red-yellow-green status reporting is a handy way to give execs an at-a-glance look at project status or progress for a predictive project, but they can often be misleading. I've worked on countless projects where the project manager reported the schedule in green status for weeks on end, only to shift status to red six weeks before the expected completion date. This type of project is often referred to as a *watermelon project* because it's green on the outside and red on the inside. There are reasons this might occur: perhaps the vendor didn't perform as expected or the team lost a key member, or any other number of incidents suddenly shifted the schedule status. This makes the red-yellow-green status more useful as an in-the-moment measure, rather than as a prediction of future performance—although don't forget for the exam that red-yellow-green status is a predictive type forecast.

Earlier in this chapter we looked at burndown and burnup charts. These are a type of capacity measure. Capacity measures are also a type of in-the-moment measure. These charts help the team see at a glance the amount of work they have completed, what they have left to do, and whether they will finish on time. The capacity is measured against their original estimates. The burnup chart we discussed earlier showed the team estimated they could complete 90 user stories in eight iterations. Keep in mind that each team has its own capacity and no two agile teams will be the same. An experienced team might have looked at the same 90 user stories and said they could finish in six iterations, whereas a team new to agile might need 10 or 12 iterations to complete the work. The story points, user stories, or other measures of work a team uses to estimate and construct a burnup or burndown chart are unique to that agile team. Tracking the velocity of the story points, or other work units used on these charts, helps the team plan the next iteration and continually improve their estimates. They can review historical performance to date on the project and adjust based on the velocity the team is experiencing to date. The continual nature of experimenting, delivering a minimum viable product, and receiving feedback helps the team to learn and improve future estimates. According to the *Agile Practice Guide* (PMI®, 2017), it takes between four and eight iterations for a team to reach a steady velocity. Each iteration will help the team learn the pace at which they can work, and the feedback received will also help them learn and improve.

Agile projects are more suited to empirical, capacity-based, and value-based measures than predictive measures. The baseline measurements we discussed earlier in this chapter are not that useful on an agile project. Many of them are predictive measures and don't describe the work completed to date.

According to the *Agile Practice Guide* (PMI®, 2017), flow-based agile methodologies like Kanban use measures such as cycle time, lead time, and response time. You'll recall that Kanban is a just-in-time, pull-based approach where work is moved from one state to another only when there is capacity to take on the work. Cycle time is the time it takes to complete the work. For example, you'll recall that the Kanban board is constructed with columns indicating the stages of work. The first column is typically a placeholder for the tasks or cards that are up and coming. For our example, let's title this column "Requested." You would start counting the cycle time for the task once it is pulled from Requested to the first work column. Don't count the time it waits in the Requested column. Cycle time will be unique for each task because each task is unique.

Cycle time can also be used to estimate iteration-based agile projects. For example, let's say the team needs to complete 90 user stories in eight iterations. If the team has completed 30 user stories at the end of iteration 4, they will likely need a total of 12 iterations to complete the work. I calculated it this way: 90 user stories − 30 user stories = 60 remaining user stories to complete. On average to date, the team is completing 7.5 user stories per iteration (30 user stories / 4 iterations). There are 60 user stories remaining; 60 / 7.5 = 8 remaining iterations. Four iterations have already occurred, plus the remaining eight means a new total of 12 iterations. You can see that cycle time helps the team see delays and recognize potential holdups in the work. Cycle time is also considered a predictive measure.

If you are measuring story points, you are measuring capacity. If you are measuring user stories or completed functionality, you are measuring finished work. Remember that working software is the goal of agile approaches.

Lead time is another flow-based measure that considers the total time it takes for the user story from initiation (and placement on the board) to the completion and delivery of the work. In this case, the time starts when the user story card is added to the Requested column. Response time is the time the user story waits before the work starts—in other words, how long it waits in the Requested column only. The *Agile Practice Guide* (PMI®, 2017) recommends diagramming cycle time and lead time on a hurricane-style chart. Hurricane-style charts show the variability in lead and cycle time. However, I'm not certain what the *Agile Practice Guide* (PMI®, 2017) is referring to with this type of chart, since there is no reference to a hurricane-style chart on the Internet or other publications that I have read. However, I would recommend knowing that hurricane-style charts display variability that's easily understood by your stakeholders.

The measures we've discussed in this section are quantitative measures. You can also use value-based measures to measure agile projects. This includes examining features or values that are not easily measured. For example, you might consider measuring customer satisfaction at times throughout the project and at the end of the project. Ask them about their impressions of the agile methodology. Did it work well for them? Did they like this approach? What would they improve or change? Don't forget to survey the team using the same questions. What worked well for the team may not have worked as well for the stakeholders, and this information will help you modify and tailor the agile process on future projects. Team morale is another way to measure the value of the agile process. If morale and camaraderie are high, the agile approach was likely a success. If team morale is low, there is conflict and tension, or the team members seem to keep to themselves, the agile process probably needs some work. You may also ask the stakeholders to measure the value of the functionality received during the iteration. Expressions like "We can't live without this feature" are a type of value-based measure.

This concludes the Monitoring and Controlling processes. Next we'll focus on the Closing process group. You're closing in on closing out the project!

Formulating Project Closeout

All good projects must come to an end, as the saying goes. Ideally, you've practiced all the topics I've talked about that have led up to this point, and you've delivered a successful project to the stakeholders and customers. You've also put some of the vital tools of project management into play—planning, executing, controlling, and communicating—to help you reach that goal.

But how do you know when a project has ended successfully? Delivering the product or service of the project doesn't mean it has been completed satisfactorily. Recall the opening chapters, where I said a project is completed successfully when it meets stakeholders' expectations and satisfies the goals of the project. During the Closing process—Close Project or Phase—you'll document the acceptance of the product of the project with a formal sign-off and file it with the project records for future reference. The formal sign-off is the way stakeholders indicate that the goals have been met, that the project scope and deliverables were achieved, and that the project meets their expectations so the project ends.

Characteristics of Closing

A few characteristics are common to all projects during the Closing processes. One is that the probability of completing the project is highest during this process and risk is lowest. You've already completed the work of the project, so the probability of not finishing the project is very low.

Stakeholders have the least amount of influence during the Closing process, whereas project managers have the greatest amount of influence. Costs are significantly lower during this process because the majority of the project work and spending has already occurred. Remember those costly S curves we talked about in the Determine Budget process? This is where they taper off as project spending comes to an end.

One last common characteristic of projects during closing is that weak matrix organizations tend to experience the least amount of stress during the Closing processes. This is because, in a weak matrix organization, the functional manager assigns all tasks (project-related tasks as well) so that the team members have a job to return to once the project is completed and there's no change in reporting structure.

All projects do eventually come to an end. You'll now examine a few of the reasons for project endings before getting into the Close Project or Phase process.

Project Endings

Projects come to an end for several reasons:

- They're completed successfully.
- They're canceled or killed prior to completion.
- They evolve into ongoing operations and no longer exist as projects.

Four formal types of project endings exist that you might need to know for the exam:

- Addition
- Starvation
- Integration
- Extinction

You'll look at each of these ending types in detail in the following sections.

Addition

Projects that evolve into ongoing operations are considered projects that end because of *addition*; in other words, they become their own ongoing business unit. An example of this is the installation of an enterprise resource planning system. These systems are business management systems that integrate all areas of a business, including marketing, planning, manufacturing, sales, financials, and human resources. After the installation of the software, these systems can develop into their own business unit because ongoing operations, maintenance, and monitoring of the software require full-time staff. These systems usually evolve into an arm of the business reporting system that no one can live without once it's installed.

A project is considered a project when it meets these criteria: it is unique, has a definite beginning and ending date, and is temporary in nature. When a project becomes an ongoing operation, it is no longer a project.

Starvation

When resources are cut off from the project or are no longer provided to the project, it's starved prior to completing all the requirements, and you're left with an unfinished project on your hands. *Starvation* can happen for any number of reasons:

- Other projects come about and take precedence over the current project, thereby cutting the funding or resources for your project.

- The customer curtails an order.

- The project budget is reduced.

- A key resource quits.

Resource starving can include cutting back or withholding human resources, equipment and supplies, or money. In any case, if you're not getting the people, equipment, or money you need to complete the project, it's going to starve and probably end abruptly.

In such cases, documentation becomes your best friend. Organizations tend to have short memories. As you move on to bigger and better projects, your memory regarding the specifics of the project will fade. Six months after the fact when someone important wonders why that project was never completed and begins the finger-pointing routine, the project documents will clearly outline the reasons the project ended early. That's one of the reasons project documentation is such an important function. I'll talk more about documenting project details shortly.

Integration

Integration occurs when the resources of the project—people, equipment, property, and supplies—are distributed to other areas in the organization or are assigned to other projects. Perhaps your organization begins to focus on other areas or other projects, and the next thing you know, functional managers come calling to retrieve their resources for other, more important things. Again, your project will come to an end due to lack of resources because they have been reassigned to other areas of the business or have been pulled from your project and assigned to another project.

The difference between starvation and integration is that starvation is the result of staffing, funding, or other resource cuts, whereas integration is the result of reassignment or redeployment of the resources.

Again, good documentation describing the circumstances that brought about the end of a project because of integration should be archived with the project records for future reference.

Extinction

This is the best kind of project end because *extinction* means the project was completed and accepted by the stakeholders. As such, it no longer exists because it had a definite ending date, the goals of the project were achieved, and the project was closed out.

🌐 Real World Scenario

Pied Piper

Jerome Reed is the project manager for Pied Piper's newest software project. His team is working on a program that will integrate the organization's human resources information, including payroll records, leave-time accruals, contact information, and so on. His top two programmers, Brett and Kathy, are heading up the coding team and are in charge of the programming and testing activities.

Pied Piper recently hired a new CIO who started working with the company just a few weeks ago. Jerome is concerned about his human resources project. It was the former CIO's pet project, but he's not sure where it falls on the new CIO's radar screen.

Jerome is in the computer room checking out the new hardware that just arrived for his project. Liz Horowitz, the director of network operations, approaches Jerome.

"That's a nice piece of hardware," Liz comments.

"It sure is. This baby is loaded. It's going to process and serve up data to the users so fast they'll be asking us to upgrade all the servers."

Liz replies, "You're right about that. I've asked Richard to burn it in and load the software."

"What software?" Jerome asks.

"You know, the new customer relationship management software. The CIO hired some vendor she has worked with before to come in and install their CRM system here. She said it was our top priority. I knew this new server was already on order, and it happens to be sized correctly for the new CRM system."

"I purchased this server for the human resources project. What am I supposed to use for that?"

Liz answers, "I've got a server over there on the bottom of the third rack that might work, or maybe you can order another one. But you should take this up with the CIO. All I know is she authorized me to use this server. She understood I was taking it from your project, so maybe she's thinking about going another direction with the human resources project.

"You probably should know I also asked to have Brett and Kathy assigned to the CRM project. Even though it's a vendor project, it still requires some of our coders. The CIO wanted the best, and they're the best we've got. It shouldn't take them long to make the changes I need, and then you can have them back for your project. In the meantime, they can give directions to your other programmers so they can keep working."

"I'm going to go see whether the CIO is in," Jerome replies.

This is a case where Jerome's project ends by integration because of the reassignment of resources. The new CIO came on board and changed the direction and focus of the project priorities, making her new project a higher priority than the previous project. As a result, Jerome's hardware and his top two resources were reassigned to the new project. Had the CIO cut the resources and equipment on the original project altogether, it would have ended because of starvation.

Closing Out the Project

The key activity of *Close Project or Phase* is concerned with completing all the activities associated with closing out the project management processes in order to officially close out the project or phase and with releasing team members. The work of the project was accepted and validated back in the Validate Scope process. In this process, you (as the project manager) will review the scope baseline and other project documents to make certain the work reflected in the documents was completed and met the objectives of the project. Remember that the project scope is measured against the project management plan, so by this point in the project, the project management plan and associated project documents should be complete and up to date. For example, perhaps some scope change requests were implemented that changed some of the characteristics of the final product. The project information you're collecting during this process should reflect the characteristics and specifications of the final product. Don't forget to update your resource assignments as well. Some team members will have come and gone over the course of the project; you need to double-check that all the resources and their roles and responsibilities are noted.

Once the project outcomes are documented, you'll request formal acceptance from the stakeholders or the customer. They're also interested in knowing whether the product or service of the project meets the goals the project set out to accomplish. If your documentation is up to date, you'll have the project results at hand to share with them.

Remember that you obtained acceptance of the deliverables back in the Validate Scope process. In this process, you'll obtain final sign-off for the project or phase itself, verifying that all the exit criteria have been met.

The Close Project or Phase process is also responsible for analyzing the project management processes to determine their effectiveness and to document lessons learned concerning them. One of the other key functions of the Close Project or Phase process is to archive all project documents for historical reference and perform administrative closure procedures for the project. This may include documenting the exit criteria for the project or phase, transferring the results of the phase or project to the operations area of the organization, and/or collecting project records and performing audits and lessons learned. You can probably guess that Close Project or Phase belongs to the Project Integration Management Knowledge Area since this process touches so many areas of the project.

Every project requires closure. According to the *PMBOK® Guide*, the completion of each project phase requires project closure as well. Before we dive into the inputs of this process, let's examine some of the administrative closure procedures you will perform as part of the project or phase closeout.

Exam Spotlight

Project closure occurs at the end of each phase of the project in order to properly document project information and keep it safe for future reference. You shouldn't wait until project completion to perform the Close Project or Phase process, but rather perform it at the end of every phase, no matter whether the project phase was completed successfully or ended for some other reason.

Administrative Closure Procedures

Administrative closure procedures involve collecting all the records associated with the project, performing activities to satisfy the exit criteria, performing a comprehensive analysis of the project success (or failure), documenting and gathering lessons learned, archiving project records, updating the organization's knowledge base, and measuring stakeholder satisfaction with the project and the processes. The project records may include legal, financial, procurement, and other administrative records. Keep in mind that when projects are performed under contract, the archiving of financial records is especially important. These records might need to be accessed if there are payment disputes, so you need to know where they are and how they were filed. Projects with large financial expenditures also require particular attention to the archiving of financial records for the same reasons. Financial information is especially useful when estimating future projects—so again, be sure to archive the information so that it's easily accessible.

All of these documents should be indexed for reference and archived in an accessible place, typically on the intranet or as part of your project management information system. Don't forget that when records are stored electronically, they are subject to your organization's retention policies, and depending on the organization, they may also be subject to statutory requirements regarding their storage, retention, and destruction. These policies will describe when it is appropriate to archive and eventually delete the records.

Administrative closure procedures also document the project team members' and stakeholders' roles and responsibilities in performing this process. According to the *PMBOK® Guide*, this should include the processes and methodologies for the following:

- Documenting approval requirements, as defined by the stakeholders, for project deliverables and changes to the deliverables.

- Assuring and confirming that the project meets the requirements of the stakeholders, customers, and sponsor. This includes documenting necessary actions to verify that the deliverables have been delivered and accepted and exit criteria have been met.

- Ensuring that all activities associated with transferring the product, service, or result to the operations area (or the next phase of the project) are complete.

- Ensuring and confirming that the exit criteria for the project are satisfied.

- Releasing resources (both human and material) from the project.

- Finalizing procurement documents, including open claims.

- Determining and measuring stakeholder satisfaction with the project and project management processes.

Regression Analysis

The three tools and techniques of Close Project or Phase are expert judgment; data analysis, including document analysis, regression analysis, trend analysis, and variance analysis; and meetings. When you're conducting administrative closure activities, which we talked about earlier, the subject matter experts can help ensure that the process is performed according to the organization's and project management standards. The data analysis tool and technique includes analyzing documents to identify lessons learned and capturing knowledge for future projects. Regression analysis helps determine why the project performed the way it did by examining different variables on the project and determining how those variables contributed to the outcomes of the project. This is for improving future projects. We've looked at trend and variance analysis in previous chapters. Trends use past performance to help determine future performance. Variance analysis examines and compares what was planned against what actually took place. Meetings in this process are used to validate that the exit criteria for the project have been met; for final reviews or validation of the deliverables, lessons learned, and closeout contracts; to measure stakeholder satisfaction; and to celebrate success (hopefully).

Close Project or Phase Final Report

Sometimes you'll work on projects where everything just clicks, your project team functions at the performing stage, the customers and stakeholders are happy, and things just fall into place according to plan. I often find it difficult to perform Close Project or Phases on projects that have progressed particularly well, just because I don't want them to end. Believe it or not, the majority of your projects can fall into this category if you practice good project management techniques and exercise those great communication skills.

The Close Project or Phase process has the following outputs: project document updates (lessons learned register); final product, service, or result transition; final report; and organizational process assets updates.

We'll look at the last three outputs in the following sections.

Final Product, Service, or Result Transition

The name of this output is somewhat misleading. This actually refers to the turnover of the product, service, or result of the project to the organization. This usually requires a formal sign-off (I'll talk about that next), and in the case of a project performed on contract, it definitely requires a formal sign-off or receipt indicating acceptance of the project.

Formal acceptance includes distributing a notice of the acceptance of the product or service of the project by the stakeholders, customer, or project sponsor to stakeholders and customers. You should require formal sign-off indicating that those signing accept the product of the project. It's important to note that you'll want financial, legal, and administrative sign-off so that you can communicate formal project closeout and transfer liability to the owner of the product of the project. You will use the communications management plan as your guide in distributing the final project report and the notice of acceptance.

> The final product, service, or result is concerned with transitioning the final product, service, or result to the organization; organizational process assets updates involve documenting and archiving formal acceptance.

Final Report

The final report, typically written by the project manager, contains a summary of what occurred on the project. It may include information on scope, cost, schedule, risk, and quality outcomes, as well as whether the benefits of the final project were achieved. Remember that the benefits were outlined in the benefits management plan way back in the Initiating phase of the project. This was one of the inputs to the Develop Project Charter process (and it's one of the project document inputs of this process). That seems like a long time ago. Since we're on memory lane, you might recall that another input of that process was the business case. The final report will also summarize whether the objectives and business needs outlined in the business plan were met.

Organizational Process Assets Updates

The organizational process assets updates output is where the formal sign-off of the acceptance of the product is documented, collected, and archived for future reference. Documenting formal acceptance is important because it signals the official closure of the project, and it is your proof that the project was completed satisfactorily.

Another function of sign-off is that it kicks off the beginning of the warranty period. Sometimes project managers or vendors will warranty their work for a certain time period after completing a project. Projects that produce software programs, for example, might be warranted from bugs for a 30- or 60-day time frame from the date of implementation or acceptance. Typically in the case of software projects, bugs are fixed for free during the warranty period. Watch out, because users will try to squeeze new requirements into the "bug" category mold. If you offer a warranty, it's critical that the warranty spells out exactly what is covered and what is not.

This is also where the other project records and files are collected and archived. This includes the project planning documents (for example, the project scope statement, budget, schedule, risk responses, and quality plan and baselines), change records and logs, issue logs, lessons learned, and so on.

Project or phase closure documents are included in this output. These include documentation showing that the project or phase is completed and that the transfer of the product of the project to the organization (or department responsible for ongoing maintenance and support) has occurred in accordance with the project plan. In the case of a phase completion, the transfer would be the official hand-off to the next phase of the project rather than to an operations or maintenance group. If your project is canceled or ends prematurely, you should document the reasons for its premature end as well as the procedures for transferring the completed and uncompleted deliverables.

Exam Spotlight

According to the *PMBOK® Guide*, the project manager is responsible for reviewing documentation from prior phases of the project, reviewing and obtaining customer acceptance documentation (formal acceptance of the deliverables occurs in the Validate Scope process, but you will collect and store the sign-off document during this process), and reviewing the contract if applicable to make certain all the requirements of the project are completed. If the requirements are not complete, you should not proceed with the Close Project or Phase process unless the project or phase has been terminated.

Historical information and lessons learned are used to document the successes and failures of the project. You should collect all the lessons learned throughout the project and hold one last comprehensive review of the project. As an example, lessons learned document the reasons specific corrective actions were taken, their outcomes, the causes of performance variances, unplanned risks that occurred, mistakes that were made and that could have been avoided, and so on.

Unfortunately, sometimes projects do fail. You can learn lessons from failed projects as well as from successful projects, and you should document this information for future reference. Most project managers, however, do not document lessons learned. The reason for this is that employees don't want to admit to making mistakes or learning from mistakes made during the project. They do not want their names associated with failed projects or even with mishaps on successful projects.

> Lessons learned can include some of the most valuable information you'll take away from a project. We can all learn from our experiences, and what better way to have even more success on your next project than to review a similar past project's lessons learned document? However, lessons learned will be there only if you document them now. I strongly recommend you not skip this step.

You and your management team will have to work to create an atmosphere of trust and assurance that lessons learned are not reasons for reprimanding or dismissing employees but are learning opportunities that benefit all those associated with the project. Lessons learned allow you to carry knowledge gained on this project to other projects you'll work on going forward. They'll also prevent repeat mistakes in the future if you take the time to review the project documents and lessons learned before undertaking your new project.

Post-implementation audits aren't an official output, but they are a good idea. They go hand in hand with lessons learned because they examine the project from beginning to end and look at what went right and what went wrong. They evaluate the project goals and determine whether the product or service of the project satisfies the objectives. Post-implementation audits also examine the activities and project processes to determine whether improvements are possible on future projects.

Organizations, or the PMO, might conduct post-implementation audits instead of lessons learned sessions. Documenting and gathering information during this procedure can serve the same function as lessons learned if you're honest and include all the good, the bad, and the ugly. Let's hope there's very little ugly.

 Real World Scenario

Cimarron Research Group

The Cimarron Research Group researches and develops organic pesticides for use on food crops. It is a medium-sized company and has established a PMO to manage all aspects of project work. The PMO consists of project managers and administrative staff who assist with information handling, filing, and disbursement.

Terri Roberts is the project manager for a project that has just closed. Terri diligently filed all the pertinent project documents as the project progressed and has requested the research files and engineering notes from the director of engineering. All information regarding the research on this project should be included with the project archives because it's important that all the information about the project be in one place.

The engineering department complies with her request but chooses to keep its own set of research records as well.

Terri's assistant has named all the project documents in accordance with the organization's records management policies, and recently sent notice of formal acceptance and approval of this project to the stakeholders, project sponsor, and management team. This notice officially closes the project. The next step is to archive the files and store them.

Closing Out the Procurements

You'll recall from Chapter 11 that closed procurements is an output of the Control Procurements process. That was where you provided formal notice to the seller that the procurement is complete and the formal acceptance and closure of the procurement occurred.

Some of the administrative functions of closing out the procurement documents occur during the Close Project or Phase process. This is where you'll confirm formal acceptance of the work described in the contract and ensure that it was completed accurately and satisfactorily. You'll update records to document the final results of the contract or agreement, close out open claims, and archive the procurement documents for future reference.

Procurement documents include the contract itself (or other procurement documents) and all the supporting documents that go along with it. These might include things such as the WBS, the project schedule, change control documents, technical documents, financial and payment records, and quality control inspection results. This information—along with all the other information gathered during the project—is filed or stored according to the organization's records management policies once the project is closed out so that anyone considering a future project of similar scope can reference what was done. Procurement documents might have specific terms or conditions for completion and closeout. You should be aware of these terms or conditions so that project closure isn't held up because you missed an important detail. If you are not administering the procurement yourself, be certain to ask your procurement department whether there are any special conditions that you should know about so that your project team doesn't inadvertently delay contract or project closure.

Exam Spotlight

For the exam, remember that closed procurements is an output of the Control Procurements process. This output is where formal notice is given to the vendor that the contract is complete. The Close Project or Phase process is where contract documents are gathered and archived. Remember that product documentation and deliverables are validated and accepted during the Validate Scope process. One more note: When projects end prematurely, the Validate Scope process is where the level of detail concerning the amount of work completed gets documented.

Closing Out an Agile Project

Closing out an agile project is similar to the closing procedures outlined in this chapter. Agile projects deliver functionality as the project progresses. That means there is a continuous hand-off of deliverables or functionality throughout the life of the project. One of the important tasks of an iteration should concern ensuring knowledge transfer regarding the functionality at the end of the iteration or flow-based work period. At the end of the project, be certain to perform a final retrospective meeting and lessons learned meeting to capture information that will be useful on future projects.

There should be a closeout process performed at the very end of the agile project (or at the end of a release in a multiphased agile project), just like a predictive project. Several steps are involved in handing off the deliverables of the project:

- Examine the criteria to successfully close the project. The criteria for each iteration are defined as the project progresses. There should also be criteria that define how to successfully close the project or release. A classic example is passing the user acceptance test. This means the software works without error and all functions perform as expected.

- Obtain final acceptance of the deliverables to confirm that the project is complete and satisfies the objectives. Sign-off might occur once user acceptance testing is conducted with no errors. Or this may also occur after a warranty period passes with no errors.

- Obtain legal, financial, procurement, and administrative closure for the project. Ensure that the contract is closed, procurements are closed, and other administrative tasks are completed.

- Ensure that all project documentation is up to date and set up archive procedures. Ensure that you've conducted a final lessons learned meeting with stakeholders and update the lessons learned register. Ensure that you've also held a final retrospective with the team and that this information is recorded in the lessons learned document.

- Prepare and distribute the final project report indicating project completion. This should be done in accordance with the communications management plan.

- Validate the readiness for transition and transfer ownership of the deliverables to the stakeholders. Ensure that stakeholders are informed ahead of time when transition will occur and that they are prepared to receive the final product.

- Ensure knowledge transfer to the end user for product continuity. The users have been participating in the project all along, but the end of the project should include formal knowledge transfer to the end users. This might include training, written manuals or documentation, one-on-one tutorials, and so on.

- Confirm your approach for knowledge transfer. You'll want to ensure that knowledge transfer is successful and that the approach you are using is helpful to the end users. For example, maybe web-based training is sufficient for general users whereas super users need in-person, hands-on training in small groups.

- Release agile team resources.

Celebrate!

I think it's a good idea to hold a celebration at the conclusion of a successfully completed project or a successfully completed phase on a large project. The project team should celebrate their accomplishment, and you should officially recognize their efforts and thank them for their participation. Any number of ideas come to mind here—a party, a trip to a ball game, pizza and sodas at lunchtime. This shouldn't be the only time you've recognized your team, as discussed during the Develop Team process, but now is the time to officially close the project and thank your team members. Even if no funds are available for a formal celebration, your heartfelt "thank you" can go a long way with the members of your team. It's tough for team members to remain disgruntled, and it's easier to revive them, when they know their efforts are appreciated.

A celebration helps team members formally recognize the project end and brings closure to the work they've done. It also encourages them to remember what they've learned and to start thinking about how their experiences will benefit them and the organization during the next project.

Releasing Project Team Members

Releasing project team members is not an official process. However, it should be noted that at the conclusion of the project, you will release your project team members, and they will go back to their functional managers or be assigned to a new project if you're working in a matrix-type organization.

You will want to keep the functional managers or other project managers informed as you get closer to project completion so that they have time to adequately plan for the return of their employees. Start letting them know a few months ahead of time what the schedule looks like and how soon they can plan on using their employees on new projects. This gives the other managers the ability to start planning activities and scheduling activity dates.

In all the excitement of wrapping up the project, we shouldn't forget about our stakeholders. In the next section, we take a look at balancing the stakeholders' interests and ensuring that we've met their expectations.

Balancing Stakeholders' Interests at Project Close

We discussed Manage Stakeholder Engagement in an earlier chapter. Although stakeholder satisfaction is the key to project success and acceptance, I want to add some closing thoughts on this topic and discuss a few new concepts.

Projects are undertaken at the request of customers, project sponsors, executive managers, and others. You'll recall that stakeholders are those who have something to gain or lose by implementing the project. As such, stakeholders have different interests and needs, and one of your jobs is to balance the needs of the stakeholders.

Customer satisfaction is probably the primary goal you're striving for in any project. If your customer is satisfied, it means you've met their expectations and delivered the product or service they were expecting. You've got a winning combination when the customer is satisfied with the product, and you've also provided excellent customer service along the way. Satisfied customers tell others about your success and will most likely use your services in the future.

One of the key ways to ensure that customer satisfaction is achieved is to apply appropriate project management techniques to your project. This includes taking the time to discover all the requirements of the project and documenting them in the scope statement. You will find that stakeholders who have a clear understanding of the requirements and have signed off on them won't suffer from faulty memory or pull the ever-famous "I thought that *was* included" technique. Take the time to define your requirements and get stakeholder sign-off. You can't forget or fudge what's written down.

At project closure, you will once again meet with the stakeholders to obtain their observations and feedback on the project processes and evaluate their satisfaction with the project results. The stakeholder management plan will be your guide for this meeting because it describes the strategies for encouraging stakeholder participation and the process for decision-making. Together, you will examine if these processes were successful and document what worked well and what needs improvement on future projects.

In the following sections, I'll discuss how to juggle the competing needs of stakeholders, how to handle the issues and problems with stakeholders, and finally, how to balance project constraints against stakeholder needs.

Competing Needs

Stakeholders come from all areas of the organization and include your customer as well. Because stakeholders do not all work in the same areas, they have competing needs and interests. One stakeholder's concern on a typical IT project might take the form of system security issues, whereas another stakeholder is concerned about ease of use. As the project manager, you will have situations in which stakeholder needs compete with each other, and you'll have to decide between them and set priorities. Sometimes you'll be able to accommodate their needs, and sometimes you'll have to choose. You want to examine the needs against the project objectives and then use your negotiation and communication skills to convince the stakeholders of priorities.

Individual stakeholders might or might not have good working relationships with other stakeholders. Because of this, office politics come into play. I advise you to stay away from the politics game but get to know your stakeholders. You'll want to understand their business processes and needs in order to make decisions about stakeholder requirements.

Dealing with Issues and Problems

Problems will occur on your project—they're part of the process. Throughout this book, I've talked about how to deal with problems and risks and how to use conflict resolution techniques in handling problems. Balancing stakeholder needs comes into play here also.

You'll have to determine alternatives that will meet the key requirements of the project without jeopardizing the competing needs of stakeholders. Once you're into the Executing processes of the project and beyond, redefining scope becomes less and less of an option. So, your responsibility is to resolve issues and determine alternative solutions to problems as they occur without changing the original objectives of the project. Enlist the help of your project team members and stakeholders during these times. Use some of the techniques such as brainstorming and the Delphi technique to find solutions.

You might also have difficulty trying to make stakeholders understand your decision or the technical nature of a problem. Again, this is where communication skills help you immensely. Take the case of a technical problem that's cropped up on your project. You should not expect your stakeholder to understand the technical aspects of rocket science if they work in the finance department, for example. It's up to you to keep the explanation at a level they can understand without loading them down with technical jargon and specifications. Keep your explanations simple, yet don't skip important details they'll need to make decisions.

Balancing Constraints

Your toughest issues will almost always center on the triple constraints: time, cost, and scope. Because of the nature of the constraints, one of these is the primary driver, and one is the least important. Keep a close eye on stakeholders who want to switch the priority of the constraints for their own purposes. Let's say the project sponsor already told you that time is the primary constraint. Another stakeholder tells you that a requirement of the project is being overlooked and quality is suffering. Be careful that the stakeholder isn't trying to divert the primary constraint from time to quality to suit their own objectives.

We have officially closed out the project and hopefully satisfied our stakeholders' expectations. We have also completed our study of all the project management processes from Initiating to Closing. But there's one more area you need to know about for the exam. We'll look at it next.

Professional Responsibility

As a certified Project Management Professional, you are required to adhere to the *PMI® Code of Ethics and Professional Conduct*. You can find a copy of this code on the PMI® website, www.pmi.org.

You should read and understand this code because you will be agreeing to adhere to its terms as part of the certification process. The *PMI® Code of Ethics and Professional Conduct* outlines four areas of focus:

- Responsibility
- Respect
- Fairness
- Honesty

Exam Spotlight

Each of these areas is described in the code, along with sections called "Aspirational Standards" and "Mandatory Standards" for each one. We will not cover every standard in this chapter, so I recommend you read them and understand them for the exam. The *PMI® Code of Ethics and Professional Conduct* notes that although it might be difficult to measure the aspirational standards, they are something we should hold ourselves and other practitioners accountable for and they are *not* optional.

There is no section on the exam that pertains solely to professional responsibility. Instead, the exam incorporates professional responsibility concepts into other questions throughout the exam. To make it easier to learn the principles in the professional responsibility area, I've broken this topic out into its own section in this chapter with questions based on these principles.

You may find at least a few exam questions regarding professional responsibility on the exam, so we'll look at each of these areas and a few others for good measure.

Responsibility

Responsibility is the act of making decisions that are for the good of the organization rather than ourselves, admitting our mistakes, being responsible for the decisions we make (or those we don't make) and the consequences that result, along with other actions. We'll look at several elements that fall within the responsibility arena.

Ensuring Integrity

As a project manager, one of your professional responsibilities is to ensure the integrity of the project management process, the product, and your own personal conduct. I've spent the majority of this book discussing how to achieve project management integrity by following the project management processes.

A product that has integrity is one that is complete and sound or fit for use.

Correctly applying the project management processes you've learned will ensure the integrity of the product. The effective execution of the Planning, Executing, and Monitoring and Controlling processes—including documenting scope, performing quality inspections, measuring performance, and taking corrective actions—will ensure that a quality product that satisfies the requirements of the stakeholders (and the project management plan) is produced. As you learned earlier, you will seek acceptance of the product from the stakeholders and customer during the Closing process group. You can ensure integrity using agile or hybrid approaches by producing small, frequent deliverables and incorporating feedback and changes during the iterations or work periods. Don't forget that all-important retrospective meeting to discuss lessons learned and how you can improve the next iteration.

Accepting Assignments

This topic, as well as many of the others discussed in these sections, crosses two of the areas noted in the *PMI® Code of Ethics and Professional Conduct*—responsibility and honesty. You should always honestly report your qualifications, your experience, and your past performance of services to potential employers, customers, PMI®, and others. You should not knowingly accept assignments that are beyond your capabilities or experience.

Be honest about what you know and what you don't know as it relates to your experience. For example, the Quality Management processes as described in the *PMBOK® Guide* are used extensively in the manufacturing industry. However, the information technology field looks at quality issues in different ways. If you're a project manager in the IT field, you've probably never used control charts and cause-and-effect diagramming techniques. Don't lead others to believe that you have used techniques you haven't used or that you have experience you don't have.

Emphasize the knowledge you do have and how you've used it in your specific industry, and don't try to fudge it with processes and techniques you've never used. Potential clients and employers would much rather work with you and provide training where you might need it than think they've got someone fully experienced with the project or industry techniques needed for a project when they don't.

If you're working on contract or you're self-employed, you have a responsibility to ensure that the estimates you provide potential customers are accurate and truthful. Clearly spell out what services you're providing and let the customer know the results they can expect at the end of the project. Accurately represent yourself, your qualifications, and your estimates in your advertising and in person.

Laws and Regulations Compliance

This might seem obvious, but as a professional, you're required to follow all applicable laws and rules and regulations that apply to your industry, organization, or project. This includes PMI® organizational rules and policies as well. You should also follow any ethical standards and principles that might govern your industry or the state or country in which you're working. Remember that rules or regulations you're used to in the United States might or might not apply to other countries, and vice versa.

Confidential Information

Many project managers work for consulting firms where their services are contracted out to organizations that need their expertise for particular projects. If you work in a situation like this, you will likely come across information that is sensitive or confidential. Again, this might seem obvious, but as part of the *PMI® Code of Ethics and Professional Conduct*, you agree not to disclose sensitive or confidential information or use it in any way for personal gain.

Often when you work under contract, you'll be required to sign a nondisclosure agreement. This agreement simply says that you will not share information regarding the project or the organization with anyone—including the organization's competitors—or use the information for your personal gain.

However, you don't have to work under contract to come into contact with sensitive or private information. You might work full-time for an organization or a government agency that deals with information regarding its customer base or citizens. For example, if you work for a bank, you might have access to personal account information. If you work for a government agency, you might have access to personal tax records or other sensitive material. It would be highly unethical and maybe even illegal to look up the account information of individuals not associated with the project at hand just to satisfy your own curiosity. In my organization, that is grounds for dismissal.

 Don't compromise your ethics or your organization's reputation by sharing information that is confidential to the organization or that would jeopardize an individual's privacy.

Company Data

Although it might seem obvious that you should not use personal information or an organization's trade secrets for personal gain, sometimes the organization has a legitimate need to share information with vendors, governmental agencies, or others. You need to understand which vendors or organizations are allowed to see sensitive company data. In some cases, you might even need to help determine which individuals can have access to the data. When in doubt, ask.

Here are some examples. Maybe the company you're working with has periodic mailings it sends to its customer base. If one of your project activities includes finding a new vendor to print the mailing labels, your organization might require the vendor to sign a nondisclosure agreement to guard the contents of the customer lists. Discovering just who should have access to this information might be tricky.

Another example involves data on citizens that is maintained by the government. You might think that because the data belongs to one agency of the government—say, the Internal Revenue Service—any other agency of the government can have access to it. This isn't the case. Some agencies are refused access to the data even though they might have good reason to use it. Others might have restricted access, depending on the data and the agency policy regarding it. Don't assume that others should have access to data because it seems logical.

Most organizations require vendors or other organizations to sign nondisclosure agreements when the vendors or others will have access to sensitive company data. It's your responsibility to ensure that the proper nondisclosure agreements are signed prior to releasing the data. The procurement department often handles this function.

Intellectual Property

You are likely to come into contact with intellectual property during the course of your project management career. Intellectual property includes items developed by an organization that have commercial value but are not tangible and copyrighted material, such as books (including this one), software, and artistic works. It might also include ideas or processes that are patented. Or it might involve an industrial process, business process, or manufacturing process that was developed by the organization for a specific purpose.

Intellectual property is owned by the business or person who created it. You might have to pay royalties or ask for written permission to use the property. Intellectual property should be treated just like sensitive or confidential data. It should not be used for personal gain or shared with others who should not have access to it. This book is an example of intellectual property, and posting it on the Internet is stealing and that's illegal. Thank you for not doing that.

Respect

Respect involves several areas as well, including the way we conduct ourselves and the way we treat others—listening to other viewpoints, conducting ourselves in a professional manner, and so on. We'll look at some of the elements of respect next.

Professional Demeanor

Acting in a professional manner is required of almost everyone who works in the business world. Although you are not responsible for the actions of others, you are responsible for your own actions and reactions. Part of acting professionally involves controlling yourself and your reactions in questionable situations. For example, a stakeholder or customer might lash out at you but have no basis for their outburst. You can't control what they said or did, but you can control how you respond. As a professional, your concern for the project and the organization should take precedence over your concern for your own feelings. Therefore, lashing out in return would be unprofessional. Maintain your professional demeanor, and don't succumb to shouting matches or ego competitions with others.

As project manager, you have a good deal of influence over your project team members. One of the items on the agenda at the project team kickoff meeting should be a discussion of where the team members can find a copy of organizational policies regarding conflict of interest, cultural diversity, standards and regulations, and customer service and standards of performance. Better yet, have copies with you that you can hand out at the meeting or have available the addresses of websites where they can find these documents.

When you see project team members acting out of turn or with less-than-desirable customer service attitudes, coach and influence those team members to conform to the standards of conduct expected by you and your organization. Your team members represent you and the project. As such, they should act professionally. It's your job as the project leader to ensure that they do.

Reporting Ethics Violations

When you hold the PMP® certification, one of the responsibilities that falls into this category is your responsibility to report violations of the PMP® code of conduct. To maintain the integrity of the profession, everyone who holds the PMP® certification must adhere to the code of conduct that makes all of us accountable to each other.

When you know a violation has occurred and you've verified the facts, notify PMI®. Part of this process—and a requirement of the code of conduct—is that you'll verify that an ethics violation has occurred (in other words, don't report bogus or unsubstantiated reports) and will assist PMI® in the investigation by supplying information, confirming facts and dates, and so on. This includes anything listed as violations in the *PMI® Code of Ethics and Professional Conduct*, such as conflicts of interest, untruthful advertising, false reporting of project management experience and credentials, appearances of impropriety, and so on. This one calls for some judgment on your part, but it's mostly based on common sense. For example, in most situations, a project manager with the PMP® designation should not have a family member working on the project team reporting to them (unless they own and run a family business).

Cultural Awareness

More and more companies compete in the global marketplace. As a result, project managers with multinational experience are increasingly in demand. This requires a heightened awareness of cultural influences and customary practices of the country where they're temporarily residing.

If you are used to working in the United States, for example, you know that the culture tends to value accomplishments and individualism. U.S. citizens tend to be informal and call each other by their first names, even if they've just met. In some European countries, people tend to be more formal, using surnames instead of first names in a business setting, even when they know each other well. Their communication style is more formal than in the United States, and although they tend to value individualism, they also value relationships, history, hierarchy, and loyalty. The Japanese, on the other hand, tend to communicate indirectly and consider themselves part of a group, not as individuals. The Japanese people value hard work and success, as most of us do.

One thing I've witnessed when working in foreign countries is U.S. citizens trying to force their own culture or customs on those with whom they're visiting or working. That isn't recommended, and it generally offends those you're trying to impress. Don't expect others to conform to your way of doing things, especially when you're in their country. You know the saying "When in Rome, do as the Romans do." Although you might not want to take that literally, the intent is good. For example, a quick kiss on both cheeks is a customary greeting in many countries. If that is the case and it's how you're greeted, respond with the same—unless you need to practice social distancing, as in the case of the COVID-19 days.

Culture Shock

Working in a foreign country can bring about an experience called *culture shock*. When you've spent years acting certain ways and expecting normal, everyday events to follow a specific course of action, you might find yourself disoriented when things don't go as you expected.

One of the ways you can avoid culture shock is to read about the country you're going to work in before going there. The Internet is a great resource for information such as this. You could also talk with coworkers who have worked in this culture and ask them for some pointers.

When in doubt about a custom or what you should do in a given situation, ask your hosts or a trusted contact from the company you'll be working with to help you. People are people all over the world, and they love to talk about themselves and their cultures. They're also generally helpful, and they will respect you more for asking what's expected rather than acting as though you know what to do when you clearly do not.

Diversity Training

Sometimes you might find yourself working with teams of people from different countries, cultures, and ethnic backgrounds. Some team members might be from one country and some from another. The best way to ensure that cultural or ethical differences do not hinder your project is to provide training for all team members.

Team-building activities are ways to build mutual trust and respect and bond team members with differing backgrounds. Choose activities that are inoffensive and ones in which everyone can participate.

Diversity training makes people aware of differences between cultures and ethnic groups, and it helps them to gain respect and trust for those on their team. Provide training regarding the project objectives and the company culture as well.

Remember that project objectives are why you are all together in the first place. Keeping the team focused on the objectives cuts across cultural boundaries and will help everyone concentrate on the project and tasks at hand rather than each other.

Respecting Your Neighbors

Americans tend to run their lives at high speed and get right down to business when working with vendors or customers. It isn't unusual for a businessperson to board a plane in the morning, show up at the client site and take care of business, and hop another flight to the next client site that night.

You'll find that this is not that common in many other countries. Often, the people you're working with will expect you to take time to get to know them and build relationships first and then proceed to business. Don't expect to do that relationship building in a few hours. It could take several days, depending on the culture. They might even want you to meet their family and spend time getting to know them. Resist the urge to get right down to business if that's not customary in the culture because you'll likely spoil the deal or damage relationships past the point of repair.

 Spend time building relationships with others. Once an atmosphere of mutual trust and cooperation is established, all aspects of project planning and management—including negotiating and problem solving—are much easier to navigate.

Perceiving Experiences

All of us see the world through our own experiences. Your experiences are not someone else's experiences; therefore, what you perceive about a situation might be very different from what others believe. Keep this in mind when it appears that a misunderstanding has occurred or that someone you're working with doesn't respond as you expect. This is especially true when you're working with someone from another country. Always give others the benefit of the doubt and ask for clarification if you think there is a problem. Put your feelings in check temporarily and remember that what you think the other person means is not necessarily as it appears.

Fairness

Fairness includes avoiding favoritism and discrimination against others, avoiding and reporting conflict of interest situations, and maintaining impartiality in a decision-making process. We'll focus on conflict of interest in this section.

Conflict of Interest

The *PMI® Code of Ethics and Professional Conduct* discusses your responsibility to report to the stakeholders, customers, or others any actions or circumstances that could be construed as a *conflict of interest*.

A conflict of interest is when you put your personal interests above the interests of the project or when you use your influence to cause others to make decisions in your favor without regard for the project outcome. In other words, your personal interests take precedence over your professional obligations, and you make decisions that allow you to personally benefit regardless of the outcome of the project. Let's look at a few examples.

Associations and Affiliations

Conflicts of interest might include your associations or affiliations. For example, perhaps your brother-in-law owns a construction company and you are the project manager who has just published an RFP. Your brother-in-law bids on the project and ends up winning the bid.

If you sit on the decision committee and don't tell anyone about your association with the winning bidder, that is clearly a conflict of interest. If you influence the bid decision so that it goes to your brother-in-law, he benefits from your position. Alternatively, if you are not so fond of your brother-in-law, you could influence the decision against him, in which case he loses the bid. Either way, this is a conflict of interest. You put your personal interests—or in this case the interests of your associations—above the project outcome. Even if you did not influence the decision in any way, when others on the project discover the winning bidder is your brother-in-law, they will assume a conflict of interest occurred. This could jeopardize the awarding of the bid and your own position as well.

The correct thing to do in this case would be to, first, inform the project sponsor and the decision committee that your brother-in-law intends to bid on the project. Second, refrain from participating on the award-decision committee so as not to unduly influence others in favor of your brother-in-law. Last, if you've done all these things and your brother-in-law still wins the bid, appoint someone else in your organization to administer the contract and make the payments for the work performed by him. Also, make certain you document the decisions you make regarding the activities performed by him and keep them with the project files. The more documentation you have, the less likely someone can make a conflict-of-interest accusation stick.

Vendor Gifts

Some professionals work in situations where they are not allowed to accept gifts in excess of certain dollar amounts. This might be driven by company policy, the department manager's policy, and so on. I have worked in organizations where it was considered a conflict of interest to accept anything from a vendor, including gifts (no matter how small), meals, or even a cup of coffee. Vendors and suppliers often provide their customers and potential customers with lunches, gifts, ball game tickets, and the like. It's your responsibility to know whether a policy exists that forbids you from accepting these gifts. It's also your responsibility to inform the vendor if they've gone over the limit and you are unable to accept the gift.

The same situation can occur here as with the brother-in-law example earlier. If you accept an expensive gift from a vendor and later award that vendor a contract or a piece of the project work, it looks like and probably is a conflict of interest. This violates PMI® guidelines and doesn't look good for you personally either.

Using the "I didn't know there was a policy" excuse probably won't save you when there's a question about conflict of interest. Make it your business to find out whether the organization has a conflict of interest policy and understand exactly what it says. Get a copy, and keep it with your files. Review it periodically. Put a note on your calendar every six months to reread the policy to keep it fresh in your mind. This is a case where not knowing what you don't know can hurt you.

Don't accept gifts that might be construed as conflicts of interest. If your organization does not have a policy regarding vendor gifts, set limits for yourself depending on the situation, the history of gift acceptance by the organization in the past, and the complexity of the project. It's always better to decline a gift you're unsure about than to accept it and later lose your credibility, your reputation, or your PMP® status because of bad judgment.

Stakeholder Influence

Another potential area for conflict of interest comes from stakeholders. Stakeholders are usually folks with a good deal of authority and important positions in the company. Make certain you do not put your own personal interests above the interests of the project when you're dealing with powerful stakeholders. They might have the ability to promote you or reward you in other ways. That's not a bad thing—but if you let that get in the way of the project or let a stakeholder twist your arm with promises of rewards, you're getting mighty close to a conflict of interest. Always weigh your decisions with the objectives of the project and the organization in mind—not your own personal gain.

Keep in mind that you might not always be on the receiving end of the spectrum. You should not offer inappropriate gifts or services or use confidential information you have at your disposal to assist others because this can also be considered a conflict of interest.

 Real World Scenario

The Golf Trip

Amanda Anderson is the project manager for a network upgrade project for her organization. This project will be outsourced to a vendor, and Amanda will manage the vendor's work. She also wrote the RFP and is a member of the selection committee.

The project consists of converting the organization's network from 100 MB Ethernet to Gigabit Ethernet. This will require replacing all the routers and switches. Some of the cabling in the buildings will need to be replaced as well. The RFP requires that the vendor who wins the bid install all the new equipment and replace the network interface cards in each of the servers with Gigabit Ethernet cards. The grand total for this project is estimated at $1.2 million.

Steve James is a vendor with whom Amanda has worked in the past. Steve is very interested in winning this bid. He drops in on Amanda one day shortly after the RFP is posted.

"Amanda, it's good to see you," Steve says. "I was in the neighborhood and thought I'd stop in to see how things are going."

"I'm great," Amanda replies. "I'll bet you're here to talk to me about that RFP. You know I can't say anything until the whole process closes."

"You bet we bid on the project. I know you can't talk about the RFP, so I won't bring it up. I wanted to chat with you about something else. My company is sending 15 lucky contestants and one friend each to Scottsdale, Arizona, for a 'conference.' The conference includes the use of the Scottsdale Golf Club (green fees paid, of course), and all the hotel and meal expenses are on us for the length of the trip. I know Scottsdale is one of your favorite places to golf, so I thought of you. What do you say?"

Amanda sits forward in her chair and looks at Steve for a minute. "That sounds fabulous, and I do love Scottsdale. But you and I both know this isn't a conference. I wouldn't feel right about accepting it."

"Come on, Amanda. Don't look a gift horse in the mouth. It is a conference. I'll be there and so will the top brass from the company. We have some presentations and demonstrations we'd like to show you while you're there, no obligation of course, and then you're free to spend the rest of the time however you'd like."

"I appreciate the offer. Thanks for thinking of me, but no thanks. I'm on the selection committee for the RFP, and it would be a conflict of interest if I attended this conference. Besides, the value of the conference is over the $100 limit our company sets for vendor gifts and meals," Amanda says.

"Okay, we'll miss you. Maybe next time."

Honesty

Honesty can include a lot of topics: reporting the truth regarding project status, being honest about your own experience, not deceiving others, not making false statements, and so on.

One of the aspects of honesty includes your truthfulness about your PMP® application and certification, your qualifications, and the continuing reports you provide to PMI® to maintain your certification. Let's look at a few others.

Personal Gain

Honesty involves not only information regarding your own background and experience but information regarding the project circumstances as well. For example, let's say you're a project manager working on contract. Part of your compensation consists of a bonus based on total project billing. Now, let's suppose your project is finishing sooner than anticipated, and this means your personal profit will decrease by $1,500. Should you stretch the work to meet the original contracted amount so that your personal bonus comes in at the full amount even though your project team is finished? I think you know the answer is of course not.

Your personal gain should never be a consideration when billing a customer or working on a project. Personal gain should never be a factor in any project decision. If the project finishes sooner than planned, you should bill the customer according to the terms of the procurement agreement. Compromising the project for the sake of your personal gain shows a lack of integrity, which could ultimately cost you your PMP® status and even your job.

Truthful Reporting

As a project manager, you are responsible for truthfully reporting all information in your possession to stakeholders, customers, the project sponsor, and the public when required. Always be up front regarding the project's progress.

> Nothing good will come of telling stakeholders or customers that the project is on track and everything looks great when in fact the project is behind schedule or several unknown risk events have occurred that have thrown the project team a curveball. I've personally witnessed the demise of the careers of project managers who chose this route.

Tell the truth regarding project status, even when things don't look good. Stakeholders will likely go to great lengths to help you solve problems or brainstorm solutions. Sometimes, though, the call needs to be made to kill the project. This decision is usually made by the project sponsor or the stakeholders based on your recommendation and predictions of future project activities. Don't skew the reporting to prevent stakeholders from making this decision when it is the best solution based on the circumstances.

Truthful reporting is required when working with the public as well. When working in situations where the public is at risk, truthfully report the facts of the situation and what steps you're taking to counteract or reduce the threats. I recommend you get approval from the organization regarding public announcements prior to reporting the facts. Many organizations have public relations departments that will handle this situation for you.

You probably remember something your mother always told you: actions speak louder than words. Always remember that you lead by example. Your team members are watching. If you are driven by high personal ethics and a strong desire for providing excellent customer service, those who work for you will likely follow your lead.

Role Delineation Study

In addition to the areas covered in the *PMI® Code of Ethics and Professional Conduct*, you should be aware of four other focus areas that PMI® discusses in its role delineation study. This study was published in PMI®'s publication *Project Management Professional (PMP®) Examination Specification*. The four focus areas are as follows:

- Ensure personal integrity and professionalism.
- Contribute to the project management knowledge base.
- Enhance personal professional competence.
- Promote interaction among team members and other stakeholders.

Throughout this chapter, I've covered most of the concepts surrounding these focus areas with the exception of contributing to the project management knowledge base. We'll look at that topic next.

Applying Professional Knowledge

Professional knowledge involves the knowledge of project management practices as well as specific industry or technical knowledge required to complete an assignment.

As a PMP® credential holder, you should apply project management knowledge to all your projects. Take the opportunity to educate others by keeping them up to date on project management practices, training your team members to use the correct techniques, informing stakeholders of the correct processes, and then sticking to those processes throughout the course of the project. This isn't always easy, especially when the organization doesn't have any formal processes in place. But once the stakeholders see the benefits of good project management practices, they'll never go back to the "old" way of performing their projects.

One way to apply professional knowledge is to become and remain knowledgeable in project management best practices techniques. I'll cover this topic next, along with how to become a PMI® education provider and the importance of having industry knowledge.

Project Management Knowledge

Project management is a growing field. Part of the responsibility when you hold the PMP® designation is to stay abreast of project management practices, theories, and techniques. You can do this in many ways, one of which includes joining a local PMI® chapter. There are hundreds of local chapters throughout the United States and in other countries as well. You can check the PMI® website to find a chapter near you.

Chapter meetings give you the opportunity to meet other project managers, find out what techniques they're using, and seek advice regarding your project. Usually, guest speakers appear at each chapter meeting and share their experiences and tips. Their stories are always interesting, and they give you the opportunity to learn from someone else's experiences and avoid making wrong turns on your next project. You might have a few stories of your own worth sharing with your local chapter. Volunteer to be a speaker at an upcoming meeting, and let others learn from your experiences.

One of the things you'll get when you join the PMI® organization and pay your yearly dues is its monthly magazine. This publication details real-life projects and the techniques and issues project managers have to deal with on those projects. Reading the magazine is a great way to learn new project management techniques or reinforce the information you already know. You might discover how to apply some of the knowledge you've already learned in more efficient ways as well.

PMI® offers educational courses through its local chapters and at the national level as well. These courses are yet another way for you to learn about project management and meet others in your field.

Education Providers

PMI® has a program that allows you or your organization to become a Registered Education Provider (REP). This enables you—once you're certified—to conduct PMI®-sanctioned project management training, seminars, and conferences. The best part is that your attendees are awarded professional development units (PDUs) for attending the training or seminar. As a PMP® credential holder, and especially as an REP, you have a responsibility to the profession and to PMI® to provide truthful information regarding the PMI® certification process, the exam applications, the PDU requirements, and so on. Keep up to date on PMI's certification process by periodically checking the website.

Industry Knowledge

Contributing and applying professional knowledge goes beyond project management experience. You likely have specific industry or technical experience as well. Part of applying your professional knowledge includes gaining knowledge of your particular industry and keeping others informed of advances in these areas.

Information technology has grown exponentially over the past several years. It used to be that if you specialized in network operations, for example, it was possible to learn and become proficient in all things related to networks. Today that is no longer the case. Each specialized area within information technology has grown to become a knowledge area in and of itself. Many other fields have either always had individual specialties or just recently experienced this phenomenon, including the medical field, bioengineering, manufacturing, and pharmaceuticals, to name a few. You need to stay up to date regarding your industry so that you can apply that knowledge effectively. Today's fast-paced advances can leave you behind fairly quickly if you don't stay on top of things.

Earlier in the book, I mentioned that as a project manager you are not required to be a technical expert, and that still holds true. But it doesn't hurt to stay abreast of industry trends and knowledge in your field and have a general understanding of the specifics of your business. Again, you can join industry associations and take educational classes to stay on top of breaking trends and technology in your industry.

Congratulations! You've learned a great deal about project management and I appreciate you hanging in there until the end. I wish you the best of luck on the exam and in your future project management endeavors.

 Real World Scenario

Project Case Study: New Kitchen Heaven Retail Store

Stakeholders have asked for an updated status on the project schedule as well as a remaining cost projection. You decide to provide several cost and schedule performance figures for the project on the status report.

"Build-out is behind schedule. They were scheduled to be completed by the 15th of January, but they aren't going to finish up until the 24th."

"What's that going to do to my schedule?" Jill asks. "I'm starting interviews for the store positions on the 16th. I hope to have that wrapped up by the 19th. As long as I have the majority of the staff hired by the 20th, we can have them stocking shelves starting the 22nd."

"Let's finish up the status of the other items, and I'll come back to that."

You've calculated some performance measurements, including earned value measures, and you show them to Jill and Dirk (all figures are in millions of dollars):

BAC = 2, PV = 1.86, EV = 1.75, AC = 1.70

SPI is 0.94 (1.75 / 1.86)

CPI is 1.03 (1.75 / 1.70)

EAC is 1.96; 1.70 + [(2 − 1.75) / (1.03 × 0.94)]

ETC is 0.30 (2 − 1.70)

"What is all this telling us?" Dirk asks.

"The cost performance index tells us we're getting a good return for the money spent on the project so far. In other words, we've experienced a $1.03 value for every dollar spent to date," you respond.

"The schedule performance index isn't as cheery, but it's not dreadful news either. This performance indicator says that work is progressing at 94 percent of what we anticipated by this point.

"The estimate at completion tells us that based on what we know today, the total project will cost $1.94 million. That's coming in under the original $2 million we had budgeted for completion, so we're on track with the project budget. The last figure is the estimated cost of the remaining work."

"It looks like we're a little behind schedule based on what you have figured here," Dirk says.

"Yes, that's correct," you reply. "That brings us back to Jill's question. I have two alternatives to propose. One, we overlap the schedule and allow Gomez's crew to complete their work while Jill's staff starts stocking shelves."

Jill says, "I don't like this option. We'll be tripping over each other, and I don't want merchandise damaged by workers who are still dragging equipment around inside the store. What's your other option?"

"We could ask Gomez to increase the crew size so that they complete on time according to the contract. We have a provision in the contract that stipulates they add crew members if it looks as though they'll miss the scheduled completion date. I will instruct the contract management department to inform Gomez that we're requiring additional crew members."

"That will do the trick," Jill says. "We need the storefront to ourselves when stocking and preparing for opening. I'm glad you had that stipulation in the contract."

You report that sign-off has been obtained for the completed deliverables to date. Quality inspections and comparisons of the deliverables to the acceptance criteria were completed to Jill and Ricardo's satisfaction on the work performed to date.

You also report that Ruth and team have completed all but two user stories for the website launch. The hybrid approach of Scrum and Kanban has worked well. The visual representation of the work remaining on the burndown chart was very useful in helping the team to gauge progress.

Project Closeout a Few Weeks Later

Dirk strolls into your office, maintaining his formal and dignified manners as always, and then sits down in the chair beside your desk.

"I just want to congratulate you on a job well done," he says. "The grand opening was a success, and the store had a better-than-expected first week. I'm impressed you were able to pull this off and get the store opened prior to the Home and Garden Show. That was the key to the great opening week."

"Thank you, Dirk. Lots of people put in a lot of hard work and extra hours to get this job done. I'm glad you're happy with the results."

"I thought the banner with our logo, 'Great Gadgets for People Interested in Great Food,' was a wonderful touch."

"That was Jill's idea. She had some great ideas that made the festivities successful. As you know, though," you continue, "we did have some problems on this project. Fortunately, they weren't insurmountable, but I think we learned a thing or two during this project that we can carry forward to other projects."

"Like what?" Dirk asks.

"We should have contracted with Gomez Construction sooner so that we didn't have to pay overtime. We had a very generous budget, so the overtime expense didn't impact this project, but it might impact the next one. And we came fairly close to having a hardware disaster on our hands. Next time, we should order the equipment sooner, test it here at headquarters first, and then ship it out to the site after we know everything is working correctly."

"Good ideas. But that's old news." Dirk continues, "Now that this project is over, I'd like to get you started on the next project. We're going to introduce cooking classes in all of our retail stores. The focus is the home chef, and we might just call the classes the Home Chef Pro series. We'll offer basic classes all the way to professional-series classes if the project is a hit. We'll bring in guest chefs from the local areas to give demos and teach some of the classes as well."

"I'm very interested in taking on this project and can't wait to get started. I'm thrilled that you want me to head this up. But I do have a few things here to wrap up before I start work on the new project," you reply.

Dirk says, "The project is over. The grand opening was a success. It's time to move on. Let Jill take over now; the retail stores are her responsibility."

"Jill has taken over the day-to-day operations. However, I've got to finish collecting the project information, close out the contract with Gomez, and make the final payment. Jake verified that all the work was completed correctly and to his satisfaction. Ruth said the hybrid-agile methodology was a perfect choice and the team learned a lot. She will continue to tailor and use agile methodologies for future projects. I will create a lessons learned document that outlines these things so that we can reference them on the next project. I also need to publish the formal acceptance notice to all of the stakeholders via email. Then, after all those things are completed, all of the project records need to be indexed and archived. I can have all that done by the end of the week and will be free starting Monday to work on requirements-gathering and the charter document for the new project."

"This is just like the planning process discussion we got into with the tree, the breakdown structure thing, and all the planning, I suspect. I do have to admit all the planning paid off." As he gets up to leave, Dirk continues, "I'll give you until the end of the week to close out this project. Come see me Monday to get started on Home Chef Pro."

Jill Overstreet thought you did such a great job of managing this project that she has offered to buy you lunch at one of those upscale, white-tablecloth-type French restaurants. The iced teas have just been delivered, and you and Jill are chatting about business.

"I'm impressed with your project management skills," she remarks. "This store opening was the best on record. And you really kept Dirk in line—I admire that. He can be headstrong, but you had a way of convincing him what needed doing and then sticking to it."

"Thanks," you reply. "I've got several years of project management experience, so many of those lessons learned on previous projects helped me out with this project. I enjoy project management and read books and articles on the subject whenever I get the chance. It's nice when you can learn from others' mistakes and avoid making them yourself."

Jill takes a long drink of tea. She glances at you over the top of her glass and pauses before setting it down. "You know," she begins, "we almost didn't hire anyone for your position. Dirk wanted to do away with the project management role altogether. He had a real distaste for project management after our last project."

"Why is that?"

"The last project manager got involved in a conflict-of-interest situation. She was working on a project that involved updating and remodeling all the existing stores. Things like new fixtures, signs, shelving, display cases, and such were up for bid—and it was a very sizable bid. Not only did she accept an all-expenses-paid weekend visit to a resort town from one of the vendors bidding on the contract, she also revealed company secrets to them, some of which leaked to our competitors."

Your mouth drops open. "I can't believe she would accept gifts like that from a vendor. Revealing company secrets is even worse. Conflict-of-interest situations and not protecting intellectual property violate the code of professional conduct that we agree to when we gain PMP® certification. I can understand why Dirk didn't want to hire another project manager. Behavior like that makes all project managers look bad."

"I'm glad you kept things aboveboard and won Dirk back over. The project management role is important to Kitchen Heaven, and I know your skills in this discipline are what made this project such a success."

"Jill, not only would I never compromise my own integrity through a conflict-of-interest situation, I would report the situation and the vendor to the project sponsor and to PMI® as an ethics violation. It's better to be honest and let the project sponsor or key stakeholders know what's happening than to hide the situation or, even worse, compromise your own integrity by getting involved in it in the first place. You have my word that I'll keep business interests above my own personal interests. I'll report anything that even looks like it would call my actions into question just to keep things honest and out in the open."

"That's good to hear," Jill replies. "Congratulations on your new assignment. Dirk and I were discussing the new Home Chef Pro project yesterday. We're venturing into new territory with this project, and I'm confident you'll do an excellent job heading it up. Dirk made a good choice."

Project Case Study Checklist

- Control Costs
  - Cost change control system
  - Performance measurement analysis
  - Forecasting
- Control Schedule
  - Schedule compression
- Validate Scope
  - Validated work results
- Control Quality
  - Assured quality requirements were met
- Close Project or Phase
  - Product verification (work was correct and satisfactory)
  - Collecting project documents
  - Disseminating final acceptance notice
  - Documenting lessons learned
  - Archiving project records
  - Product verification (work was correct and satisfactory), notification to vendor that work is complete, gathering and archiving procurement information

- Adhering to the PMI® Code of Ethics and Professional Conduct
  - Ensuring personal and professional integrity
  - Not placing personal gain above business needs
  - Avoiding conflict of interest situations
  - Truthfully reporting questionable situations and maintaining honesty
  - Protecting intellectual property

Understanding How This Applies to Your Next Project

Earned value management is a tool you can't live without on a waterfall project, particularly the cost and schedule variance and the cost and schedule performance indexes. The size and complexity of the project will dictate how often you should run the performance measurements. The mantra of stakeholders everywhere is "on time and on budget." Therefore, controlling the project budget and the schedule will likely be two of your most time-consuming project management tasks. Use the tools we discussed in this chapter to keep yourself and your stakeholders informed of what's happening regarding these two important areas. If cost or schedule changes must occur, it's imperative you communicate what happened, why it happened, and if it's expected to happen again and that you realign stakeholders' expectations with the new forecasted estimates.

Control Scope is absolutely essential for all projects. Time and again I've seen changes to scope end up pitting stakeholders against the project team because the requirements weren't defined adequately in the first place and because neither party clearly understood what was being requested. Scope changes can kill a project by significantly delaying the finish date or by so drastically modifying the original objective of the project that it no longer resembles what it set out to accomplish. I try to keep the questions regarding scope change simple, as in, "Do you absolutely have to have this to meet the objective of this project?" That isn't always easy for people to understand because we often confuse wants with needs. So, I might come up with an analogy they can relate to—something like this: "Let's say your one and only culinary skill is boiling water. Do you really need a designer stove with dual-fuel options and a built-in warming oven to boil water? Wouldn't a simple store brand work in that case?" If you think scope changes are likely on your project, consider a hybrid or agile development methodology.

Closing out the project should always include a formal sign-off (from the project sponsor and key stakeholders at a minimum) that the work of the project is acceptable and complete. You've probably experienced, as I have, stakeholders with short-term memory lapses. For example, I'll get to the end of the project and find that a stakeholder has a list of additional requirements longer than their arm. Some of this goes back to ensuring that requirements are accurately defined during the Planning processes and that ongoing communication regarding project status, along with feedback from the stakeholders, is occurring on a regular basis. Another key here is making certain you are managing stakeholder expectations throughout the project. You might indeed have identified all the requirements and accurately documented the objective of the project. However, if the stakeholder had expectations that weren't captured or that took shape after the work of the project began, you could end up with a very unhappy stakeholder on your hands and, potentially, a failed project. In my experience, stakeholder expectations tend to stray midway through the project. That's because they are beginning to see nearly completed deliverables and new possibilities for the product are developing that weren't thought of during the Planning stage. It's similar to buying a new house based on blueprints and then going on that first walk-through when the framing is finished. It's during walk-through that you think, "Gee, I thought that closet was bigger than this." Tune into what your stakeholders are saying and get at the basis of their questions (dig deep) so that there are no hidden expectations and you're able to manage any new expectations that pop up.

The professional responsibility section of this chapter can easily be boiled down to the Golden Rule: "Do unto others as you would have them do unto you." I wish I could tell you that everything I talked about here is practiced by all project managers everywhere. Unfortunately, you've likely read, as I have, the endless stories about corporate execs acting in their own best interests rather than those of their organizations. Project managers make the news as well, especially when they're working on projects that involve public funds or charitable organizations. In my humble opinion, ignoring the practices and advice in this chapter isn't worth the damage it might cause to my organization, to the project, or to my career.

I'd also add that if your first thought on a new project is what a great résumé builder it's going to be for you and how you'll likely score that next big promotion after the project is complete, you've started off on the wrong foot. Although I'm the first to admit big projects (when they are executed well and satisfy the stakeholder expectations) *are* résumé builders, it's the wrong reason to take on a project. Consider your experience level and how and what you'll be able to contribute to the project to make it a success. It's okay if the project is a stretch for you—you can't grow your experience without taking on more complex projects as your career progresses. But also be wise enough to know when you might be in over your head.

There's no substitute for integrity and honesty when conducting your projects. Once you've tarnished your integrity, whether intentionally or not, it's almost impossible to regain the trust of your stakeholders and management staff. Because of this, I'm never afraid of telling anyone, "I don't know," but I always follow it up with, "But I'll find out." You're a project manager, not a miracle worker. No one expects you to have all the answers any more than they expect you to perform every single task on the project.

I had the unfortunate experience of being instructed to lie about project status on a previous job and to purposely withhold project information from oversight boards. I spent a few sleepless nights worrying about where I'd find my next job because I immediately disobeyed those orders and reported the truth. I knew it would cost me my job. But I also knew it was better for me to lose the job than to compromise my integrity. As it turned out, I found a new job quickly. In retrospect, if I had not left the position when I did, my reputation would have taken a hit—not because of anything I had done but through association with the people on the project who compromised their integrity when they shouldn't have.

I hope you've found this study guide helpful both for your studies for the PMP® exam and for your next project. Thank you for spending some time with me in the pages of this book. I wish you the best of luck in your project management endeavors.

Summary

We covered a lot of material in this chapter, and we also closed out the Monitoring and Controlling process group.

Control Costs involves managing changes to project costs. It's also concerned with monitoring project budgets to prevent unauthorized or incorrect costs from getting included in the cost baseline. Control Costs uses tools and techniques such as earned value management (CV, SV, CPI, SPI), forecasting (ETC, EAC, and TCPI), variance analysis, trend analysis, and EVM to monitor costs.

Control Schedule involves determining the status of the project schedule, determining whether changes have occurred or should occur, and influencing and managing schedule changes.

Validate Scope involves validating and accepting work results. Control Scope is concerned with controlling changes to project scope.

Agile projects use several different measures to monitor and measure progress. Definition of done (DoD) is a checklist of elements that ensures the deliverable is ready for the customer to use. Definition of ready (DoR) describes the specifics of the tasks planned for the iteration before the team begins work. Empirical measures are usually expressed as deliverables, functionality, or features. Flow-based projects measure cycle time, lead time, and response time. Cycle time is the time it takes to complete work on a task from the time work starts. Lead time is the time it takes for a task to go from request to completion. Response time is the time a task waits before work starts.

Project closure is the most often neglected process of all the project management processes. The Close Project or Phase process is the only process in the Closing process group. The four most important tasks of closure are as follows:

- Checking the work for completeness and accuracy
- Documenting formal acceptance
- Disseminating project closure information
- Archiving records and lessons learned

Close Project or Phase involves analyzing the project management processes to determine their effectiveness and to document lessons learned concerning them. It also involves archiving all project documents for historical reference and performing administrative closure procedures for the project. Documenting the formal acceptance of the project product is an important aspect of project closure as well, because this ensures that the project scope, compared to the project management plan, is complete and accurate and meets the exit criteria. Ultimately, you want to assure that the stakeholder or customer is satisfied with the work of the project and that it meets their needs.

Projects come to an end in one of four ways: addition, starvation, integration, or extinction. Addition is when projects evolve into their own business unit. Starvation happens because the project is starved of its resources. Integration occurs when resources are taken from the existing project and dispersed back into the organization or assigned to other projects. Extinction is the best ending because the project was completed, accepted, and closed.

Close Project or Phase is performed at the end of each phase of the project as well as at the end of the project. Close Project or Phase involves documenting formal acceptance and disseminating notice of acceptance to the stakeholders, customer, and others. All documentation, including procurement documentation, gathered during the project and collected during this process is archived and saved for reference purposes on future projects.

Lessons learned document the successes and failures of the project and of the procurement processes. Many times lessons learned are not documented because staff members do not want to assign their names to project errors or failures. You and your management team need to work together to assure employees that lessons learned are not exercises used for disciplinary purposes but benefit both the employee and the organization. Documenting what you've learned from past experiences lets you carry this forward to new projects so that the same errors are not repeated. It also allows you to incorporate new methods of performing activities that you learned on past projects.

Closing an agile project involves many of the same tasks you'll use to close a predictive project. Ensure the criteria have been met to close the project, obtain final acceptance, perform administrative closure procedures, ensure the documentation is up to date, prepare and distribute the final project report, validate the readiness for transition to the users, ensure knowledge transfer to the end user, confirm your approach for knowledge transfer, and release team members.

Project management professionals are responsible for reporting truthful information about their PMP® status and project management experience to prospective customers, clients, employers, and PMI®. As a project manager, you're responsible for the integrity of the project management process and the product. In all situations, you are responsible for your own personal integrity.

Personal integrity means adhering to an ethical standard. As part of your PMP® designation, you'll be required to adhere to the *Code of Ethics and Professional Conduct* established by PMI®. Part of this code involves avoiding putting your own personal gain above the project objectives.

As a professional, you should strive to maintain honesty in project reporting. You're required to abide by laws, rules, and regulations regarding your industry and project management practices. You should also report any instance that might appear to be a conflict of interest. It's always better to inform others of an apparent conflict than to have it discovered by others and have your methods called into question after the fact.

You will likely come across confidential information or intellectual property during your project management experiences. Respect the use of this information, and always verify who might have permission to access the information and when disclosures are required.

Stakeholders have competing needs and business issues and as such will sometimes cause conflict on your project. You will be required to balance the needs and interests of the stakeholders with the project objectives.

Many project managers today are working in a global environment. It's important to respect and understand the cultural differences that exist and not try to impose your cultural beliefs on others. Culture shock is an experience that occurs when you find yourself in an unfamiliar cultural environment. Training is a good way to provide project team members with relationship management techniques regarding cultural and ethnic differences.

Exam Essentials

Name the purpose of the Control Costs process.   The Control Costs process is concerned with monitoring project costs to prevent unauthorized or incorrect costs from being included in the cost baseline.

Be able to describe earned value management techniques.   Earned value management (EVM) monitors the planned value (PV), earned value (EV), and actual costs (AC) expended to produce the work of the project. Cost variance (CV), schedule variance (SV), cost performance index (CPI), and schedule performance index (SPI) are the formulas used with the EVM technique.

Name the purpose of the Validate Scope process.   The purpose of Validate Scope is to determine whether the work is complete and whether it satisfies the project objectives. Validate Scope should be performed for both predictive and agile projects when a project is canceled.

Be able to describe product verification.   Product verification confirms that all the work of the project was completed accurately and to the satisfaction of the stakeholder. This is performed at the end of an iteration or flow-based work period on agile projects.

Be able to name the primary activity of the Close Project or Phase process.   The key activity of this process is concerned with completing all the activities associated with closing out the project management processes in order to officially close out the project or phase. It also ensures that all project work is complete.

Be able to describe when the Close Project or Phase process is performed.   Close Project or Phase is performed at the close of each project phase and at the close of the project.

Be able to define the purpose of lessons learned.   The purpose of lessons learned is to describe the project successes and failures and to use the information learned on future projects.

Be able to name several measures used on agile projects.   Measure include the following: definition of done (DoD) defines exit criteria for the iteration; definition of ready (DoR) is entry criteria for the iteration; capacity measures (such as burndown and burnup charts) measure velocity; empirical measures are usually expressed as deliverables, functionality, or features; cycle time is a flow-based measure that calculates the time it takes for a task to complete work; lead time is a flow-based measure that calculates the time it takes for a task to complete work from request to completion; and response time is a flow-based measure that calculates the time a task waits to start work.

Be able to describe the steps to close out an agile project.   Examine criteria to close; obtain final acceptance of deliverables; obtain legal, financial, procurement, and administrative closure; update and archive all project documents; hold a final retrospective and a final lessons learned meeting; prepare and distribute final project report; validate readiness for transition; ensure knowledge transfer; confirm approach for knowledge transfer; and release team members.

Be able to name the publication that describes the ethical standards to which PMP® **credential holders are required to adhere.**   The ethical standards PMP® credential holders are required to adhere to are described in the *PMI® Code of Ethics and Professional Conduct.*

Describe the areas in which PMP® credential holders must apply professional knowledge.   PMP® credential holders must apply professional knowledge in the areas of project management practices, industry practices, and technical areas.

Know the key activity that ensures customer satisfaction.   The key activity that ensures customer satisfaction is documenting project requirements and meeting them.

Review Questions

You can find the answers to the review questions in Appendix A. Be sure to download the Bonus Exams and Bonus Questions so that you'll have a broader exposure and more experience answering questions related to the topics in this chapter.

1. You are working on a project that was proceeding well until a manufacturing glitch occurred that requires corrective action. It turns out the glitch was an unintentional enhancement to the product, and the marketing people are absolutely crazy about its potential. The corrective action is canceled, and you continue to produce the product with the newly discovered enhancement. As the project manager, you know that a change has occurred to the product scope because the glitch changed the characteristics of the product. Which of the following statements is true?

   A. Changes to product scope should be reflected in the project scope.

   B. Changes to product scope should be documented in the scope management plan.

   C. Changes to product scope will result in cost changes.

   D. Changes to product scope are a result of corrective action.

2. You are a project manager for a manufacturing firm that produces Civil War–era replicas and memorabilia. You discover a design error during a test production run on your latest project. Time is a critical constraint on this project. The stakeholders are anxious to get this new product to market and their bonuses are on the line if this project doesn't come in on time. Which of the following is the most likely response to this problem?

   A. Reduce the technical requirements so that the error is no longer valid.

   B. Go forward with production and ignore the error.

   C. Go forward with production, but inform the customer of the problem.

   D. Develop alternative solutions to address the error.

3. You are using a burnup chart to visually provide information about project progress to the agile team. Which of the following are true regarding a burnup chart?

   A. This is a type of capacity measure.

   B. This shows the completed work for the iteration or project.

   C. Accuracy depends on the accuracy of the team's estimates.

   D. It can't tell you if you are working on the right things.

   E. A, B, D

   F. A, B, C, D

   G. A, C, D

4. You are a project manager for Dakota Software Consulting Services. You're working with a major retailer that offers online and in retail stores. You're using a performance review to examine elements such as actual start and end dates for schedule activities. You also used SV and SPI calculations to determine the impact of schedule variations, which are more

than the stakeholders will tolerate. Rumors have started and stakeholders have heard there might be a schedule delay. They've asked for a full report on the situation. Which of the following are true regarding this scenario? (Choose two.)

A. You should ensure you know the facts about the schedule variances and how much the schedule might be delayed before reporting anything to the stakeholders.

B. You should act as quickly as possible to implement corrective actions so that the schedule is brought back in line with the plan and delays are reduced as far as possible.

C. Examine float variances because the activities with the least amount of float have the potential to cause the biggest schedule delays.

D. Delays to the critical path tasks may cause delays to the project completion date, so examine any variances in ascending order of critical path tasks.

5. You are the project manager for a top-secret software project for an agency of the U.S. government. Getting top-secret clearances for contractors takes quite a bit of time and waiting for clearances could jeopardize the implementation date. Your mission—should you choose to accept it—is to complete the project using internal resources. Your agile programmers are 80 percent of the way through the programming and testing when your agency appoints a new executive director. Your programmers are siphoned off this project to work on the executive director's hot new project before they finish this project. Which of the following are true regarding this question?

A. This project ended due to starvation.

B. You should perform the Validate Scope process which documents the level and degree of completion.

C. The DoD, such as describing the functionality, purpose, and how the functionality should be performed, helps determine if the functionality produced at the end of the iteration is ready for the customer to use.

D. The Validate Scope process documents the correctness of the work according to stakeholders' expectations and is not needed for an agile project.

6. You are a project manager for Dutch Harbor Consulting. Your latest project involves the upgrade of an organization's operating system on 236 servers. You performed this project under contract. You are closing out the project and the procurement and know that you should document and file which of the following?

A. Administrative documents

B. The written notice that the contract is complete

C. Formal acceptance of the work of the project

D. Product verification of the work of the project

E. A, B, D

F. C, D

G. B, C, D

7. You are a project manager for Laurel's Theater Productions. Your new project is coming in over budget and requires a cost change through the cost change control system. You know which of the following statements are true regarding Control Costs?

A. A description of how cost changes should be managed and controlled is found in the cost management plan.

B. Approved cost changes are reflected in the cost baseline.

C. EVM is used to determine the cost performance that must be realized for the remaining work of the project to meet the BAC goal.

D. This equation, EAC = BAC / cumulative CPI, is used to forecast an estimate at completion assuming future project performance will be the same as past performance.

E. A, B, C, D

F. A, B, D

8. Which of the following might require re-baselining of the cost baseline?

A. Corrective action

B. Revised cost estimates

C. Updates to the cost management plan

D. Budget updates

9. What are the performance measurements for the Control Schedule process?

A. SV = (EV − PV) and SPI = (EV / PV)

B. SV = (EV − AC) and SPI = (EV / AC)

C. SV = (EV − BAC) and SPI = (EV / BAC)

D. SV = (PV − EV) and SPI = (PV / EV)

10. This measurement is the value of the work that has been completed to date compared to the budget.

A. PV

B. AC

C. EV

D. EAC

11. You are a contract project manager for a wholesale flower distribution company. Your project is to develop a website for the company that allows retailers to place their flower orders online. You will also provide a separate link for individual purchases that are ordered, packaged, and mailed to the consumer directly from the grower's site. This project involves coordinating the parent company, growers, and distributors. You are preparing a performance review and have the following measurements at hand: PV = 300, AC = 200, and EV = 250. What do you know about this project?

A. The EAC is a positive number, which means the project will finish under budget.

B. You do not have enough information to calculate CPI.

C. The CV is a negative number in this case, which means you've spent less than you planned to spend as of the measurement date.

D. The CV is a positive number in this case, which means you're under budget as of the measurement date.

12. A negative result from an SV calculation means which of the following?

A. PV is higher than EV.

B. PV equals 1.

C. EV is higher than PV.

D. EV is higher than AC.

13. You are a contract project manager for a wholesale flower distribution company. Your project is to develop a website for the company that allows retailers to place their flower orders online. You will also provide a separate link for individual purchases that are ordered, packaged, and mailed to the consumer directly from the grower's site. This project involves coordinating the parent company, growers, and distributors. You are preparing a performance review and have the following measurements at hand: PV = 300, AC = 200, and EV = 250. What is the CPI of this project?

A. 0.80

B. 1.25

C. 1.5

D. 0.83

14. You have accepted project performance to date and assume future work (ETC) will be performed at the budgeted rate. If BAC = 800, ETC – 275, PV = 300, AC = 200, EV = 250, and CPI = 1.25, what is the EAC?

A. 640

B. 750

C. 600

D. 550

15. You know that EAC = 375, PV = 300, AC = 200, and EV = 250. You expect the work of the project to continue as planned. What is the ETC?

A. 300

B. 125

C. 175

D. 50

16. You expect future project performance to be consistent with the project performance experienced to date for this work component. If BAC = 800, ETC = 275, PV = 300, AC = 200, EV = 250, and CPI = 1.25, what is the EAC?

A. 640

B. 750

    C. 600

    D. 550

17. You know that BAC = 500, PV = 325, AC = 275, CPI = 0.9, and EV = 250, and you are using actual costs to date and assuming ETC uses the budgeted rate. What is Variance at Completion?

    A. 25

    B. −52

    C. 52

    D. −25

18. You know that BAC = 2500, PV = 1250, AC = 1275, EV = 1150, and that you are experiencing typical cost variances. What is ETC?

    A. 1467

    B. 2625

    C. 1500

    D. 2778

19. You are a contract project manager working with the State of Bliss. Your latest project involves rewriting the Department of Revenue's income tax system. As project manager, you have taken all the appropriate actions regarding confidentiality of data. One of the key stakeholders is a huge movie buff, and she has the power to promote you into a better position at the conclusion of this project. She's reviewing some report data that just happens to include confidential information regarding one of her favorite movie superstars. What is the most appropriate response?

    A. Report her to the management team.

    B. Request that she immediately return the information, citing conflict of interest and violation of confidential company data.

    C. Do nothing, because she has the proper level of access rights to the data and this information showed up unintentionally.

    D. Request that she immediately return the information until you can confirm that she has the proper level of access rights to the data.

20. This tool and technique of the Close Project or Phase process looks at project variables and how they contributed to project outcomes. This is for the purpose of improving performance on future projects.

    A. Variance analysis

    B. Trend analysis

    C. Regression analysis

    D. Document analysis

Appendices

Appendix

A

Answers to Review Questions

Chapter 1: Building the Foundation

1. C. The Project Management Institute (PMI)® is the industry-recognized standard for project management practices.

2. B. Projects exist to create a unique product, service, or result. The new color line is not a unique product. A minor change has been requested (to add a new color), indicating that this is an ongoing operations function. Some of the criteria for projects are that they are unique, temporary with definitive start and end dates, and considered complete when the project goals are achieved.

3. A. This is a project. The product line is new, which implies that this is a unique product—it hasn't been done before. You can discern a definite start and end date by the fact that the new appliances must be ready by the spring catalog release.

4. B, D. You can't know the characteristics and features without consulting with the stakeholders. Progressive elaboration is the process of determining the characteristics and features of the product of the project. Progressive elaboration is carried out via steps in detailed fashion. The project was already approved, as stated in this question, so the business value should have already been defined. The project life cycle planning will start in the Planning process group of the project. You are still in the Initiating process.

5. B. This came about because of an organizational need. Staff members were spending unproductive hours producing information for the management report that wasn't consistent or meaningful.

6. C. Project management brings together a set of tools and techniques to organize project activities. Project managers are the ones responsible for managing the project management processes.

7. A, C, D. There are three ways phases can be performed in a multiphased project: sequential, iterative, and overlapping. Iterative is when more than one phase is being performed at the same time, and overlapping occurs when one phase starts before the prior phase completes. All phases should perform a phase gate review.

8. A, C, D. Option B describes the waterfall, or predictive, process.

9. B. When one business develops new pricing structures or new products, or offers more for the same, competing businesses must do something similar in order to stay competitive. This is the definition of competitive forces.

10. D. Portfolios are collections of projects and/or programs. The projects or programs do not have to be directly related or interdependent to reside within the portfolio.

11. D. The processes, in order, are Initiating, Planning, Executing, Monitoring and Controlling, and Closing.

12. C. The Initiating process group is where stakeholders have the greatest ability to influence outcomes of the project. Risk is highest during this stage because of the high degree of unknown factors.

13. B, C, D. The three types of PMOs are supportive, controlling, and directive.

14. C. The seven wastes are associated with Kaizen, a Lean methodology.

15. E, F. Scrum and Kanban are considered pull systems because user stories are moved from one point in the process to the next, thereby freeing up space to pull other user stories from the backlog.

16. B. Six Sigma relies heavily on statistical data.

17. A, C, D. The agile methodologies require close contact with the stakeholders, and this provides continuous feedback to the project team throughout the project. In a waterfall approach, stakeholders typically have a lot of contact and involvement with the team at the beginning of the project, and this involvement tapers off toward the end of the project. Option E also describes an agile methodology approach. Waterfall, or predictive, methodologies do not have iterative reviews and changes are rigorously controlled and managed.

18. B, C. The *Agile Practice Guide* (PMI®, 2017). focuses on value to the customer, not in measuring processes or the quality of deliverables. Success is measured in incremental steps.

19. A-3, B-1, C-4, D-2. Project and program management focuses on performing the projects in the right way, whereas portfolio management focuses on working on the right projects and programs at the right time. OPM aligns projects with the organization's strategic business objectives. The PMO is a centralized unit that oversees the management of projects throughout the organization.

20. C. This describes a hybrid development life cycle. Early on, requirements might be gathered in detail, and as the project progresses, the team reverts to an agile approach to deliver functionality incrementally.

Chapter 2: Assessing Project Needs

1. A. OPAs are internal to the organization. EEFs can be internal or external to the organization but they are always outside the control of the project team. EEFs may include government or industry regulations and can therefore drive compliance requirements for the project.

2. D, E. According to the *PMBOK® Guide*, the project charter should be issued by the project sponsor or the project initiator. Once the charter is approved, the project manager will use the goals of the project to set a clear vision and mission for the project and continually inform the team and the stakeholders of the objectives and mission of the project. It's also used to help align stakeholder expectations with the objectives of the project. Project charters are typically required when using a predictive methodology, and they may be also be used in a hybrid methodology approach.

3. B, C. Integration is managed and performed by the project manager when using an agile methodology. The agile team is responsible for planning, delivery, and control of the product. Integration focuses on coordinating all aspects of the project, not just those mentioned in Option D. Integration can be used with any project management methodology, not just predictive.

4. B, D. This describes the Project Schedule Management Knowledge Area, which involves the following processes: Plan Schedule Management, Define Activities, Sequence Activities, Estimate Activity Durations, Develop Schedule, and Control Schedule. The question also describes a predictive methodology. The question describes a step-by-step approach by finishing the first activity (estimating) before creating a schedule. In agile, these activities are performed together.

5. A. The benefits management plan is created early in the project and it is updated iteratively throughout the life of the project. The business case is reviewed at each phase gate.

6. C. The project charter is the document that names the project manager.

7. B. Historical information on projects of a similar nature can be helpful when initiating new projects. They can help in formulating project deliverables and identifying constraints and assumptions and will be helpful later in the project Planning processes as well.

8. C. Procurement Management generally occurs at the beginning of the project in an agile methodology. You may have procurements at some point in the project and they may become a iteration activity, but they are not performed before, during, and after every iteration.

9. A. Option B describes the project exit criteria. Option C is not true, and Option D is partially true. The project manager does not dictate what the approval requirements are. The stakeholders and sponsor will come to agreement on the success criteria, and the project manager may or may not have a voice in the decision.

10. B. Projects with NPV greater than 0 should be given an accept recommendation.

11. D. Projects with the highest IRR value are favored over projects with lower IRR values.

12. A. Net present value (NPV) assumes reinvestment is made at the cost of capital.

13. C. Year 1 and 2 inflows are each $100,000 for a total of $200,000. Year 3 inflows are an additional $300,000. Add one more quarter to this total, and the $575,000 is reached in three years and three months, or 39 months.

14. D. IRR assumes reinvestment at the IRR rate and is the discount rate when NPV is equal to 0.

15. C. The purpose of the business case is to understand the business need for the project and determine whether the investment in the project is worthwhile. This may include analysis using benefit measurement methods. The benefits, how they're measured, and how they're obtained is contained in the benefits management plan.

16. A, D, F. The results of the needs assessment are documented in the business case. The Develop Project Charter resides in the Project Integration Management Knowledge Area.

17. B. Project B has a payback period of 21 months; $50,000 is received in the first 12 months, with another $75,000 coming in over each of the next three quarters, or nine months.

18. C. Payback period does not consider the time value of money and is, therefore, the least precise of all the cash flow analysis techniques.

19. C. The project should be kicked off with a project charter that authorizes the project to begin, assigns the project manager, and describes the project objectives and purpose for the project. Doing so ensures that everyone is working with the same purposes in mind.

20. A, B, C, E. There is not a requirement to print project documents. They should be available for future reference as historical information for future projects.

Chapter 3: Delivering Business Value

1. D. A project is considered successful when it achieves its objectives and stakeholder needs and expectations are met.

2. B, C. Conflicts between stakeholders should always be resolved in favor of the customer. This question emphasizes the importance of identifying your stakeholders and their needs as early as possible in the project.

3. A, C, D. The level of authority the project manager has is determined by the organizational structure, interactions with various management levels, and the project management maturity level of the organization.

4. A, B, E. Advantages for employees in a functional organization are that they have only one supervisor and a clear chain of command exists. Organizational structures and culture are independent from the development or life-cycle methodology you'll use to manage the project (such as predictive, adaptive, or hybrid). Adaptive and hybrid methodologies can be challenging because resources on these teams are self-organized and self-directed, which could be difficult in a functional organization. The project manager will have to work with functional managers to obtain resources for the project in this structure.

5. D. DMAIC stands for define, measure, analyze, improve, and control. The define phase is where the project goals are established, stakeholders are identified, and the project charter is written.

6. E. All of the options are true. Option A describes the EEFs as an input to the Identify Stakeholder process. All of these factors should be taken into consideration when assessing stakeholders.

7. A, B. The project manager is responsible for ensuring corrective actions are taken. Change requests bring about changes, not corrective actions. However, a corrective action may bring about a change request.

8. E. All of these are true. The business need that brought about the project is twofold, improving customer satisfaction and a business need. Customers were complaining according to the question and the business need involved implementing a new system that would in turn improve satisfaction scores. Customer satisfaction is often measured using surveys. The project manager is responsible for championing the business value for the project throughout its life cycle. An agile approach would have allowed for this change to be incorporated into a future iteration and could have averted a two-month delay.

9. A-1, B-4, C-2, D-6, E-5, F-3. The Salience model is a tool and technique in the Identify Stakeholder process that allows you to categorize stakeholders according to power, urgency, and legitimacy.

10. B. Business values are those values that will lead to short- and long-term benefits for the organization.

11. A, B, D. This question describes the stakeholder register. You may have used the power/interest grid to classify your stakeholders, but this tool does not contain other information such as contact information.

12. A-5, B-1, C-2, D-4, E-3. This question describes the various titles project managers may have in different adaptive methodologies.

13. E. Business value involves bringing short- and -long term benefits to the organization. Creating a new car model and implementing a new accounting system are project goals, not a business value.

14. C, D, E. The project manager is responsible for understanding the urgency required to deliver the business value of the project. The project sponsor defines the urgency, but the project manager is responsible for delivering the project according to the urgency driving the business value. Urgency is subjective and is defined by the project sponsor.

15. A, C. You should examine and report on business value throughout the life of the project, you should always consider delivering business value incrementally whenever possible, and it's the project manager's responsibility to champion the business value of the project (not expecting others to periodically review the business value statement in the project charter).

16. C. Key performance indicators are a way to measure business value. It is the project manger's responsibility to measure and report on value. This is objective, not subjective, like asking the project sponsor what they think.

17. B. The business value network can include internal resources, employees, and stakeholders, not just external resources.

18. A, D. The minimum viable product is a component of work, or a task, that's been broken down into the lowest tangible feature, function, or result possible. The product owner is responsible for determining if business value has been achieved. The project team does not manage or prioritize the backlog; the product owner does. The Scrum master is a facilitator. Team members do not report to the Scrum master, and they do not assign tasks to team members.

19. D. All of the options are true regarding the minimum viable product.

20. B. Adjustments, changes, and corrections can be made at any time during the iteration.

Chapter 4: Developing the Project Scope

1. A, B, D. Option C describes the project scope statement.

2. B, C, E. Options A and D may have been considered when developing the project documents so far, but they are not reasons to hold a kickoff meeting.

3. B, E. The scope management plan describes how project scope will be defined and validated, how the scope statement will be developed, how the WBS will be created and defined, and how project scope will be managed and controlled. Project scope is measured against the project management plan, whereas product scope is measured against the product requirements. It is based on the approved project scope. The project scope management plan defines, maintains, and manages the scope of the project.

4. A, D. These four decision-making techniques belong to the Collect Requirements process and are part of the decision-making tool and technique in this process. They help stakeholders make decisions and come to agreement on the requirements of the project.

5. B, C, E. The project scope statement further elaborates the project deliverables and documents the product scope description, acceptance criteria, and project exclusions. It serves as a basis for future project decisions. It is an agreement between the project team and the customer on the precise work of the project. This question describes a hybrid approach so you could use either a project scope statement or a product backlog to compile user stories, which are the deliverables and requirements for the project. Options A and D describe the scope management plan.

6. D. The requirements traceability matrix links requirements to their origin and traces them throughout the project. Option A describes the requirements management plan, not the requirements document. Option B is partially true, with the exception of the first statement. Requirements documents do not have to be formal or complex. Option C refers to the project scope statement, not the requirements.

7. F. The scope baseline consists of the approved project scope statement, the WBS, and the WBS dictionary.

8. A, B. Brainstorming and lateral thinking are not decision-making techniques. They are used to help generate free form ideas and create information that can later be decided on.

9. A. You could use each product as a level one entry on the WBS, so option A is correct, but you may choose to construct the WBS differently. Option C is not correct because rolling wave planning is the process of fully elaborating near-term WBS work packages and elaborating others, like the third product in this question, at a later time when all information is known.

10. C. Poor scope definition might lead to cost increases, rework, schedule delays, and poor morale. Option C describes the project scope management plan.

11. D. Each element in the WBS is assigned a unique identifier called a code of accounts identifier. Typically, these codes are associated with a corporate chart of accounts and are used to track the costs of the individual work elements in the WBS.

12. B, C, D. The work package level is the lowest level in the WBS and facilitates resource assignment and cost and time estimates. The work package level on an agile project is the user story. In this question, the work package level contains four subprojects, so it would not be used to create the activity list. The activity list will be created from the work package level for each WBS created for each subproject.

13. C. The primary constraint is time. Since the trade show demos depend on project completion and the trade show is in late September, the date cannot be moved. The budget is the secondary constraint in this example.

14. E. Decomposition subdivides the major deliverables into smaller components. It is a tool and technique of the Create WBS process and is used to create a WBS. In an agile methodology, user stories are the work package level of the WBS and are documented in the product backlog. They are pulled from there into the spring backlog at the beginning of the iteration and further decomposed into tasks.

15. C, D. Acceptance criteria are documented in the project scope statement in a predictive methodology and in the user stories for an adaptive methodology.

16. D. The primary constraint is quality. If you made the assumption as stated in options A, B, and C, you assumed incorrectly. Clarify these assumptions with your stakeholders and project sponsors.

17. C. This is an example of an assumption. You've used this vendor before and haven't had any problems. You're assuming there will be no problems with this delivery based on your past experience.

18. B, C. This describes a regulatory standard and/or compliance issue, which are part of the organization's EEFs. Constraints restrict the actions of the project team.

19. A. The project came about because of a business need. The phones have to be answered because that's the core business. Upgrading the system to handle more volume is a business need. An assumption has been made regarding vendor availability. Always validate your assumptions.

20. B. The steps of decomposition include identify major deliverables, organize and determine the structure, identify lower-level components, assign identification codes, and verify correctness of decomposition.

Chapter 5: Creating the Project Schedule

1. C, D. Activity lists can be created using any tool and technique the agile team wants to use, not just the ones found in the Define Activities process. The iteration planning meeting is where the user stories are broken into activities, but a high-level breakdown may also occur early on in the project to determine project duration estimates. These estimates may be based on story points or iterations.

2. A, C, D. According to the *PMBOK® Guide*, the schedule management plan should be completed no matter what life cycle methodology you're using.

3. C. Reserve analysis takes schedule risk into consideration and adds a percentage of time or additional work periods to the estimate to prevent schedule delays.

4. A. Parametric estimating uses an algorithm or formula that multiplies a known element—such as the quantity of materials needed—by the time it takes to install or complete one unit of materials. The result is a total estimate for the activity. In this case, 10 servers multiplied by 16 hours per server gives you a 160-hour total duration estimate.

5. A, B, D. The activity list is a component of the project schedule, not the WBS. The activity list includes all the project activities, an identifier, and a description of the activity. The activity list is an output of the Define Activities process.

6. D. This is an example of a mandatory dependency, also known as hard logic. Mandatory dependencies are inherent in the nature of the work. Discretionary dependencies, also called preferred logic, preferential logic, and soft logic, are defined by the project management team.

7. A-2, B-4, C-1, D-3. The foundational principles of the *Agile Manifesto* include a maximum focus on delivering value to the customer with minimum focus on the process.

8. D. The project schedule should be easily accessible by all stakeholders, and you should notify them when there are updates. The original schedule baseline should be saved so that you can compare future changes to it.

9. D. Finish-to-start (FS) is the most commonly used logical relationship in PDM and the default relationship in most project management software packages.

10. C. CPM calculates a single early and late start date and a single early and late finish date for each activity. Once these dates are known, float time is calculated for each activity to determine the critical path. The other answers contain elements of PERT calculations.

11. B. The only information you have for this example is activity duration; therefore, the critical path is the path with the longest duration. Path A-D-E-H with a duration of 34 days is the critical path.

12. D. The only information you have for this example is activity duration, so you must calculate the critical path based on the durations given. The duration of A-B-C-E-H increased by 3 days, for a total of 35 days. The duration of A-F-G-H and A-F-G-E-H each increased by 3 days. A-F-G-E-H totals 36 days and becomes the new critical path.

13. D. You calculate the critical path by adding together the durations of all the tasks with zero or negative float. The critical path can be compressed using crashing techniques.

14. C. The calculation for PERT is the sum of optimistic time plus pessimistic time plus four times the most likely time divided by 6. The calculation for this example is as follows: (48 + 72 + (4 * 60)) / 6 = 60.

15. D. You calculate the standard deviation by subtracting the optimistic time from the pessimistic time and dividing the result by 6. The calculation for this example is as follows: (72 − 48) / 6 = 4.

16. D. There is a 95 percent probability that the work will finish within plus or minus two standard deviations. The expected value is 500, and the standard deviation times 2 is 24, so the activity will take from 476 to 524 days.

17. B. This is known as the basis of estimates.

18. A. Crashing the schedule includes tasks such as adding resources to the critical path tasks or speeding up deliveries of materials and resources.

19. C. Resource leveling is used for overallocated resources and allows for changes to the schedule completion dates. Crashing and fast tracking are schedule compression techniques that shorten the schedule. Resource smoothing techniques will not allow for changes to the critical path or project end date, and since you are concerned about not overusing this resource, lengthening the schedule is a better option.

20. B, C. Velocity is used to determine how long it will take to complete the work of the iteration. It measures the speed with which the team progresses, not capacity. Kanban boards display work based on capacity. Velocity is time bound.

Chapter 6: Developing the Project Budget and Engaging Stakeholders

1. B, C. The communications management plan, not the stakeholder engagement plan, defines and manages the flow of project information. The communications management plan also documents the types of information needs the stakeholders have, when the information should be distributed, how the information will be delivered, and how communications will be monitored and controlled throughout the project. The communications management plan considers the organizational structure and stakeholder requirements. The stakeholder engagement plan captures the strategies needed to engage stakeholders throughout the project and to assist in decision-making and project execution.

2. A, D. The cost management plan, an output of the Plan Cost Management process, is developed by considering the project scope, schedule, resources, and the approved project charter. The level of precision and accuracy and the units of measure and control thresholds are included in the cost management plan.

3. D. Control thresholds are variance thresholds (typically stated as a percentage of deviation from the baseline) used for monitoring cost performance.

4. C. The cost management plan is established using the WBS and its associated control accounts. A control account can be placed at any level of the WBS and is used for earned value measurement calculations regarding project costs.

5. A. Scope definition is the key component of determining cost estimates. It should be completed early in the project because costs are more easily influenced at the beginning of the project.

6. B. Three-point estimating can improve activity cost estimates because it factors in estimation uncertainty and risk.

7. B, D. Option A describes a review meeting and option C describes a retrospective meeting.

8. E. All of the options are true regarding mentoring stakeholders and ensuring buy-in for the project.

9. A, B, D. Future period operating costs are considered ongoing costs and are not part of project costs.

10. A, C, D. Stakeholders are classified in the stakeholder register. The neutral classification is when stakeholders neither support nor resist the project. Resistant stakeholders are not supportive of the project and may actively resist engaging. Neither of these classifications describes an adaptive methodology.

11. B, C, D. Funding limit reconciliation concerns reconciling the funds to be spent on the project, with funding limits placed on the funding commitments for the project.

12. B. The cost baseline is displayed as an S curve because of the way project spending occurs. Spending begins slowly, picks up speed until the spending peak is reached, and then tapers off as the project winds down.

13. B, C. Options A, D, and E describe the iteration-based daily stand-up. Option E asks whether any roadblocks prevent you from doing your work and, in a flow-based approach, this question would ask whether there are bottlenecks or roadblocks in the workflow.

14. B, C, E. Review meetings are not decision-making meetings. Option D describes the team members present at the retrospective meeting.

15. D. There are 36 channels of communication, yourself plus eight stakeholders. The formula is 9 (9 − 1) / 2 = 36. Lines of communication are considered when using the communications requirements analysis tool and technique.

16. A, C, E. Options B and D describe elements of the stakeholder engagement plan.

17. C. Acknowledgment means the receiver has received the message but does not mean that they agree with the message.

18. A-2, B-5, C-1, D-4, E-3. Each methodology has its own focus, benefits, and planning methods.

19. D. XP focuses on frequent cycles and delivering software to the customer when the customer needs it.

20. A, C. Osmotic communication is a form of polite eavesdropping where talking is occurring in the background, within earshot of team members who may pick up important project information. It is associated with the Crystal method.

Chapter 7: Identifying Project Risks

1. B. Ambiguity risk is addressed using gap analysis or benchmarking. Document what you know and identify what is uncertain or not understood and work with experts outside of the organization to fill in the gaps in knowledge.

2. B, D, E. The purpose of Perform Qualitative Risk Analysis is to determine what impact the identified risk events will have on the project and the probability they'll occur. It also puts risks in priority order according to their effects on the project objectives and assigns a risk score for the project. Options A and C describe the Perform Quantitative Risk Analysis process.

3. C, D. The risk management plan details how risk management processes will be implemented, monitored, and controlled throughout the life of the project. The risk management plan does not include strategies for overall project risks or responses to risks or triggers. Responses to risks are documented in the risk register as part of the Plan Risk Responses process.

4. C. Continuous probability distributions graphically display the probability of risk to the project objectives as well as the time or cost elements.

5. D. The data representation tool and technique called a bubble chart is a way to graphically display three factors of risk characteristics.

6. C. Criticality analysis determines the risk impact to critical path tasks on the project schedule.

7. E. All of the options are benefits of using an agile approach to manage projects with high degrees of uncertainty.

8. B. This risk event has the potential to save money on project costs, so it's an opportunity, and the appropriate strategy to use in this case is the exploit strategy.

9. D. This risk event has the potential to save money on project costs. Sharing involves using a third party to help ensure that the opportunity occurs.

10. A. The best answer is A. Triggers are warning signs of an impending risk event.

11. C, D. The probability and impact matrix multiplies the probability by the impact to determine a risk score. Using this score and a predetermined matrix that's defined by the organization or key stakeholders, you determine whether the score is a high, medium, or low designation. This also helps in determining which risks need detailed response plans.

12. A, C, D. Monte Carlo analysis is a simulation technique that computes project costs many times in an iterative fashion.

13. C. Mitigation attempts to reduce the impact of a risk event should it occur. Making plans to arrange for the leased equipment reduces the consequences of the risk.

14. B, E. Risk appetite and risk threshold are the components of risk attitude.

15. A. This question describes the risk threshold levels of the stakeholders. Risk triggers are recorded in the risk register during the Plan Risk Responses process. The risk of buying a machine from a new supplier would pose a threat to the project, not an opportunity. Interviewing might have been used, but this question wasn't describing the Identify Risks process.

16. B. The question describes sensitivity analysis, which is a tool and technique of the Perform Quantitative Risk Analysis process. Tornado diagrams are often used to display sensitivity analysis data.

17. B. This question describes a prompt list, used in the Identify Risks process as a tool and technique to help the team determine risks for the project.

18. D. The RBS describes risk categories, and the lowest level can be used as a prompt list to help identify risks. Risk owners are not assigned from the RBS but typically are assigned as soon as the risk is identified.

19. A-2, B-3, C-4, D-1. Risk processes are performed in each iteration, or workflow process, in an adaptive approach. Each approach has a different objective, but they all perform the risk processes repeatedly.

20. C, D, E. Product specification, production capability, and process suitability are the three characteristics of uncertainty.

Chapter 8: Planning and Procuring Resources

1. D. Make-or-buy analysis is determining whether it's more cost effective to purchase the goods or services needed for the project or more cost effective for the organization to produce them internally.

2. C. Firm fixed-price contracts have the highest risk to the seller and the least amount of risk to the buyer. However, the price the vendor charges for the product or service will compensate for the amount of risk they're assuming.

3. E. Either the buyer or the seller can write the SOW. Sometimes the buyer will write the SOW and the seller might modify it and send it back to the buyer for verification and approval.

4. D. According to the *Agile Practice Guide* (PMI®, 2017), when using a hybrid approach to manage the project you should use a governing agreement such as a master service agreement contract.

5. F. All of the options are true and describe the Plan Resource Management process.

6. A, B, D. Source selection criteria can be based on price alone when there are multiple vendors who can readily supply the goods or services. The question states that only three vendors make the machine, which means source selection criteria should be based on more than price. Sole source is used when you have justification for engaging only one vendor in the bidding process. In this case, there are three vendors who can meet the specifications, so a sole source is not recommended.

7. A. A constraint can be anything that limits the option of the project team. Organizational structures, collective bargaining agreements, and economic conditions are all constraints that you might encounter during this process. Location and logistics should be considered when documenting the resource management plan, but they are not a constraints.

8. B. Fixed-price contracts can include incentives for meeting performance criteria, but the question states the vendor helping with the programming task will be reimbursed for their costs and, depending on your satisfaction with their results, may receive an additional award. This describes a cost plus award fee contract.

9. A, D. RACI stands for responsible, accountable, consult, and inform.

10. B. Plan Procurement Management can directly influence the project schedule, and the project schedule can directly influence this process.

11. G. A value-driven structure is not conducive to agile methodologies because it emphasizes meeting a milestone rather than producing value-driven deliverables.

12. E. Fit-for-use reviews should occur at least once every two weeks.

13. C, D. Philip Crosby devised the zero defects theory, meaning do it right the first time. Proper Plan Quality Management leads to less rework and higher productivity. Joseph Juran's fitness for use says that stakeholders' and customers' expectations are met or exceeded.

14. A. W. Edwards Deming conjectured that the cost of quality is a management problem 85 percent of the time and that once the problem trickles down to the workers, it is outside their control.

15. F. The benefits of meeting quality requirements are increased stakeholder satisfaction, lower costs, higher productivity, and less rework. Lower risk is not associated with meeting quality requirements.

16. A. Internal failure costs are costs associated with not meeting the customer's expectations while you still had control over the product. This results in rework, scrapping, and downtime.

17. A, C, D. Quality metrics are an output of the Plan Quality Management process and not part of the quality management plan.

18. D. This is an example of design of experiments.

19. B. Six Sigma is a measurement-based strategy that focuses on process improvement and variation reduction by applying Six Sigma methodologies to the project.

20. C. Teaming agreements are not a named input of the Plan Procurement Management process, but they are an input to all of the Planning processes collectively.

Chapter 9: Developing the Project Team

1. B Corrective action brings anticipated future project outcomes back into alignment with the project management plan. Because an important deadline that depends on a positive outcome of this test is looming, the equipment is exchanged so that the project plan and project schedule are not impacted.

2. A. The most difficult aspect of the Direct and Manage Project Work process is coordinating and integrating all the project elements. The clue to this question is in the next-to-last sentence.

3. F. All of the options are true.

4. F. Training is an important element on agile projects. The project manager or team facilitator should ensure that all of the steps outlined in the options are fulfilled so that training is successful.

5. D. Work performance data includes elements such as schedule status, the status of deliverables completion, lessons learned, resource utilization, KPIs, actual durations of activities that are complete, technical performance measures, and more.

6. B, D. Backlog refinement and impact mapping are two techniques you can use to help the team understand how user stories fit into the big picture of the overall project.

7. A. This question describes a defect repair that was discovered and implemented. Defect repairs come about as a result of approved changes. Approved changes are implemented in the Direct and Manage Project Work process. The question implies that the change request was approved because the problem was corrected and the rest of the production run went smoothly.

8. B. Teams in the norming stage of team development exhibit affection and familiarity with one another and make joint decisions.

9. C. The introduction of a new team member will start the formation and development of the team all over again with the forming stage.

10. D. Delegating is a situational leadership style that is used when team members have performed the task before and need little to no input from the manager. The supporting style requires some input from the leader. An autocratic manager makes all the decisions with no input from the team, and the laissez-faire leader lets the team make all the decisions and has little involvement with the team. Neither the autocratic nor laissez-faire leadership style is a named style in the Blanchard Situational Leadership II Model.

11. B. The Expectancy Theory says that people are motivated by the expectation of good outcomes. The outcome must be reasonable and attainable.

12. A, D. Colocation would bring your team members together in the same location and allow them to function more efficiently as a team. At a minimum, meeting in a common room, such as a war room, for all team meetings would bring the team closer together. If this isn't possible, you could also consider virtual workspaces that allow the team to stay in contact with one another through videoconferencing, teleconferencing, screen sharing, and voice communication.

13. G. Options A–D describe teams that are new to agile methodologies. The question describes a hybrid methodology because the planning was performed with a waterfall approach and development will use a Scrum methodology.

14. A. Referent power is power that is inferred on a leader by their subordinates as a result of the high level of respect for the leader.

15. B. Theory Y managers believe that people will perform their best if they're provided with the proper motivation and the right expectations.

16. F. All of the options describe agile techniques to maximize the speed at which the team can work.

17. E. This technique is called *360-degree review*. It's part of the project performance appraisals tool and technique of the Manage Team process and is used for individual performance reviews, not team assessments.

18. A, B. This question refers to preassignments, which are a tool and technique of the Acquire Resources process. Resources promised as part of the project proposal should be noted in the project charter. Preassignment can pertain to human or physical resources. Team members promised as part of the project proposal should be noted in the project charter.

19. E. Option D describes diversity. Inclusion is ensuring that all team members have the ability to participate on the team and voice their opinions and concerns, and that they are collaborative, supportive, and respectful of one another.

20. A-2, B-5, C-4, D-1, E-3. Testing at all levels is a concept used in Extreme Programming, and other agile methodologies, to expose issues and problems early in the coding process.

Chapter 10: Sharing Information

1. D. Procurement documentation includes bids, quotations, RFIs, IFBs, RFPs, RFQs, and so on as you learned in the Plan Procurement Management process. They are an input of the Conduct Procurements process.

2. A, B, D, E. Receivers filter information through cultural considerations, knowledge of the subject matter, language abilities, geographic location, emotions, and attitudes.

3. E. All of the options are true. Engaged and informed stakeholders help increase the chances of success and it's the project manager's responsibility to ensure they have access to all project artifacts and information.

4. B. Groups of 5 to 11 participants make the most accurate decisions.

5. A, B, C. Lessons learned (which is what this question describes) are useful for activities and processes for the current project as well as future projects. Lessons learned are similar to retrospective meetings on agile projects. Review meetings are held to demonstrate functionality or a prototype on agile projects.

6. B, C, E. The complexity of the contract will determine how extensive the contract negotiations will be, not the procurement officers. Weighted systems are a type of proposal evaluation technique. Seller rating systems should not be used as the sole evaluation technique when evaluating and selecting vendors. They should be combined with other techniques.

7. F. Bidder conferences are also known as vendor conferences, prebid conferences, and contractor conferences, according to the *PMBOK® Guide*.

8. A, D. Qualified sellers lists are a component of the organizational process assets input of Conduct Procurements. Their purpose is to provide information about the sellers.

9. B. Option A is a decision tree diagram. Option B is a cause-and-effect diagram—also called fishbone diagrams or Ishikawa diagrams. They show the relationships between the causes and effects of problems. Option C is a flowchart diagram.

10. C. Fait accompli is a tactic used during contract negotiations where one party convinces the other that the particular issue is no longer relevant or cannot be changed.

11. D. Customer collaboration is more highly valued than contract negotiation, according to the *Agile Manifesto*.

12. D. The acceptance of work results happens later during the Validate Scope process, not during a quality audit.

13. A. Process analysis is a tool and technique in the Manage Quality process and involves examining constraints, problems, and activities that don't add value to create process improvements.

14. C. The data representation tool and technique of Manage Quality includes: affinity diagrams, cause-and-effect diagrams, flowcharts, histograms, matrix diagrams, and scatter diagrams. This question refers to tree diagrams that are other data representation tools you could use in this process, including PDPCs, interrelationship digraphs, tree diagrams, prioritization matrices, and activity network diagrams.

15. C. Face-to-face meetings are the most effective means for resolving stakeholder issues, provided these meetings are practical.

16. B. Status meetings are to report on the progress of the project. They are not for demos or show-and-tell. Option C is not correct because stakeholders are not concerned about the content of the technical documentation; they need to know that a qualified technician has reviewed the technical documentation and that the documentation task is accurate and complete.

17. F. All of the options outline ways you can help reduce conflict on a project. Team ground rules should be established early on in the project and should include stakeholder expectations.

18. A. The project manager alone is responsible for managing stakeholder expectations.

19. B, E. Smoothing (also known as accommodating) and withdrawal (also known as avoidance) are both lose-lose techniques. Forcing is a win-lose technique. Reconcile is where neither side wins or loses. Problem solving is a win-win technique.

20. A-1, B-5, C-4, D-2, E-3. Scaling agile frameworks is a technique used to scale agile practices to the organization and incorporate multiple teams using agile methodologies.

Chapter 11: Measuring and Controlling Project Performance

1. D. The Monitor and Control Project Work process involves reporting actual project results and how they compare to the project management plan, analyzing performance data, monitoring change requests, and more.

2. A, C, D. You are performing the Monitor and Control Project Work process and should recommend a change request that can take the form of a corrective action. Preventive actions reduce the possibility of negative impacts from risk events and do not apply to this situation. Lessons learned should not be used to document employee performance. However, you could document the missed activity and why it was missed without naming names.

3. D. Negotiation, using alternative dispute resolution (ADR), is the preferred method of settling claims or disputes in the Control Procurements process.

4. C, D, E. Option A is incorrect because it is the project manager's responsibility to manage stakeholder expectations regarding project goals. The vendor may assist with this, but the responsibility for this rests with the project manager. Option B is not correct; the meanings are reversed in the option. Inspections are structured reviews of the work performed by the contractor, and audits are structured reviews of the procurement process.

5. B. Contracts end in default typically due to vendor nonperformance, or when the vendor fails to perform according to the terms of the contract. In this case, the vendor performed the work but did not meet the requirements. Contracts that end for cause are due to violations of the terms of the contract. This question describes poor performance, not contract violations. Options A and D are also viable options if you want to continue to work with this vendor.

6. B. According to the *PMBOK® Guide*, the Control Procurements process is closely coordinated with the Direct and Manage Project Work, Control Quality, Perform Integrated Change Control, and Monitor Risks processes.

7. E. Status review meetings should contain relevant information, and they should be tailored to the needs of the stakeholders.

8. E. All of the options are true. Option D is important when you sense stakeholder engagement is waning. Ask stakeholders what will be helpful to them and what else you can do to help keep them engaged.

9. F. You may engage in corrective actions in this case, not preventive actions. Preventive actions reduce the possibility of negative impacts from risk events and do not apply to this situation. They should have been applied before the situation occurred.

10. B. The Delphi method, technology forecasting, scenario building, and forecast by analogy are all in the judgmental methods category of forecasting.

11. D. Scatter diagrams display the relationship between an independent and dependent variable over time.

12. B. Workarounds are unplanned responses. They deal with negative risk events as they occur. As the name implies, workarounds were not previously known to the project team. The risk event was unplanned, so no contingency plan existed to deal with the risk event, and thus it required a workaround.

13. F. Change requests are submitted through other processes like the Monitor and Control Project Work process, and they are reviewed, tracked, managed, analyzed, and documented in this process.

14. G. Change control systems are documented procedures that describe how to submit change requests. They track the status of the change requests, document the management impacts of change, track the change approval status, and define the level of authority needed to approve changes.

15. E. All of the options are true and describe aspects of the Control Quality process. Inspections are a tool and technique of this process that consider tolerable results and attributes.

16. C. Change control boards (CCBs) review change requests and have the authority to approve or deny them. Their authority is defined by the configuration control and change control process.

17. A. The three activities associated with configuration management are configuration identification, configuration status accounting, and configuration verification and auditing.

18. B, D. You are using a control chart. Control charts depict if the process is in control or not in control. This process is in control because it is within three standard deviations.

19. C. In this question, there is no response plan, so you'll need to use reserves to deal with the risk. The risk auditors should be informed about what happened with this risk, but not as a first course of action. Option B is a tool and technique used in Control Quality. Testing and product evaluation is not the first step you should take in this situation, but it may be performed later. Option D is part of the Monitor and Control Project Work process and perhaps should be considered, but not until you deal with the risk first.

20. E. This question describes an acquisition. Acquisitions usually involve large companies buying or acquiring smaller companies and the smaller company ceases to exist. A merger occurs when two organizations agree to form a new business entity.

Chapter 12: Controlling Work Results and Closing Out the Project

1. A. Changes to product scope should be reflected in the project scope.

2. D. The best answer to this problem is to develop alternative solutions to address the design error. Reducing technical requirements might be an alternative solution, but it's not one you'd implement without looking at all the alternatives. Ignoring the error and going forward with production will result in an unsatisfactory product for the customer.

3. G. A burndown chart shows the work remaining for the iteration; a burnup chart shows the completed work.

4. B, C. Option A is not correct. You should always report the truth, even if you don't yet know the full picture. This question states the rumors have already started so stakeholders are already wondering what's going on. It's better to get ahead of the rumors and let them know there is an issue and that you are working on a full report. Option D is not correct because critical path tasks will *always* cause delays to the project completion date, not "may" cause delays.

5. A, B. Option C describes definition of ready (DoR), not definition of done (DoD). DoD ensures that the deliverable or functionality produced at the end of the iteration is ready for the customer to use. DoD measures the iteration, not the deliverable. Examples include integration testing is complete and code review is complete. Option D describes the Control Quality process.

6. F. Written notice of contract closure is part of the Control Procurements process. This process involves the project manager documenting the formal acceptance of the work of the contract.

7. F. To-complete performance index determines the cost performance that must be realized for the remaining work of the project to meet a goal such as BAC or EAC.

8. D. Budget updates might require cost re-baselining.

9. A. Schedule variance is (EV – PV) and schedule performance index is (EV / PV).

10. C. Earned value is referred to as the value of the work that's been completed to date compared to the budget.

11. D. The CV is a positive number and is calculated by subtracting AC from EV as follows: 250 – 200 = 50. A positive CV means the project is coming in under budget, meaning you've spent less than you planned as of the measurement date.

12. A. The SV calculation is EV – PV. If PV is a higher number than EV, you'll get a negative number as a result.

13. B. CPI is calculated as follows: EV / AC. In this case, 250 / 200 = 1.25.

14. B. When you accept project performance to date and assume future ETC work will be performed at the budgeted rate, EAC is calculated as follows: AC + (BAC – EV). Therefore, the calculation for this question looks like this: 200 + (800 – 250) = 750.

15. C. The correct formula for ETC for this question is as follows: EAC – AC. Therefore, ETC is as follows: 375 – 200 = 175.

16. A. When project performance is expected to behave like past performance, EAC is calculated as follows: EAC = BAC / CPI. Therefore, the calculation for this question looks like this: 800 / 1.25 = 640.

17. D. You first have to calculate EAC in order to calculate VAC. EAC for variances that are atypical is AC + (BAC – EV). So, our numbers are 275 + (500 – 250) = 525. VAC is calculated this way: BAC – EAC. Therefore, 500 – 525 = –25. Our costs are not doing as well as anticipated.

18. C. You must first calculate CPI in order to calculate ETC. CPI is EV / AC. We have 1150 / 1275 = .90. ETC with typical cost variances is (BAC –EV) / CPI. Our numbers are (2500 – 1150) / .90 = 1500.

19. C. As project manager, you have taken all the appropriate actions regarding confidentiality of data; this question indicated that you did that. In this case, option D is not correct because it implies that you did not verify ahead of time that the stakeholder had the proper levels of approval to use the data.

20. C. Regression analysis examines project variables and how they contributed to project outcomes for the purpose of improving performance on future projects.

Appendix
B

Process Inputs and Outputs

Throughout this book, I've discussed the inputs and outputs to the PMI® processes. In this appendix, you'll find the inputs, tools and techniques, outputs, and Knowledge Areas of the project management processes listed by process in the order in which they appear in the text. I think you'll appreciate the convenience of having all this information in one location. Enjoy!

Initiating Processes

Table B.1 lists the inputs, tools and techniques, outputs, and Knowledge Areas for the Initiating process group.

TABLE B.1   Initiating processes

Process name	Inputs	Tools and techniques	Outputs	Knowledge Area
Develop Project Charter	Business documents: Business case Benefits management plan	Expert judgment	Project charter	Integration
	Agreements	Data gathering: Brainstorming Focus groups Interviews	Assumption log	
	Enterprise environmental factors	Interpersonal and team skills: Conflict management Facilitation Meeting management		
	Organizational process assets	Meetings		

Process name	Inputs	Tools and techniques	Outputs	Knowledge Area
Identify Stakeholders	Project charter	Expert judgment	Stakeholder register	Stakeholder
	Business documents: Business case Benefits management plan	Data gathering: Questionnaires Brainstorming	Change requests	
	Project management plan: Communications management plan Stakeholder engagement plan	Data analysis: Stakeholder analysis Document analysis	Project management plan updates: Requirements management plan Communications management plan Risk management plan Stakeholder engagement plan	
	Project documents: Change log Issue log Requirements documentation	Data representation: Stakeholder mapping / representation	Project documents updates: Assumption log Issue log Risk register	
	Agreements	Meetings		
	Enterprise environmental factors			
	Organizational process assets			

Planning Processes

Table B.2 lists the inputs, tools and techniques, outputs, and Knowledge Areas for the processes in the Planning process group.

TABLE B.2   Planning processes

Process name	Inputs	Tools and techniques	Outputs	Knowledge Area
Develop Project Management Plan	Project charter	Expert judgment	Project management plan	Integration
	Outputs from other processes	Data gathering: Brainstorming Checklists Focus groups Interviews		
	Enterprise environmental factors	Interpersonal and Team Skills: Conflict management Facilitation Meeting management		
	Organizational process assets	Meetings		
Plan Scope Management	Project charter	Expert judgment	Scope management plan	Scope
	Project management plan: Quality management plan Project life cycle description Development approach	Data analysis: Alternatives analysis	Requirements management plan	

Process name	Inputs	Tools and techniques	Outputs	Knowledge Area
	Enterprise environmental factors	Meetings		
	Organizational process assets			
Collect Requirements	Project charter	Expert judgment	Requirements documentation	Scope
	Project management plan: Scope management plan Requirements management plan Stakeholder engagement plan	Data gathering: Brainstorming, Interviews Focus groups Questionnaires and surveys Benchmarking	Requirements traceability matrix	
	Project documents: Assumption log Lessons learned register Stakeholder register	Data analysis: Document analysis		
	Business documents: Business case	Decision-making: Voting Multicriteria decision analysis		
	Agreements	Data representation: Affinity diagrams Mind mapping		

TABLE B.2   Planning processes *(continued)*

Process name	Inputs	Tools and techniques	Outputs	Knowledge Area
	Enterprise environmental factors	Interpersonal and Team Skills: Nominal group Observation/ conversation Facilitation		
	Organizational process assets	Context diagram		
		Prototypes		
Define Scope	Project charter	Expert judgment	Project scope statement	Scope
	Project management plan: Scope management plan	Data analysis: Alternatives analysis	Project documents updates: Assumption log Requirements documentation Requirements traceability matrix Stakeholder register	
	Project documents: Assumption log Requirements documentation Risk register	Decision-making: Multicriteria decision analysis		
	Enterprise environmental factors	Interpersonal and team skills: Facilitation		
	Organizational process assets	Product analysis		

Process name	Inputs	Tools and techniques	Outputs	Knowledge Area
Create WBS	Project management plan: Scope management plan	Expert judgment	Scope baseline	Scope
	Project documents: Project scope statement Requirements documentation	Decomposition	Project documents updates: Assumption log Requirements documenta-tion	
	Enterprise environmental factors			
	Organizational process assets			
Plan Schedule Management	Project charter	Expert judgment	Schedule management plan	Schedule
	Project management plan: Scope management plan Development approach	Data analysis: Alternatives anal-ysis		
	Enterprise environmental factors	Meetings		
	Organizational process assets			

TABLE B.2   Planning processes *(continued)*

Process name	Inputs	Tools and techniques	Outputs	Knowledge Area
Define Activities	Project management plan: Schedule management plan Scope baseline	Expert judgment	Activity list	Schedule
	Enterprise environmental factors	Decomposition	Activity attributes	
	Organizational process assets	Rolling wave planning	Milestone list	
		Meetings	Change requests	
			Project management plan updates: Schedule baseline Cost baseline	
Sequence Activities	Project management plan: Schedule management plan Scope baseline	Precedence diagramming method (PDM)	Project schedule network diagrams	Schedule
	Project documents: Activity attributes Activity list Assumption log Milestone list	Dependency determination and integration	Project documents updates: Activity attributes Activity list Assumption log Milestone list	

Process name	Inputs	Tools and techniques	Outputs	Knowledge Area
	Enterprise environmental factors	Leads and lags		
	Organizational process assets	Project management information system (PMIS)		
Estimate Activity Resources	Project management plan: Resource management plan Scope baseline	Expert judgment	Resource requirements	Resource
	Project documents: Activity attributes Activity list Assumption log Cost estimates Resource calendars Risk register	Bottom-up estimating	Basis of estimate	
	Enterprise environmental factors	Analogous estimating	Resource breakdown structure	
	Organizational process assets	Parametric estimating	Project documents updates: Activity attribute Assumption log Lessons learned register	

TABLE B.2   Planning processes *(continued)*

Process name	Inputs	Tools and techniques	Outputs	Knowledge Area
		Data analysis: Alternatives analysis		
		Project management information system (PMIS)		
		Meetings		
Estimate Activity Durations	Project management plan: Schedule management plan Scope baseline	Expert judgment	Duration estimates	Schedule
	Project documents: Activity attributes Activity list Assumption log Lessons learned register Milestone list Project team assignments Resource breakdown structure Resource calendars Resource requirements Risk register	Analogous estimating	Basis of estimates	

Process name	Inputs	Tools and techniques	Outputs	Knowledge Area
		Parametric estimating	Project documents updates:	
			Activity attributes	
			Assumption log	
			Lessons learned register	
		Three-point estimating		
		Bottom-up estimating		
		Data analysis:		
		Alternatives analysis		
		Reserve analysis		
		Decision-making		
		Meetings		
Develop Schedule	Project management plan:	Schedule network analysis	Schedule baseline	Schedule
	Schedule management plan			
	Scope baseline			

TABLE B.2   Planning processes *(continued)*

Process name	Inputs	Tools and techniques	Outputs	Knowledge Area
	Project documents:	Critical path method	Project schedule	
	Activity attributes			
	Activity list			
	Assumption log			
	Basis of estimates			
	Duration estimates			
	Lessons learned register			
	Milestone list			
	Project schedule network diagrams			
	Project team assignments			
	Resource calendars			
	Resource requirements			
	Risk register			
	Enterprise environmental factors	Resource optimization	Schedule data	
	Organizational process assets	Data analysis: What-if scenarios Simulation	Project calendars	
		Leads and lags	Project management plan updates: Schedule management plan Cost baseline	

Process name	Inputs	Tools and techniques	Outputs	Knowledge Area
		Schedule compression	Project documents updates:	
			Activity attributes	
			Assumption log	
			Duration estimates	
			Lessons learned register	
			Resource requirements	
			Risk register	
		Project management information system (PMIS)		
		Agile release planning		
Plan Cost Management	Project charter	Expert judgment	Cost management plan	Cost
	Project management plan:	Data analysis:		
	Schedule management plan	Alternatives analysis		
	Risk management plan			
	Enterprise environmental factors	Meetings		
	Organizational process assets			

TABLE B.2   Planning processes *(continued)*

Process name	Inputs	Tools and techniques	Outputs	Knowledge Area
Estimate Costs	Project management plan: Cost management plan Quality management plan Scope baseline	Expert judgment	Cost estimates	Cost
	Project documents: Lessons learned register Project schedule Resource requirement Risk register	Analogous estimating	Basis of estimates	
	Enterprise environmental factors	Parametric estimating	Project documents updates: Assumption log Lessons learned register Risk register	
	Organizational process assets	Bottom-up estimating Three-point estimating Data analysis: Alternatives analysis Reserve analysis Cost of quality		

Process name	Inputs	Tools and techniques	Outputs	Knowledge Area
		Project management information system (PMIS)		
		Decision-making: Voting		
Determine Budget	Project management plan: Cost management plan Resource management plan Scope baseline	Expert judgment	Cost baseline	Cost
	Project documents: Basis of estimates Cost estimates Project schedule Risk register	Cost aggregation	Project funding requirements	
	Business documents: Business case Benefits management plan	Data analysis: Reserve analysis	Project documents updates: Cost estimate Project schedule Risk register	
	Agreements	Historical information review		
	Enterprise environmental factors	Funding limit reconciliation		

TABLE B.2 Planning processes *(continued)*

Process name	Inputs	Tools and techniques	Outputs	Knowledge Area
	Organizational process assets	Financing		
Plan Stakeholder Engagement	Project charter	Expert judgment	Stakeholder engagement plan	Stakeholder
	Project management plan: Resource management plan Communications management plan Risk management plan	Data gathering: Benchmarking		
	Project documents: Assumption log Change log Issue log Project schedule Risk register Stakeholder register	Data analysis: Assumption and constraint analysis Root cause analysis		
	Agreements	Decision-making: Prioritization/ranking		
	Enterprise environmental factors	Data representation: Mind mapping Stakeholder engagement assessment matrix		
	Organizational process assets	Meetings		

Process name	Inputs	Tools and techniques	Outputs	Knowledge Area
Plan Communications Management	Project charter	Expert judgment	Communications management plan	
	Project management plan: Resource management plan Stakeholder engagement plan	Communication requirements analysis	Project management plan updates: Stakeholder engagement plan	
	Project documents: Requirements documentation Stakeholder register	Communication technology	Project documents updates: Project schedule Stakeholder register	
	Enterprise environmental factors	Communication models		
	Organizational process assets	Communication methods		
		Interpersonal and team skills: Communication styles assessment Political awareness Cultural awareness		
		Data representation: Stakeholder engagement assessment matrix		
		Meetings		

TABLE B.2   Planning processes *(continued)*

Process name	Inputs	Tools and techniques	Outputs	Knowledge Area
Plan Risk Management	Project charter	Expert judgment	Risk management plan	Risk
	Project management plan: Any component	Data analysis: Stakeholder analysis		
	Project documents: Stakeholder register	Meetings		
	Enterprise environmental factors			
	Organizational process assets			
Identify Risks	Project management plan: Requirements management plan	Expert judgment	Risk register	Risk
	Schedule management plan			
	Cost management plan			
	Quality management plan			
	Resource management plan			
	Risk management plan			
	Scope baseline			
	Schedule baseline			
	Cost baseline			

Process name	Inputs	Tools and techniques	Outputs	Knowledge Area
	Project documents: Assumption log Cost estimates Duration estimates Issue log Lessons learned register Requirements documentation Resource requirements Stakeholder register	Data gathering: Brainstorming Checklists Interviews	Risk report	
	Agreements	Data analysis: Root cause analysis Assumption and constraint analysis SWOT analysis Document analysis	Project documents updates: Assumption log Issue log Lessons learned register	
	Procurement documentation	Interpersonal and team skills: Facilitation		
	Enterprise environmental factors	Prompt lists		
	Organizational process assets	Meetings		
Perform Qualitative Risk Analysis	Project management plan: Risk management plan	Expert judgment	Project documents updates: Assumption log Issue log Risk register Risk report	Risk

TABLE B.2 Planning processes *(continued)*

Process name	Inputs	Tools and techniques	Outputs	Knowledge Area
	Project documents: Assumption log Risk register Stakeholder register	Data gathering: Interviews		
	Enterprise environmental factors	Data analysis: Risk data quality assessment Risk probability and impact assessment Assessment of other risk parameters		
	Organizational process assets	Interpersonal and team skills: Facilitation Risk categorization Data representation: Probability and impact matrix Hierarchical charts Meetings		
Perform Quantitative Risk Analysis	Project management plan: Risk management plan Scope baseline Schedule baseline Cost baseline	Expert judgment	Project documents updates: Risk report	Risk

Process name	Inputs	Tools and techniques	Outputs	Knowledge Area
	Project documents:	Data gathering:		
	Assumption log	Interviews		
	Basis of estimates			
	Cost estimates			
	Cost forecasts			
	Duration estimates			
	Milestone list			
	Resource requirements			
	Risk register			
	Risk report			
	Schedule forecasts			
	Enterprise environmental factors	Interpersonal and team skills:		
		Facilitation		
	Organizational process assets	Representations of uncertainty		
		Data analysis:		
		Simulations		
		Sensitivity analysis		
		Decision tree analysis		
		Influence diagrams		
Plan Risk Responses	Project management plan:	Expert judgment	Change requests	Risk
	Resource management plan			
	Risk management plan			
	Cost baseline			

TABLE B.2   Planning processes *(continued)*

Process name	Inputs	Tools and techniques	Outputs	Knowledge Area
	Project documents: Lessons learned register Project schedule Project team assignments Resource calendars Risk register Risk report Stakeholder register	Data gathering: Interviews	Project management plan updates: Schedule management plan Cost management plan Quality management plan Resource management plan Procurement management plan Scope baseline Schedule baseline Cost baseline	
	Enterprise environmental factors	Interpersonal team skills Facilitation	Project documents updates: Assumption log Cost forecast Lessons learned register Project schedule Project team assignments Risk register Risk report	
	Organizational process assets	Strategies for threats		

Process name	Inputs	Tools and techniques	Outputs	Knowledge Area
		Strategies for opportunities		
		Contingent response strategies		
		Strategies for overall project risk		
		Data analysis: Alternatives analysis Cost–benefit analysis		
		Decision-making: Multicriteria decision analysis		
Plan Procurement Management	Project charter	Expert judgment	Procurement management plan	Procurement
	Business documents: Business case Benefits management plan	Data gathering: Market research	Procurement strategy	
	Project management plan: Scope management plan Quality management plan Resource management plan Scope baseline	Data analysis: Make-or-buy analysis	Bid documents	

TABLE B.2 Planning processes *(continued)*

Process name	Inputs	Tools and techniques	Outputs	Knowledge Area
	Project documents:	Source selection analysis	Source selection criteria	
	Milestone list			
	Project team assignments			
	Requirements documentation			
	Requirements traceability matrix			
	Resource requirements			
	Risk register			
	Stakeholder register			
	Enterprise environmental factors	Meetings	Make-or-buy decisions	
	Organizational process assets		Independent cost estimates	
			Change requests	
			Project documents updates:	
			Lessons learned register	
			Milestone list	
			Requirements documentation	
			Requirements traceability matrix	
			Risk register	
			Stakeholder register	

Process name	Inputs	Tools and techniques	Outputs	Knowledge Area
			Organizational process assets updates	
Plan Resource Management	Project charter	Expert judgment	Resource management plan	
	Project management plan: Quality management plan Scope baseline	Data representation: Hierarchical-type charts Responsibility assignment matrix Text-oriented formats	Team charter	
	Project documents: Project schedule Requirements documentation Risk register Stakeholder register	Organizational theory	Project documents updates: Assumption log Risk register	
	Enterprise environmental factors	Meetings		
	Organizational process assets			
Plan Quality Management	Project charter	Expert judgment	Quality management plan	Quality

TABLE B.2   Planning processes *(continued)*

Process name	Inputs	Tools and techniques	Outputs	Knowledge Area
	Project management plan: Requirements management plan Risk management plan Stakeholder engagement plan Scope baseline	Data gathering: Benchmarking Brainstorming Interviews	Quality metrics	
	Project documents: Assumption log Requirements documentation Requirements traceability matrix Risk register Stakeholder register	Data analysis: Cost–benefit analysis Cost-of-quality	Project management plan updates: Risk management plan Scope baseline	
	Enterprise environmental factors	Decision-making: Multicriteria decision analysis		
	Organizational process assets	Data representation: Flowcharts Logical data model Matrix diagrams Mind mapping	Project documents updates: Lessons learned register Requirements traceability matrix Risk register Stakeholder register	

Process name	Inputs	Tools and techniques	Outputs	Knowledge Area
		Test and inspection planning	Project documents updates	
		Statistical sampling		
		Meetings		

Executing Processes

Table B.3 lists the inputs, tools and techniques, outputs, and Knowledge Areas for the processes in the Executing process group.

TABLE B.3   Executing processes

Process name	Inputs	Tools and techniques	Outputs	Knowledge Area
Direct and Manage Project Work	Project management plan: Any component	Expert judgment	Deliverables	Integration
	Project documents: Change log Lessons learned register Milestone list Project communications Project schedule Requirements traceability matrix Risk register Risk report	Project management information system (PMIS)	Work performance data	

TABLE B.3 Executing processes *(continued)*

Process name	Inputs	Tools and techniques	Outputs	Knowledge Area
	Approved change requests	Meetings	Issue log	
	Enterprise environmental factors		Change requests	
	Organizational process assets		Project management plan updates: Any component	
			Project documents updates: Activity list Assumption log Lessons learned register Require- ments documentation Risk register Stakeholder register	
			Organizational process assets updates	

Process name	Inputs	Tools and techniques	Outputs	Knowledge Area
Acquire Resources	Project management plan: Resource management plan Procurement management plan Cost baseline	Decision-making: Multicriteria decision analysis	Physical resource require-ments	Human Resource
	Project docu-ments: Project schedule Resource calendars Resource requirements Stakeholder register	Interpersonal and team skills: Negotiation	Project team assignments	
	Enterprise environmental factors	Pre-assignment	Resource calendars	
	Organizational process assets	Virtual teams	Change requests Project management plan updates: Resource management plan Cost base-line	

TABLE B.3 Executing processes *(continued)*

Process name	Inputs	Tools and techniques	Outputs	Knowledge Area
			Project documents updates:	
			Lessons learned register	
			Project schedule	
			Resource breakdown structure	
			Resource requirements	
			Risk register	
			Stakeholder register	
			Enterprise environmental factors updates	
			Organizational process assets updates	
Develop Team	Project management plan: Resource management plan	Colocation	Team performance assessments	Human Resource

Process name	Inputs	Tools and techniques	Outputs	Knowledge Area
	Project documents: Lessons learned register Project schedule Project team assignments Resource calendars Team charter	Virtual teams	Change requests	
	Enterprise environmental factors	Communication technology	Project management plan updates: Resource management plan	
	Organizational process assets	Interpersonal and team skills: Conflict management Influencing Motivation Negotiation Team building	Project documents updates: Lessons learned register Project schedule Project team assignments Resource calendars Team charter	
		Recognition and rewards	Enterprise environmental factors updates	

TABLE B.3   Executing processes *(continued)*

Process name	Inputs	Tools and techniques	Outputs	Knowledge Area
		Training	Organizational process assets updates	
		Individual team assessments		
		Meetings		
Manage Team	Project management plan: Resource management plan	Interpersonal and team skills: Conflict management Decision-making Emotional intelligence Influencing Leadership	Change requests	Human Resource
	Project documents: Issue log Lessons learned register Project team assignments Team charter	Project management information system	Project management plan updates: Resource management plan Schedule baseline Cost baseline	
	Work performance reports		Project documents updates: Issue log Lessons learned register Project team assignments	

Process name	Inputs	Tools and techniques	Outputs	Knowledge Area
	Team performance assessments		Enterprise environmental factors updates	
	Enterprise environmental factors			
	Organizational process assets			
Implement Risk Response	Project management plan: Risk management plan	Expert judgment	Change requests	Risk
	Project documents: Lessons learned register Risk register Risk report	Interpersonal and team skills: Influencing	Project documents updates: Issue log Lessons learned register Project team assignments Risk register Risk report	
	Organizational process assets	Project management information system (PMIS)		

TABLE B.3 Executing processes *(continued)*

Process name	Inputs	Tools and techniques	Outputs	Knowledge Area
Conduct Procurements	Project management plan:	Expert judgment	Selected sellers	Procurement
	Scope management plan			
	Requirements management plan			
	Communications management plan			
	Risk management plan			
	Procurement management plan			
	Configuration management plan			
	Cost baseline			
	Procurement documents:	Advertising	Agreements	
	Lessons learned register			
	Project schedule			
	Requirements documentation			
	Risk register			
	Stakeholder register			

Process name	Inputs	Tools and techniques	Outputs	Knowledge Area
	Procurement documentation	Bidder conferences	Change requests	
	Seller proposals	Data analysis: Proposal evaluation	Project management plan updates:	
			Requirements management plan	
			Quality management plan	
			Communications management plan	
			Risk management plan	
			Procurement management plan	
			Scope baseline	
			Schedule baseline	
			Cost baseline	

TABLE B.3 Executing processes *(continued)*

Process name	Inputs	Tools and techniques	Outputs	Knowledge Area
	Enterprise environmental factors	Interpersonal and team skills: Negotiation	Project documents updates: Lessons learned register	
			Requirements documentation	
			Requirements traceability matrix	
			Resource calendars	
			Risk register	
			Stakeholder register	
	Organizational process assets		Organizational process updates	
Manage Quality	Project management plan: Quality management plan	Data gathering: Checklists	Quality reports	Quality
	Project documents: Lessons learned register Quality control measurements Quality metrics Risk report	Data analysis: Alternatives analysis Document analysis Process analysis Root cause analysis	Test and evaluation documentations	

Process name	Inputs	Tools and techniques	Outputs	Knowledge Area
	Organizational process assets	Decision-making: Multicriteria decision analysis	Change requests	
		Data representation: Affinity diagrams Cause-and-effect diagrams Flowcharts Histograms Matrix diagrams Scatter diagrams	Project management plan updates: Quality management plan Scope base-line Schedule baseline Cost base-line	
		Audits	Project documents updates: Issue log Lessons learned reg-ister Risk register	
		Design for X		
		Problem solving		
		Quality improvement methods		
Manage Project Knowledge	Project management plan: All components	Expert judgment	Lessons learned register	Integration

TABLE B.3   Executing processes *(continued)*

Process name	Inputs	Tools and techniques	Outputs	Knowledge Area
	Project documents: Lessons learned register Project team assignments Resource breakdown structure Source selection criteria Stakeholder register	Knowledge management	Project management plan updates: Any component	
	Deliverables	Information management	Organizational process assets updates	
	Enterprise environmental factors	Interpersonal and team skills: Active listening Facilitation Leadership Networking Political awareness		
	Organizational process assets			
Manage Communications	Project management plan: Resource management plan Communications management plan Stakeholder engagement plan	Communication technology	Project communications	Communications

Process name	Inputs	Tools and techniques	Outputs	Knowledge Area
	Project documents:	Communication methods	Project management plan updates:	
	Change log		Communications management plan	
	Issue log		Stakeholder engagement plan	
	Lessons learned register			
	Quality report			
	Risk report			
	Stakeholder register			
	Work performance reports	Communication skills:	Project documents updates:	
		Communication competence	Issue log	
		Feedback	Lessons learned register	
		Nonverbal	Project schedule	
		Presentations	Risk register	
			Stakeholder register	
	Enterprise environmental factors	Project management information system (PMIS)	Organizational process assets updates	
	Organizational process assets	Project reporting		
		Interpersonal and team skills:		
		Active listening		
		Conflict management		
		Cultural awareness		
		Meeting management		
		Networking		
		Political awareness		
		Meetings		

TABLE B.3   Executing processes *(continued)*

Process name	Inputs	Tools and techniques	Outputs	Knowledge Area
Manage Stakeholder Engagement	Project management plan: Communications management plan Risk management plan Stakeholder engagement plan Change management plan	Expert judgment	Change requests	Stakeholder
	Project documents: Change log Issue log Lessons learned register Stakeholder register	Communication skills: Feedback	Project management plan updates: Communications management plan Stakeholder engagement plan	
	Enterprise environmental factors	Interpersonal and team skills: Conflict management Cultural awareness Negotiation Observation/ conversation Political awareness	Project documents updates: Change log Issue log Lessons learned register Stakeholder register	
	Organizational process assets	Ground rules Meetings		

Monitoring and Controlling Processes

Table B.4 lists the inputs, tools and techniques, outputs, and Knowledge Areas for the Monitoring and Controlling group processes.

TABLE B.4   Monitoring and Controlling processes

Process name	Inputs	Tools and techniques	Outputs	Knowledge Area
Monitor and Control Project Work	Project management plan: Any component	Expert judgment	Work performance reports	Integration
	Project documents: Assumption log Basis of estimates Cost forecasts Issue log Lessons learned register Milestone list Quality reports Risk register Risk report Schedule forecasts	Data analysis: Alternatives analysis Cost–benefit analysis Earned value analysis Root cause analysis Trend analysis Variance analysis	Change requests	
	Work performance information	Decision-making	Project management plan updates: Any component	

TABLE B.4 Monitoring and Controlling processes *(continued)*

Process name	Inputs	Tools and techniques	Outputs	Knowledge Area
	Agreements	Meetings	Project documents updates: Cost forecast Issue log Lessons learned register Risk register Schedule forecasts	
	Enterprise environmental factors			
	Organizational process assets			
Control Procurements	Project management plan: Requirements management plan Risk management plan Procurement management plan Change management plan Schedule baseline	Expert judgment	Closed procurements	Procurement

Process name	Inputs	Tools and techniques	Outputs	Knowledge Area
	Project documents: Assumption log Lessons learned register Milestone list Quality reports Requirements documentation Requirements traceability matrix Risk register Stakeholder register	Claims administration	Work performance information	
	Agreements	Data analysis: Performance reviews Earned value analysis Trend analysis	Procurement documentation updates	
	Procurement documentation	Inspection	Change requests	
	Approved change requests	Audit	Project management plan updates: Risk management plan Procurement management plan Schedule baseline Cost baseline	

TABLE B.4   Monitoring and Controlling processes *(continued)*

Process name	Inputs	Tools and techniques	Outputs	Knowledge Area
	Work performance data		Project documents updates:	
			Lessons learned register	
			Resource requirements	
			Requirements traceability matrix	
			Risk register	
			Stakeholder register	
	Enterprise environmental factors		Organizational process assets updates	
	Organizational process assets			
Monitor Communications	Project management plan:	Expert judgment	Work performance information	Communications
	Resource management plan			
	Communications management plan			
	Stakeholder engagement plan			
	Project documents:	Project management information system (PMIS)	Change requests	
	Issue log			
	Lessons learned register			
	Project communications			

Process name	Inputs	Tools and techniques	Outputs	Knowledge Area
	Work performance data	Data analysis: Stakeholder engagement assessment matrix	Project management plan updates: Communications management plan Stakeholder engagement plan	
	Enterprise environmental factors	Interpersonal and team skills: Observation/ conversation	Project documents updates: Issue log Lessons learned register Stakeholder register	
	Organizational process assets	Meetings		
Perform Integrated Change Control	Project management plan: Change management plan Configuration management plan Scope baseline Schedule baseline Cost baseline	Expert judgment	Approved change requests	Integration
	Project documents: Basis of estimates Requirements traceability matrix Risk report	Change control tools	Project management plan updates: Any component	

TABLE B.4   Monitoring and Controlling processes *(continued)*

Process name	Inputs	Tools and techniques	Outputs	Knowledge Area
	Work performance reports	Data analysis: Alternatives analysis Cost–benefit analysis	Project documents updates: Change log	
	Change requests	Decision-making: Voting Autocratic decision-making Multicriteria decision analysis		
	Enterprise environmental factors	Meetings		
	Organizational process assets			
Monitor Stakeholder Engagement	Project management plan: Resource management plan Communications management plan Stakeholder engagement plan	Data analysis: Alternatives analysis Root cause analysis Stakeholder analysis	Work performance information	Stakeholder
	Project documents: Issue log Lessons learned register Project communications Risk register Stakeholder register	Decision-making: Multicriteria decision analysis Voting	Change requests	

Process name	Inputs	Tools and techniques	Outputs	Knowledge Area
	Work performance data	Data representation: Stakeholder engagement assessment matrix	Project management plan updates: Resource management plan Communications management plan Stakeholder engagement plan	
	Enterprise environmental factors	Communication skills: Feedback Presentations	Project documents updates: Issue log Lessons learned register Risk register Stakeholder register	
	Organizational process assets	Interpersonal and team skills: Active listening Cultural awareness Leadership Networking Political awareness Meetings		
Monitor Risks	Project management plan: Risk management plan	Data analysis: Technical performance analysis Reserve analysis	Work performance information	Risk

TABLE B.4 Monitoring and Controlling processes *(continued)*

Process name	Inputs	Tools and techniques	Outputs	Knowledge Area
	Project documents: Issue log Lessons learned register Risk register Risk report	Audits	Change requests	
	Work performance data	Meetings	Project management plan updates: Any component	
	Work performance reports		Project documents updates: Assumption log Issue log Lessons learned register Risk register Risk report	
			Organizational process assets updates	
Control Costs	Project management plan: Cost management plan Cost baseline Performance measurement baseline	Expert judgment	Work performance information	Cost

Process name	Inputs	Tools and techniques	Outputs	Knowledge Area
	Project documents: Lessons learned register	Data analysis: Earned value management Variance analysis Trend analysis Reserve analysis	Cost forecasts	
	Project funding requirements	To-complete performance index (TCPI)	Change requests	
	Work performance data	Project management information system (PMIS)	Project management plan updates: Cost management plan Cost baseline Performance measurement baseline	
	Organizational process assets		Project documents updates: Assumption log Basis of estimates Cost estimates Lessons learned register Risk register	

TABLE B.4 Monitoring and Controlling processes *(continued)*

Process name	Inputs	Tools and techniques	Outputs	Knowledge Area
Control Schedule	Project management plan: Schedule management plan Schedule baseline Scope baseline Performance measurement baseline	Data analysis: Earned value analysis Iteration burndown chart Performance reviews Trend analysis Variance analysis What-if scenario analysis	Work performance information	Schedule
	Project documents: Lessons learned register Project calendars Project schedule Resource calendars Schedule data	Critical path method	Schedule forecasts	
	Work performance data	Project management information system (PMIS)	Change requests	
	Organizational process assets	Resource optimization	Project management plan updates: Schedule management plan Schedule baseline Cost baseline Performance measurement baseline	

Process name	Inputs	Tools and techniques	Outputs	Knowledge Area
		Leads and lags	Project documents updates:	
			Assumption log	
			Basis of estimates	
			Lessons learned register	
			Project schedule	
			Resource calendars	
			Risk register	
			Schedule data	
		Schedule compression		
Control Quality	Project management plan:	Data gathering:	Quality control measurements	Quality
	Quality management plan	Checklists		
		Check sheets		
		Statistical sampling		
		Questionnaires and surveys		
	Project documents:	Data analysis:	Verified deliverables	
	Lessons learned register	Performance reviews		
	Quality metrics	Root cause analysis		
	Test and evaluation documents			
	Approved change requests	Inspection	Work performance information	
	Deliverables	Test/product evaluations	Change requests	

TABLE B.4   Monitoring and Controlling processes *(continued)*

Process name	Inputs	Tools and techniques	Outputs	Knowledge Area
	Work performance data	Data representation: Cause-and-effect diagrams Control charts Histograms Scatter diagrams	Project management plan updates: Quality management plan	
	Enterprise environmental factors	Meetings	Project documents updates: Issue log Lessons learned register Risk register Test and evaluation documents	
	Organizational process assets			
Validate Scope	Project management plan: Scope management plan Requirements management plan Scope baseline	Inspection	Accepted deliverables	Scope
	Project documents: Lessons learned register Quality reports Requirements documentation Requirements traceability matrix	Decision-making: Voting	Work performance information	

Process name	Inputs	Tools and techniques	Outputs	Knowledge Area
	Verified deliverables		Change requests	
	Work performance data		Project documents updates:	
			Lessons learned register	
			Requirements documentation	
			Requirements traceability matrix	
Control Scope	Project management plan:	Data analysis: Variance analysis Trend analysis	Work performance information	Scope
	Scope management plan			
	Requirements management plan			
	Change management plan			
	Configuration management plan			
	Scope baseline			
	Performance measurement baseline			

TABLE B.4 Monitoring and Controlling processes *(continued)*

Process name	Inputs	Tools and techniques	Outputs	Knowledge Area
	Project documents:		Change requests	
	Lessons learned register			
	Requirements documentation			
	Requirements traceability matrix			
	Work performance data		Project management plan updates:	
			Scope management plan	
			Scope baseline	
			Schedule baseline	
			Cost baseline	
			Performance measurement baseline	
	Organizational process assets		Project documents updates:	
			Lessons learned register	
			Requirements documentation	
			Requirements traceability matrix	

Process name	Inputs	Tools and techniques	Outputs	Knowledge Area
Control Resources	Project Management Plan:	Data Analysis:	Work performance information	Resource
	Resource management plan	Alternatives analysis	Change requests	
	Project Documents:	Cost–benefit analysis	Project management plan updates:	
	Issue log	Performance reviews	Resource management plan	
	Lessons learned register	Trend analysis		
	Physical resource assignments	Problem solving	Schedule baseline	
	Project schedule	Interpersonal and team skills:	Cost baseline	
	Resource breakdown structure	Negotiation	Project documents updates:	
	Resource requirements	Influencing	Assumption log	
	Risk register	Project management information systems	Issue log	
	Work performance information		Lessons learned register	
	Agreements		Physical resource assignments	
	Organizational process assets		Resource breakdown structure	
			Risk register	

Closing Processes

Table B.5 lists the inputs, tools and techniques, outputs, and Knowledge Areas for the processes in the Closing process group.

TABLE B.5   Closing processes

Process name	Inputs	Tools and techniques	Outputs	Knowledge Area
Close Project or Phase	Project charter	Expert judgment	Project documents updates: Lessons learned register	Integration
	Project management plan All components	Data analysis: Document analysis Regression analysis Trend analysis Variance analysis	Final product, service, or result transition	
	Project documents: Assumption log Basis of estimates Change log Issue log Lessons learned register Milestone list Project communications Quality control measurements Quality reports Requirements documentation Risk register Risk reports	Meetings	Final report	

Process name	Inputs	Tools and techniques	Outputs	Knowledge Area
	Accepted deliverables		Organizational process assets updates	
	Business documents:			
	Business case			
	Benefits management plan			
	Agreements			
	Procurement documentation			
	Organizational process assets			

Index

C

CA (control Account), 302
case studies, Kitchen Haven, 89–93
Change Management, 669
chart of accounts, 302
Close Project or Phase process, 53, 836–837
  Agile projects, 724–725
  characteristics of closing, 713–714
  closing out procurements, 723
  closing out projects, 717–718
    administrative closure procedures, 718–719
    final report, 719–722
    regression analysis, 719
  project endings
    addition, 714
    extinction, 715–716
    integration, 715
    starvation, 715
  stakeholder interests, 725–727
Closing process group, 22, 24–25, 836–837
CMMI (Capability Maturity Model
    Integration), 463
Collect Requirements process, 179,
    785–786
  benchmarking, 186
  brainstorming, 186
  context diagram, 189
  data analysis, 186
  data representation, 187
  decision-making, 186–187
  document gathering, 180–181
    assumption log, 181–182
    assumptions, 181–182
    constraints, 182–185
  documenting requirements, 185–189
  expert judgment, 185
  focus groups, 186
  gathering requirements, 185–189
  interpersonal skills, 187–188
  interviews, 186
  product backlog, 199
  prototypes, 189
  questionnaires, 186
  requirements
    documentation, 189–191
    traceability matrix, 191–192

surveys, 186
  team skills, 187–188
  user stories, 188
communication, 323–324
  Agile teams, 333–335
  communication needs, 325–331
  documenting plan, 331–333
  planning, 324–325
communication skills
  active listening, 589
  communication competence, 587–588
  communication model, 586–587
  feedback, 588
  forms of communication, 587
  information exchange, 585–586
  methods of communication, 587
  nonverbal, 588
  presentations, 588
  reporting, 588–589
Communications Management Plan, 170
compliance, 409
Conduct Procurements process, 814–816
  Agile projects, 573–574
  procurement agreements, 571–572
    contract life cycles, 572–573
  proposal evaluation
    advertising, 564
    analytical techniques, 567
    BATNA (best alternative to a negotiated
      agreement), 568
    bidder conferences, 565
    contingent agreements, 568–569
    negotiating strategy, 568–571
    screening systems, 566
    seller rating systems, 566–567
    weighting systems, 566
conflict resolution skills, 589–594
constraints, 183–184
continuous improvement, 462–463. See
    also Kaizen
contracts
  cost-reimbursable, 431
    CPAF (cost plus award fee), 432
    CPFF (cost plus fixed fee), 431
    CPIF (cost plus incentive fee), 431–432
    CPPC (cost plus percentage of cost), 432
  dynamic scope, 444

E

Q